APPLICATIONS OF SYSTEMS APPROACHES AT THE FARM AND REGIONAL LEVELS

Volume 1

## Systems Approaches for Sustainable Agricultural Development

VOLUME 5



*Aims and Scope*

The book series *Systems Approaches for Sustainable Agricultural Development* is intended for readers ranging from advanced students and research leaders to research scientists in developed and developing countries. It will contribute to the development of sustainable and productive systems in the tropics, subtropics and temperate regions, consistent with changes in population, environment, technology and economic structure.

The series will bring together and integrate disciplines related to systems approaches for sustainable agricultural development, in particular from the technical and the socio-economic sciences, and presents new developments in these areas.

Furthermore, the series will generalize the integrated views, results and experiences to new geographical areas and will present alternative options for sustained agricultural development for specific situations.

The volumes to be published in the series will be, generally, multi-authored and result from multi-disciplinary projects, symposiums, or workshops, or are invited. All books will meet the highest possible scientific quality standards and will be up-to-date. The series aims to publish approximately three books per year, with a maximum of 500 pages each.

*The titles published in this series are listed on p. viii.*

# Applications of Systems Approaches at the Farm and Regional Levels

# Volume 1

Proceedings of the Second International Symposium on Systems Approaches for Agricultural Development, held at IRRI, Los Baños, Philippines, 6–8 December 1995

Edited by

P. S. TENG
*International Rice Research Institute, Manila, Philippines*

M. J. KROPFF
*Wageningen Agricultural University, Wageningen, The Netherlands*

H. F. M. ten BERGE
*Research Institute for Agrobiology & Soil Fertility (AB-DLO), Wageningen, The Netherlands*

J. B. DENT
*University of Edinburgh, Edinburgh, Scotland, UK*

F. P. LANSIGAN
*University of the Philippines, Los Baños, Philippines*

H. H. van LAAR
*Wageningen Agricultural University, Wageningen, The Netherlands*

IRRI

KLUWER ACADEMIC PUBLISHERS
DORDRECHT / BOSTON / LONDON
in cooperation with
International Rice Research Institute
International Consortium for Agricultural Systems Applications

Library of Congress Cataloging-in-Publication Data is available.

ISBN 0-7923-4285-2 (vol. 1)
ISBN 0-7923-4286-0 (vol. 2)
ISBN 0-7923-4287-9 (set)

---

Published by Kluwer Academic Publishers,
P.O. Box 17, 3300 AA Dordrecht, The Netherlands.

Kluwer Academic Publishers incorporates
the publishing programmes of Martinus Nijhoff,
Dr W. Junk, D. Reidel, and MTP Press.

Sold and distributed in the U.S.A. and Canada
by Kluwer Academic Publishers,
101 Philip Drive, Norwell, MA 02061, U.S.A.

In all other countries, sold and distributed
by Kluwer Academic Publishers Group,
P.O. Box 322, 3300 AH Dordrecht, The Netherlands.

*Printed on acid-free paper*



Printed and bound in Great Britain by
Hartnolls Limited, Bodmin, Cornwall

# Contents

# Systems Approaches for Sustainable Agricultural Development

1. Th. Alberda, H. van Keulen, N.G. Seligman and C.T. de Wit (eds.): *Food from Dry Lands.* An Integrated Approach to Planning of Agricultural Development. 1992
ISBN 0-7923-1877-3
2. F.W.T. Penning de Vries, P.S. Teng and K. Metselaar (eds.): *Systems Approaches for Agricultural Development.* Proceedings of the International Symposium (Bangkok, Thailand, December 1991). 1993
ISBN 0-7923-1880-3; Pb 0-7923-1881-1
3. P. Goldsworthy and F.W.T. Penning de Vries (eds.): *Opportunities, Use, and Transfer of Systems Research Methods in Agriculture to Developing Countries.* Proceedings of an International Workshop (The Hague, November 1993). 1994
ISBN 0-7923-3205-9
4. J. Bouma, A. Kuyvenhoven, B.A.M. Bouman, J.C. Luyten and H.G. Zandstra (eds.): *Eco-regional Approaches for Sustainable Land Use and Food Production.* Proceedings of a Symposium (The Hague, December 1994). 1995
ISBN 0-7923-3608-9
5. P.S. Teng, M.J. Kroppf, H.F.M. ten Berge, J.B. Dent, F.P. Lansigan and H.H. van Laar (eds.): *Applications of Systems Approaches at the Farm and Regional Levels.* 1996
ISBN 0-7923-4285-2
6. M.J. Kroppf, P.S. Teng, P.K. Aggarwal, J. Bouma, B.A.M. Bouman, J.W. Jones and H.H. van Laar (eds.): *Applications of Systems Approaches at the Field Level.* 1996
ISBN set volume 5 & 6: 0-7923-4287-9; ISBN 0-7923-4286-0

KLUWER ACADEMIC PUBLISHERS – DORDRECHT / BOSTON / LONDON

# Preface

The world has indeed become a smaller place because of modern communication and information technologies. Agriculture's stakeholders are constantly reminded of the interrelatedness between agriculture and other sectors of national economies and the global economy. Often, this reminder is a shocking one, especially in industrialized countries, as agriculture is relegated to a secondary role due to perceived farm product surpluses. At the same time, it has become fashionable to talk about the complex issues facing agriculture in the post 'Green Revolution' era in developing countries, especially in fora to discuss the future of agriculture in relation to food security, food sufficiency and the environment. Warnings of impending food crises due to continuing high population growth rates in developing countries, reduction in agricultural land, degradation of existing granaries, competing uses of water, scarcity of farm labour and the negative aspects of global climate change were frequent in the mass media in 1995, a year which saw a temporary shortage of rice, the staple food in Asia. Yet, from an holistic systems perspective, these two apparently conflicting situations, highlighted by modern communication and information media in a seemingly similar time frame, are not necessarily at odds, and should not be so. They provide an opportunity to work towards a globally food secure situation based on sustainable natural resource management and social justice. It was towards this end that the theme of this book, and the symposia series, was conceived.

Systems approaches for agricultural development are needed if rational strategies for incorporating the role of agriculture into national development are to be made. Mathematical models and computer simulation provide objective tools for applying science to determine and evaluate options for resource management at field, farm and regional scales. Yet, these tools will not be very useful unless they incorporate social and economic dimensions. The second SAAD symposium attempted to foster this link between the biophysical and social sciences in the selection of keynote papers and oral presentations. This is a further reflection of how systems approaches have definitely moved beyond the research mode into the application mode. The large number and quality of interdisciplinary research teams in different parts of the globe, working to determine land-use options that will meet multiple goals and yet sustain natural resource bases, is a key indicator of this 'coming of age'.

At the farm level, where trade-off decisions between processes and products (commodities) feature strongly, much progress is also evident in the development of systems-based tools for decision making. The continuing distinction from, and yet links back to, parts of the older 'farming systems' approach with its strong sociological slant, is a promising development that will lead to more relevant applications. Also at the SAAD symposium, many computer software programs which had potential for farm-level decision support were demonstrated. We see this role of computer technology to implement systems approaches at the farm level as an increasing trend in all parts of the world.

This book contains a subset of the papers presented at the Second International Symposium on *Systems Approaches for Agricultural Development (SAAD)*, held at the International Rice Research Institute (IRRI), Los Baños, Philippines, 6–8 December 1995. The papers in this book address issues at the farm and regional levels while the companion volume, similarly named, contains papers which address issues at the field level. Together, their contents reflect the objectives of SAAD, which were:

- to review the status of applications of systems research and modelling in agricultural research, with special focus on countries where agricultural development is facing major challenges; and
- to promote international collaborative activities and to increase awareness of the opportunities for using systems approaches as a tool in research and planning.

We believe these objectives were met by the more than 200 scientists from 32 countries, when they came together in December 1995.

Production of this volume would not have been possible without the tireless efforts of a support team made up of Ms. J.M.C.A. 'PeeWee' Cabrera, Ms. R. 'Say' Badrina and Ms. Viring Lantican. Thanks are also due to the Communication and Publication Services of IRRI for editorial assistance.

The second SAAD symposium was guided by an International Steering Committee comprising J.T. Ritchie, J.B. Dent, J.W. Jones, H.F.M. Ten Berge, M.J. Kropff and P.S. Teng, while local organization at Los Baños was done by T.P. Tuong, S. Pandey, K.G. Cassman, G.J.D. Kirk and P.S. Teng. The SAAD symposium was sponsored by IRRI, the International Consortium for Agricultural Systems Applications (ICASA) and by the Simulation and System Analysis in Rice Production (SARP) project. We thank IRRI for its financial, logistical and moral support, without which the symposium and this book would not have been possible. The following persons actively contributed to the logistical support during the symposium: Ms. Viring Lantican, Ms. Susan Telosa, Ms. J.M.C.A. 'PeeWee' Cabrera, Ms. Anaida Ferrer, Ms. Cecille Lopez, Ms. Perly Villamayor and Mr. Benjie Nunez.

Lastly, we hope that the two books from this second international symposium on systems approaches will be further testimony of the vision and philosophy, provided by the late Professor C.T. De Wit, Wageningen, The Netherlands, which has guided many applications of the approach.

P.S. Teng
M.J. Kropff
H.F.M. Ten Berge
J.B. Dent
F.P. Lansigan
H.H. Van Laar

PART I

# Farm level

# Opportunities for systems approaches at the farm scale

J.W. JONES[1], P.K. THORNTON[2] and J.W. HANSEN[1]
[1]*University of Florida, Agricultural and Biological Engineering Department, P.O. Box 110570, Gainesville, FL 32611-0570, USA*
[2]*International Fertilizer Development Center, P.O. Box 2040, Muscle Shoals, AL 35662, USA*

**Key words:** farm-scale systems approach, farming systems, farming systems research, simulation, farm model, crop model, hierarchy

**Abstract**
Considerable progress has been made in the development and application of models at the field scale. However, suitability of a particular management technology for a farm cannot be determined by crop, soil and weather factors alone. Resource constraints cannot be determined at the field scale, and neither can risks to the economic or nutritional well-being of the household. Systems approaches at the farm scale must take account of biophysical characteristics such as soil, weather, crops and animals, characteristics of the farm and the farm household, and the social, economic, political and ecological environment of the farm. To make significant advances in farm-scale systems analysis, a 'bottom-up' approach is required. Three levels of analysis provide a practical framework for future efforts. The unconstrained production level relies on enterprise simulation and simple budgeting models to aggregate farm production, estimate profit and compute the resources required to manage a farm in a particular way. The resource-constrained production level considers internal farm constraints in assessing what the farm with its existing limitations can attain. At the adaptive production level, a farm decision model considers the characteristics and preferences of the farm household. These three levels of farm system analysis parallel levels of analysis at the field scale that continue to provide guidance in the development and use of crop models. This framework may lead to more generic farm-scale models, definitions of minimum data sets for their use and wider application of systems approaches at the farm scale.

## Introduction

The title of this paper suggests that opportunities exist for applying systems approaches to farming systems, as if this were a new idea. On the contrary, numerous scientific studies have analysed farming systems for one purpose or another for several decades. There are three major reasons for focusing on this topic today: (i) the nature of contemporary problems and issues, (ii) the need for knowledge about agricultural systems by a wide range of users and (iii) the inadequacy of traditional methods for supplying the information to those who need it when they need it.

The need to understand agricultural systems and their interaction with other segments of society is greater today than ever before. Population projections suggest a doubling of food demand by the middle of the twenty-first century (Islam 1995). In addition, if certain climate change scenarios come to pass, agricultural production in some regions, particularly in the tropics, could suffer significantly (Rosenzweig and Parry 1994). Many examples could be cited where local or regional land and water resources have been degraded (e.g. Campbell et al. 1994). The lack of new



*P.S. Teng* et al. *(eds.): Applications of Systems Approaches at the Farm and Regional Levels, 1–18.*

land that can be brought into agricultural production will limit our ability to meet anticipated food demands. These issues pose major threats to food security. How can productivity be increased to meet the growing demands for food while ensuring the sustainability of agriculture and the environment for future generations? Which technologies are efficient, economically viable and environmentally appropriate for a farmer endowed with specific resources? How can public policy be formed to help ensure that agriculture will meet current and future needs? There is considerable debate about how to introduce sustainability and natural resource management into research and development programmes at all scales from field to farm, watershed and ecosystem. Research must result in an understanding of agriculture's interactions with other segments of society if such issues are to be addressed adequately.

Decision makers need information supplied by research to make informed choices about new agricultural technologies and to make and implement policies that can enhance food production and sustainability. Many decision makers at various levels, including policy makers, businesses that supply agricultural inputs and market products, and the farmer, impact on how food is produced and how resources are managed. The effects of many of their decisions are hard to characterize. Ultimately, however, it is the farmer who makes the final choices about acceptance of new methods or technologies. If agriculture is to meet the needs of society as a whole – producing food and other products while protecting natural resources – it must first meet the needs of farmers who manage the land. Furthermore, policy makers need to understand the impacts of their decisions on the economic well-being of the farm household and also on the natural resource base and the regional or national economy. The users of information generated through research are not only farmers but, increasingly, decision makers at all levels in the private and public sectors.

Traditional agricultural research conducted along reductionist lines is not adequate for providing the understanding necessary to address the broad issues outlined above, nor for packaging the information in a way that a wide variety of users can easily access and interpret. Field- or plot-scale research is not sufficient. Economic risk cannot be adequately assessed at the field level. There may be many resource constraints to farm production that cannot be dealt with at the field level. A systems approach is essential, therefore, for understanding sustainability and the effects of changing external influences on farming systems. A systems approach requires the analyst to define system boundaries and system components, to develop models of the components, and to integrate them for studying the behaviour of the entire system (Rabbinge et al. 1994). A farm consists of the farm household, an area of land, a set of resources and various production enterprises. It belongs to a food industry system (through its enterprises), to an ecological system (through its basis in land, water, and air resources) and to an sociopolitical system (through its household). The boundary conditions of a farm are the set of policies, prices and constraints, as well as the climatic conditions, social conditions and ecological surroundings.

The greatest difficulty in using a systems approach is the proper formulation and development of a model of the system. Ideally, a farm model would describe the decision-making process of the farmer in response to biophysical conditions of the farm, the goals of the farmer and the boundary conditions of the farm. It would also describe the biophysical and economic behaviour of the farm through time in response to its internal characteristics and environment. One could then conduct experiments on the farm model to infer behaviour of the real farm. The model might be used to test hypotheses about farm behaviour under different internal and external conditions. Such an ideal model would be extremely useful. Researchers could characterize a farm, simulate a new technology and determine the chances of its adoption. They could use the model to help the farmer understand how the new technology could fit into their existing production system and how it would affect various aspects of the farm. Such a model would also be useful to policy makers who might be considering legislation to enhance the economic stability of farmers, to promote production of a commodity important to the economy or to protect natural resources. In short, the model would be a tool for experimentation to allow decision makers to understand the consequences of their decisions at the farm level before making them. Unfortunately, such ideal models do not exist, and the requirements for developing something even remotely close to the ideal model seem daunting and overwhelming. What can be done?

In this paper, we first present a perspective of previous approaches to farm-level systems research. We then outline a framework for farm-scale analysis based on several levels; for each level, we discuss model requirements and appropriate interpretation and use. This provides a practical pathway for progress by building on what has been done in the past and integrating the biophysical with the socioeconomic. In taking this approach, we admit a bias towards the biophysical aspects of farming systems. However, it is appropriate to emphasize the possibilities of merging the significant recent advances in field-scale modelling with the wider perspectives that predominate in the economics literature. A few critical efforts fuelled rapid development in the use of the systems approach at the field scale; opportunities for similar advances at the farm scale may exist, and could perhaps be fuelled in the same way.

## A perspective on previous research

### *Modelling Farming Systems*

Early research at the whole-farm scale emphasized farm budget analyses. Before the 1950s, agricultural economists were developing procedures for accounting and production economics at the farm scale (Lee 1983). Much of this early work has evolved into a large number of computer software packages. Klein and Narayanan (1992) referred to 122 farm analysis software packages developed in the USA and Canada during the 1980s. These packages dealt with farm accounting, tax management, financial analysis and farm business planning. Most of this early work was designed specifically for farmers' use, and computer technology evolved so that farmers now widely use this type of application.

In the late 1950s and 1960s, research emphasis in the US shifted towards the use of Linear Programming (LP) as a tool, largely credited to the influence of Heady's work (Heady 1952; Lee 1983). With this tool, economists analysed farm growth, response to policies, cost minimization and minimum resource requirements for specified farm income, assuming profit maximization behaviour by the farmer. They extended single period LPs to cover multiple planning years for more realistic analyses (Klein and Narayanan 1992). Mathematical programming procedures were also developed and applied to risk and uncertainty. The early study of Camm (1962), various quadratic risk programming procedures (outlined in Anderson et al. 1977) and the mean absolute deviation (MOTAD) approach (Hazell 1971), are examples. Simulation was also introduced as a tool for more flexible analysis of complex farms (Halter and Dean 1965). These models were aimed at improving operation of farms.

During the early 1970s, a shift occurred in the USA and Canada towards the development and use of macroeconomic models for policy analysis. During this time, there were major concerns about regional and national agricultural policy issues, including international trade. Some studies were conducted to determine how to aggregate farm models to regional and national results. However, these did not seem practical at the time. Emphasis on modelling at the farm scale during this time concentrated on econometric household models that could apply to developing agriculture (for example, Barnum and Squire 1979).

In the early 1980s, some efforts were directed back towards farm-scale analyses (e.g. Baum and Schertz 1983). The macro models did not adequately allow one to understand the likely effects of macro policy and market conditions on individual farmers. Hybrid LP-simulation models such as FLIPSIM (Richardson and Nixon 1985) were introduced, as well as multi-year, stochastic farm simulation models (Crawford and Milligan 1982). These and other models such as Beck and Dent (1987) allowed more comprehensive interactions among farm components, and sequential decisions to mimic farm progress over multiple time periods more realistically. Taylor (1983) suggested that more complementary modelling efforts between macro and micro (farm) levels were needed, and recommended a bottom-up approach for modelling farming systems. Nelson (1983) pointed out that a major limitation of farm models was the lack of data. He recommended more efforts to develop and integrate data for farm analyses. Unfortunately, little progress has been made towards this goal. In some ways, this is not surprising because so many models and analytical techniques are available, each with different data requirements.

Several studies using farm-scale models for a range of applications have been published recently. Lee et al. (1995) used a farm model to help understand the behaviour of subsistence farmers in Western Samoa after government efforts to stimulate production of exportable crops had only limited success. Systems approaches have been used to study economic risk due to drought for livestock farms (e.g. Stafford-Smith and Foran 1992) and cropping farms (e.g. Klein et al. 1989; Kingwell et al. 1992). Because of concern for sustainability of natural resources and the environment, several recent studies have considered both environmental effects and economics, often in a multiple goal programming framework. For example, Fraisse et al. (1995) developed a generic farm-scale model to study the effects of dairy farm management on nutrient loading into surface and groundwater. This model used water quality simulation models (Leonard et al. 1987) in a Geographic Information System (GIS) framework to compute farm-scale material flows and nutrient losses in water that leaves the farm. This model was based on earlier work by Negahban et al. (1993), who developed a dairy model for use by the state water management authorities in Florida. Farm models in The Netherlands have focused more on economics and the trade-offs between profits and pollution. For example, Berentsen and Giesen (1995) used an LP framework to study dairy farms and the incorporation of a levy on nitrogen losses. De Koeijer et al. (1995) also combined economic and environmental pollution effects in a model to study mixed crop and livestock farming. Another type of hybrid farm model involves using biophysical simulation models to derive input–output coefficients that are then used in a mathematical programming model. Examples of this approach are Veloso et al. (1994), who used such an approach to study feasible cropping plans for newly opened agricultural land in the savannahs of Brazil, and Herrero et al. (1995), who integrated a cattle simulator with multiple criteria programming models to devise management strategies for grazing dairy systems in Costa Rica. Examples of each of these types of farm model are numerous for crops, less so for livestock systems.

We can make some generalizations concerning farm-scale modelling and analysis, although like all good generalizations, exceptions abound. First, the main purpose of most of the recent studies has been to gain improved understanding of farm systems regarding response to policies, resource requirements, farm growth potential and, most recently, environmental impacts of farming practices. Second, farmers are generally not involved much in model development and use, except perhaps by providing information through surveys. Third, there is usually a new model for each study, with little acknowledgement of the potential for using the same model for different farms or different types of analysis. FLIPSIM is an exception, as are recent models that emphasize biophysical interactions at the farm scale rather than economics, such as the GIDM (Generic Interactive Dairy Model) model (Fraisse et al. 1995), FSS (Farming System Simulator; Hansen 1995) and the sheep farm model of Cacho et al. (1995). Fourth, most farm-scale models have been developed with a bias towards economics and limited consideration of the biophysical components. Many of these farm studies have relied on historical data or surveys to provide estimates of production responses. Few efforts have attempted to merge biophysical models of crop and animal production with economic and environmental considerations at a whole-farm scale, although the approach has been suggested in the past (Dent and Thornton 1988). An exception is the farm model of Edwards-Jones and MacGregor (1994) constructed around crop simulation models; household decisions are modelled using sets of rules and a socioeconomic data base. The testing of many of these farm models, however, presents problems that we have scarcely started to address.

### *Farming Systems Research*

How are the farm modelling efforts referred to above related, if at all, to Farming Systems Research (FSR) approach? FSR shares with most farm-level modelling work the objective of understanding and improving existing farming systems. Its focus, in contrast, is on farmer participation and on-farm research. The general methodology of FSR has been to describe an existing farming system, identify constraints impeding farmers from reaching their goals, and then develop methods to overcome those constraints and test them on the same farm (Anderson 1985; Norman and Collinson 1985). FSR has had a major impact on promoting interdisciplinary approaches to overcome constraints at the farm level. Ideally, this approach would lead to identification of factors beyond the farm gate that may affect the appropriateness of a particular technology and its acceptance by the farmer. The farmer works directly with the FSR teams and participates in all aspects of the research.

Achievements of the approach have not matched expectations (Dent 1993; Stroosnijder and Van Rheenen 1993). Menz and Knipsheer (1981) pointed out one of the basic problems: FSR results are location-specific. Findings for one farm cannot be implemented satisfactorily over large geographical areas without modifications to local needs and conditions. This is the same problem that Nix (1984) identified with most field-scale research; it does not facilitate the transfer of production technology to other locations with different edaphic and climatic conditions. To overcome this problem at the field scale, Nix (1984) recommended a systems approach in which minimum data requirements are defined for assessing results among different locations. This seems like a simple recommendation, but it has profound implications. One cannot compare experiments conducted in two locations unless the same observations are made in both. Such data could be used with field-scale models to assess the performance of production practices in new areas where experiments have not been conducted but where the minimum weather and soil data are available. This framework guided the former IBSNAT project and led to major progress in the effective development of a systems approach to the management of field crops (Uehara and Tsuji 1993).

If one's goal is to balance required increases in food production with environmental protection – the challenge proposed by this symposium – then the FSR approach has other limitations. FSR does not adequately address the important consideration of risk in the form of either temporal variability or system degradation or enhancement. The static nature of the FSR approach is analogous to single season experiments at specific field sites. Norman and Collinson (1985) suggested that most FSR successes resulted more from the ability of existing farming systems to adapt to constraints than from the ability of FSR to relieve the constraints. Factors far from the farm gate cause many of the constraints that farmers face in developing countries. Site-specific, on-farm research simply fails to address such issues adequately. FSR lacked, and still lacks, a methodology to effectively address policy issues that constrain farming systems.

Various authors have suggested using socioeconomic (Anderson 1985; Dillon and Virmani 1985) and biophysical models (Dent and Thornton 1988) within the general framework of FSR to overcome the location specificity problem. McCown et al. (1994) suggested that the term operational research be added to the FSR approach outlined by Dillon and Virmani (1985). The operational research component would incorporate models of farms and their enterprises to help researchers, farmers and policy makers understand the best options for a farm, given its internal characteristics and socioeconomic and political setting. They proposed combining on-farm research, in close cooperation with farmers, with the more formal systems approach that uses socioeconomic and biophysical models. Although this is a challenge, it would provide a framework for analysis, tools that could be used across locations, an approach to provide information to both farmers and policy makers, and a structured agenda for interdisciplinary research.

FSR is still in an early stage of scientific development, characterized by an emphasis on observation and site-specific, trial-and-error experiments. If it is to evolve into a more mature scientific endeavour, then practitioners must develop theories, quantify relationships and build models that capture the major behavioural characteristics of farming systems. The people that Röling (1994) referred to in a dazzling oxymoron as 'post-Newtonian biophysical scientists' will doubtless consider this conclusion anathema or nonsense, or both. Nevertheless, the simple fact is that at the farm level we have not yet been sufficiently creative.

*Biophysical modelling*

A systems approach at the farm scale requires quantitative estimates of enterprise performance for the different management options being considered. Some economic models use average yields of crops and livestock based on historical data. Others add a random component to reflect uncertainties. Statistical production function models are sometimes used to relate crop yields to resources such as water or nitrogen. While any of these approaches could provide the estimates needed for profit calculations in the farm models, they are highly restrictive because they require data for the management practices of interest collected under the soil and weather conditions of the farm being studied. These approaches are inadequate if farm models are to be used across locations with different soil and weather conditions, or if they are to be used to examine crops or production technologies that are new to a region. Furthermore, decision makers need to know more about the performance of an enterprise than simply yield. Soil erosion, nutrient leaching, or pesticide run-off, for example, may be of equal importance. Such variables depend on the characteristics of the site, and are difficult and expensive to measure. Biophysical models are needed for simulating enterprise performance for the conditions of the farm under study.

Biophysical models of crop and livestock production have been under development for over 25 years. The efforts of several research groups around the world (e.g. Uehara and Tsuji 1993; Ten Berge 1993; Jones 1993) have resulted in considerable progress in the development and widespread testing and use of crop models. The International Consortium for Agricultural Systems Applications (ICASA) is a new effort to promote the efficient and effective use of such models. Field-scale models will play an important role in future farm-scale models. These types of models simulate growth of the plant or animal (often on a daily basis), timing of physiological events, economic yield and yield components for given soil, weather and manage-

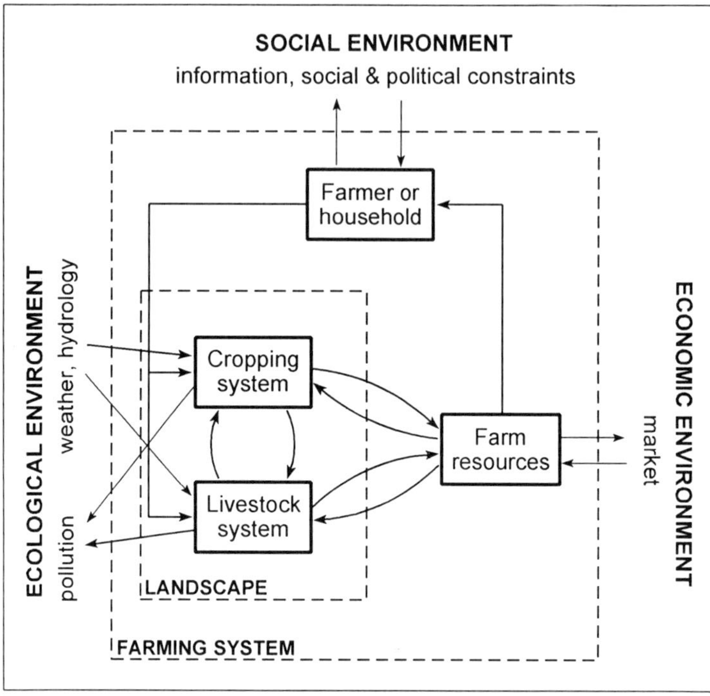


*Figure 1.* Schematic of a farming system showing crop and livestock production systems on the farm landscape, resources used and produced by farm activities, a household with consumption and decision-making characteristics, and the interface to the ecological, economic and social environment of the farm.

ment conditions. They may also estimate soil erosion, nutrient leaching and other factors important to sustainable production of a farm. Simulated outputs may be used in a decision module. Figure 1 shows a simple schematic of how crop and livestock models can be used in a whole-farm model.

Although it is not a new idea, progress in combining biophysical and economic models for farm-scale analyses has been strictly limited for at least four reasons:

(1) The availability of crop and livestock models, and researchers' confidence in them, has been slow to develop.

(2) Suitable data sets have been scarce and costly to obtain.

(3) Computer hardware and software difficulties have only recently been overcome. Powerful computers are widely available now, whereas they were not widely available ten years ago. Programming difficulties remain, however. New programming methods are needed to allow crop models to run in parallel for several fields and in parallel with management operations and resource accounting at the farm level. Object-oriented programming provides a structure for major advances in this area, but the inertia in agricultural system modelling still lies in procedural programming languages, and little progress has been made to overcome limitations in existing models.

(4) Crop and livestock models do not include all factors that affect production in the real world. They usually over-predict performance because they do not account for pest, disease or weed effects. Researchers will continue to improve these models, yet they are useful even at their current stage of development. Various research efforts are now demonstrating the feasibility and value of combining biophysical and economic models at the farm scale. Although the development of farm-scale models that include biophysical models is costly, such models have the potential to be used widely because of the generic nature of the biophysical models.

## Levels of analysis and model requirements

One major lesson learned in applying systems approaches at the field level of analysis is that a single crop model will not suffice for all purposes. Some research objectives are oriented towards understanding the processes and interactions that occur in a crop field. Others attempt to determine yield potential for different areas and to identify gaps between potential and actual yields. Yet other studies are concerned with defining optimal water and fertilizer management or predicting yields at several sites. While the models used in such studies have varying levels of detail, models clearly do not have to include all factors to be useful for some purposes.

A useful characterization of crop models for use at the field scale is due to De Wit (Penning de Vries and Van Laar 1982). Initially, four levels of models were defined, but these were later simplified to three levels (Rabbinge 1993). First is the potential yield level; crop growth is dependent on weather factors alone, such as temperature, solar radiation and perhaps day length and $CO_2$ concentrations in the atmosphere. This level of model includes basic crop growth processes such as photosynthesis, respiration, tissue growth and development. Its main use is to understand how these factors affect the potential production of a crop in a given climate when water and fertilizers are in adequate supply and no pest damage occurs. A second level of detail is characterized as attainable yield. At this level, the model has components that describe water and nitrogen limitations to production. Attainable yield thus depends on water and nitrogen supplied by management and by rainfall and soil organic matter. Many existing crop models are able to describe crop growth and yield with these assumptions (Jones et al. 1994b). The third level of detail is the actual yield level that attempts to explain all of the factors that influence growth in a field, including pest damage and mismanagement. This characterization of levels of detail has been useful in providing both a practical pathway for model development and a conceptual framework for model applications. It has also served to determine data requirements for model development, considering the need for more comprehensive data sets as the level of model detail increases.

We can define a similar characterization for farm-scale models. This will help in several ways. First, it will provide a pathway for evolution of less comprehensive into more comprehensive models. Most crop models evolved from potential yield models to attainable yield models that include soil water and nitrogen balances in

*Table 1.* Suggested levels of farm-scale models with associated inputs, model components and outputs.

| Farm production level | Inputs | Model components | Outputs |
|---|---|---|---|
| 1. *Unconstrained* | Endogenous: fields, enterprises<br><br>Exogenous: technology packages, weather, prices<br><br>Initial conditions: physical characteristics | Soil, crop, livestock, ecological, budgeting, resource accounting<br><br>Decisions: none | Production, costs, profits, resources needed, non-market outputs, risk |
| 2. *Resource-constrained* | Endogenous: fields, enterprise options, technology options, labour, capital, equipment<br><br>Exogenous: weather, prices<br><br>Initial conditions: physical characteristics, resources (labour, capital, etc.) | Soil, livestock, ecological, budgeting, resource accounting<br><br>Decisions: enterprise mix, resource use conflicts, household consumption | Production, costs, profits, resources needed, non-market outputs, risk, enterprise mix, technology, constraints |
| 3. *Adaptive* | Endogenous: fields, enterprise options, technology options, labour, capital, equipment, farmer characteristics<br><br>Exogenous: weather, prices, social factors<br><br>Initial conditions: physical characteristics, resources | Soil, crop, livestock, ecological, budgeting, resource accounting<br><br>Decisions: enterprise mix, resource use conflicts, household consumption, technology adoption, farm expansion/contraction | Production, costs, profits, resources needed, non-market outputs, risk, enterprise mix, technology, constraints |

order to simulate production under given weather, irrigation, and nitrogen fertilizer and residue management. Second, it will help provide a basis for defining minimum data requirements for applying farm-scale models across a range of farming conditions. Finally, it will give users a framework for analysis, helping to ensure that they interpret model results with full acknowledgement of the limitations and assumptions of the model.

We suggest three levels of analysis for farm system models: unconstrained, resource-constrained and adaptive. The unconstrained level assumes that the crops grown and the production practices used in each field are fixed. Crop models are run for each field and resulting production is computed based on the size of each field. The farm model includes a budgeting component that calculates the total resources required, costs and profit. If the farm has livestock, appropriate biophysical models would also simulate the livestock and their interactions with crop enterprises. A major assumption at this level is that resources other than land are not limiting. This is hardly ever the case, so how would such a model be useful? Its main use would be to compare the potential of different ways of managing the farm under its weather and soil conditions. This potential can be expressed as biological production, net returns, resources required or environmentally important outputs. The farming system could be simulated for an extended period to examine long-term impacts of soil degradation or price trends. The simulation could be replicated with the same initial conditions but different sequences of weather and prices to examine the impact of these sources of risk on farm viability or on the potential acceptability of a particular technology to farmers. Table 1 summarizes the model components, inputs, and categories of output suggested for each level of detail. Hansen et al. (1996) present data requirements for this and other levels of analysis.

At the first level of analysis, we focus on the ecological components of a farm while largely ignoring the economic and social components. The economic component becomes important at the resource-constrained level. At this level, we assume that the farm has a particular set of resources. The question addressed at this level is, what is attainable from an economic point of view if one or more farm resources are fixed. A farm model at the resource-constrained level would include a decision model to optimally allocate resources based on profit maximization. The decision model would select from a set of alternatives the production option that maximizes profits without violating constraints. No attempt is made to model behaviour based on characteristics or preferences of the particular farmer. Such analysis would provide decision and policy makers with an upper economic limit of what is feasible for a particular set of management options and resource limitations. Results could suggest ways to alleviate limitations and how they would affect farm production.

At the adaptive level of analysis, we attempt to deal rigorously with the social component of a farming system besides the ecological and economic components. A farm model at this level includes a decision model that attempts to describe behaviour based on farm household characteristics and preferences in order to gain understanding of factors that influence decisions such as acceptance of new technology and choice and management of crop and livestock enterprises. It would need to deal with strategic and tactical decisions, so that the farm itself could evolve in response to its changing environment. This level is the most general and by far the most challenging of the three proposed. To date we have made little progress in understanding the decision-making processes of farmers.

We summarize the distinguishing characteristics of the three levels as follows. The unconstrained production level describes the biophysical potential of a prescribed set of management practices without any consideration of farm constraints other than land. The resource-constrained production level identifies the economically optimal production for a set of management options, given the internal farm resource constraints and biophysical conditions. The adaptive production level adds social considerations: farm goals (which may be far more complicated than profit maximization), and farmer decision making, taking into account internal and external constraints and biophysical conditions of the farm. We can make considerable progress at the unconstrained and resource-constrained levels before achieving satisfactory capabilities for modelling the decision maker. In theory, the analyst could progress from unconstrained to resource-constrained to adaptive levels in a building-block approach, similar to that used by many crop modelling efforts.

Of course many existing models of farms do not fit very nicely into this classification. Nevertheless, we believe the classification is useful because the advancement of farm-scale models must include adequate representation of the biophysical responses, and indeed different levels of assumptions can be made regarding the farmer as decision maker. It builds on the considerable progress already made in modelling crops and livestock systems. It emphasizes a minimum data set approach (Hansen et al. 1996). It provides a framework for interpreting farm analyses. Finally, it provides a framework for interdisciplinary cooperation between biophysical scientists and economists for addressing issues at the farm scale.

## Examples of farm models for characterizing sustainability

*Unconstrained level of analysis*

As an example of a farm model at the unconstrained production level, we describe research to quantify and diagnose constraints to sustainability of small farms in South America (Hansen 1995; Thornton et al. 1995; Hansen and Jones 1996). A farm model (FSS) developed for this research simulates various enterprises in the fields of a farm using biophysical simulation models, and aggregates resource use, costs and profits for the whole farm. Although FSS simulates household consumption and interactions among resources, it does not attempt to model the decision-making behaviour of the farmer. FSS uses the biophysical models contained in DSSAT v3 (Jones 1993; Tsuji et al. 1994). These crop-scale models were combined using an object-oriented framework to characterize the farm, its fields, resources and enterprises (Hansen 1995). The farm model controls the operation of each of the crop models based on a management plan that the user has completely specified. The plan includes the sequence and management of crops on each field on the farm. The farm model requires the same data as the DSSAT v3 crop models (Jones et al. 1994a), plus data for commodity prices and for resource accounting (Hansen et al. 1996). FSS simulates the growth of crops in each field through time and summarizes resource use and yields for each field, crop and season combination. A farm scenario is simulated over several years to evaluate cumulative effects of farm management and external environment (such as weather and price variability) on resource stores, changes in soil properties and farm wealth. The multiple year scenario is replicated using different weather and price time series, so that probability statements can be made about the behaviour of the farm in terms of uncertainty and risk.

If one defines minimum thresholds of farm resources such as wealth, the probability that the farm can 'sustain' can be estimated as the proportion of simulated sequences in which farm resources remain above the minimum threshold (Hansen and Jones 1996). The outputs through time for one farm plan can be compared with another to determine which would be more 'sustainable' (i.e. more likely to remain above thresholds for a given length of time). Sensitivity analyses can also be carried out to examine the effects of different internal or external factors on farm behaviour and 'sustainability'. It allows researchers to better understand the potential impacts of different farm plans before investing large amounts of resources in testing them on farms. Hypotheses about constraints to sustainability can be tested by sensitivity analysis. Hansen (1995) has examples of these types of analyses for a farm in the Cauca Region of southwest Colombia.

Farm simulation results can be presented in several useful ways. A box plot with minimum, 25th, 50th, 75th percentiles and maximum values of some relevant state variable, such as wealth, provides a view of how risk changes through time. Sustainability expressed as the probability of not violating a system threshold integrates the effects of levels, trends, variability and autocorrelation, and relates them to goal levels. Another way to provide a visual representation of results is by using GIS. GIS provides a way to store spatial information about different fields of the farm, such as soil characteristics and simulated yields, resources used, erosion and chemical leaching, and to present results as maps. Several farm-scale GIS decision support systems have been developed with biophysical models of crop production (Engel et al. 1995) and environmental quality (Negahban et al. 1993; Fraisse et al. 1995). These DSSs organize spatial and attribute data of the farm, provide a way for users to specify production practices on any of the fields of the farm, simulate the biophysical responses and display results. Such systems easily aggregate results over different fields. The various approaches for presenting results of farm-scale analyses are equally appropriate for the other levels of analysis.

*Extending the approach to the resource-constrained level*

The addition of a decision model and internal farm constraints could extend this approach to simulate resource-constrained farm production. One difficulty is that each crop model has been developed to run for an entire season, whereas the models must run in parallel to allow for tactical decisions throughout the season to account for internal resource conflicts and other management constraints. This problem has practical solutions. Kelly (1995) developed an approach that used a multiple stage, non-linear decision model along with the DSSAT v3 crop simulation models to study the sustainability of a peanut–pasture–corn farm in North Florida. The decision model considered constraints to labour, machinery and to nutrient leaching. He developed the farm model with a recursive structure to optimize practices based on information available at each decision period, taking into account the constraints and expected performance of the enterprises. He also used different years of weather and price information to study variability in farm performance. As expected, results showed that internal constraints would shift optimal farm plans. Another example is the farm model described by Edwards-Jones and MacGregor (1994) that addresses the parallelism problem by iteration; if resource conflicts arise, it repeats each growing season with small changes in resource use until the farm plan can operate wholly within the constraints imposed by the finite resource levels of the farm household. The problem of identifying and resolving farm-level resource conflicts due to competing enterprises would be greatly simplified if crop models were restructured to allow all fields to be simulated in parallel each simulation day.

*Adaptive level of analysis*

We are not aware of many examples of adaptive production farm-level model that incorporate biophysical models. Efforts such as the work by Dillon et al. (1989), which included biophysical models, internal farm constraints, resource accounting and a decision model that considered farmer risk preferences, approach this level of analysis. The main reasons for so little progress at this level are presumably the complexity of the task and the fact that our lack of mechanistic understanding of the household decision-making process. This explains why current decision models must use some simple, empirical objective functions. If such objective functions were better predictors of human behaviour, then we could largely side-step the whole sociocultural problem. Unfortunately, this is not so. It is not at all clear how we should go about improving their predictive power.

## Discussion

Dent and Thornton (1988), nearly eight years ago, listed several problems with incorporating biological models into FSR. They were then discussing farm-level models at the adaptive production level, although they did not express it in those terms. The first item they discussed was the provision of valid biophysical models. The approach to farm-level modelling that we are proposing is critically dependent on such models. The situation has not improved much since the late 1980s. Although we have more models for more crops, like farm models these are often created *ab initio* with little consideration for what could be reused or recycled. This means that most biophysical models are poorly tested against good, independent data. Despite decades of field research at institutions throughout the world, where are the good data sets to test the models over even a single season? The situation for crop sequences and rotations is even worse. The implications of this are profound; even the basic building-blocks for farm models lack credibility.

Two things might help to change this situation. First, the creation of good data standards could have an enormous impact. Such work is neither scientifically interesting nor particularly publishable, but it is a critical component of increased application of the systems approach. Unless we have good, validated biophysical models, the types of farm models we are proposing would have little value. This suggests giving more attention to testing the capabilities of existing models. For example, do they really do a good job of simulating nitrogen transformations in the soil and uptake by the plant? If we are serious in wanting to look at sustainability issues, we must validate their sequencing capabilities. We have scarcely even started on this. Of course, working on new modules for the crop models is more interesting, but unless existing modules are properly tested, it is wasted effort. The second thing that would help the data set situation is equally mundane: sifting and collating the enormous quantities of field data that exist to generate data sets, wherever possible, that can be used for various purposes (including model testing). Adding value to previously collected research results sitting on dusty shelves could have an immense impact. Many experiments could never be incorporated into meaningful data sets because of various inadequacies, but there must also be many that could.

The second problem area that Dent and Thornton (1988) identified was variability: the variability of the real system compared with the variability expressed in the model; variability between on-farm and on-station yields (i.e. the so-called 'yield gap'), and variability in the system itself, particularly spatial variability at the micro scale. System versus model variability has received almost no attention over the last eight years. It is part of the wider validation question of what is an acceptable model. While validation is difficult to characterize philosophically, in practice it seems to cause few problems (although if we did better testing of our models, it might become more of an issue). The yield gap is a problem of system characterization and input data: if the crop models are good enough and we can characterize the farmer's field and the experiment station's plot, then no constraints should exist to simulating on-farm yield levels. This has now been well demonstrated. The spatial variability component has received much attention recently. Precision agriculture has grown up around its existence, and in high-input systems valuable information concerning spatial variability is available. Again, this is largely a problem of input data. In summary, these variability issues are not of great consequence, despite the fears expressed in 1988.

The third major problem area outlined in Dent and Thornton (1988) was the socioeconomic framework: 'The building of a whole farm model requires that decision making in a constraint-ridden production system can be represented in some algorithmic way ... it is better to use an inaccurate but internally-consistent framework than a woefully inadequate one.' In retrospect, internal consistency is not really worth much, and in fact we are still left with woeful inadequacy. The situation has changed somewhat, however. As noted above, we need considerably more work towards understanding the decision-making complex if we are to make much progress (Edwards-Jones and MacGregor 1994) and, equally important, if such models are going to be amenable to objective validation. The socioeconomic framework, therefore, still presents a considerable problem.

In practical terms, more energy should be spent in building and testing farm-level models because this is really the only way in which their inadequacies will become known. Such efforts are bound to suggest new ways of dealing with decision making at the farm level. For now, better characterization of the farming systems under study is one practical way to improve the applicability and relevance of farm-level models. Another is simply to build better links between economic and biophysical component models. If we handle the major impacts of human actions on physical and economic variables in the modelled system credibly, then decision makers themselves may use and experiment with farm-level models to generate the information that they need to make informed decisions consistent with their preferences and objectives.

It is hard to escape the conclusion that, despite the difficulty of modelling the human agency in agricultural systems, the major constraint to useful farm-level models is still inadequate biophysical models. Despite the increasing gravity of the world's food security and natural resource situation, the systems approach is still not well integrated in teaching, research and development. Opportunities for applying the systems approach at the farm scale, and to agricultural systems in general, will continue to be limited until we achieve much better integration in generating data and turning them into useful information, thereby increasing the knowledge base.

## References

Anderson J R (1985) Assessing the impact of farming systems research: Framework and problems. Agric. Admin. 20:225–235.

Anderson J R, Dillon J L, Hardaker J B (1977) Agricultural decision analysis. The Iowa State University Press, Ames, IA, USA. 344 p.

Anderson J R, Dillon J L, Hardaker J B (1985) Socioeconomic modelling of farming systems. Pages 77–88 in Remenyi J V (Ed.) Agricultural systems research for developing countries. Australian Centre Proceedings No. 11, Canberra, Australia.

Barnum H N, Squire L (1979) A model of an agricultural household: Theory and evidence. Occasional Paper No. 27, World Bank, Washington DC, USA. 107 p.

Baum K H, Schertz L P (Eds.) (1983) Modeling farm decisions for policy analysis. Westview Press, Boulder, CO, USA. 418 p.

Beck A C, Dent J B (1987) A farm model for policy analysis in an extensive pastoral production system. Aust. J. Agric. Econ. 31(1):29–44.

Berentsen P B M, Giesen G W J (1995) An environmental-economic model at farm level to analyse institutional and technical change in dairy farming. Agric. Syst. 49(2):153–175.

Cacho O J, Finlayson J D, Bywater A C (1995) A simulation model of grazing sheep. II. Whole farm model. Agric. Syst. 48(1):27–50.

Camm B M (1962) Risk in vegetable production on a fen farm. Farm Econ. 10:89–98.

Campbell K L, Graham W D, Botcher A B (Eds.) (1994) Environmentally sound agriculture: Proceedings of the Second Conference. ASAE, St. Joseph, MI, USA. 578 p.

Crawford E W, Milligan R A (1982) A multi-year, stochastic, farm simulation model for Northern Nigeria: An experimental design approach. Am. Agr. Econ. Assoc. 64:728–737.

De Koeijer T J, Renkema J A, Van Mensvoort J J M (1995) Environmental-economic analysis of mixed crop-livestock farming. Agric. Syst. 48:515–530.

Dent J B (1993) Potential for systems simulation in farming systems research? Pages 325–339 in Penning de Vries F W T, Teng P S, Metselaar K (Eds.) Systems approaches for agricultural development. Kluwer Academic Publishers, Dordrecht, The Netherlands.

Dent J B, Thornton P K (1988) The role of biological simulation models in farming systems research. Agric. Admin. Ext. 29:111–122.

Dillon C R, Mjelde J W, McCarl B A (1989) Biophysical simulation in support of crop production decisions: A case study in the Blacklands Region of Texas. S. J. Agr. Econ. 21(1):73–85.

Dillon J L, Virmani S S (1985) The farming systems approach. Pages 507–532 in Muchow R C (Ed.) Agro-research for the semi-arid tropics. Northwest Australia University Queensland Press, Brisbane, Australia.

Edwards-Jones G, MacGregor M J (1994) The necessity, theory and reality of developing models of farm households. Pages 338–352 in Dent J B, MacGregor M J (Eds.) Rural and farming systems analysis: European perspectives. CAB International, Wallingford, UK.

Engel T, Hoogenboom G, Jones J W (1995) AEGIS/WIN – A Windows interface combining GIS and crop simulation models. ASAE Paper No. 95–3244. ASAE, St. Joseph, MI, USA. 12 p.

Fraisse C W, Campbell K L, Jones J W, Boggess W G (1995) Generic Interactive Dairy Model (GIDM): User's and developer's manual. Research Report AGE 95–3. Agricultural and Biological Engineering Department, University of Florida, Gainesville, FL, USA. 112 p.

Halter A N, Dean G W (1965) Use of simulation in evaluating management policies under uncertainty: Application to a large scale ranch. J. Farm Econ. 147:557–573.

Hansen J W (1995) A systems approach for characterizing farm sustainability. PhD dissertation, Agricultural and Biological Engineering Department, University of Florida, Gainesville, FL, USA. 265 p.

Hansen J W, Jones J W (1996) A systems framework for characterizing farm sustainability. Agricultural Systems (in press).

Hansen J W, Thornton P K, Jones J W, Jacobson B M (1996) Towards data standards for enterprise and farm-level analysis. Pages 229–245 in Teng P S, Kropff M J, Ten Berge H F M, Dent J B, Lansigan F P, Van Laar H H (Eds.) Applications of systems approaches at the farm and regional levels. Kluwer Academic Publishers, Dordrecht, The Netherlands (this volume).

Hazell P B R (1971) A linear alternative to quadratic and semivariance programming for farm planning under uncertainty. Am. J. Agr. Econ. 53:53–62.

Heady E O (1952) Economics of agricultural production and resource use. Prentice-Hall, Englewood Cliffs, NJ, USA. 850 p.

Herrero M, Fawcett R H, Dent J B (1995) Integrating simulation models to optimize nutrition and management for dairy farms: A methodology. In The Study of Livestock Farming Systems in research and development framework: Research, socio-economics and the land manager. European Association for Animal Production (EAAP) Publications (in press).

Islam N (Ed.) (1995) Population and food in the early twenty-first century: Meeting future food demands of an increasing population. IFPRI Occasional Publication, Washington DC, USA.

Jones J W (1993) Decision support system for agricultural development. Pages 459–471 in Penning de Vries F W T, Teng P S, Metselaar K (Eds.) Systems approaches for agricultural development. Kluwer Academic Publishers, Dordrecht, The Netherlands.

Jones J W, Hunt L A, Hoogenboom G, Godwin D C, Singh U, Tsuji G Y, Pickering N B, Thornton P K, Bowen W T, Boote K J, Ritchie J T (1994) Input and output files, Volume 2–1. Pages 1–94 in Tsuji G Y, Uehara G, Balas S (Eds.) DSSAT v3. University of Hawaii, Honolulu, HI, USA.

Jones P G, Thornton P K, Hill P (1994) Agrometeorological models: Crop yield and stress indices. Proceedings, EU/FAO Expert Consultation on crop yield forecasting methods. Villefranche-sur-Mer, France (in press).

Kelly T C (1995) A bioeconomic systems approach to sustainability analysis at the farm level. PhD thesis, Department of Food and Resource Economics, University of Florida, Gainesville, FL, USA. 210 p.

Kingwell R S, Morrison D A, Bathgate A D (1992) The effect of climatic risk on dryland farm management. Agric. Syst. 39:153–175.

Klein K K, Narayanan S (1992) Farm level models: A review of developments, concepts and applications in Canada. Can. J. Agr. Econ. 40:351–368.

Klein K K, Kulshreshtha S N, Klein S A (1989) Agricultural drought impact evaluation model: Description of components. Agric. Syst. 30:117–138.

Lee D J, Tipton T, Leung P (1995) Modeling cropping decisions in a rural developing country: Multiple-objective programming approach. Agric. Syst. 49(2):101–111.

Lee J E (1983) A perspective on the evolution and status of micro modeling in Agricultural Economics. Pages 1–9 in Baum K H, Schertz L P (Eds.) Modeling farm decisions for policy analysis. Westview Press, Boulder, CO, USA.

Leonard R A, Knisel W G, Still D A (1987) GLEAMS: Groundwater loading effects of agricultural management systems. Trans. ASAE 30(5):1403–1418.

McCown R L, Cox P G, Keating B A, Hammer G L, Carberry P S, Probert M E, Freebairn D M (1994) The development of strategies for improving agricultural systems and land-use management. Pages 81–96 in Goldsworthy P R, Penning de Vries F W T (Eds.) Opportunities, use, and transfer of systems research methods in agriculture to developing countries. Kluwer Academic Publishers, Dordrecht, The Netherlands.

Menz K M, Knipsheer H C (1981) The location specificity problem in farming systems research. Agric. Syst. 7:95–103.

Negahban B, Fonyo C, Boggess W G, Jones J W, Campbell K L, Kiker G, Hamouda E, Flaig E, Lal H (1993) A GIS-based decision support system for regional environmental planning. Pages 169–178 in Heatwole C D (Ed.) Application of advanced information technologies: Effective management of natural resources. Proceedings of the 18–19 June 1993 Conference, Spokane, Washington. ASAE, St. Joseph, MI, USA.

Nelson T R (1983) Specification and use of farm growth and planning models in farm management and extension. Pages 367–374 in Baum K H, Schertz L P (Eds.). Modeling farm decisions for policy analysis. Westview Press, Boulder, CO, USA.

Nix H A (1984) Minimum data sets for agrotechnology transfer. Pages 181–188 in Proceedings of the International Symposium on Minimum Data Sets for Agrotechnology Transfer. International Crops Research Institute for the Semi-Arid Tropics Center, Patancheru, India.

Norman D, Collinson M (1985) Farming systems research in theory and practice. Pages 16–30 in Remenyi J V (Ed.) Agricultural systems research for developing countries. Australian Centre for International Agricultural Research Proceedings No. 11, ACIAR, Canberra, Australia.

Penning de Vries F W T, Van Laar H H (Eds.) (1982) Simulation of plant growth and crop production. Simulation Monographs. Pudoc, Wageningen, The Netherlands. 308 p.
Rabbinge R (1993) The ecological background of food production. Pages 2–29 in Crop protection and sustainable agriculture. Ciba Foundation Symposium 177. John Wiley and Sons, Chichester, UK.
Rabbinge R, Leffelaar P A, Van Latesteijn H C (1994) The role of systems analysis as an instrument in policy making and resource management. Pages 67–79 in Goldsworthy P R, Penning de Vries F W T (Eds.) Opportunities, use, and transfer of systems research methods in agriculture to developing countries. Kluwer Academic Publishers, Dordrecht, The Netherlands.
Richardson J W, Nixon C J (1985) FLIPSIM V: A general firm level policy simulation model. B-1528. Department of Agricultural Economics, Texas A&M University, College Station, Tex., USA. 179 p.
Röling N (1994) Platforms for decision making about ecosystems. Pages 385–393 in Fresco L O, Stroosnijder L, Bouma J, Van Keulen H (Eds.) The future of the land: Mobilising and integrating knowledge for land use options. John Wiley and Sons, Chichester, UK.
Rosenzweig C, Parry M L (1994) Potential impacts of climate change on world food supply. Nature 367:133–138.
Stafford-Smith M, Foran B (1992) An approach to assessing the economic risk of different drought management tactics on a South Australian pastoral sheep station. Agric. Syst. 39(1):83–106.
Stroosnijder L, Van Rheenen T (1993) Making farming systems a more objective and quantitative research tool. Pages 341–353 in Penning de Vries F W T, Teng P S, Metselaar K (Eds.) Systems approaches for agricultural development. Kluwer Academic Publishers, Dordrecht, The Netherlands.
Taylor C R (1983) Complementarities between micro- and macro-systems simulation and analysis. Pages 63–72 in Baum K H, Schertz L P (Eds.) Modeling farm decisions for policy analysis. Westview Press, Boulder, CO, USA.
Ten Berge H F M (1993) Building capacity for systems research at national agricultural centers: SARP's experience. Pages 515–538 in Penning de Vries F W T, Teng P S, Metselaar K (Eds.) Systems approaches for agricultural development. Kluwer Academic Publishers, Dordrecht, The Netherlands.
Thornton P K, Hansen J W, Knapp E B, Jones J W (1995) Designing optimal crop management strategies. Pages 333–351 in Bouma J, Kuyvenhoven A, Bouman B A M, Luyten J C, Zandstra H G (Eds.) Eco-regional approaches to develop sustainable land use and food production. Kluwer Academic Publishers, Dordrecht, The Netherlands.
Tsuji G Y, Uehara G, Balas S (Eds.) (1994) DSSAT v3. University of Hawaii, Honolulu, HI, USA. 491 p.
Uehara G, Tsuji G Y (1993) The ISBNAT Project. Pages 505–513 in Penning de Vries F W T, Teng P S, Metselaar K (Eds.) Systems approaches for agricultural development. Kluwer Academic Publishers, Dordrecht, The Netherlands.
Veloso R F, McGregor M J, Dent J B, Thornton P K (1994) Técnicas de modelagem de sistemas aplicadas em planejamento agrícola dos cerrados. Pesq. Agropec. Bras. Brasília 29(12):1877–1887.

# Systems approaches to farm management in variable environments

A. PATANOTHAI
*Faculty of Agriculture, Khon Kaen University, Khon Kaen 40002, Thailand*

**Key words:** systems approaches, farm management, farming systems research, farmer participation, simulation models

**Abstract**
Under variable environments, farm management decisions involve a series of complex issues, since factors affecting the farm households are numerous, including not only biophysical, but also socioeconomic factors which interact with each other in a complex manner. There is also a significant diversity among farm systems in the biophysical environments and socioeconomic circumstances in which they operate. The concern on sustainability adds more complexity to the issues. Systems approaches to resolving these issues are, thus, essential. Farming Systems Research (FSR) has been the approach widely used over the past two decades. Although FSR has contributed significantly to the systems movement, it fell far short of expectations. The approach, however, still provides a logical framework for the development of farm management technologies that fit the farmers' circumstances. In recent years, there has been significant progress in three broad areas of systems research methodologies: the development of conceptual and analytical tools for qualitative analysis; the participatory methodologies; and quantitative modelling and use of Geographic Information Systems (GIS). If what is available were to be used more widely and effectively, much improvement could be made to FSR. However, much more research is needed, particularly on simulation models for combination of crops and enterprises and the inclusion of a socio-economic dimension in the models. Problems still exist in evaluating alternative practices in the context of the whole-farm system with multiple enterprises, multiple goals and objectives, and land fragmentation. Much more attention needs to be paid to farmers' decision-making criteria and processes. The focus of FSR only at the farm household level and with individual farmers appears to be inadequate for natural resource management, as many of the interventions have to be implemented at a higher level and with groups of farmers. Finally, research should aim at generating a range of management alternatives for the farmers to choose from rather than a fixed set, and farmers' participation is still essential throughout the research process.

## Introduction

Variable environments make agricultural production risky. In developing countries, most farmers not only have limited resources, but the environments in which they operate also are, for the most part, variable and diverse. Over time, these farmers have developed coping strategies to minimize risks. These include diversification of enterprises and adjustment of management practices. As a consequence, small farmers in developing countries generally operate multiple enterprises, some for subsistence and some for commercial goals, and are often engaged in off-farm and non-farm activities. These create multiple goals and priorities. Although soils and climatic conditions are the major determinants of farm production, farm manage-



*P.S. Teng* et al. *(eds.): Applications of Systems Approaches at the Farm and Regional Levels, 19–29.*

ment practices are influenced by input and output markets and government policies. They are also bound by culture, beliefs and community regulations. These factors interact with each other in a complex manner, making farming management decisions quite complicated. Farming systems are also dynamic, and so are the economic conditions. These make agricultural development difficult since it has to deal with several complex issues. Thus, the need for systems approaches has long been recognized. The issues now are even more complicated as the current concern about sustainability calls for a shift of attention from production to natural resource management, while free markets and integration of international economies put pressure towards increased productivity and economic efficiency to be competitive in the absence of subsidies, tariffs or protection from imports. Systems approaches are, thus, essential to deal with these complex issues. This paper broadly examines the trends in the development of systems approaches for improving farm management practices under variable environments.

## Farming Systems Research: contributions and limitations

Farming Systems Research (FSR) is, perhaps, the most widely used systems approach to agricultural research and development. It evolved around the mid-1970s in the post-Green Revolution era with the concern that technologies generated by the conventional research approach, which had contributed to a substantial increase in food production, were not benefiting resource-poor farmers who normally live in the unfavourable environments. The problem has its roots in the lack of understanding by scientists of the conditions under which small farmers operate. It was realized that a broader approach that takes into account the real farm conditions and farmers' circumstances was needed (Tripp et al. 1990).

The basic principle of FSR is that the process of technology generation and adaptation should be responsive to the needs and requirements of the targeted farming systems, the farmers' objectives and the conditions under which specific populations must practise agriculture. Based on this principle, it follows that a systems approach is necessary because any given agricultural enterprise is a final product of a complex and dynamic interplay of numerous components and processes that take place both within and outside the farm: the farmers' circumstances are location-specific and so is the research and extension process; a diagnosis phase is required in the overall process of technology generation and transfer in order to understand the set of conditions that will eventually determine the adoption rate and the final impact of the whole effect; and on-farm adaptation and testing of technology would enhance the probability of success of any given innovation. Farmer participation is indispensable if the improved technologies are to be responsive to the farmers' priorities and objectives (Berdegue 1993).

Over the past two decades, FSR has been widely employed by international agricultural research centres and national agricultural research and extension institutions (IARCs 1987; Merrill-Sands et al. 1989), and has evolved an immense range of activities (Simmonds 1985; Merrill-Sands 1986). Despite the many variations and expressions, all approaches have stages for description and diagnosis, design, testing and dissemination. They may vary in emphasis on the stages and in the scope of enterprises, area or time coverage (Fresco 1984). In the early years, FSR was predominantly focused on a predetermined crop (Byerlee and Collinson 1980), but has shifted to a broader enterprise coverage in later years. Cropping systems research is now accompanied by on-farm animal research (Amir and Knipsheer 1989), rice-fish research (Lightfoot 1990) and other multi-enterprise approaches.

In spite of so much interest and investment, the tangible results of FSR have been modest and have fallen short of expectations. Tripp (1991a) listed four factors contributing to the limited success of FSR. One is the institutional context in which FSR has operated in developing countries. In many cases, national agricultural research institutions conducting FSR are poorly funded and organized, with the FSR framework frequently imposed upon these institutions by outside funding agencies with specific interests and agenda. The second is the quality of FSR carried out by national and international institutions. Much research under the name of FSR has been poorly conceived and implemented. The third is the isolation of FSR from other parts of the research organization. Consequently, FSR has been denied access to the technical support it needs, preventing it from contributing to the identification and definition of research priorities at the national level. Finally, FSR has often been carried out with inadequate links to extension services and little consideration of the policy environment.

Orientation of FSR activities towards resource-poor farmers living in marginal areas, in itself, limits the success of FSR. With such orientation, most projects have faced one or more of the following conditions: unfavourable soils, climate and topography; lack of credit; underdeveloped or non-existent markets for both inputs and agricultural products: weak farmers' organizations; economic policies that discriminate against agriculture as a whole and against poor farmers in particular; lack of interest on the part of national agricultural institutions which feel that their resources would yield more benefits if invested in more favourable areas; and the attitude of many farmers who consider innovations to be risky (Berdegue 1993).

Baker (1993) considered the inability to deal with agricultural policy as a critical weakness of FSR. He pointed out that the orientation of FSR to a strategy of modifying technologies rather than modifying farmer circumstances and to focus on problem solving, minimum investments in data collection and quick results created a strong bias against longer-term investments in data generation for policy analysis and reform. Consequently, there have been several missed opportunities in which FSR teams could have provided needed empirical information and a farm-level perspective to policy design, implementation and monitoring.

Despite all these shortcomings, FSR has contributed significantly to the way that agricultural research is organized and conceived. It has changed the perspective of researchers, enabling them to view farms in a holistic manner and to see farm family welfare as dependent on a wide range of variables. As a consequence of this perspective, agricultural researchers have broadened their criteria for assessing the acceptability of new technology. The farming systems perspective has also encouraged a growing acceptance of a problem-oriented approach to planning agricultural research, and has fostered a more realistic look at the targeting of agricultural research programmes. Another major contribution of FSR has been the development of new methods for adaptive agricultural research. These include informal techniques for data gathering, the design, management and evaluation of on-farm experiments, and the involve-ment of farmers in managing experiments. FSR has also contributed to significant improvements in the management of agricultural research. These include the move towards more interdisciplinary research and a more location-specific approach to research (Tripp 1991b).

## Trends in the development of methodology

Over the past decade, there has been significant progress in systems research methodologies for agricultural development. Much of it was derived from attempts to improve the effectiveness and efficiency of FSR. The trends in methodological development can be broadly grouped into three clusters: the development of conceptual and analytical tools for qualitative analysis, participatory methodologies, and quantitative modelling and the use of Geographic Information Systems (GIS).

### *Conceptual and analytical tools for qualitative analysis*

Analysing complex farming systems is a formidable task for agricultural researchers who normally do not have training or background in systems concepts and methodologies. The fact that everything is related to everything often makes them feel they are playing with a jigsaw puzzle. The use of conceptual models helps them conceptualize the interactions and processes of the various components and, thus, helps organize their own thinking. The concepts of systems boundary and systems hierarchy help them visualize the organization and structure of the different systems, while that of systems properties allows for an understanding of the combined effects of interactions and processes of the lower-level components. The application and use of these concepts and principles by agricultural researchers and extension workers represent significant progress towards the use of systems approaches in agricultural research and development.

Analytical procedures have been developed to provide guidelines for analysis of complex agricultural systems. One approach which, including its variants, appears to have been used widely, particularly in Asian countries, is agroecosystems analysis (Conway 1985). This approach focuses on the identification and understanding of key functional relationships that have great influences on systems properties. The basic assumption is that significant improvement in systems performances can be achieved by changing only few key management decisions. Therefore, it is not necessary or desirable to know everything about the system of concern in order to produce a realistic and useful analysis. The procedure involves defining systems boundaries and assessing systems properties by examining patterns in space, time, flows and decisions that allow for integration of biophysical and socioeconomic information. Important management decisions are expressed in terms of key questions upon which guidelines or working hypotheses for research and development are formulated. The procedure extensively uses simple models such as maps, transects, graphs, flow and bar diagrams, decision trees and Venn diagrams.

Several conceptual tools have also been developed to help the examination of interactions and functional relationships among various biophysical and socio-economic factors, and the interrelations among systems components and subsystems. Examples of these can be seen in Conway (1987), Limpinuntana (1987) and Lightfoot et al. (1991, 1993).

Another approach developed for planning on-farm research involves the analysis of problems and causes, and their interrelationships. The procedure includes identification of problems limiting the productivity of the farm, ranking of these problems, identification of causes of the problems, analysis of interrelationships among problems and causes, identification of possible solutions to the problems and evaluation of possible solutions. The procedure is described in detail in Tripp and Woolley (1989).

All of these tools aim to help researchers gain in-depth understanding of complex farming systems, and facilitate analyses to determine research priorities and appropriate interventions. The models used, however, are essentially qualitative models, and the analyses, for the most part, are qualitative.

*Farmer participatory methodologies*

The importance of farmer participation in technology development has long been recognized. This comes from a realization that farmers know best their local environments, their available resources and constraints, and their needs. Most importantly, they are the ones who will decide to accept or reject the technology.

In FSR, farmer participation is an essential element, and is supposed to be included in all stages of the research process, from diagnosis to on-farm testing and evaluation of technologies. Over the years, there have been improvements in methods and techniques to enhance farmer participation in a meaningful and productive way. However, the form of farmer participation still has generally been limited to the consultative type, with farmers acting as informants and evaluators of relatively developed technology. There has not been sufficient involvement of farmers in making decisions together with researchers about what to test and how to test it (Ashby 1991).

One of the approaches which has enhanced farmer participation in FSR is Rapid Rural Appraisal (RRA; Beebe 1985). The key feature of this approach is the informal interview of farmers and key informants. The interview may be done on the spot in an open or non-structured manner, but follows certain guidelines. The use of RRA together with agroecosystem analysis and various conceptual and qualitative analytical tools has helped researchers gain better understanding of the interactions and processes in the farming systems, leading to significant improvements in the diagnosis phase of FSR. The techniques also have been used widely in research conducted with other objectives (Khon Kaen University 1987). However, the type of farmer participation is essentially a consultative type.

Attempts to involve farmers in a more active role have resulted in methods for analysis by farmers. Chambers (1993) noted two streams of innovation contributing to methods for analysis by farmers: Farmer Participatory Research (FPR), sometimes characterized as 'farmer-first', and Participatory Rapid Appraisal (PRA). These methods overlap with, and have antecedents in applied anthropology, FSR and RRA. These methods are evolving and, in many cases, have been used in combination. Chambers (1993) classified these applications into two groups: one more verbal and the other more visual. In the more verbal mode, analysis and communication are through discussion, often supported by observation and demonstration. In the more visual mode, analysis and communication are based on participatory diagramming. Participatory visual methods described include participatory mapping and modelling, analysis of aerial photographs, matrix scoring and ranking, farmers' flow, linkage, causal diagramming, seasonal analysis and trend analysis.

Another development in participatory methodologies is the move towards a more active role of farmers in the diagnosis, design and experimentation at their own initiative. This development coincided with the shift of research focus towards natural resource management. An example is provided by Lightfoot and Noble (1993) which showed that the use of participatory methods has resulted in farmers' experimenting and developing their own unique rice-fish systems. Lightfoot et al. (1993) presented the details of farmer participatory procedures for managing and monitoring sustainable farming systems. The procedures allow households to experiment on natural resource management, improve their natural resource management skills and provide researchers with analytical frameworks for monitoring and assessing impact on households and natural resources.

*Quantitative modelling and the use of geographic information systems (GIS)*

Parallel with the development in qualitative modelling and farmer participatory methodology, there has also been development towards the use of quantitative models in agricultural research. This approach has gained much interest in recent years, especially with the advances in computer technology. Over the past decade, substantial efforts have been spent in the development and application of quantitative models, particularly simulation models, and significant progress has been made. Systems simulation techniques have now been developed for use in research and decision making at different levels in the hierarchy of agricultural systems: at the level of crop, farm unit and farming systems, and at national and regional levels for land-use planning. Currently, simulation methods are used widely for the study of a great variety of issues in agricultural research and development, and in natural resources management (Goldsworthy and Penning de Vries 1994).

Two groups of quantitative models can be distinguished, namely mathematical economic models and biophysical simulation models. Among the mathematical economic models, linear programming models have been used more often for farm-level analysis. For example, at ICARDA, linear programming models have been used to analyse the whole-farm economics of production systems. *Ex ante* analyses were done to compare the potential profitability of crop-crop and crop-pasture rotations, integrated with livestock, in different environments. Sensitivity analyses were also used to help identify research goals and potential adoption domains (Harris et al. 1994). Recently, Stroosnijder and Van Rheenen (1994) proposed the use of Interactive Multiple Goal Linear Programming (IMGLP) for farming systems analysis. This mathematical programming technique links biotechnical and socioeconomic components of farming systems, and can be used to explore development options, taking into account technological innovations and risk acceptance at farm level. With IMGLP, various goals can be taken into account, and their trade-offs can be illustrated.

Among the biophysical simulation models, crop models appeared to have received the greatest attention. Simulation models are now available for several crops including maize, wheat, groundnut, rice, drybean, sorghum and millet (Jones 1993). Recently, several models have also been developed for multiple cropping systems (Caldwell and Hansen 1993), integrated aquaculture systems (Zweig 1992) and tree crop-grazing livestock farming systems (Dahlan and Shahar 1992).

Another development is the use of GIS as a tool for spatial analysis of various components of farming systems. This tool greatly facilitates the description and diagnosis activities of FSR. It is also being used to complement simulation methods for geographic extrapolation of research findings. This combination of methods offers completely new opportunities for the international agricultural research centres and the more advanced and better-endowed national agricultural research systems in targeting their research (Goldsworthy and Penning de Vries 1994).

The currently available models and systems approaches still have limitations. One of these is their inability to deal adequately with long-term trends. Other elements that limit the level of confidence of systems models include the limited availability and quality of data, the limits of our understanding of the physical, biological and socioeconomic processes involved, the unpredictability of human behaviour and the evolution of the context in which problems are seen. Also, most of the models have been designed originally for application in a developed country, whereas the objectives of users in developing countries are often different. Consequently, the models available will not always provide answers to questions that are being asked by the national agricultural research systems (Goldsworthy and Penning de Vries 1994).

### Improving and broadening the scope of FSR

The qualitative models, farmer participatory methodologies and quantitative models all have their own strengths and weaknesses. In combination, these will complement each other. Although FSR has several drawbacks, it still provides a logical framework for agricultural research and development and natural resource management. These methodologies, if used appropriately in the FSR framework, will certainly improve the effectiveness of FSR. In fact, many of them have already been incorporated in the FSR methodologies (Zandstra 1991).

Most of the qualitative models and farmer participatory methodologies have been developed for improving the effectiveness of FSR, and have already been widely used. Simulation models, introduced recently in agricultural research, have great potential for problem identification, priority setting, assessing alternative management strategies and extrapolation of technologies. Most importantly, they can reduce the time and resources of actual experimentation. Examples illustrating the values of simulation models in FSR can be seen in Dent (1993).

The use of quantitative models also can broaden the scope of FSR in assessing a broader range of management strategies and in dealing with policy. McCown et al. (1994) described an approach which uses improved agricultural production simulation models and economic decision models to be able to compare options that are not currently available or feasible for resource-poor farmers. The approach is an expansion of the FSR framework in which a more analytical planning stage is added. The usefulness of the approach was demonstrated using experiences from work in semi-arid Kenya where investment in soil enrichment in fertility-depleted cropland was deterred by the unreliability of rainfall. The approach was also used to compare the risks and returns of two policy strategies, i.e. migration of farmers to drier areas versus investment in increasing the soil fertility of existing croplands. They concluded that the best prospects for developing better policy and management strategies lie with skilful use of 'hard' systems tools within a 'soft' systems philosophical framework.

With the current concern on sustainability and the shift of research emphasis from production to natural resource management, there has also been a shift of research focus from a farm level to a landscape level. Many of the principles, and the philosophy of FSR, are still applicable, and combinations of qualitative, quantitative and farmer participatory methods are also used. The framework of the Sustainable Agriculture and Natural Resource Management Collaborative Research Support Program (SANREM CRSP; Hargrove et al. 1994) represents this type of integrated systems approach. In this approach, the complex processes within and between the individual ecosystems of a toposequence which transect two or more agroecological zones are described, covering social, physical and biological dimensions. Improved management techniques are designed with farmer participation, and their appropriateness is evaluated by farmers and other users of natural resources in terms of agricultural, environmental, economic and social sustainability. The approach also includes a mix of experimental and systems analysis techniques, focused on a 'demand-driven' research agenda, aimed at developing, testing and implementing improved strategies for sustainable agriculture and natural resource management. The framework of SANREM CRSP is quite comprehensive, covering a broad range of goals and objectives, but provides a menu of research areas which individual projects can draw upon.

## Concluding remarks

The multiple goals and enterprises of farm households, and the diversity and variability of the biophysical and socioeconomic conditions, make farm management difficult.

Issues involved are complex. Progress in systems methodologies provides a bright prospect for dealing with these complex issues. Despite its limited success, FSR still provides a logical framework and sound basic principles. If the available systems methods and techniques are used widely and well implemented, the effectiveness of FSR could be greatly enhanced. The scope of FSR could also be broadened to include institutional and policy issues. However, much more work is needed, particularly on simulation models for multiple enterprises and the inclusion of the socioeconomic dimension in the models.

Any alternative management practice or set of practices will involve a change in only a part of the farm system. Evaluation of its effect on the whole-farm system is still a problem because farmers differ in their resources, enterprises, goals and objectives. In order to better evaluate management alternatives, much more understanding is needed on farmers' decision-making criteria and processes.

The focus of FSR only at the farm household level and with individual farmers appears to be inadequate for natural resource management, because effects of management practices often occur across boundaries of several farms. In these cases, interventions will have to be made at a higher hierarchical level and with groups of farmers. Focus on both farm level and landscape level would be more appropriate.

Research in the past has considered farmers almost exclusively as 'adopters' of technologies. Empirical evidence has indicated that farmers themselves do experiment with new alternatives all the time. What they are using has been tried and modified to fit their specific resources and environments. They actually are more of the 'adapter' than the 'adopter' (Ashby 1991). This reflects a diversity among farm households in which fine tuning of technologies to fit the individual households will not be possible or desirable. This suggests that research should aim at producing a 'prototype' rather than a 'fine-tuned' technology, as well as a range of management alternatives for farmers to choose from rather than a fixed set. It also points out the need to involve farmers from the early phase of the research process. As formal methodologies are more and more developed, there is also a tendency for researchers to move away from farmers. The point to emphasize here is that, whatever progress has been made in systems analysis and simulation, it is still far from predicting the real world and farmers' behaviour. Therefore, farmers' participation is still essential throughout the research process.

## References

Amir P, Knipsheer H C (1989) Conducting on-farm animal research: Procedures and economic analysis. Winrock International, Morillton, AR, USA, and International Development Research Centre, Ottawa, Canada. 244 p.

Ashby J A (1991) Adopters and adapters: The participation of farmers in on-farm research. Pages 273–286 in R. Tripp (Ed.) Planned change in farming systems: Progress in on-farm research. John Wiley and Sons, Chichester, UK.

Baker D (1993) Inability of farming systems research to deal with agricultural policy. J. Farming Syst. Res. Ext. 4(1):67–86.

Beebe J (1985) Rapid rural appraisal: The critical first step in a farming systems approach to research. Networking Paper No. 5. Farming Systems Support Project, University of Florida, Gainesville, FL, USA.

Berdegue J A (1993) Challenges of farming systems research and extension. J. Farming Syst. Res. Ext. 4(1):1–9.
Byerlee D K, Collinson M P (1980) Planning technologies appropriate to farmers: Concepts and procedures. International Maize and Wheat Improvement Center (CIMMYT), Mexico. 71 p.
Caldwell R M, Hansen J W (1993) Simulation of multiple cropping systems with CropSys. Pages 397–412 in Penning de Vries F W T, Teng P S, Metselaar K (Eds.) Systems approaches for agricultural development. Kluwer Academic Publishers, Dordrecht, The Netherlands.
Chambers R (1993) Methods for analysis by farmers: The professional challenge. J. Farming Syst. Res. Ext. 4(1):87–101.
Conway G R (1985) Agroecosystem analysis for research and development. Winrock International Institute for Agricultural Development, Bangkok, Thailand. 111 p.
Conway G R (1987) Rapid rural appraisal and agroecosystem analysis: A case study from Northern Pakistan. Pages 228–254 in Proceedings of the 1985 International Conference on Rapid Rural Appraisal, 2–5 September 1985. Rural Systems Research and Farming Systems Research Projects, Khon Kaen University, Khon Kaen, Thailand.
Dahlan J, Shahar M S (1992) Simulation models for quantitative decisions in crop-livestock integrated farming systems. J. Asian Farm. Syst. Assoc. 1:351–360.
Dent J B (1993) Potential for systems simulation in farming systems research. Pages 325–339 in Penning de Vries F W T, Teng P S, Metselaar K (Eds.) Systems approaches for agricultural development. Kluwer Academic Publishers, Dordrecht, The Netherlands.
Fresco L O (1984) Comparing anglophone and francophone approaches to farming systems research and extension. Networking Paper 1. Farming Systems Support Project, University of Florida, Gainesville, FL, USA.
Goldsworthy P R, Penning de Vries F W T (Eds.) (1994) Opportunities, use, and transfer of systems research methods in agriculture to developing countries. Kluwer Academic Publishers, Dordrecht, The Netherlands. 366 p.
Hargrove W L, Bonner J W, Kanemasu E T, Neely C L, Dar W (1994) The SANREM CRSP: A framework for integration of system analysis methods in a sustainable agriculture and natural resource management research and development agenda. Pages 233–246 in Goldsworthy P R, Penning de Vries F W T (Eds.) Opportunities, use, and transfer of systems research methods in agriculture to developing countries. Kluwer Academic Publishers, Dordrecht, The Netherlands.
Harris H D, Nordblom T L, Rodriguez A, Smith P (1994) Experience of the use of system analysis in ICARDA. Pages 295–302 in Goldsworthy P R, Penning de Vries F W T (Eds.) Opportunities, use, and transfer of systems research methods in agriculture to developing countries. Kluwer Academic Publishers, Dordrecht, The Netherlands.
IARCs – International Agricultural Research Centers (1987) Proceedings of the Workshop on Farming Systems Research, 17–21 February 1986. International Crops Research Institute for the Semi-Arid Tropics, Patancheru, India.
Jones J W (1993) Decision support systems for agricultural development. Pages 459–471 in Penning de Vries F W T, Teng P S, Metselaar K (Eds.) Systems approaches for agricultural development. Kluwer Academic Publishers, Dordrecht, The Netherlands.
Khon Kaen University (1987) Proceedings of the 1985 International Conference on Rapid Rural Appraisal, 2–5 September 1985, Rural Systems Research and Farming Systems Research Projects, Khon Kaen University, Khon Kaen, Thailand.
Lightfoot C (1990) Integration of aquaculture and agriculture: A route to sustainable farming systems. Naga, International Center for Living Aquatic Resources Management, ICLARM Q. 13(1):9–12.
Lightfoot C, Noble R (1993) A participatory experiment in sustainable agriculture. J. Farming Syst. Res. Ext. 4(1):11–13.
Lightfoot C, Noble R, Morales R (1991) Training resource book on a participatory method for modelling bioresource flows. International Center for Living Aquatic Resources Management, Manila, Philippines. 30 p.
Lightfoot C, Dalsgaard J P, Bimbo M, Fermin F (1993) Farmer participatory procedures for managing and monitoring sustainable farming systems. J. Asian Farm. Syst. Assoc. 2(1):67–87.
Limpinuntana V (1987) Conceptual tools for RRA in agrarian society. Pages 144–173 in Proceedings of the 1985 International Conference on Rapid Rural Appraisal, 2–5 September 1985, Rural Systems Research and Farming Systems Research Projects, Khon Kaen University, Khon Kaen, Thailand.
McCown R L, Cox P G, Keating B A, Hammer G L, Carberry P S, Probert M E, Freebairn D M (1994) The development of strategies for improved agricultural systems and land-use management. Pages

81–96 in Goldsworthy P R, Penning de Vries F W T (Eds.) Opportunities, use, and transfer of systems research methods in agriculture to developing countries. Kluwer Academic Publishers, Dordrecht, The Netherlands.

Merrill-Sands D (1986) Farming systems research: Clarification of terms and concepts. Exp. Agric. 22: 87–103.

Merrill-Sands D, Ewell P, Biggs S, McAllister J (1989) Issues in institutionalizing on-farm client-oriented research: A review of experiences from nine national agricultural research systems. Q. J. Int. Agric. 25(3/4):279–300.

Simmonds N W (1985) Farming systems research: A review. World Bank Technical Paper No. 43, Washington DC, USA.

Stroosnijder L, Van Rheenen T (1994) Making farming system analysis a more objective and quantitative research tool. Pages 341–353 in Goldsworthy P R, Penning de Vries F W T (Eds.) Opportunities, use, and transfer of systems research methods in agriculture to developing countries. Kluwer Academic Publishers, Dordrecht, The Netherlands.

Tripp R (1991a) The limitations of on-farm research. Pages 247–256 in Tripp R (Ed.) Planned change in farming systems: Progress in on-farm research. John Wiley and Sons, Chichester, UK.

Tripp R (1991b) The farming systems research movement and on-farm research. Pages 3–16 in Tripp R (Ed.) Planned change in farming systems: Progress in on-farm research. John Wiley and Sons, Chichester, UK.

Tripp R, Woolley J (1989) The planning stage of on-farm research: Identifying factors for experimentation. International Maize and Wheat Improvement Center (CIMMYT), Mexico and International Center for Tropical Agriculture (CIAT), Cali, Colombia. 85 p.

Tripp R, Anandajayasekeram P, Byerlee D, Harrington L (1990) Farming systems research revisited. Pages 384–399 in Eicher C K, Staatz J M (Eds.) Agricultural development in the third world. The Johns Hopkins University Press, Baltimore and London.

Zandstra H G (1991) Approaches in farming systems research and extension: Problems and improvements. J. Asian Farm. Syst. Assoc. 1(1):101–111.

Zweig R D (1992) Ecological-economic computer simulation models for the design and analysis of integrated aquaculture systems. J. Asian Farm. Syst. Assoc. 1:337–350.

# Interdisciplinary tool for optimizing agricultural production and erosion: a conceptual model

L.G.J. BOERBOOM, I.D. FLITCROFT and E.T. KANEMASU

*Department of Crop and Soil Sciences, University of Georgia, 3111 Miller Plant Sciences Building, Athens, GA 30602, USA*

**Key words:** multiple objective optimization, non-point source pollution modelling, expert systems, erosion control, interdisciplinarity

**Abstract**

What are the areas in a catchment where soil conservation practices can be most effectively implemented? To be able to answer that question, erosion rates and farm profitability from different land uses need to be optimized according to physical economic and social constraints, which differ for the interest groups in and outside a catchment. The Interdisciplinary Tool for Optimizing Productivity and Erosion (ITOPE), a conceptual model at this point, offers the opportunity for hydrologists, economists and anthropologists to integrate their knowledge about the erosion problem. It optimizes farm profitability and soil loss and quantifies the economic impact of soil conservation measures for subcatchments, given physical, social and economic constraints, by linking a hydrologic model, a multiple criteria optimization model and an expert system. It should serve as a tool for regional planners, which ensures that social and economic boundary conditions for each of the interest groups in a watershed are considered. It should also serve as a negotiation tool for groups inside and outside the watershed, to allocate the costs and benefits of soil conservation. The development of the conceptual model has led to the observation of interdisciplinary methodological opportunities.

## Introduction

As with many environmental concerns, tackling soil erosion problems demands an interdisciplinary and intersectoral approach (Goldsworthy et al. 1994). The interests of stakeholders, within and between economic sectors, each with their own motivations, options and limitations, require that solutions encompass spatial and intergenerational equity (De Groot 1992). For this purpose, Multiple Criteria Optimization (MCO) is a set of powerful algorithms that determine the optimal allocation of natural resources (Janssen 1992). Given a number of constraints which determine a set of feasible solutions, the algorithms maximize or minimize objective functions (Steuer 1989). The tool presented in this paper uses MCO to allocate crops and soil conservation methods optimally in a watershed, with the goal of balancing farm profitability and soil loss. In addition to MCO, an expert system is introduced to overcome a limitation of MCO and hydrologic models.

## Overcoming the limitation of multiple criteria optimization

Three different perspectives on decision making reveal the same limitation of MCO: the assumption of rational economic behaviour (Luenberger 1995). The first per-



*P.S. Teng* et al. *(eds.): Applications of Systems Approaches at the Farm and Regional Levels, 31–43.*

spective is the choice to analyse decision making by deduction or induction. Economists explain decision making from theory (deduction), whereas anthropologists and sociologists observe economically irrational decisions that contradict the theory with their inductive approach, in both western and non-western societies (Herskovits 1941; Knight 1941; Schneider 1989; Fairweather and Keating 1994).

Cognitive models are a second perspective to perceive irrational economic behaviour. Nazarea-Sandoval (1995) shows that, although there are cultural commonalities in communities, cognitive models are differentiated by socioeconomic class and gender. The differences in cognitive models lead to different agricultural decision making, in which non-economical factors, for instance gastronomical considerations, play an important role. Fairweather and Keating (1994) do not search for differentiation by class or gender, but look at different management styles as they are shaped by farmers' personality. Röling (1994) observed that neither MCO nor ecosystem models account for the different cognitive models that stakeholders have on the use of natural resources. He suggests several research areas where humanities and sciences should interact, such as development of information systems which reflect stakeholders' views of natural resource use, development of land-use negotiation tools instead of land-use planning tools (Brinkman 1994) and using products from science as tools for social learning. We attempt to respond to this research agenda.

The third perspective is expressed by Simon (1992), who explains that people cannot handle the overload of information in a decision process, but nevertheless make decisions, which makes them 'satisficers' – people who try to attain a level of satisfaction – not economic maximizers. In other words, they accept a local instead of a global maximum or minimum, because they cannot handle the whole search space to find solutions.

Our objective is to develop a planning tool that optimizes land use in terms of farm profitability and soil loss on a catchment and subcatchment scale. It should serve as a tool for regional planners, which ensures that social and economic boundary conditions for each of the interest groups in a watershed are considered. It should also serve as a negotiation tool for those groups and groups outside the watershed, to allocate the costs and benefits of soil conservation. Although the tool applies an MCO algorithm, we intend to circumvent its limitation by integrating the MCO model and a non-point source pollution model with an expert system. The expert system contains social and economic boundary conditions as experienced by farmers.

Here we present only the first phase of the project, which is the development of the conceptual model, called Interdisciplinary Tool for Optimizing Productivity and Erosion (ITOPE). It is being developed and will be tested within the multidisciplinary institutional framework of the Sustainable Agriculture and Natural Resources Management Collaborative Research Support Program (SANREM CRSP). To assist land-users with the development of more sustainable land uses, information generated in different disciplines should interact to solve environmental problems which reveal themselves in many, closely interacting disciplinary dimensions. The model can serve as an interdisciplinary tool for hydrologists, economists and anthropologists to improve information-intensive land- and water-use planning with respect to farm

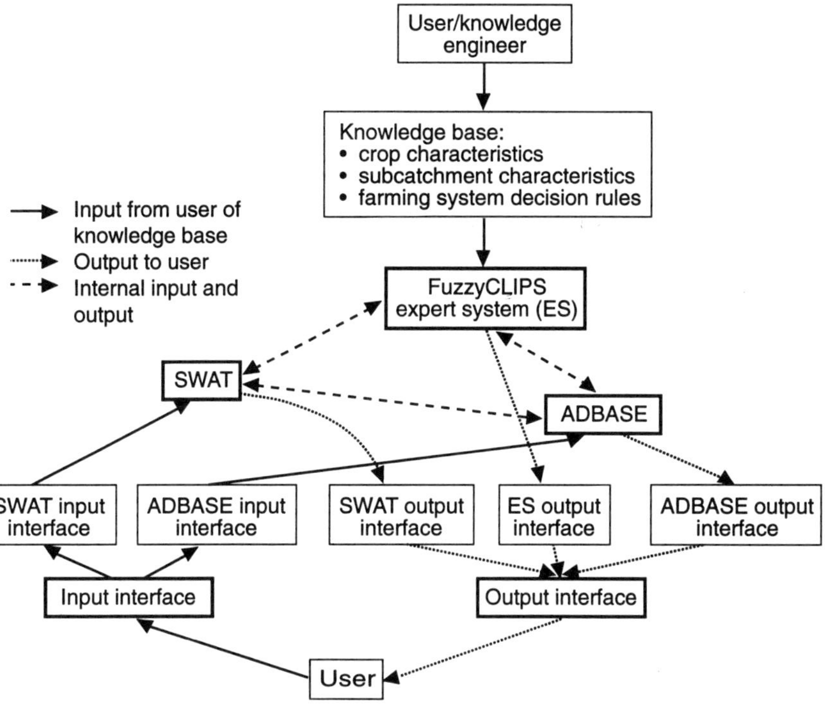


*Figure 1.* ITOPE's three model components: SWAT (Soil and Water Assessment Tool), ADBASE (Adjacent Efficient Basis algorithm) and FuzzyCLIPS, and interfaces.

profitability and erosion control. The study site is the 50000 ha Manupali watershed (8° 05′ N, 124° 55′ E), mainly in the municipality of Lantapan, on the island of Mindanao in the Philippines.

## ITOPE components

Figure 1 shows the three model components: SWAT, ADBASE and FuzzyCLIPS. All three components can be run autonomously or as one model. ITOPE is being developed in a Linux environment on a 486/66 personal computer for portability to the Philippines. The model uses the interfaces that exist for SWAT and ADBASE. Additional input and output interfaces are created.

## Hydrology component

SWAT, the Soil and Water Assessment Tool (Arnold et al. 1996), is the hydrologic component of ITOPE. It is a non-point source pollution model, which simulates sediment, nutrient and pesticide transport through basins. SWAT is a distributed parameter, continuous time model, which operates on a daily time step and can simulate many consecutive years. It divides a basin into subwatershed or grid cells, which it assumes to be homogeneous, except that it allows for spatially non-specific differentiation of crops within a subbasin or grid cell. This permits simulation of farming systems in the subbasins and thereby links farm and watershed scales. Nevertheless, the *de facto* modelling scale unit is the catchment, not the farm.

SWAT consists of modules for weather generation, surface runoff, lateral flow, groundwater flow, transmission losses, sediment and chemical routing through ponds, reservoirs and streams, and for irrigation. Sediment yield is estimated with the Modified Universal Soil Loss Equation and is routed down the basin reaches and reservoirs. In the crop growth model, intercepted energy and a biomass conversion parameter determine potential growth. Maturation is determined with the number of heat units received by the crop. Actual yield is determined by a harvest index, and water and nutrient stress.

The hydrologic model is linked to the Geographic Resources Analysis Support System (GRASS), a public domain raster Geographic Information System (GIS), which serves as an input and output user interface (Srinivasan and Arnold 1994). SWAT contains a USDA soils data base and a crop data base for 46 crops, of which 25 are vegetable crops. SWAT originates from the Simulator for Water Resources in Rural Basins, SWRRB (Arnold et al. 1990) and its validity appears to be reasonably good in the USA for monthly and annual water discharge (Arnold et al. 1995). For the Philippine site, it will be tested with Indonesian USLE parameters (Hamer 1981) and validated with water and sediment discharge measurements from the whole Manupali catchment and two of its subcatchments (±100 ha).

## Multiple criteria optimization component

### *Explanation of the technique*

Multiple objective linear programming (MOLP), the linear variant of MCO, is used in this model. The two most important elements of MOLP are the objectives and the constraints. Two or more objective functions are maximized or minimized. The maximization and minimization are restricted by constraints, which determine the region of feasible solutions (Steuer 1989).

An example of a problem that could be solved with MOLP would be a farmer who has to choose how to attribute land to two crops, corn and potato. The farmer's objectives are to maximize both profit and crop yield, which will both get a weight of 0.5. The constraints are that: the farm is 7 ha; only 300 work days per year are available, consisting of the farmer and eldest son, who helps out when possible; and at least 1 ha of those very tasty potatoes are needed for annual festivities. Knowing the prices, yields, and labour requirements of the crops, the optimal allocation could be determined.

The multiple criteria optimization component consists of ADBASE, the Adjacent Efficient Basis algorithm, together with its auxiliary programs Lambda and Filter, which are used to perform the augmented weighted Tchebycheff procedure (Steuer 1989). The procedure (Figure 2) identifies the boundary of the feasible region (*Z*),

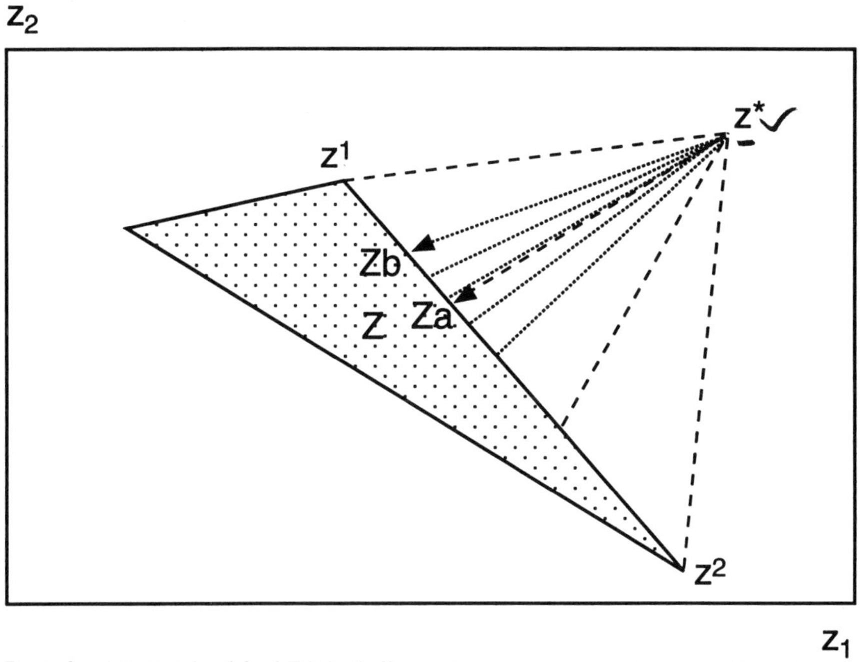


*Figure 2.* Augmented weighted Tchebycheff procedure: $z_1$ and $z_2$ are objective function values, e.g. profit and crop yield. $Z$ is the feasible region, $z^*$ is the point from which the boundary of $Z$ between extreme points $z^1$ and $z^2$ is sampled. Broken line is the first evenly disperse set offered, dotted line is the second evenly disperse set offered. *Za* and *Zb* are selected points by the decision maker.

which is defined by the constraints that apply to the optimization problem. For the illustration given above, it would maximize profit ($z_1$) and crop yield ($z_2$). The maximum solutions would be located on the line $\{z^1, z^2\}$, but where? First a point ($z^*$) outside $Z$ is selected with a vector larger than any of the vectors constituting $Z$. From $z^*$, the first evenly disperse set of points on line $\{z^1, z^2\}$ is generated (broken lines) and offered to the decision maker, who in this case selects point *Za* as the most favourable combination of objectives $z_1$ and $z_2$. In the next iteration, a second disperse set is generated (dotted lines), which is closer to the point *Za* selected by the decision maker. This time the decision maker selects *Zb*. The process continues until the user is satisfied. In ITOPE, the user can interact with ADBASE for each of the farming systems in each subbasin or use an example interaction from sessions with farmers in the Manupali watershed.

The method is attractive, because, unlike other algorithms, it can converge to final solutions other than the corners of the feasible region in both linear and non-linear multiple objective programmes. Furthermore, it does not require the decision maker to weight objectives as the farmer in the above example needed to do. Weighting is difficult, especially for problems in higher dimensions, i.e. with many crops, and if several objective functions are involved, for which the final maximized and minimized values are not known *a priori*.

*Implementation in ITOPE*

Each of the subcatchments is to be characterized by one farming system, which is defined by the category of crops that are grown, e.g. vegetables, grains and fibre, with or without livestock, and by the socioeconomic status of the farm household. For each subcatchment, an optimization problem is to be defined. Here, we review the objectives and then the crops that will be modelled in the MCO component of ITOPE.

Input cost minimization, monetary output maximization, soil loss and nutrient loss minimization are the objectives in the MOLP. The first interviews that have been carried out (several informal and eight formal semi-structured), suggest that vegetable farms in the Manupali watershed are not concerned only with profit, but also with the input costs required to achieve these profits. Some farmers wait for what they call a 'jackpot harvest' which enables them to invest and change to higher-input vegetable farming. We introduced input cost minimization as one of the objectives, to account for the willingness to invest in an extra unit of output. Combinations of financial output and input, and soil loss generated by the augmented weighted Tchebycheff procedure would be presented to the farmer to weigh the relative importance of these objectives. We intend to include minimization of nutrient losses, as an on-site effect of soil conservation, which is an appealing indicator to farmers. It also is an objective that coincides with the objective of reduction of downstream sediment yield.

The maximum number of objectives for each subcatchment is four, because of computational time requirements (Steuer 1994). Two farming systems may be attributed to one subcatchment if only one erosion objective and one profitability objective, accompanied by input constraints, are used for each farming system. A simulation of 100 subwatersheds, with 30 crops, four objectives and ten constraints would take approximately five minutes on a 486/66 personal computer.

ADBASE will potentially simulate the 46 crops in SWAT's data base and allow for user-added crops. The model is being developed to run its calculations on a yearly basis with two or three crops potentially on the same area. During the development stage the number of crops will be limited to seven. The seven crops are corn, tomatoes, potatoes, asparagus, pigeon peas, pineapple and tea. Corn is the staple food in Lantapan; tomatoes have recently been introduced and crop management is controlled by local industry; potatoes are a high-value crop that most farmers are eager to grow, but which has several disease problems; and the other four are high-value crops that are being tested as hedgerow soil conservation measures (AVRDC 1995).

## Expert system component

*Explanation of the technique and examples in agriculture*

Expert systems are computer programs that represent and reason with knowledge of some specialist subject to solve problems or give advice, replacing experts or assisting decision makers. They are written in symbolic languages, which enables them to process qualitative information and, unlike MCO, solve problems by heuristic methods. They consist of a knowledge base, which represents domain-specific knowledge, and an inference engine, which performs the reasoning (Jackson 1990). The knowledge base contains facts and rules, which are interpreted by the inference engine. The two main forms of operation for an inference engine are to conclude from facts (forward chaining) or to prove the conditions for a hypothesis to be true (backward chaining).

Plant (1993) discusses a number of expert system applications in agriculture and resource management. Many agricultural applications are concerned with crop management, pesticide application or irrigation scheduling. GIS, remote sensing tools and expert systems have been linked for resource management applications (Friedl et al. 1988; Coulson et al. 1991). Expert systems have been used less to decide which crops to grow. Nevo et al. (1994) list characteristics of the crop selection process which favour the use of an expert system, and address the limitation of MCO. The crop selection problem is sufficiently complex to require experts, and has a major cognitive component. Crop decisions require reasoning with uncertain or incomplete data, have to be made under conditions of risk such as crop or market failure and are often based on heuristics.

*Implementation in ITOPE*

ITOPE will have three knowledge bases: a crop characteristic, a subcatchment characteristic and a decision knowledge base. These knowledge bases are matched. They are still being constructed and only some examples can be presented.

The crop-characteristic knowledge base contains facts about the crops, as perceived by farmers, and importantly, the explanation they give for these facts. Such facts and explanations could include labour needs for transportation, requirements to plant along permanent streams to benefit from temperature effects, minimum subsistence, or gastronomic requirements; for instance, preference for tasty upland rice varieties. The subcatchment characteristics knowledge base could contain facts about farming systems, elevation, indigenous soil classification, stream length and type, area, etc. Topographic information is extracted from SWAT's digital elevation model and road maps. The inference engine looks at the facts, in a forward chaining fashion, and determines which rules of the decision knowledge base apply.

A hypothetical inference could be that the engine sees the facts of a vegetable being vulnerable to transportation damage, having high transportation cost, but with a low probability of crop failure. The catchment is 'far' (a term that is well handled with fuzzy logic) from the road, and contains a vegetable farming system with a certain percentage of middle socioeconomic class farmers and the balance of low socioeconomic class farmers. The inference engine concludes that it needs to apply the rules of the middle and low class, respectively. From the crop and catchment conditions, it may conclude that a middle-class farmer would still grow the crop but only at 5% of the area of his farm, possibly causing serious erosion, but the low-class farmer may decide not to grow the crop at all.

Apart from the knowledge base and the inference engine, the expert system has an explanatory facility, to explain its conclusions and trace the rules that were used. The explanation facility is particularly important if ITOPE is to be used to enable different interest groups in a watershed to interact to find feasible soil conservation strategies (Duchessi and O'Keefe 1992).

The expert system is implemented in FuzzyCLIPS (KSL 1994), an expert system development environment. It is an extended version of CLIPS, the C Language Integrated Production System, and allows representation and manipulation of fuzzy facts and rules (Zadeh 1965). FuzzyCLIPS is selected because it creates the opportunity to model risk behaviour with fuzziness factors.

Fuzziness will not be realized in the first version of ITOPE, but the simulation would probably be improved if the degree of membership of a crop to the set of suitable crops or the set of unsuitable crops would be known. Priorities could be given to crops with higher membership values for the set of suitable crops.

### Introducing the overall model

Expert systems and multiple criteria optimization have been linked, for instance to provide heuristic short-cuts in the algorithm of multiple criteria optimization (MacLeod and Reeves 1993), or to simulate crop decision making (Nevo et al. 1994). The latter application relies on expertise to obtain production estimates. A primary requirement for expert system development is that there is an expert available to extract knowledge from. However, it is difficult to find experts for the erosion process. Farmers generally pay more attention to their particular environment and may provide expertise about their farm plots and their direct surroundings, but there is no expert who regularly monitors all subcatchments and has an overview for a whole catchment. To solve this problem, SWAT is used to estimate the contribution of the different subbasins to the whole-basin sediment load and to simulate crop production. Figure 3, which shows ITOPE's program control, explains the information streams between the subcomponents in Figure 1.

### Local versus global optimization

ITOPE operates in two phases, first for subcatchment optimization (phase I), then for catchment optimization (phase II). The 'tragedy of the commons' typically exhibits local optimality versus global suboptimality. This is simulated in phase I, where for each subcatchment the revenues are maximized and inputs and soil loss are minimized. Erosion constraints are not defined in this phase, and the importance that farmers in the subcatchments attach to soil loss among the other objectives is reflected in the augmented weighted Tchebycheff procedure. The procedure should first be run without the erosion objective to get a baseline decision set, because erosion has presumably never been a quantified decision criterion for farmers, and if presented with these figures, they may make decisions that do not reflect their usual decision making.

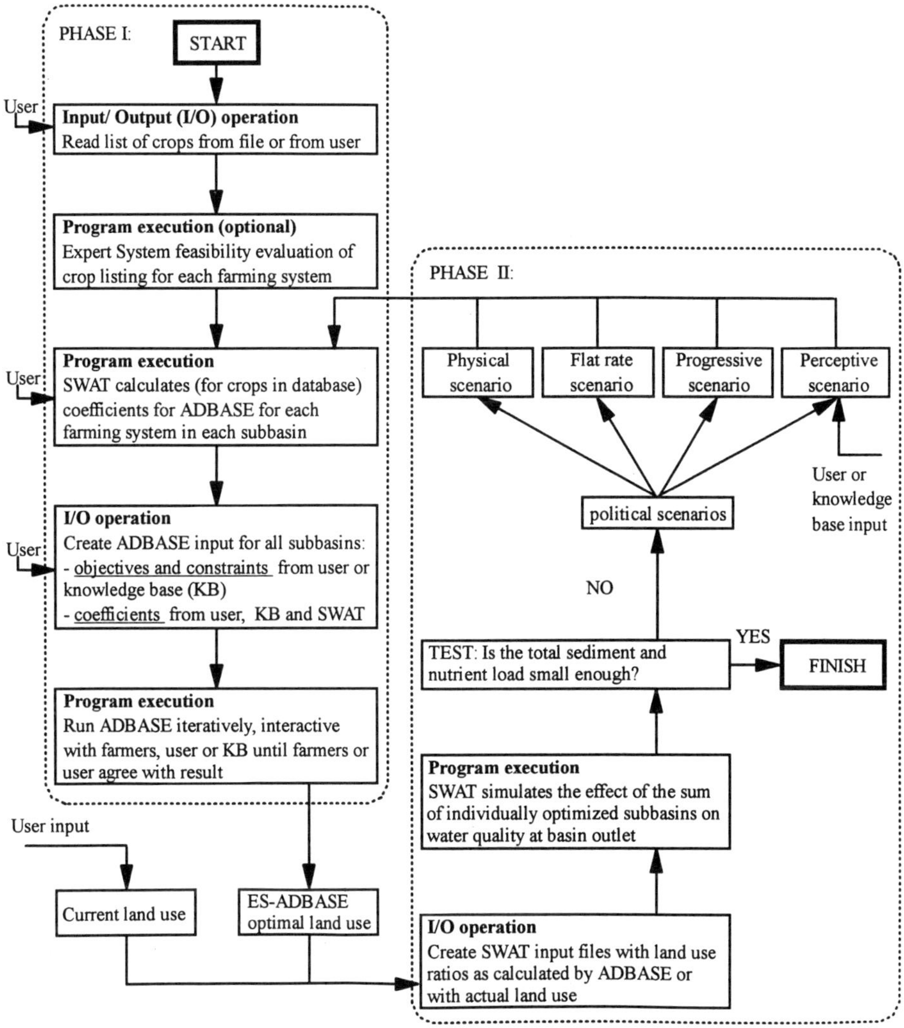


*Figure 3.* ITOPE's program control structure.

If catchment erosion rates exceed downstream water quality thresholds, erosion constraints are introduced in phase II, which are implemented in a new iteration of the multiple objective program. The user selects one of four political scenarios, which are executed by FuzzyCLIPS, in particular because of the more qualitative fourth scenario. The physical scenario attributes higher constraints to catchments with higher erosion rates. The flat rate scenario applies the same reduction rate to all subcatchments. The progressive scenario restricts erosion rates for catchments where farm profitability is higher. The perceptive scenario could apply constraints to those areas where awareness is highest, or to those areas that people perceive as sources of sediment. This latter, more qualitative scenario will not be developed for the Manupali watershed setting because input data are not available.

### Program control

In the first step of phase I, the user selects the crops from the SWAT data base, that he or she would like to work with and provides information for additional crops, which are not in SWAT's data base. Then, as the second step, the crops are evaluated by the expert system for their feasibility in each subcatchment. A list of crops, which farmers consider as feasible, is generated for each subcatchment from the list that the user started within step one. In the third step of phase I, SWAT calculates erosion and crop production rates for each crop that FuzzyCLIPS selected. For each crop in a subcatchment, the rates are calculated, assuming full coverage of the subcatchment. The user sets the number of years of simulation to generate average annual rates. Crop and input price information is multiplied with crop production rates and inputs, to calculate input and output objective function coefficients. In the fourth step, the input files for the augmented weighted Tchebycheff procedure are created. Objective function coefficients are taken from step three. Constraints are defined by the expert system or by the user. The Tchebycheff procedure is executed in step five, which results in an attribution of area for each crop in each subcatchment. Either this land-use distribution or the actual land use is processed in phase II.

Then, in step six, SWAT input files are created, using the actual or optimal land-use map. In the seventh step, SWAT calculates erosion rates for the catchment as a whole, with the input files generated in step six. If catchment sediment load exceeds a water quality threshold, one of the political scenarios is executed and subcatchments are identified to be restricted in sediment load (step eight). The iterative process continues, until a target sediment value for the whole catchment is achieved. For each catchment, the costs or benefits will be presented by comparison with the first optimization.

### Opportunities elicited by ITOPE

A few insights have appeared from the development of ITOPE. First, artificial intelligence and anthropology can benefit from each other's knowledge elicitation approaches. When an expert system is developed, a single recognized expert is asked to participate in knowledge elicitation methods. Anthropologists, in contrast, strive for elicitation of knowledge of communities. This knowledge may be either common or expertise knowledge for those communities, but it all certainly is expertise for outsiders. An example of an anthropological application of expert systems technology is the indigenous soil classification system described by Furbee (1989).

Research on a watershed level requires interaction with many spatially dispersed interest groups. Financial resources are spread out and interaction with communities is shallower. In order to obtain in-depth knowledge, elicitation techniques applied to a single expert, as practised in expert systems development, may balance shallow interaction with a community. Consensus theory (Romney et al. 1986; Boster and Johnson 1989) can identify the experts. Knowledge of one expert has to be tested against knowledge of other experts with the same management style to make it represent communal knowledge, for instance using focus groups (Krueger 1988). Consensus theory may be interesting for the artificial intelligence discipline in its exploration of expert systems with multiple experts (Liou 1992). What is interesting about the application described in this paper is that it simulates multiple experts, working in the same domain – farmers in one interest group – as well as multiple experts in different domains – spatially different areas and different interest groups (Dieng et al. 1992). For anthropologists, the application implies a focus on both heterogeneity in a watershed and homogeneity in a community.

Second, from an economic perspective, the maximization procedure has become subject to environmental and social boundary conditions introduced by a deterministic hydrologic model and reasoning from heuristics. The procedure is used as a deductive assumption that farmers certainly consider maximization of profit important, but given social and physical and mental constraints.

Third, from a hydrologist's viewpoint, the integration of the three tools provides an explicit social framework for the hydrologic model. It will give guidance in planning projects and prevent the development of unsuitable plans, both from an on-site and off-site perspective. Besides, it recognizes that solving erosion problems is a social and political problem. Consequences of erosion can be translated to social and political implications.

And fourth, last and most importantly, farmers and regional planners would have a tool that would assist them in discussing or negotiating costs and benefits from soil conservation strategies. The tool could reflect the constraints that farmers face, which not only become explicit to planners and other interest groups, but also determine simulation outcome.

## Acknowledgements

This research is a component of the Sustainable Agriculture and Natural Resource Management Collaborative Research Support Program (SANREM CRSP) supported by US AID Grant No. LAG-4198-A-00-2017-00 and by contributions from participating institutions. The CRSP accession number is JA-93-0002.

## References

Arnold J G, Williams J R, Nicks A D, Sammons N B (1990) SWRRB, a basin scale simulation model for soil and water resources management. Texas A&M University Press, College Station, Tex., USA. 142 p.

Arnold J G, Williams J R, Maidment D R (1995) Continuous-time water and sediment routing model for large basins. J. Hydrol. Eng. 121(2):171 – 183.

Arnold J G, Williams J R, Srinivasan R, King K W (1996) SWAT, soil and water assessment tool. United States Department of Agriculture, Agricultural Research Service, Grassland, Soil and Water Research Laboratory, Temple, Tex., USA. 102 p.

AVRDC – Asian Vegetable Research and Development Center (1995) Progress report 1994. AVRDC, Shanhua, Tainan, Taiwan (ROC). 520 p.

Boster J S, Johnson J C (1989) Form or function: A comparison of expert and novice judgements of similarity among fish. Am. Anthropol. 91:866–889.

Brinkman R (1994) Recent developments in land use planning, with special reference to FAO. Pages 11–21 in Fresco L O, Stroosnijder L, Bouma J, Van Keulen H (Eds.) The future of the land: Mobilising and integrating knowledge for land use options. John Wiley and Sons, New York, USA.

Coulson R N, Lovelady C N, Flamm R O, Spradling S L, Saunders M C (1991) Intelligent geographic information systems for natural resource management. Quantitative Methods Landscape Ecol. 82: 153–172.

De Groot W T (1992) Environmental science theory, concepts and methods in a one-world, problem-oriented paradigm. Elsevier, Amsterdam, The Netherlands. 583 p.

Dieng R, Giboin A, Tourtier P, Corby O (1992) Knowledge acquisition for explainable multi-expert, knowledge-based design systems. Pages 298–317 in Siekmann J (Ed.) Lecture notes in artificial intelligence. Springer-Verlag, Berlin, Germany.

Duchessi P, O'Keefe R M (1992) Contrasting successful and unsuccessful expert systems. Eur. J. Oper. Res. 61:122–134.

Fairweather J R, Keating N C (1994) Goals and management styles of New Zealand farmers. Agric. Syst. 44:181–200.

Friedl M A, Estes J E, Star J L (1988) Advanced formation-extraction tools in remote sensing for earth science application: AI and GIS. AI Appl. 2(2-3):17–30.

Furbee L (1989) A folk expert system: Soils classification in the Colca Valley, Peru. Anthropol. Q. 62(2):83–102.

Goldsworthy P R, Eyzaguirre P, Boerboom L (1994) The institutional and organizational implications of integrating natural resource management and production-oriented research. Pages 139–152 in Goldsworthy P R, Penning de Vries F W T (Eds.) Opportunities, use, and transfer of systems research methods in agriculture to developing countries. Kluwer Academic Publishers, Dordrecht, The Netherlands.

Hamer W I (1981) Soil conservation consultant report. Part A: A soil degradation assessment methodology, Technical note No. 7. Center for Soil Research, Bogor, Indonesia. 50 p.

Herskovits M J (1941) Economics and anthropology, a rejoinder. J. Political Econ. 49:269–278.

Jackson P (1990) Introduction to expert systems. 2nd Edition. Addison-Wesley, Wokingham, UK. 526 p.

Janssen R (1992) Multi objective decision support for environmental management. Kluwer Academic Publishers, Dordrecht, The Netherlands. 232 p.

Knight F H (1941) Anthropology and economics. J. Political Econ. 49:247–268.

Krueger R A (1988) Focus groups: A practical guide for applied research. Sage Publications, Newbury Park, CA, USA. 197 p.

KSL – Knowledge Systems Laboratory (1994) FuzzyCLIPS version 6.02: User's guide. National Research Council, Canada. 71 p.

Liou Y I (1992) Collaborative knowledge acquisition. Expert Syst. Appl. 5:1–13.

Luenberger D G (1995) Microeconomic theory. McGraw-Hill, New York, USA. 486 p.

MacLeod K R, Reeves G R (1993) AXIS: A framework for interactively combining structured algorithms and knowledge based systems. Comput. Oper. Res. 20(6):613–623.

Nazarea-Sandoval V D (1995) Local knowledge and agricultural decisionmaking in the Philippines, class, gender, and resistance. Cornell University Press, Ithaca, NY, USA. 226 p.

Nevo A, Oad R, Podmore T H (1994) An integrated expert system for optimal crop planning. Agric. Syst. 45:73–92.

Plant R E (1993) Expert systems in agriculture and resource management. Technol. Forecast. Soc. Change 43:241–257.

Röling N (1994) Platforms for decisionmaking about ecosystems. Pages 385–393 in Fresco L O, Stroosnijder L, Bouma J, Van Keulen H (Eds.) The future of the land: Mobilising and integrating knowledge for land use options. John Wiley and Sons, New York, USA.

Romney A K, Weller S C, Batchelder W H (1986) Culture as consensus: A theory of culture and informant accuracy. Am. Anthropol. 88:313–338.

Schneider H K (1989) Economic man: The anthropology of economics. 2nd Edition. Sheffield Publishing Company, Salem, WI, USA. 278 p.

Simon H A (1992) Rational choice and the structure of the environment. Pages 39–54 in Egidi M, Marris R (Eds.) Economics, bounded rationality, and the cognitive revolution. Edward Elgar Publisher, Brookfield, VT, USA.

Srinivasan R, Arnold J G (1994) Integration of a basin-scale water quality model with GIS. Water Res. Bull. 30(3):453–462.

Steuer R E (1989) Multiple criteria optimization: Theory, computation and application. 2nd Edition. Krieger Publishing, Malabar, FL, USA. 546 p.

Steuer R E (1994) Random problem generation and the computation of efficient extreme points in multiple objective linear programming. Comput. Organ. Appl. 333–347.

Zadeh L A (1965) Fuzzy sets. Inf. Control 8:118–153.

# The economics of wheat production in the rice-wheat system in Nepal

M. BHATTARAI[1] and S. PANDEY[2]

[1]*Agricultural Projects Services Centre (APROSC), P.O. Box 1440, Kathmandu, Nepal*
[2]*International Rice Research Institute, P.O. Box 933, 1099 Manila, Philippines*

**Key words:** supply response, linear programming, infrastructure, rice-wheat system

**Abstract**

Wheat is grown in over 0.5 million ha as a part of the rice-wheat system in Nepal. Despite its importance, the productivity of wheat in Nepal remains low. The economics of wheat production in a typical rice-wheat area of Nepal is examined using a linear programming model. In order to assess the effect of irrigation and marketing infrastructure on the economics of wheat production and elasticity of wheat supply, representative models for four locations with differential access to the infrastructure were developed. Wheat production was found to be economically more profitable in locations with better access to irrigation and marketing infrastructure. The price elasticity of supply of wheat, derived by parametric variation of the wheat price in the linear programming model, was found to be dependent on the level of infrastructure development. Farmers in areas with access to irrigation and markets are found to be much more responsive to changes in wheat prices than farmers without access to such infrastructure. The effectiveness of price policies in increasing wheat production thus appears to be conditional on the level of infrastructure development.

## Introduction

Wheat is one of the major winter crops of Nepal, occupying over 0.5 million ha and ranking third in importance in terms of area planted. Despite the expansion of the wheat area by over 500 percent between 1960 and 1990, the productivity of wheat has remained stagnant at around 1.2 ton $ha^{-1}$. A host of technological, economic and policy factors has kept the productivity of wheat at a low level (Bhattarai 1994; Fujisaka et al. 1994). Although high-yielding varieties have been adopted in most wheat-producing areas of Nepal, the application of chemical fertilizers on wheat remains at a very low level of about 15 kg of nutrients per hectare (Gill 1995; Upadhyaya 1996). Although the rate of return to wheat research in Nepal is high at around 70% (Morris et al. 1992), area expansion is the main contributing factor.

The major objective in this paper is to investigate the effects of irrigation and marketing infrastructure on the economics of wheat production and its supply response to price changes. Being mostly a cash crop, its production and supply can be expected to be governed to a large extent by market prices. Changes in price policies are often advocated to encourage adoption of yield-increasing wheat technologies. The effectiveness of price policy in encouraging output expansion depends critically on the supply response. The study attempts to throw light on infrastructure factors that constrain the effectiveness of price policy in increasing wheat production in Nepal.



*P.S. Teng* et al. *(eds.): Applications of Systems Approaches at the Farm and Regional Levels, 45–52.*

*Table 1.* Different production environments (locations) selected for the farm level analysis, Morang District, Nepal.

| Locations | Village name (locality) | Distance from wholesale market | Availability of irrigation |
|---|---|---|---|
| A | Damabandi | 2 km | Assured |
| B | Mangalbare | 2 km | Unassured |
| C | Amardaha W.No. 6 | 22 km | Assured |
| D | Amardaha W.No. 6 | 20 km | Unassured |

## Rice-wheat production systems of Nepal

Wheat is a winter-season crop planted after the harvest of rice. In Nepal, wheat is grown mainly in the flat plains (Terai) and mid-hills. The Terai is the main granary of Nepal and is characterized by a moist subtropical climate with temperatures ranging between 15°C and 32°C. Wheat in Terai is grown under both rainfed and irrigated conditions with moderate fertilizer application and the main constraints to yield increase are late planting, variability in irrigation supply and disease pressure (Fujisaka et al. 1994).

The mid-hills are characterized by a warm to cool temperate climate, with temperatures ranging between 5°C and 20°C. Wheat is grown mostly under partial irrigation in valleys and terraced fields on the hills. Both input usage and yields are lower than in the Terai. At the national level, hills and Terai contribute 40 and 60% of the total wheat supply (MOA 1994), respectively.

## The study area and survey design

The study area was chosen to represent a typical rice-wheat production area in the eastern part of the Terai. In the Morang district, wheat is grown in over 20 000 ha out of the total cultivated area of 105 000 ha. Nearly 42% of the total cultivated area is partially or fully irrigated. The average productivity of wheat is around 2 ton $ha^{-1}$ (MOA 1994).

To examine the effect of irrigation and marketing infrastructure on the economics of wheat production, four locations (villages) with differential access to marketing facilities (judged by the distance to market) and irrigation were purposefully selected from the Morang district (Table 1). Thirty-five farmers from each location were randomly selected for the survey, which included rice as well as a range of winter crops including wheat. Cost and production data for 1992–93 were obtained from farmers by using structured interview schedules. In addition to this, qualitative data regarding constraints to wheat production were generated by using the key-informant survey approach.

## Analytical method

The response of agricultural output to output price changes (i.e. supply response) can be obtained by econometric estimation using historical data or by parametric variation of prices in a model of farm households (Shumway and Chang 1977; Hazel and Norton 1986). As the current study is based on cross-sectional data (i.e. farm data for only one year) and seeks to assess the impact of structural variables such as marketing and irrigation infrastructure, estimation based on the second approach was judged to be more appropriate. Accordingly, a deterministic linear programming (LP) model for a representative rice-wheat farmer in each of the four villages was constructed.

The basic structure of the LP model is presented in Table 2. Rice is the only activity considered in the rainy season. Wheat, grass peas, mustard and lentil are alternative crops grown in the winter season. Wheat can be planted early (early October to mid-November) or late (mid-November to late December), with the late planted wheat suffering a yield reduction of 10% because of higher temperature during the grain filling stage. Both early and late wheat compete with rice because of peak labour demand during harvesting and threshing of rice and land preparation for wheat sowing. Working capital constraints were specified in the model with a maximum limit on borrowing. Similarly, land renting was allowed with a maximum limit. Maximum limits on other winter crops and a minimum limit on wheat for subsistence requirements were also specified. The farmer was assumed to allocate land to various activities in order to maximize the total gross margin. The detailed justification of the structure of the programming model, the parameters used and calibration can be found in Bhattarai (1994).

## Results and discussion

### *Micro-economics of wheat production*

The average costs of production of wheat in these four locations are presented in Table 3. The average wheat yield is 2.2 ton $ha^{-1}$ in location A, which has access to relatively assured low-cost irrigation from canals. Due to the proximity to market, the farm-gate price is 18% higher here than in market-inaccessible locations. Accordingly, the gross margin is highest at Rs 4400 $ha^{-1}$. As wheat is an important cash crop, almost 80% of the area is planted to wheat in this location (Table 4). Only about 5% of land is fallowed in winter (Table 5).

The land-use pattern is quite different in locations B, C and D. Wheat occupies a relatively smaller area in these locations. Lentil, mustard and other winter crops, which are used mainly for domestic consumption, are more popular. Also, the extent of fallowing is higher in locations with less wheat. Due to unavailability of cheap irrigation and/or limited access to market, farmers substitute rainfed crops for wheat. These alternative crops are more attractive due to their lower cash cost, as well as gross margins, which are comparable to that of wheat (Table 5). For example, the ratio of gross returns to cash cost is 1.67 for wheat, but for lentil, mustard and grass peas, the ratios are 6.8, 2.97 and 4.86, respectively (Table 5). The cash cost of wheat production is also higher in market accessible locations (i.e. A and B) than in market inaccessible locations (C and D).

*Table 2.* Structure of the basic LP model for an average farm, Morang District, Nepal, 1993.

| | RHS rows/type | Unit | Activities included in LP matrix: Early wheat | Late wheat | Lentil | Grass peas | Mustard | Rice | Labour hiring Early season | Labour hiring Mid-season | Plough hiring Early season | Plough hiring Mid-season | Capital borrow. Main season | Capital borrow. Winter season | Land renting Main season | Land renting Winter season |
|---|---|---|---|---|---|---|---|---|---|---|---|---|---|---|---|---|
| I Objective function (max.) | | | + | + | + | + | + | + | – | – | – | – | – | – | – | – |
| II Resources constraints | | | | | | | | | | | | | | | | |
| 1 Land | | | | | | | | | | | | | | | | |
| 1.1 Main season | ≥ | ha | | | | | | + | | | | | | | – | |
| 1.2 Winter land type 1 | ≥ | ha | + | + | + | + | + | 0 | | | | | | | | – |
| 1.3 Winter land type 2 | ≥ | ha | | 0 | + | + | | | | | | | | | | |
| 2 Farm family labour | | | | | | | | | | | | | | | | |
| 2.1 Early season | ≥ | No | + | 0 | + | + | + | + | – | | | | | | | |
| 2.2 Mid-season | ≥ | No | + | + | + | + | + | + | | – | | | | | | |
| 3 Family plough (animal days) | | No | | | | | | | | | | | | | | |
| 3.1 Early season | ≥ | No | + | 0 | + | 0 | + | + | | | – | | | | | |
| 3.2 Mid-season | ≥ | No | + | + | 0 | 0 | 0 | + | | | | – | | | | |
| 4 Working capital (family) | | | | | | | | | | | | | | | | |
| 4.1 Main season | ≥ | Rs | | | | | | + | | | | | – | | | |
| 4.2 Winter season | ≥ | Rs | + | + | + | + | + | 0 | | | | | | – | | |
| 5 Maximum limit of credit available for winter season | ≥ | Rs | | | | | | | | | | | | + | | |
| 6 Maximum limit on rented land | | | | | | | | | | | | | | | | |
| 6.1 Main season | ≥ | ha | | | | | | | | | | | | | + | |
| 6.2 Winter season | ≥ | ha | | | | | | | | | | | | | | + |
| 7 Maximum limit of lentil | ≥ | ha | | | + | | | | | | | | | | | |
| 8 Maximum limit of grass peas | ≥ | ha | | | | + | | | | | | | | | | |
| 9 Minimum mustard area | ≤ | ha | | | | | + | | | | | | | | | |
| 10 Minimum wheat area | ≤ | ha | + | + | | | | | | | | | | | | |

*Table 3.* Costs and returns of wheat production in the Morang District, Nepal, 1993.

| Item | Unit[1] | Location A | Location B | Location C | Location D | Average |
|---|---|---|---|---|---|---|
| 1 Material input costs | $Rs\,ha^{-1}$ | 3315 | 4191 | 2806 | 3418 | 3433 |
| 2 Total labour cost[2] | $Rs\,ha^{-1}$ | 4485 | 4924 | 5578 | 5067 | 5014 |
| 2.1 Family supplied | $Rs\,ha^{-1}$ | 1979 | 3413 | 4876 | 4101 | 3593 |
| 2.2 Hired | $Rs\,ha^{-1}$ | 2506 | 1511 | 702 | 966 | 1421 |
| 3 Total variable (cash) costs (1+2.2) | $Rs\,ha^{-1}$ | 5821 | 5702 | 3508 | 4384 | 4854 |
| 4 Average yield | $t\,ha^{-1}$ | 2.19 | 1.74 | 1.67 | 1.80 | 1.86 |
| 5 Farm-gate price | $Rs\,t^{-1}$ | 4675 | 4675 | 3959 | 3959 | 4317 |
| 6 Gross return | $Rs\,ha^{-1}$ | 10 238 | 8135 | 6612 | 7126 | 8030 |
| 7 Total cost | $Rs\,ha^{-1}$ | 7800 | 9115 | 8384 | 8485 | 8447 |
| 8 Gross margin (6−3) | $Rs\,ha^{-1}$ | 4417 | 2432 | 3104 | 2742 | 3175 |

[1]US $1 = Rs 50 at 1994 exchange rate.
[2]Total labour cost includes both human labour and animal draft power (ploughing) and family labour valued at the market wage rate.

*Table 4.* Land-use pattern in winter season, Morang District, Nepal, 1993–94 (*Source:* Bhattarai 1994).

| | Percentage area under | | |
|---|---|---|---|
| Location | Wheat | Lentil and mustard | Fallow |
| A | 80 | 15 | 5 |
| B | 15 | 60 | 25 |
| C | 22 | 62 | 16 |
| D | 10 | 58 | 32 |

*Table 5.* Per hectare costs ($Rs\,ha^{-1}$) and returns for various crops in the Morang District, Nepal, 1993.

| | Crops | | | |
|---|---|---|---|---|
| Item | Rice | Wheat | Lentil | Mustard |
| 1 Total material inputs cost | 1717 | 3433 | 360 | 1610 |
| 2 Total human labour cost | 3231 | 2474 | 1493 | 1659 |
| 2.1 Family supplied | 921 | 1489 | 1119 | 1244 |
| 2.2 Hired | 2309 | 985 | 374 | 415 |
| 3 Animal ploughing cost | 2900 | 2540 | 1247 | 2871 |
| 3.1 Family supplied | 2228 | 2101 | 1247 | 2871 |
| 3.2 Hired | 672 | 436 | 0 | 0 |
| 4 Total cash (variable) cost (1+2.2+3.2) | 4698 | 4854 | 733 | 2025 |
| 5 Total production cost (1+2+3) | 7848 | 8446 | 3100 | 6140 |
| 6 Average yield ($ton\,ha^{-1}$) | 2.53 | 1.86 | 0.49 | 0.52 |
| 7 Average farm-gate price ($Rs\,ton^{-1}$) | 4420 | 4317 | 10 157 | 11 657 |
| 8 Average value of by-products | 970 | 48 | 0 | 0 |
| 9 Gross return (6×7+8) | 12 153 | 8078 | 4979 | 6007 |
| 10 Gross margin (9−4) | 7455 | 3224 | 4246 | 3982 |
| 11 Net return | 4305 | −371 | 1867 | −133 |
| 12 Gross return per unit of cash cost (ratio) | 1.59 | 1.67 | 6.80 | 2.97 |

*Note:* US $1 = Rs 50 at 1994 exchange rate.

*Table 6.* Comparison of existing and model-predicted wheat acreage at current wheat prices.

| Locations | Existing wheat area (ha) | Model-predicted wheat area (ha) | Current wheat acreage to land holding (%) |
|---|---|---|---|
| A | 1.77 | 2.32 | 80 |
| B | 0.40 | 0.40 | 15 |
| C | 0.40 | 0.58 | 22 |
| D | 0.33 | 0.4 | 10 |

*Cost comparison of wheat versus rice cultivation*

The cost of material input for wheat is almost twice that of rice (Table 5), but rice is a more labour-intensive crop due to the requirement of transplanting and manual weeding. As these operations have to be completed within a short period, rice farmers rely much more on hired labour than do wheat farmers. The cost of hired labour for rice production is more than 50% of the total variable cash cost.

Although land preparation costs for both rice and wheat are similar (around Rs 4700-4800 per hectare), animal labour for land preparation for rice is mostly family-owned. In the case of wheat, the need to prepare land rapidly soon after the harvest of rice forces farmers to use hired draft animals. Therefore, any delay in rice harvesting and unavailability of draft power in time for land preparation can constrain the expansion of the wheat area.

*Wheat supply response*

The supply elasticity, which measures the percentage change in the quantity supplied for a 1% change in price, was derived by altering the price of wheat, other things remaining the same, and observing the change in the optimal wheat area in the LP model. Optimal wheat areas obtained from the model were then regressed on prices to obtain the supply elasticity. As the price response to yield changes was not modelled, the supply elasticities presented are area elasticities. Farmers may expand the area as well as increase yield by applying more inputs in response to price increases. The area response estimated in this study thus provides only a lower limit estimate of the supply elasticity.

An attempt at validating the LP model was made by comparing the existing wheat area with the model-predicted wheat area in each location. For locations B, C and D, the model predictions are close to the actual values (Table 6). For location A, model prediction is about 20% higher. Overall, the model simulates the farmer allocation of land to wheat reasonably closely.

The supply elasticities, estimated in the neighbourhood of the current wheat price (i.e. –5 to +5%), are presented in Table 7. Within each category of market access, areas with more assured irrigation have a higher elasticity of supply. For example, within the market accessible category, location A has a higher elasticity than location B. Similarly, within the market inaccessible category, location C has a higher elasticity

*Table 7.* Farm level price elasticity of wheat supply (acreage).

| Production environment | Location | Price elasticity (i) | Price elasticity (ii) |
|---|---|---|---|
| Market accessible assured irrigated area | A | 0.66 | 1.98 |
| Market accessible unassured irrigated area | B | 0 | 1.22 |
| Market inaccessible assured irrigated area | C | 0.50 | 2.56 |
| Market inaccessible unassured irrigated area | D | 0 | 0.27 |

(i) Price elasticities estimated by considering the price variation of −5% to +5% from the current wheat price level.
(ii) Price elasticities estimated by considering the price variation of −15% to +25% from the current wheat price level.

than location D. Thus, farmers' responsiveness to wheat price changes seems to be determined largely by access to irrigation.

In locations B and D, the supply response is nonexistent as the expansion of wheat area (within the range examined) is severely constrained by limited access to irrigation (in location B), as well as by poor marketing accessibility (in location D).

When the range of price variation in the model is widened, the basic conclusions remain the same – although the numerical magnitude of supply elasticities increases. For example, in the range of price variation from −15 to +25 percent of the current price, the effects of market access and irrigation in determining supply response become more pronounced (Table 7).

## Concluding remarks

The low price elasticity in the neighbourhood of the current wheat price indicates that price policy may not be an effective tool in manipulating wheat supply, especially in areas with limited access to irrigation and marketing infrastructure. As wheat production is economically more profitable to farmers in areas with better access to market and availability of cheap irrigation, better targeting of public sector intervention is desirable.

In market accessible areas with irrigation facilities, where wheat is a major crop, further expansion of wheat production may require yield-increasing technology since land constraints restrict the expansion of acreage in these locations. Provision of improved technology at the farmers' level, more effective extension and adequate input supplies are some of the ways to increase wheat production in these locations.

In areas with limited market access, where the current area of wheat is relatively low, policies that encourage area expansion such as rural road improvement and provision of market infrastructure and better market information are likely to be more effective than yield-increasing technologies *per se* in increasing wheat production. In the past, efforts at extending improved agricultural technologies in these areas have not been successful because of limited incentives to wheat production.

## References

Bhattarai M (1994) Wheat supply response in Nepal: Econometric and programming approaches. MS thesis, Department of Agricultural Economics, University of the Philippines Los Baños, Philippines. 165 p.

Fujisaka S, Harrington L, Hobbs P R (1994) Rice-wheat in South Asia: System and long term priorities established through diagnostic research. Agric. Syst. 46:169–187.

Gill G J (1995) Fertilizers and food grains in the Nepal Terai. Winrock International, Nepal. 25 p.

Hazel P B R, Norton D (1986) Mathematical programming for economic analysis in agriculture. Macmillan Publishing Company, New York, USA. 400 p.

MOA – Ministry of Agriculture (1994) Agricultural statistics. Kathmandu, Nepal.

Morris M, Dubin H J, Pokhrel T (1992) Returns to wheat research in Nepal. CIMMYT Economic Working Paper 92–04. International Maize and Wheat Improvement Center, Mexico. 47 p.

Shumway C R, Chang A A (1977) Linear programming versus positively estimated supply functions: An empirical and methodological critique. Am. J. Agric. Eco. 59(2):344–357.

Upadhyaya H K (1996) Setting priorities for rice research in Nepal. In Evenson R, Herdt R, Hossain M (Eds.) Setting rice research priorities in Asia. International Rice Research Institute, P.O. Box 933, Manila, Philippines (in press).

# Modelling long-term effects of crop residue management for sustainable farming systems

M. BRUENTRUP[1], J.P.A. LAMERS[1] and L. HERRMANN[2]

[1]*Institute of Agricultural Economics and Social Sciences in the Tropics (490), University of Hohenheim, 70593 Stuttgart, Germany*

[2]*Institute of Soil Science and Land Evaluation, University of Hohenheim, 70593 Stuttgart, Germany*

**Key words:** economy, modelling, dynamic linear programming soil fertility, mulching, Niger

**Abstract**

Field research has shown a significant yield increase with the use of millet crop residues (CR) as a soil mulch. The short-term profitability of this management practice is, however, low. This paper examines the long-term effects of mulching on soil fertility, the decreasing accessibility of fallow and, consequently, the reduction in the opportunity of restoring the soil fertility through a fallow using a multi-period dynamic linear programming model. Depending on an assumed discounting rate for future incomes, the long-term evaluation indicates that a mulch application can be economically profitable compared to the tested alternative uses such as mulching at lower rates and the trade of stover. Four scenarios were analysed which focused on different land: labour ratios, discount rates, prices of CR, and a specific combination of mulch and working capital. The importance of fallow for long-term fertility management is shown, as well as the importance of a mulch application with millet stover in the absence of sufficient fallow. The optimal combination of fallow, and stover producing and consuming activities to master the mulch shortage, are presented. However, all scenarios show that with CR mulching alone, future soil degradation cannot be prevented.

## Introduction

One of the most important problems in the West African Sahel (WAS) is feeding the rapidly growing population under a continuously declining soil fertility. The Republic of Niger, one of the largest countries in this area, exemplifies the major issues: whereas the total cultivated area increased during the last two decades, the grain yields per hectare of the major staple food pearl millet (*Pennisetum glaucum* (L.) R. Br.) decreased (FAO 1993). Farmers were unable to compensate for the decline in per hectare production by an extension of the cultivated area. In addition, due to population growth rates above 3%, increasing quantities of food had to be imported (FAO 1993).

Mineral fertilizer has been identified for increasing yields substantially but, in spite of intensive propagation and high subsidies, the fertilizer application rate has remained extremely low. In 1992, for instance, total fertilizer use by farmers in Niger was only 0.3 kg $ha^{-1}$ of arable land (FAO 1993). Given the problems encountered with mineral fertilizers, organic fertilization was advocated by researchers. A broadcast mulch of millet crop residues (CR) reduced wind erosion and increased yields (Rebafka 1993; Michels 1994; Buerkert 1995).



*P.S. Teng* et al. *(eds.): Applications of Systems Approaches at the Farm and Regional Levels, 53–64.*

Short-term economic analyses of the comparative advantages of five different CR uses showed that with a broadcast mulch of 2000 kg ha$^{-1}$ returns to land were indeed higher (Lamers and Bruentrup 1995; Bruentrup et al. 1996). Yet, the use of CR as livestock feed yielded the highest returns to total labour, whereas burning 2000 kg CR ha$^{-1}$ gave the highest returns per unit weeding time. When these alternatives were compared by linear programming models for a typical farm in western Niger, all optimal solutions suggested a multi-purpose use of CR. Since farmers' current multi-purpose use of CR coincided with its optimal use (Lamers and Bruentrup 1995), it is argued that farmers use CR very rationally, even when it is taken from the fields or burned.

Yet, these household model calculations consider only the short-term effects of mulching and ignore influences on soil fertility in the medium and long run. In addition, the opportunity to restore the soil fertility through a fallow, a traditional pillar of fertility management, is neglected. Fallow, CR management and mulching practices are in a sensitive equilibrium that is influenced by a variety of economic characteristics of the farm households, such as consumption requirements, land tenure and liquidity, as well as labour availability and the ecological environment in which households operate.

This paper compares the long-term economics of a CR mulch to a complete removal of CR for a representative farm in western Niger, to answer the question of whether to take CR from, or to leave CR in, the field. The analyses are conducted from a farming systems point of view, neglecting social costs and benefits by off-farm effects of decreasing fertility and erosion.

## Methodology

### *Model structure and the representative farm household*

The main analytical tool in this analysis is a dynamic multi-period linear programming (LP) model which was used to simulate the effects of CR removal and its sale versus a mulch of CR at a rate of 2000 kg ha$^{-1}$ in the succeeding year on any farmers' field (CR transfer). In each year, over a period of 16 years, CR and fallow activities were combined on the available land of each of seven soil fertility levels. The farm organization was restricted to guarantee a certain annual self-sufficiency grain production and a minimum annual cash income in normal terms. Self-produced grain consumption was not monetarily valued since the objective of a household is likely not to be the maximization of the value of consumed grain, but to satisfy energetic requirements under certain consumption preferences. The cash income in each year, which is calculated as the sold output (grain and stover) minus the variable inputs (hired labour, purchased stover), was discounted; the LP model maximizes the discounted income. The discount rate was set at 10% (see also section Scenario 3). Including production, transfer and market operations, the model consisted of 819 activities and 481 constraining variables.

Since data on model parameters such as CR prices, land fertility distribution and

*Table 1.* Characteristics of a representative farm in western Niger (*Sources:* estimates based on household surveys from Baidu-Forson (1988) and McIntire et al. (1989, ILCA, Addis Ababa, Ethiopia, pers. commun.)).

| Feature | Amount |
|---|---|
| Total farm size including fallow (ha) | 15 |
| Family size (person) | 11 |
| – of which working members (%) | 36 |
| Subsistence requirements (kg millet $person^{-1}$ $yr^{-1}$) | 200 |
| – of which self-sufficiency level (%) | 90 |
| Total working capacity (d $person^{-1}$) | 200 |
| – of which during the weeding period (d $person^{-1}$) | 45 |

time preference rates are variable, scarce or even unknown, emphasis was given particularly to extensive sensitivity analyses. Different scenarios of labour, cash and land constraints for the farm household were evaluated.

No alternative uses of CR and no risk were considered in the model for several reasons: (i) comparisons of CR for mulching and alternative uses are discussed elsewhere: uses such as burning and feeding are still economically superior to mulching under the actual conditions of western Niger (Lamers and Bruentrup 1995; Bruentrup et al. 1996); (ii) long-term data on these alternative uses as well as on mulching on fields of different soil fertility with repetitions over several years to catch climatic risks were not available; and (iii) assuming cash income maximization (with the mentioned restrictions) provided insight into an extreme case of economic rationality, including risk considerations under conditions of few management alternatives. Thus, the long-term profitability of mulching most probably would be lower than calculated here. In any case, for long-term soil fertility management, which is the main topic of this paper, the application of CR as mulching would assure the maximum effect. However, the economic assessment of millet CR mulching techniques is comprehensive, even without the previously cited analyses.

The characteristics of the representative household which are based on surveys in western Niger villages are summarized in Table 1. Since weeding operations were found to be the decisive labour bottleneck in the farming systems (McIntire et al., 1989, International Livestock Center for Africa (ILCA), Addis Ababa, Ethiopia, pers. commun.) for this period, the household labour capacity and the requirements of the production activities were considered separately in the models. The dynamic LP model is designed to obtain annually a minimum cash income of 25 000 FCFA and to transfer a maximum of 30% of the income above this minimum in the following year for hiring labour and/or purchasing stover. The upper limit to finance these latter activities is set at 25 000 FCFA. It was assumed that grain surplus production can be sold, but not stored over the next rainy season.

Benefits and costs in the base case are valued at constant market prices. The price for grain was 80 FCFA $kg^{-1}$ (OPVN 1992), for stover 11.8 FCFA $kg^{-1}$ (own market survey) and for labour 700 FCFA $d^{-1}$ (7 h $d^{-1}$). Prices and revenues are at the rate prior to the devaluation of the FCFA in January 1994, i.e. $1 US equalled about 270 FCFA.

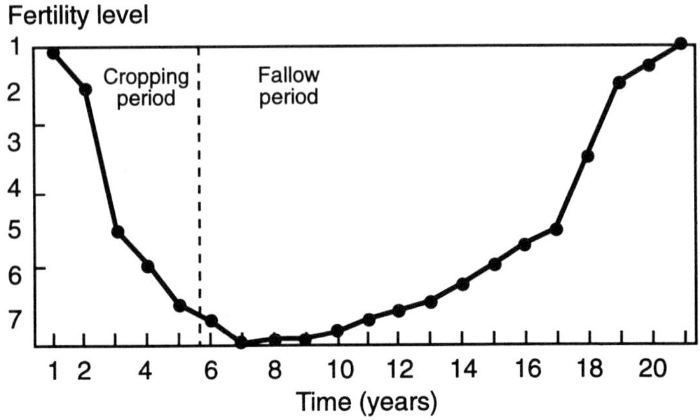


*Figure 1.* Development of soil fertility in a crop/fallow rotation under conditions in Niger.

*Assessment of yields and soil fertility dynamics for different crop residue uses*

Soil fertility here means the average production capacity (measured by crop yield) of a soil at a certain moment without external nutrient input. The information on millet grain and stover yields for different CR treatments and soil fertility levels is extremely heterogeneous, and partially contradictory (Feil et al. 1995). Thus, generalizations, which will be discussed below, were made to derive expected parameters for the LP model.

Existing data suggest that a constant cultivation of seven years is possible before the chemical degradation of the most common soil in the WAS (arenosol) is so pronounced that the land has to be transformed into fallow for fertility restoration (Rebafka 1993). Without any nutrient supplement, the average crop yield and fertility development versus time take the shape of S-curves (Figure 1). This shape is explained by the easily available nutrients (organic and inorganic) after fallow and the dependence of nutrient delivery mainly on the decomposition of more stable fractions after those are depleted (Hebel 1995). CR as well as fertilizer applications can increase yields substantially but they do not slow down the decrease in soil fertility.

The fertility under fallow follows the inverse way. Under Sahelian conditions the development of spontaneous vegetation after a cropping period is slow. Due to the low soil cover, nutrient losses due to erosion must be taken into account (Roose 1993). As soon as the vegetation is sufficiently in place, biomass yield and, therefore, nutrient uptake and storage in fallow plants, increase exponentially until a certain equilibrium is reached. Based on our data, the time ratio between degradation and recovery was estimated to be 1 : 2.

In the scenarios examined, a yield of 700 kg grain directly after a fallow was assumed, which corresponds to an average yield of 355 $kg\,ha^{-1}$ if zero mulch is made and millet cultivation is equally distributed across the seven soil fertility levels. The average on-farm millet yield in western Niger is about 350–500 $kg\,ha^{-1}$ (ICRISAT

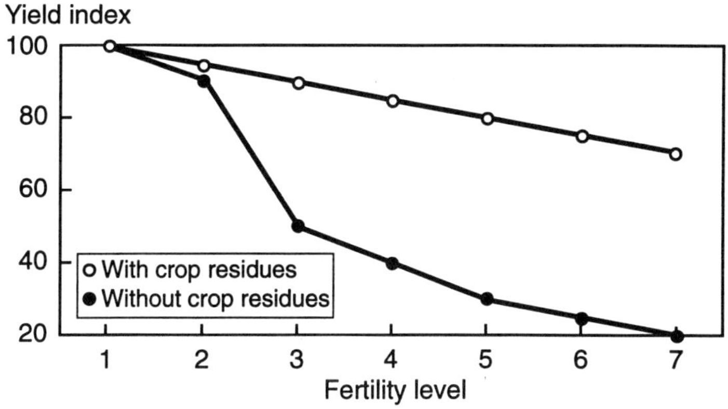


*Figure 2.* Yield indices for different soil fertility levels and crop residue treatments (index 100 is obtained directly after fallow).

1992). A constant millet stover: grain ratio of 0.75: 0.25 was assumed for the surveyed millet cultivar CIVT (Composite Inter Variétal de Tarna). Relative yields for both CR uses and different soil fertility levels (Figure 2) are based on results from experiments conducted between 1983 and 1992 at the ICRISAT Sahelian Center, Niamey, Niger (Rebafka 1993; Buerkert 1995). Due to the lack of nutrient addition in the zero mulching version, the relationship of relative yield versus fertility follows the model shown in Figure 1. The yield of the mulching activity also declines with decreasing soil fertility, but less than without a mulch. The lower limit was assumed to be 75% of the maximum yield index after fallow, whereas without CR application it was set at 20%.

For the first year of the model, it was assumed that the fields (one-third of total land) as well as fallow land were equally distributed over all fertility levels. The initial CR quantity available for the farm in model-year 1 that can be used for mulching or sale in the first year is 4718 kg corresponding to the yield from 5 ha without mulching, which was a typical value from previous analyses (Lamers and Bruentrup 1995). The transfer of CR as a mulch into a successive year is linked to a loss of 30% due to wind blow, termite activities, etc.

To reflect a dynamic fertility management and not a fixed field: fallow ratio, it was assumed that a field must not be necessarily used until the lowest fertility level is reached, but can be left in fallow before reaching this stage. Also, a piece of fallow can be taken into cultivation before reaching the maximum fertility level. To consider the low rate in fertility increase at the initial stage of fallow, a fallow field needs two years to recover before it increases in soil fertility and then always two years to gain the next highest fertility level. Under the conditions of the Sahel, little effort is necessary to clear a fallow, and since this can be done in the off-season, no additional labour time was considered for this activity. The result of the last year is an artefact of the LP formulation because a fertility decline after year 16 has no impact on the total gross margin and was, therefore, omitted in Figures 3–5.

Labour times for the zero mulch and high mulch situations are reported elsewhere (Lamers and Bruentrup 1995). It was assumed that the general labour input for soil preparation, sowing, and thinning and weeding did not change with the different soil fertility levels. In order to consider different quantities of labour needed for harvest, the harvest labour was taken to be 1 hour for harvesting 40 kg grain and 180 kg of stover. For purchase and sale activities of CR, 8 h work per 150 kg was assumed for cutting, collecting, and loading and transporting produce to and from a nearby market. For the internal transfer of CR on the farm, 7 h work per 150 kg was considered average. It was further taken for granted that the CR were protected against wild animals and cattle.

## Results and discussion

### *The base case*

Selected indicators of the optimal solution of the LP model under the assumptions in the base case are shown in Figures 3–5. In the first years, when the discounting factors place important weight on the cash income, the optimal solution shows high interannual variations in the farm organization, particularly in annual nominal cash income (Figure 3). In the initial phase of the evaluation period, liquidity constraints and the higher preference for immediate consumption (expressed by the discounting rate) hinder farmers to transfer cash into the next year. Only after about year 6 has the cash income stabilized and cash transfers are used to invest in hired labour for weeding and general activities (Figure 3). Grain is also sold, but the internal organization continues to be variable. Approximately 3.5–4 ha are cultivated each year (Figure 4). The main activities are zero mulching on soils of high fertility (levels 1 and 2), and in some years on levels 3 and 4 (Figure 5). High mulching only occurs in the first years on fields with low fertility level (level 7), where the marginal benefits from mulch application are the highest of all treatments (not shown in figure). However, at the price of 11.8 FCFA $kg^{-1}$ stover, it is more profitable for farmers to sell this material. The fertility management shows that, increasingly, fields of intermediate fertility levels are left in fallow and are recultivated in the higher fertility level 1, after six years of fallow. In later years, fallow of fertility level 2 is also used.

Previous empirical research in the same region shows that the mixed farming systems in western Niger are characterized by limited family labour and low cash inputs (McIntire et al. 1989; ILCA, Addis Ababa, Ethiopia, pers. commun.). Most of the cropping budget is spent to hire manpower for weeding. Additional cash income is earned with off-farm activities particularly during the dry season (Baidu-Forson 1988). Lower yields imply a lower investment capacity of farmers because cash reserves are normally depleted by the purchase of food (McIntire et al. 1989, ILCA, Addis Ababa, Ethiopia, pers. commun.). Higher grain yields would increase both the potential for satisfying household food demands and their investment capacity. The access to cash is thus of major importance for the development of sustainable farming systems.

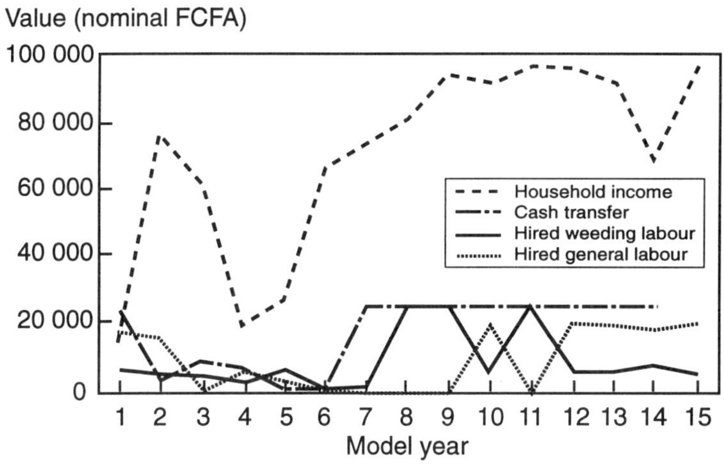


*Figure 3.* Selected economic parameters of the base case (10 000 FCFA = US $37).

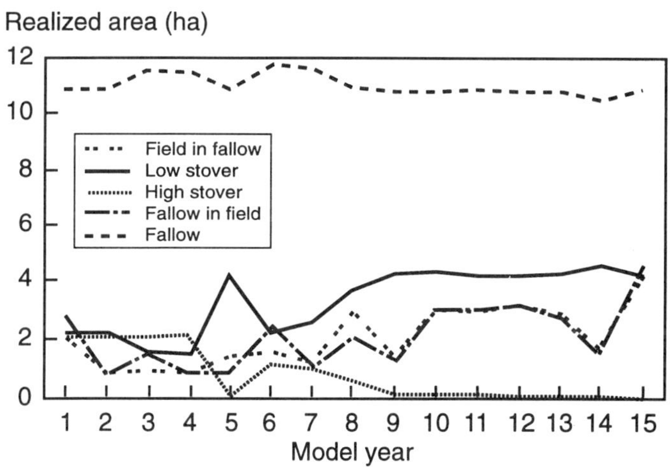


*Figure 4.* Summary of field activities of the base case (across all fertility levels).

*Scenario 1: changing family size.* The average population density in Niger is about 4.5 people per square kilometre, but this is without much meaning since three-quarters of the population live on less than 12% of the national territory where agricultural production is suitable. Given the predictions in population, land will become scarcer, larger quantities of food will be needed, but also a labour force will become increasingly available. The changing land : labour ratio is a basic reason for changing cropping patterns of farming systems in the absence of appropriate technologies. The effect of population pressure on CR management is shown in Table 2 by changing family size.

With increasing population density, the cash income from agricultural activities

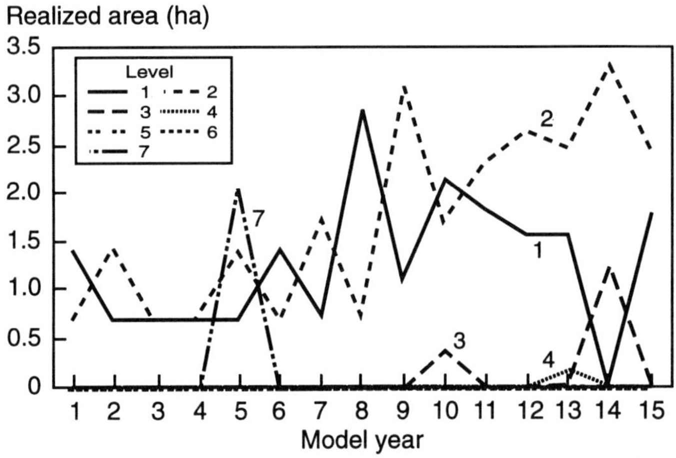


*Figure 5.* Cultivation without stover use on soils of different fertility levels.

*Table 2.* Sensitivity analysis with varying family size (CR = Crop Residues; US $1 = 270 FCFA (January 1994)).

| Family size (Person per household) | Model results | | | | |
|---|---|---|---|---|---|
| | Cropped surface[a] (ha) of: | | Total income[b] (FCFA) | Fertility index[c] of: | |
| | zero CR | 2 t CR | | total area | fallow |
| 25 | – | – | not feasible | – | – |
| 20 | 67.5 | 40.2 | 410738 | 3.79 | 6.00 |
| 15 | 60.6 | 21.7 | 606025 | 3.30 | 4.50 |
| 11[d] | 55.2 | 10.6 | 714299 | 2.58 | 3.18 |
| 9 | 53.2 | 5.0 | 755591 | 2.53 | 3.50 |

[a] Sum of annual areas cropped with zero or high CR (the total maximum area cropped by any treatment including fallow would be 15 ha × 16 years = 240 ha).
[b] Sum of the discounted annual cash gross margin.
[c] Fertility index laying between 1 (very high) and 7 (very low) according to the fertility levels in Figure 1.
[d] This represents the situation in the base case.

would decrease in favour of grain self-consumption. With higher population, increasingly land would be used for cropping, and the area devoted to high mulching would increase disproportionately. Because mulching activities are mainly realized on soils of low fertility, soils would be depleted to very low fertility levels before being left in fallow. Consequently, low soil fertility, here indicated by the increasing fertility index, is particularly found on fallow land. The increase in the partitioning of land through a growing population would destabilize the balance between fallow and crop land. Although in some areas land is still available, allowing entire fields to lay fallow and restore fertility, in other regions of Niger farmers currently fallow only parts of their field (within-field fallow). If the population increases to the equivalent of more than 20 persons per family having access to the 15 ha, no feasible solution would be possible under the given assumptions of grain and cash needs (Table 2).

*Table 3.* Sensitivity analysis with varying stover prices (CR = Crop Residues; US$1 = 270 FCFA (January 1994)).

| Stover price (FCFA kg$^{-1}$) | Model results | | | | |
|---|---|---|---|---|---|
| | Cropped surface (ha) of: | | Total income (FCFA) | Fertility index of: | |
| | zero CR | 2 t CR | | total area | fallow |
| 15 | 58.1 | 6.3 | 832585 | 2.40 | 3.37 |
| 11.8[a] | 55.2 | 10.6 | 714299 | 2.58 | 3.18 |
| 9 | 58.8 | 12.0 | 623023 | 2.90 | 4.06 |
| 6 | 55.8 | 23.9 | 551366 | 2.96 | 4.50 |
| 3 | 55.0 | 28.7 | 556938 | 2.98 | 4.58 |
| 0.1 | 52.9 | 34.5 | 640585 | 2.91 | 4.37 |

[a] This represents the situation in the base case, for more details see Table 2.

Agriculture can, thus, provide no more food for this amount of people. In such a situation, the rural population would be forced to outmigrate if no local off-farm employment opportunities are available.

In contrast, a reduction in family size translates into lower grain consumption needs. This would more than compensate for the lower labour capacity and, consequently, cash income would increase. The total area cropped would be slightly reduced, but particularly high mulching activities would be halved because of their higher labour demand. Nevertheless, the higher proportion of land in fallow would increase final soil fertility compared to the solution in the base case.

*Scenario 2: changing stover price.* The model results affirm that the stover price plays a key role in the use of stover as a mulch (Table 3). Whereas with high stover prices little land is mulched in the first years, a decrease of the stover price would drastically increase the mulching practice. In this case, high amounts of stover would be purchased and applied on infertile land. The projected effect on soil fertility is surprising: At very high stover prices the total soil fertility would not decrease, but increase. Stover is used as a tradable product, which increases the profitability of high stover producing activities (zero mulch on fertile land) and, thus, the fallow activities providing such land. With decreasing stover prices, stover consuming activities would be more profitable, meaning the cultivation of low fertility land.

Recent studies demonstrated that households in Niger, particularly in the proximity of urban centres where demand and prices for crop residues are higher than in rural areas, may increase their income through the sale of stover (Baidu-Forson 1994). The income from CR sales can become important, particularly during years of poor grain harvests since the stover:grain ratio is not fixed, but varies in wide ranges. Thus, high stover yields may in some years compensate for low grain yields. Because farmers are aware of the detrimental effects of a continuous removal of all stover from the same field, they take CR from different fields each year, or leave low amounts of stover each year on the field.

Low stover market prices may be realistic in rural areas, but the farm-internal value of stover to feed animals, as indicated by its shadow price, is unlikely to drop

*Table 4.* Sensitivity analysis with varying discount rate (CR=Crop Residues; US$1=270 FCFA (January 1994)).

| Discount rate | Model results | | | | |
|---|---|---|---|---|---|
| | Total surface (ha) of: | | Total income (FCFA) | Fertility index of: | |
| | zero CR | 2t CR | | total area | fallow |
| 0.05 | 55.9 | 11.4 | 1 129 173 | 2.35 | 3.21 |
| 0.1[a] | 55.2 | 10.6 | 714 299 | 2.58 | 3.18 |
| 0.2 | 65.6 | 4.5 | 373 461 | 3.66 | 4.50 |
| 0.3 | 72.3 | 1.4 | 264 826 | 3.77 | 4.55 |
| 0.4 | 72.5 | 0.5 | 206 295 | 4.11 | 4.50 |
| 0.5 | 66.9 | 0.0 | 168 239 | 3.64 | 4.10 |

[a] This represents the situation in the base case, for more details see Table 2.

below the real stover price. In short-term LP models which included the use of stover as animal feed during the dry season (assuming a herd size of 3 TLU on 10 ha of own land without taking fallow into account), as well as high mulching, the shadow price of stover was in all cases above the market price of 11 FCFA per kilogram (Lamers and Bruentrup 1995; Bruentrup et al. 1996).

*Scenario 3: changing discount rate.* By changing the rate to discount future cash income, a clear relationship between the propensity for actual consumption (which corresponds to high discount rates) and soil fertility development is revealed: the more future income is discounted, the lower the final fertility level will become (Table 4).

The preference for income reduces the cash transfers for hiring labour and the use of high mulching in the first years. Fallow is recultivated at a lower fertility level, before the field has regained maximum fertility. The appropriate discount rate and future income will heavily depend on the rate of time preference for consumption by farm families (consumption rate of interest) and the alternative uses of capital (marginal productivity of capital) (Markandya and Pearce 1988). Little is known about the propensity of consumption of small farmers, but most authors believe that the personal discount rate is rather high for low-income families (Lutz et al. 1994). The choice of the appropriate mulching practice will crucially depend on the capital costs and farmers' time preferences. In 1991, the real rate of interest in Niger was about 7% for savings, for borrowing about 11% (World Bank 1993). But formal capital markets in West Africa are weak and, consequently, only a few farmers have access to the official capital sector. It may well be that an alternative use of capital than investing in soil fertility would give higher returns and would, therefore, be more economical.

*Summary of analyses of other scenarios.* Other scenarios showed that a boost of CR production could not substantially increase the income of farmers since the family labour capacity is limited. Only if the transfer of cash between years is increased, would income be substantially augmented. In this case, additional initial cash income would result in a return to capital of about 33%. However, such a substantial increase in cash seems unrealistic given the urgent cash needs of households for immediate consumption.

## Conclusions and outlook

In this paper, the consequences of generally predicted developments in the WAS are simulated, such as an increasing population pressure on arable land, increased preference rate for current income or higher stover prices. The results obtained through dynamic linear programming indicate that some of developments provide incentives for farm households to implement a soil mulch with stover at a rate of 2000 kg ha$^{-1}$, but none can prevent soil degradation.

In the last decade, several technologies based on low external inputs have been developed to increase soil fertility and reduce soil degradation in small-scale farming systems in the WAS. Researchers have studied the use of locally available millet crop residues which increase yields. The results suggest that the efficiency of traditional cropping techniques can only be marginally increased by improving these low external input techniques. Moreover, it is less the absolute population growth that endangers the ecological stability of agriculture than the pressure to feed the growing population with their own resources only (supposing that no technological progress such as mineral fertilization occurs). To develop sustainable farming systems in the WAS, biomass production must increase.

From an agronomic viewpoint, the use of organic and mineral fertilizers is imperative. Yet, Sahelian farmers apply suboptimal rates due to the limited availability of organic fertilizers, high prices and their inaccessibility, or do not use them at all due to unreliable rainfall or a lack of capital. Without the use of mineral fertilizers, it is unlikely that the production in the Sahel can be increased and soil fertility maintained. Easy credit availability should accompany inorganic and organic soil improvement strategies to overcome bottlenecks in the farmers' capacities to hire labour caused by increased labour requirements or invest in labour-saving technologies. This calls for an immediate intervention of policy makers at the national and international level. However, the farmers' perspective should continue to be the focal point of research. The economic feasibility of such technology is considered a necessary although not a sufficient criterion for its adoption, particularly as long as governments in the WAS are too weak effectively to protect natural resources by other means such as regulations and laws.

## Acknowledgements

This study was supported by the German Research Foundation (Deutsche Forschungsgemeinschaft) under program SFB 308.

## References


Baidu-Forson J (1988) Characteristics of farm households and economy in western Niger villages: Some evidence from baseline surveys. Resource Management Program, Economics Group, Progress Report No. 87, International Crops Research Institute for the Semi-Arid Tropics, Hyderabad, India.

Baidu-Forson J (1994) Uses and economic value of crop residues and herbaceous plants to rural households in western Niger. Resource Management Program, Progress Report No. 5, International Crops Research Institute for the Semi-Arid Tropics, Sahelian Centre, Niamey, Niger.

Brüntrup M, Lamers J P A, Heidhues F, Von Oppen M (1995) The economic evaluation of the use of millet crop residues for wind erosion control and the competition with alternative uses in a farming systems perspective. Pages 243–256 in Bürkert A, Von Oppen M, Allison B (Eds.) Wind erosion in Niger: Consequences and control measures in millet based farming systems. Kluwer Academic Publishers, Dordrecht, The Netherlands.

Buerkert A (1995) Effects of crop residues, phosphorus, and spatial soil variability on yield and nutrient uptake of pearl millet (*Pennisetum glaucum* L.) in Southwest Niger. Ulrich Grauer Verlag, Stuttgart, Germany. 273 p.

FAO – Food and Agriculture Organization (1993) AGROSTAT-PC. Computerized information series. Statistics land use. FAO of the United Nations, Rome, Italy.

Feil P R, Lamers J P A, Herrmann L (1995) Knowledge transfer in the field: Solving crop residue problems in Niger. J. Technol. Transfer 20(1):31-41.

Hebel A (1995) Einfluß der organischen Substanz auf die räumliche und zeitliche Variabilität des Perlhirse-Wachstums auf Luvic Arenosolen des Sahel (Sadoré/Niger). Hohenheimer Bodenkundliche Hefte 24, University of Hohenheim, Stuttgart, Germany. 219 p.

ICRISAT – International Crops Research Institute for the Semi-Arid Tropics (1992) Pages 81–82 in West African programs annual report 1991. ICRISAT, Niamey, Niger.

Lamers J P A, Bruentrup M (1995) Comparative advantage of single and multipurpose uses of millet stover in Niger. Agric. Syst. 50(3):273–285.

Lutz E, Pagiola S, Reiche C (1994) The costs and benefits of soil conservation: the farmers' viewpoint. World Bank Res. Observer 9(2):273–295.

Markandya A, Pearce D (1988) Environmental considerations and the choice of the discount rate in developing countries. The World Bank Environmental Department Working Paper No. 3, Washington DC, USA. 79 p.

Michels K (1994) Wind erosion in the Southern Sahelian zone. Extent, control, and effects on millet production. Ulrich Grauer Verlag, Stuttgart, Germany. 99 p.

OPVN – Office des Produits Vivriers du Niger (1992) Marchés des céréales et du niébé au Niger, 1990/1992. OPVN, Système d'Information sur les Marchés (S.I.M.), Niamey, Niger. 24 p.

Rebafka F P (1993) Deficiency of phosphorus and molybdenum as major growth limiting factors of pearl millet and groundnut on an acid sandy soil in Niger, West Africa. Hohenheimer Bodenkundliche Hefte 9, University of Hohenheim, Stuttgart, Germany. 159 p.

Roose E (1993) Capacité des jachères à restaurer la fertilité des sols pauvres en zone soudano-sahélienne d'Afrique occidentale. Pages 233–244 in Floret C, Serpantié G (Eds.) La Jachère en Afrique de l'Ouest, Editions de l'ORSTOM, Paris, France.

World Bank (1993) World development report 1993. Oxford University Press, Washington DC, USA. 389 p.

# The impact of agrarian policies on sustainable land use

R. RUBEN[1], G. KRUSEMAN[1], H. HENGSDIJK[2] and A. KUYVENHOVEN[1]
[1]*Department of Development Economics, Wageningen Agricultural University, P.O. Box 8130, 6700 EW Wageningen, The Netherlands*
[2]*DLO-Research Institute for Agrobiology and Soil Fertility, P.O. Box 14, 6700 AA Wageningen, The Netherlands*

**Key words:** farm household modelling, agrarian policy, sustainable land use, Mali

**Abstract**
Agrarian policies for sustainable land use require the selection of appropriate policy instruments to influence resource allocation decisions by farmers. A farm household modelling approach is applied to identify microeconomic supply response reactions to various policy measures. Production and consumption decisions are simultaneously analysed, permitting an appraisal of adjustments in land use in accordance with farm household welfare objectives. Four types of farm household are identified according to their resource endowments and objective functions to acknowledge different directions of supply response reactions. The modelling framework is based on the linkage of data sets from agrotechnical and socio-economic sources. Actual and alternative (sustainable) cropping and livestock activities are defined according to soil and weather conditions, production techniques and soil management practices, and their consequences for the nutrient and carbon balance. Farm household resource endowments and expenditures condition allocation procedures. Dynamic properties are incorporated into the model through the savings and investment module, while a time discount rate accounts for risk behaviour. Food and labour balances are identified for the appraisal of market interactions and exchange among farm types. Policy consequences of this modelling approach are illustrated for the case of Koutiala Cercle in southern Mali. The effects of modifications in market prices of food and cash crops, input prices, transaction costs, credit supply and land policy are demonstrated, and trade-offs between objectives of farm household welfare and sustainable land use are acknowledged.

## Introduction

Analysis of sustainable land use requires an analytical framework that takes into account the technical possibilities and limitations for land use given the resource base, and possible adjustments in the production structure of farm households due to modifications in the market or institutional environment (Reardon and Vosti 1992). The appraisal of suitable policy instruments to modify the socioeconomic conditions in which farm households take their allocation decisions regarding land use allows possible trade-offs between farm household welfare and agroecological sustainability criteria to be identified. Policy options can be identified that permit reconciliation of both objectives, as well as different supply response reactions due to the level of market integration and risk attitudes (Kruseman et al. 1995).

This paper presents an analytical framework for the assessment of the impact of agrarian policies on farm household decisions regarding land use and welfare. A recursive Linear Programming and Farm Household Modelling approach was used



*P.S. Teng* et al. *(eds.): Applications of Systems Approaches at the Farm and Regional Levels, 65–82.*

to analyse response reactions of farm households in the short and medium term (1–5 years) to well-defined changes in the production conditions. The methodology enables the *ex ante* appraisal of feasible production systems, and supports policy makers in selecting appropriate instruments to attain agrarian development objectives pursued. While options for sustainable land use are usually explored at the regional level, the present contribution offers an operational methodology for the evaluation of policy instruments to induce producers decisions towards more sustainable land use at farm level. Integration of both approaches generates new insights and information which improves policy formulation with respect to land use in the Sudan-Sahelian region (Kuyvenhoven et al. 1995).

The farm household model was used for the assessment of the impact of policy instruments in the Koutiala Cercle in southern Mali. Due to high population growth, pressure on natural resources has rapidly increased and traditional fallow periods are strongly reduced. Traditional *transhumant* grazing systems meet difficulties due to the increasing number of animals and the competition with arable farmers, affecting the availability of rangelands. Regional development is constrained by decreasing soil fertility and limited access to productivity-enhancing techniques. As further options for extensive growth become limited, more attention is being given to improve factor productivity and input efficiency (Van Keulen and Breman 1990). Technical options for intensification of land use and suitable policy measures to induce farmers to adopt more sustainable production systems must be identified. This study identifies suitable instruments that permit farmers to modify their production system in more desired directions.

## Agricultural production structure and development in the Koutiala Cercle

The Cercle of Koutiala is in southern Mali and comprises 9100 km$^2$. The population consists of 46 320 farm households (DNSI 1991b) and has an annual growth rate of 3.3%. Agricultural production in Koutiala Cercle is based on rainfed food crops (maize, millet, sorghum and cowpea), cash crops (cotton and groundnuts) and livestock. Average annual rainfall is about 780 mm. Soils can be characterized as loamy sand (50% of all soils), gravelly (40% of all soils) and clay depression (10% of all soils) with low organic matter contents. Runoff and erosion of the loamy and gravelly soils are considered major problems (Berckmoes et al. 1990).

The cultivated area of food crops increased by 3% during recent decades, while cotton area has remained more stable. Food security levels are reasonably beyond minimum requirements, and locally produced surpluses of maize and millet are transferred to other regions. Animal numbers more than doubled during the last decade, causing an increased pressure on rangeland. Use of cotton revenue to buy livestock caused an increase in stocking rate to 0.32 TLU ha$^{-1}$ (Tropical Livestock Units) beyond the carrying capacity of 0.13–0.15 TLU ha$^{-1}$ (Bosma et al. 1993; World Bank 1994).

Cropping activities account for 35% of current land use, while pastures occupy more than 62% of available land (see Table 1). Fallow is now almost eliminated and

*Table 1.* Cercle of Koutiala: agrarian production structure (1992–93) (*Source:* DNSI 1993).

| Activity | Area (ha) | Area (%) | Employment ($10^3$ labour days) | Employment (%) |
|---|---|---|---|---|
| *Cropping activities* | | | | |
| Millet/sorghum | 154054 | 17.3 | 9705.4 | 38.0 |
| Rice | 2190 | 0.3 | 175.0 | 0.7 |
| Maize | 36329 | 4.1 | 2361.4 | 9.3 |
| Cowpea | 24831 | 2.8 | 1117.4 | 4.4 |
| Cotton | 59057 | 6.6 | 8799.5 | 34.5 |
| Peanut | 21045 | 2.4 | 1031.2 | 4.0 |
| Other crops | 13203 | 1.4 | n.d. | n.d. |
| *Livestock* | 533891 | 62.3 | 1964.7 | 7.7 |
| *Other* | 24900 | 2.8 | 364.1 | 1.4 |
| Total | 889500 | 100 | 25518.9 | 100 |

land has become a scarce factor. Current agricultural activities guarantee full absorption of the regional labour force, mainly in crop production. More than 90% of the economically active population is working in agriculture. Family work represents the major part of all labour, but wage labour is gradually becoming more important (DNSI 1991b).

Intensification of cropping systems can be attained through an improved access to animal traction and better use of fertilizers. Integration of livestock and cropping activities, including the use of crop residues (fodder) for high-quality feed, and manure for organic matter and nutrient recycling in cropping activities, allows for mutual benefits. Even in such integrated systems, fertilizers are prerequisite to sustain production. Crop diversification is also considered as a risk-reducing strategy (Reardon et al. 1988), while investment in livestock represents an additional insurance mechanism.

The Malian Government defined the following specific development objectives for Koutiala Cercle (MAEE 1992): (i) stabilization of cultivated area under cotton and intensification of cotton production techniques, permitting area expansion only in the southern part of the region; (ii) improvement of integration between cropping and livestock activities; (iii) control of erosion and soil conservation; (iv) intensification of livestock production systems, i.e. increase of value added per animal; (v) crop and income diversification; and (vi) promotion of local village management of natural resources and decentralization of productive services towards village associations.

## Approach

We combined two modelling approaches: for the analysis of the production side a Multiple Goal Linear Programming (MGLP) framework was applied to determine activity and technology choice, while a Farm Household Modelling (FHM) approach was used for the assessment of impact of modifications in prices on farm profits, factor allocation and land use. The result was a recursive modelling framework in which econometrically specified behavioural relations were used to guide the pro-

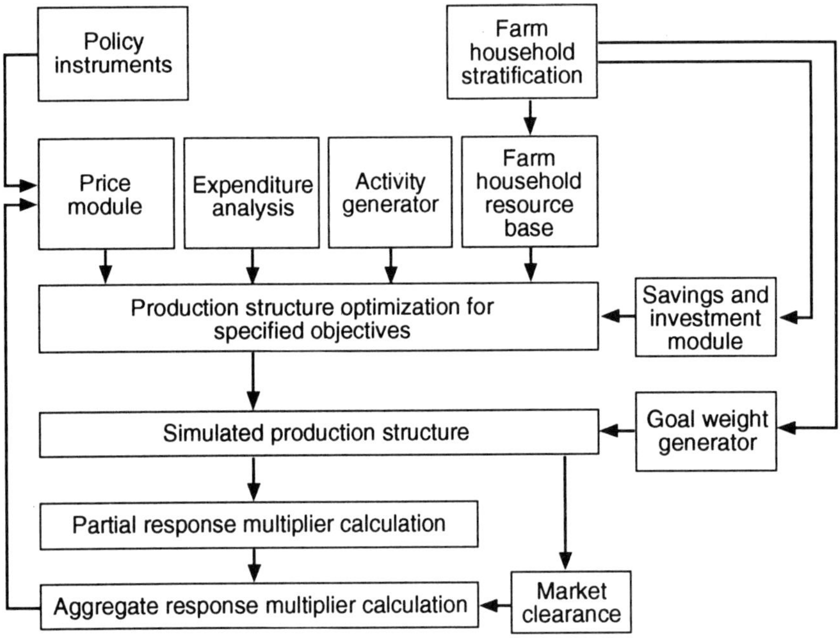


*Figure 1.* Structure of the farm household model.

cedures for linear programming optimization of the production structure for different types of farm households (Ruben et al. 1994; Kuyvenhoven et al. 1995). This set-up of the modelling framework is based on the linkages between production and consumption decisions within farm households (Singh et al. 1986; De Janvry et al. 1991).

The selection of production activities takes place in a stepwise optimization procedure, while the objective function is based on a continuously specified, negative, exponential consumption-utility function (Hazell and Norton 1986), and on a criterion of 'full income' that takes into account monetarized changes in soil quality. Separability of production and consumption decisions is assumed, as food and labour requirements are met by either farm production or market exchange, taking into account transaction costs.

The basic structure of the farm household model is schematically presented in Figure 1. The model consists of six separate modules for (i) objectives and goal weights, (ii) farm household resource endowments, (iii) production activities, (iv) prices, (v) expenditures, and (vi) savings and investments. This modular structure permits the linkage of data sets from agrotechnical sources (input-output coefficients) and from socioeconomic sources (budget surveys, farming systems research, market surveys based on e.g. crop growth simulation models). Simple adjustments of individual modules can be intro-duced without modifications in the model structure.

Farm households are characterized by their different initial resource endowments, their attitudes towards savings (savings coefficient) and their risk perceptions (expressed in a time discount rate). The combined effect results in a different allocation of income among expenditure categories. Dynamic properties are incorporated into the model through the savings and investment module, permitting adjustment of the resource base in subsequent years. The different modules are combined into a multi-objective optimization procedure and outcomes are derived with respect to (changes in) the selected production structure, related agroecological soil nutrient and carbon balances, and socioeconomic balances for labour and food.

Effects of modifications in commodity prices, transaction costs (e.g. public investment in infrastructure), credit supply and land policy are determined at farm level. Results of the modelling approach are presented in the form of *response multipliers*, indicating a percentage adjustment in factor allocation (land use, labour intensity, input use) as induced by discrete parametric changes in market prices. As for certain production factors (capital, land), no effective markets exist, price dummies are introduced into the model permitting an evaluation of the possible impact of market development policies. For a number of other commodities (cotton, livestock, fertilizers, cotton cake), prices are defined exogenously or through institutional mechanisms.

*Objectives and goal weights*

Model optimization takes place for two specified objectives: net expected full income and consumption utility. Consumption utility is derived from the expenditure module (see below). Full income is defined as the difference between net revenue and nutrient losses valued at different discount rates (Van Der Pol 1992). Each farm household type may select a different allocation of savings, preferring acquisition of livestock (in case of risk avoidance), purchase of equipment (in case of risk neutrality) or investment in land improvement (in case of risk acceptance). The model determines the optimal production structure for the weighted combination of both goals, and calculates the value of net revenue and the availability of savings.

The modelling framework is used for two subsequent optimization procedures: (i) allocation of available savings (generated under normal weather conditions) based on full income maximization and expected market prices, and (ii) definition of the modified production structure based on a weighted combination of both goals for specified prices and a probabilistic approach to weather conditions.

A specific procedure has been used to generate goal weights, that permits specification of the relative importance of each objective for all farm types. Following Romero (1993), these goal weights are determined through a confrontation of the calculated values of the specified objectives under partial optimization as generated in a base run with actual production activities and current prices, with the actual objective values as derived from farm survey data for farms with the same resource endowments (DRSPR 1992; Brons et al. 1994). Calibration of the model with relative goal weights of 0.75 for consumption utility and 0.25 for full income (including monetary value of nutrient losses) provides a reasonably good fit for the model.

*Table 2.* Characteristics of the identified farm households (*Source:* CMDT 1994).

| Criterion | Farm type A | Farm type B | Farm type C | Farm type D |
|---|---|---|---|---|
| Family size (persons) | 25.1 | 11.9 | 8.5 | 5.5 |
| Labour force (persons) | 11.8 | 5.7 | 3.9 | 2.5 |
| Land area (ha) | 17.8 | 10.1 | 5.8 | 3.3 |
| Cattle (TLU) | 23.1 | 3.0 | 0.6 | 0.1 |
| Oxen | 5.8 | 2.7 | 1.0 | 0.2 |
| Ploughs (units) | 4.2 | 2.2 | 0.9 | 0.1 |
| Sowers | 1.0 | 0.3 | 0.1 | 0.0 |
| Number (N) | 9092 | 7905 | 2383 | 401 |

*Farm household resource endowments*

Measurement of farm-level reactions to modifications in policy incentives requires a specification of farm households according to homogeneity in expected response to policy interventions, taking into account their characteristics with respect to (i) resource endowments and equipment level (Brons et al. 1994), (ii) production systems and farm management practices (Bosma et al. 1993) and (iii) farm performance (Leesberg et al. 1990; Van Der Pol 1993). Farm households are defined according to their initial resource endowments of land, labour force, livestock and equipment, and defined parameters for savings and risk behaviour. Land for cropping activities is limited to the available land with prevailing soil characteristics. Soil types are distributed over the farm households proportional to the regional soil-type distribution. Annual cotton area is limited to a quarter of the farm area to limit the occurrence of soil-borne diseases. Use of hired labour is constrained by limits on working capital.

In the Koutiala Cercle, farm households are classified by the regional development agency Compagnie Malienne pour le Développement des Textiles into four farm household types (CMDT 1994). This classification is mainly based on the availability of animal traction, as prospects for agricultural intensification depend to a large extent on access to animal traction and related equipment for timely land preparation and to increase labour productivity. Other variables, like farm size, labour availability, farm size and land use, are highly correlated with animal traction as shown by Brons et al. (1994). Cotton production influences the overall farm strategy because it is the major farm income source used to finance inputs for other crops. Access to fertilizers and cotton cake is rationed according to the cotton area, but other crops make use of the residual effect of fertilizer nutrients in the soil. Cotton revenues are used to finance expansion of livestock activities. Crop rotation requirements imply that cotton can only be grown on a quarter of the farm area.

Four types of farm households are identified with characteristics as shown in Table 2. Farm types A and B, the largest farm households, represent the major share (86%) of the number of farms in Koutiala. They combine food and cash crop production, and reach food security through their own production, while income growth is attained through investment in cattle and/or intensification of cropping activities (Van Der Pol 1993). The smaller households with less equipment (types C and D) have less diversified production systems and depend on purchase of food to guarantee self-sufficiency. These households also rely on hiring animal traction (Leesberg et al. 1990; Toulmin 1994).

*Production activities*

In the production activity module, well-defined mixtures of inputs are converted into agricultural outputs. The latter include e.g. harvestable yields for human consumption, and crop residues for fodder purposes, but also side-effects of the production process such as soil erosion. Input requirements include labour for cultivation practices, animal traction, implements, fertilizers and manure. For labour and animal traction, specific time periods are distinguished to account for peak activities. Crop activities are defined for millet, sorghum, maize, cotton, cowpea and groundnut, while livestock activities are defined for cattle, goats, sheep and oxen used for animal traction. Moreover, pasture activities are defined which quantify the agricultural output of natural rangeland in Koutiala.

Different crop activities are defined according to (i) soil type, (ii) rainfall type, (iii) production techniques, (iv) utilization of crop residues and (v) type of anti-erosion measures (Hengsdijk et al. 1996). Five soil types are identified that differ in texture, available rooting depth and presence of gravel. These characteristics influence the water availability and, thus, to a large extent the production potential of crops. Based on rainfall data for 1950–1981, three types of rainfall regimes are identified, representing the 10% driest, the middle 45% and the wettest 45% of the rainfall years. Three types of crop residue strategies are defined for harvesting crop residues, grazing by livestock and use as mulch. Finally, four production techniques are defined for each crop based on the application of fallow periods and the use of animal traction and implements.

Two types of arable crop activities are defined: activities that represent the current production techniques in Koutiala, and agroecologically sustainable activities that are technically feasible but not yet widely applied in the region. Current activities are in general characterized by negative C and nutrient balances and high soil erosion losses. Alternative activities are defined in such a way that the annual decomposition of soil organic matter is compensated by a supply of crop residues, manure or fallow periods, while the nutrient output of the system is compensated by the application of manure and, if necessary, additional fertilizers.

Agroecological sustainability is operationalized in the crop production activities in two criteria: (i) a carbon balance for crop activities monitoring the annual organic C supply to the soil and the organic matter decomposition in the soil, and (ii) a nutrient balance for N, P and K monitoring the annual supply and withdrawal of these nutrients from the system. Alternative production activities are defined with erosion measures to reduce soil loss and soil cultivation measures to improve the surface water storage capacity, increasing the amount of water available for crop production.

For livestock activities, sustainability implies a stable herd of each type of animal. The production of pasture activities equals a fraction of the total annual biomass production that does not affect the exploitation of rangelands in the long run (Breman and De Ridder 1991). For each animal type, four production levels with different feed rations are defined, each combined with two marketing strategies, i.e. ages at which offspring are sold. For pasture activities, three different exploitation regimes are defined for grazing: only during the rainy season, year-round grazing and grazing during the dry season.

*Prices*

The model makes use of expected market prices, based on a weighted average of farm-gate prices during the last three years, with coefficients of expectation set at levels of 0.5 for year $(t-1)$ and 0.25 for years $(t-2)$ and $(t-3)$, respectively. Transaction costs are defined as a margin between farm-gate and market price, subject to available transport infrastructure and market information. Hedonic price analysis is applied for the estimation of these margins at a 50% level.

Prices for cotton and livestock depend on the world market, where cereals are locally traded, but demand is rather inelastic with respect to income due to high marginal budget shares (Delgado et al. 1994). Statistical records of market prices are only available for tradable commodities (CMDT 1993). Prices for non-tradable commodities (land) and the terms of trade for reciprocal exchange transactions (family labour, animal traction) are based on implicit prices (Goetz 1992).

*Expenditures*

The relationship between household income and consumption utility is estimated directly from a cross-sectional budget survey (DNSI 1991a) for Mali. Marginal utility of consumption for different expenditure categories (cereals, meat, milk, non-agricultural commodities) is converted into utility making use of a negative exponential function (Ruben et al. 1994). Linearization of this function is carried out using the convex combination constraint (Hazell and Norton 1986).

Rural incomes and consumption patterns are sensitive to prices, not just from the supply side, but also from the demand side (Reardon 1993). As cereal consumption represents between 34% and 77% of the household budget, reproduction costs of labour and animal feed costs depend on the level of grain prices. High cereal prices will be favourable for net supply households, and lower prices shall yield positive welfare effects for net demand households.

*Savings and investment*

The savings and investment module permits a dynamic analysis of farm household behaviour, allowing for a separate specification of production and investment decisions. Production decisions are based on the availability of resources and specified fixed and transitory savings coefficients. Fixed savings vary among farm household types and are dependent on the income level. Transitory savings represent 80% of the difference between the expected income under 'average' rainfall conditions, and the effective income under real weather conditions.

Investment decisions are modelled in a separate way. Transitory savings are used for the purchase of livestock as a buffer stock out of precautionary motives (consumption smoothing) to maintain expenditures in periods of adverse weather conditions (Deaton 1990; Rosenzweig and Wolpin 1993). Fixed savings are used to assure reproduction of the resource base in the long run through the purchase of additional livestock, the purchase of equipment or investment in anti-erosion measures. Returns for alternative allocations of savings are discounted and compared between farm households.

## Development goals and policy instruments

Socioeconomic development prospects for Mali are highly dependent upon the performance of the agricultural sector. Objectives of employment creation, poverty alleviation, food security, foreign exchange earnings and government finance receive high attention (Ministère du Plan 1988). Moreover, control on resource degradation is necessary to maintain the medium- and long-term agricultural production potential. Different objectives with probable conflicting consequences result in trade-offs among these objectives. Prioritization of objectives and assessment of the impact of policy instruments are required to identify options for sustainable land use.

Economic restructuring in Mali gives priority to economic recovery within the framework of food self-sufficiency. Therefore, incentives are required to guarantee regular revenues to farmers and to mobilize rural savings for the finance of regional development, while maintaining crop production. Decentralization of allocation decisions is promoted in order to increase flexibility in resource use and improve market response reactions. Structural adjustment is directed towards demand management with restrictive budgetary and monetary policies, and progressive price liberalization to improve supply response. Major instruments include market development to enhance productivity in the private sector, and decentralized resource management at lower levels of public sector absorption (World Bank 1994). While the rural sector is acknowledged as the major engine for economic growth, effectiveness of agrarian policies depends on the functioning of factor and commodity markets. Available policy instruments to influence economic development and sustainable land use in Koutiala Cercle include (MAEE 1992): price policies on factor and commodity markets (e.g. cereals, cotton, fertilizers), market development (reduction of transaction costs, improved credit supply) and property rights regulation (land or head tax).

Efficiency and sustainability of agricultural production systems could be enhanced through selective price policies. Both input and output prices for food and cash crops are subject to policy reform, within the framework of market liberalization. Modifications of market prices directly influence cost/benefit relations of agricultural activities, and also have an indirect effect on net revenue for each farm type through different discount rates.

Effective supply response at the farm household level depends also on the development of the marketing system. Government investment in infrastructure will support the functioning of commodity markets. The development of factor markets for capital (banking system as alternative insurance mechanism) and savings offers alternative outlets for capital investment, and a control on excessive livestock expansion.

To control nutrient depletion on common property resources, a clear definition of rights on common pasture land and forest areas (*droit foncier*) is required. Specific mechanisms for land allocation and distribution can be defined within the framework of territorial planning, taking into account both efficiency (resource allocation) and welfare (income distribution) aspects.

## Policy instruments and response reactions

The farm household model described was used for the assessment of the potential impact of market policies and institutional reform on land use and welfare at farm household level. Resource allocation was analysed as a result of production, consumption and investment decisions; trade-offs between welfare effects and sustainability criteria were addressed simultaneously. Differences among farm types with respect to their net demand or supply situation on local cereals and labour markets were highlighted, since the elasticity of supply response depends on the effective functioning of these markets (De Janvry et al. 1991).

Three different model simulations were used to determine the size and direction of the response multipliers: (i) base scenario to define the initial production structure under current prices and with actual activities; (ii) technology scenario under current prices but including actual and alternative activities; and (iii) policy scenario with different modifications in prices and institutional constraints to assess the impact of different instruments on adjustments in land use and farm household welfare.

### *Base scenario*

Table 3 gives the results of the base-run for each farm type under current market prices and with actual activities, as determined by the initial resource endowments (see Table 2) and the selected objectives for optimization. All farm households dedicate a substantial proportion of resources to cash crop production (cotton and groundnut), and only farm types A and B are able to maintain a net cereal surplus. Cowpea is selected to guarantee livestock feed requirements. Fallow almost disappears with the sole exception of a small portion of land in farm type A. Reliance on common pastures is dependent on the herd size and available biomass from cropping activities. Production activities that are selected under the base-run include mainly cotton to the maximum of the rotation constraint, in addition to cereals, especially sorghum and some millet. Millet production is concentrated in farm types A and B. Farm types C and D produce groundnuts for sale in order to provide additional income to buy food and, thus, compensate for their negative food balance.

Objective values for utility and full income decrease from farm types A and B

*Table 3.* Base run: objective values, resource balances and initial production structure.

| Indicator | Unit | Farm type A | Farm type B | Farm type C | Farm type D |
|---|---|---|---|---|---|
| Consumption utility | Utility per capita | 611 501 | 598 481 | 546 688 | 525 893 |
| Full income | FCFA per capita | 68 681 | 53 306 | 32 921 | 25 416 |
| Net revenue | FCFA per capita | 81 493 | 61 419 | 36 546 | 26 848 |
| Savings | FCFA per capita | 16 299 | 6142 | 1827 | 269 |
| Carbon balance | kg ha$^{-1}$ | −1127 | −1193 | −1323 | −1182 |
| N-balance | kg ha$^{-1}$ | −42 | −48 | −47 | −44 |
| P-balance | kg ha$^{-1}$ | 2 | 0 | −2 | −1 |
| K-balance | kg ha$^{-1}$ | −19 | −20 | −10 | −10 |
| Erosion | Mt soil loss ha$^{-1}$ | 34 | 42 | 50 | 48 |
| Labour balance | Peak period labour use (%) | 95 | −5 | 93 | 90 |
| Cereal balance | kg cereals capita$^{-1}$ | 133 | 246 | −79 | −118 |
| Food self-subsistence | Self-sufficiency rate (%) | 90 | 97 | 71 | 58 |
| Cereals | ha | 10.8 | 6.6 | 2.1 | 1.2 |
| Cash crops | ha | 4.5 | 2.5 | 2.5 | 1.4 |
| Cowpea | ha | 1.6 | 1.0 | 1.3 | 0.8 |
| Fallow | ha | 1.0 | 0.0 | 0.0 | 0.0 |
| Pastures (common land) | ha | 132.0 | 23.0 | 11.0 | 3.0 |
| Livestock | Annual growth rate (%) | 7 | 5 | 2 | 4 |

*vis-à-vis* farm types C and D. Net revenues and fixed savings follow the same tendency, but the difference between net revenue and full income is less on smaller farm types due to the higher time discount rate for soil depletion on these types of farm. Nutrient balances for N, P and K and carbon balance are negative under the selected production structure, and all farms suffer serious soil erosion. Farm types C and D are net buyers of cereals and reach a low level of cereal self-sufficiency, due to limitations of land and preference for cash crop production. The labour balance, i.e. the share of available household labour required for agricultural activities during the peak period, is only negative for farm type B. Finally, livestock has a positive growth rate for all farm types, as savings are used for precautionary livestock purchase.

*Technology scenario*

Introduction of alternative (sustainable) cropping activities into the model under current prices permits only a partial improvement of the nutrient and carbon balances, while net returns increase between 15% and 29% (Table 4). Although alternative activities are technically more efficient, an important part of current activities is still maintained as economically efficient options, especially on smaller farm types C and D. The supply of alternative techniques may lead to substitution of crops and/or adjustment of technology choice. Farm types A and B can rely on the first strategy

*Table 4.* Response multipliers for the technology scenario.

| Indicator | Farm type A | Farm type B | Farm type C | Farm type D |
|---|---|---|---|---|
| Consumption utility | 2 | 4 | 4 | 2 |
| Full income | 34 | 39 | 36 | 18 |
| Net revenue | 18 | 28 | 29 | 15 |
| Carbon balance | 65 | 44 | 32 | 33 |
| N-balance | 59 | 48 | 29 | 22 |
| Erosion | −28 | −33 | −18 | −37 |
| Labour balance | 53 | −42 | −79 | 48 |
| Food balance | −122 | −16 | 92 | −10 |
| Alternative activities | 61 | 53 | 33 | 21 |

*Note:* Response multipliers indicate the percentage change in the value of the goal indicators compared to the base run. For positive nutrient and carbon balances the percentage change should be higher than 100%.

and increase their cash crop production as their food balance is already positive, while food-deficit farm types C and D give preference to improve cereal production. For all households there is a shift in cotton production from actual to alternative technologies, while differences in costs are not very substantial, and benefits in terms of carbon balance are important. Farm types C and D maintain a high proportion of actual activities, because their higher time discount rates render positive effects on the carbon balance less important. While farm type A shifts towards extensive cereal production, farm types B and D give priority to more intensive production technologies. Moreover, farm type A shifts towards sorghum production, and farm types B and C shift towards millet production that makes labour requirements in the peak periods coincide with cotton production. Changes in cropping patterns with respect to cereals can be explained by a number of factors of which differences in labour availability is the most important.

*Policy scenario*

Making alternative production techniques available can thus be considered as an option to enhance resource use efficiency, but economic incentives for adoption will still be rather limited, as adjustment costs may fully offset marginal income increase. Therefore, additional price policies could be evaluated that permit a more substantial improvement in net revenues. For this purpose, policy simulations were executed with the following instruments: (i) increase in food prices, (ii) increase in cotton prices, (iii) decrease in transaction costs (due to infrastructure investment), (iv) fertilizer price subsidy, (v) credit market development, (vi) land taxation and (vii) head tax on common pasture use. Response reactions for net farm revenue and carbon balance are presented in Table 5 for each farm type as a percentage change in the value of the indicator for a 1% upward or downward change in prices.

Several price policy measures were found to permit an improvement of net revenues at farm household level, but elasticity of these reactions differed among farm types. An increase of output prices for cotton or cereals has a general positive impact, while input prices and transaction costs yield substantially lower supply

*Table 5.* Response multipliers for different price instruments.

| Indicator | Level | Food price | Cotton price | Transaction costs | Fertilizer price | Credit supply | Land tax | Head tax |
|---|---|---|---|---|---|---|---|---|
| Net revenue | Farm type A | 0.05 | 0.05 | 0.03 | 0.01 | 0.01 | −0.01 | 0.00 |
| | Farm type B | 0.04 | 0.05 | 0.01 | 0.02 | 0.00 | −0.01 | 0.00 |
| | Farm type C | 0.04 | 0.06 | −0.02 | 0.01 | 0.01 | −0.01 | 0.00 |
| | Farm type D | 0.09 | 0.02 | 0.01 | 0.02 | 0.10 | 0.00 | 0.00 |
| Carbon balance | Farm type A | 0.11 | 0.01 | 0.06 | −0.01 | 0.03 | −0.12 | 0.00 |
| | Farm type B | −0.02 | −0.01 | −0.08 | −0.02 | 0.00 | −0.15 | 0.00 |
| | Farm type C | 0.07 | 0.06 | 0.06 | −0.03 | 0.02 | −0.02 | 0.00 |
| | Farm type D | 0.01 | 0.11 | 0.06 | 0.00 | 0.03 | 0.00 | 0.00 |

*Note:* Response multipliers indicate the percentage change in goal indicators compared to the technology scenario, due to a 1% change in prices.

response reactions. Most response reactions are due to small shifts in cropping patterns, both in terms of shifts between crop groups (cereals, cash crops, fodder crops) and in terms of shifts between cereal crops, and substitution of technologies within crop groups.

Increase in food prices has a general positive effect on the food balance, due to the substitution of cash crops by cereals and/or the selection of more efficient cereal production techniques. The most substantial increase is reached in farm type C, where further intensification and specialization in millet production occurs. Household type B was already the top of the league with respect to food balance, and increasing food prices only strengthens that position. Household type A intensifies production of cereals in response to food price increases.

Fertilizer subsidies encourage farm types C and D to adopt more alternative cotton activities. For farm type C, the potentially positive effect on the carbon balance is offset by the choice of more soil-depleting cereal activities. For farm type D, the shift towards alternative technology is so small that the effect is negligible. The negative impact of fertilizer price subsidies on the carbon balance can be understood as a disincentive to improve input efficiency at farm household level.

Market development policies provoke somewhat stronger reactions for several farm types. Reduction of transaction costs is especially relevant for the more commercially oriented farm types. The negative effect on the carbon balance in farm type B is caused by the adoption of more intensive cereal production techniques. Improved credit facilities only provoke major reactions in farm type D, which uses additional resources to increase the cereal area and reduce livestock, both with adverse effect for the carbon balance. For all other farm types, the availability of internal savings proved to be a sufficient device for consumption smoothing purposes.

Different policies regarding properties rights proved to have a limited or even negative impact on the carbon balance. Introduction of a head tax (250 FCFA $TLU^{-1}$; $1 US = 500 FCFA) does not lead to adjustment of livestock production, as objectives of consumption smoothing and manure requirements outweigh short-term revenue effects. Introduction of a land tax (250 FCFA $ha^{-1}$ pastures) results in a marginal decrease in farm household revenues, but negatively affects the carbon balance as reliance on crop residues for livestock feeding increases. Moreover, the effect of a land tax on the use of common pastures leads to less intensive use of these resources. The secondary effect is a shift towards the use of fodder available at farm level, leading to adverse effects on the carbon balance.

Improvement of the carbon balance can be reached through modification of the relative price relation between cereals and cash crops. Only farm type B suffers a lower carbon balance when food or cotton prices increase due to limited labour availability that inhibits the selection of more intensive cropping techniques. Fertilizer price subsidies will lead in all cases to adverse effects on the carbon balance, as the increasing use of fertilizers is accompanied by higher extraction rates and, thus, incentives for efficient use diminish.

Complete equilibrium of the carbon balance cannot be reached with any of the indicated policy instruments. The calculated effects on the carbon balance through policy intervention are relatively small compared to the impact of technology. Policy interventions cause small shifts in allocation efficiency, whereas technology change as defined in this model leads to substantial shifts in technical efficiency. In situations where differences between actual and alternative activities are rather substantial, the welfare consequences of intermediate technology options should be evaluated carefully.

## Conclusions

The major policy problem in intensification of African sub-Saharan agriculture is low supply response reactions, causing price policies to be largely ineffective at enhancing economic growth and sustainable land use (Delgado et al. 1994). Therefore, it has been argued that structure policies (rural infrastructure, credit systems, land policies, etc.) are required to promote technological innovation.

Technical options for intensification of cropping systems can be attained through better access to animal traction and more use and improved efficiency of fertilizers. Integration of livestock and cropping systems based on the use of crop residues (fodder) for high-quality fodder, and manure for organic matter and nutrient recycling in the cropping system, yields mutual benefits. However, even in such integrated systems, the supply of additional fertilizers is a prerequisite to sustain production.

The effectiveness of incentives for farmers to adopt these options depends on the tradability of different commodities and related supply-response reactions. Non-tradable items are actually demand constrained, but marginal budget shares spent on locally traded cereals are rather high. Consequently, multipliers of regional income growth due to increased spending on locally tradable goods are important. These demand linkages permit, in turn, adoption of production techniques that favour factor productivity and sustainable land use. Aggregate supply response is, however, also dependent on the availability of incentive goods (Bevan et al. 1987). Therefore, higher cash crop prices only lead to increased production if sufficient consumer goods are available to absorb additional income. Perverse responses may occur due to shortages in supply.

Variation in performance among farm households can be explained by differences in available resources and environmental constraints. The impact of price policies on agricultural income depends finally on the net supply or demand position of rural households (Budd 1993). If priority is given to self-sufficiency, food prices have only a marginal impact on household income, although distribution effects can be important. Households that are net buyers of cereals would benefit from lower food prices, but net buyers are harmed by such policies (Weber et al. 1988).

Finally, allocation priorities depend on the available alternatives offered by local markets and institutions. When rural households become more integrated into factor and product markets, both for income generation and for input purchase, options for intensification of farming systems increasingly depend on opportunity costs and risks. Off-farm employment can be considered as such an additional device to enhance supply response, representing an alternative mechanism to finance input purchase and capital investments that may favour farm intensification. When rural credit markets do not function adequately, off-farm labour becomes a substitute for lending (Fafchamps 1993; Reardon et al. 1994).

The presented farm household modelling approach offers an analytical framework for the appraisal of the impact of different policy incentives on farm household welfare and agroecological sustainability. The present model takes into account multiple objectives and differences in resource endowments among farm household types. Technical relations between cropping and livestock activities at farm level, as well as market relations for inputs and commodities, are explicitly integrated into a recursive framework. The results of the modelling exercise are especially suited to identify prospects for price policies, as well as institutional reform to enhance technology adoption and diffusion.

Exploration of the possibilities of technological change indicates that, with full knowledge of alternative (sustainable) technology and in the absence of transition costs, nutrient and carbon balances improve somewhat but soil depletion cannot be prevented completely. Soil mining practices are thus allocationally efficient, given the farm household resources, their goals and their subjective time discount rate. Modification of output prices tends to yield higher response reactions than input price adjustment. Both fertilizer subsidies and instalment of a head tax for livestock lead to diminishing incentives for efficient resource use. Structural policies addressing transaction costs and financial markets seem to offer better prospects to enhance tradability and reinforce intersectoral growth linkages, permitting higher supply response.

These results offer compatible outcomes with other research concerning the structural determinants for the intensification of production systems, involving both characteristics at farm household level, as well as regional parameters. Following Delgado et al. (1994), the positive impact of factors like transaction costs, off-farm income and the availability of credit are confirmed. With respect to price policies, different driving forces are identified that stimulate intensification of land use. Higher food output prices offer incentives to improve factor productivity for producers who are net suppliers. Input subsidies tend to reduce incentives for efficient allocation.

Intensification can thus be associated with both higher and lower food prices, but especially with more stability in income patterns in order to reduce the costs of smoothing consumption.

Some major refinements can be introduced into the modelling framework. As the calculated response multipliers reflect short- and medium-term reactions, adjustment costs are not taken into account. Different time lags in adjustment behaviour can be expected for consumption and investment. Moreover, the procedures for the adoption of technological innovations are strongly linked with farm size and resource endowments, while modern literature also points to a broad number of other household characteristics (e.g. education, gender, age) and copying procedures that determine innovation adoption. Linkages with non-agricultural sector and relations with other regions can be taken into account, in order to acknowledge the possible contributions of remittances and off-farm employment. Finally, market prices are considered as exogenous variables for individual producers. Aggregation towards the regional level would require specification of the supply and demand conditions to derive prices endogenously. However, the aggregate response would be less sensitive to prices, as market failures inhibit adjustment within certain farm types (De Janvry et al. 1991).

## Acknowledgements

The DLV research project is financed by The Netherlands Ministry of Agriculture, Nature Management and Fisheries (LNV) and the Directorate-General for International Cooperation (DGIS) of the Ministry of Foreign Affairs, through the DLO-Research Institute for Agrobiological and Soil Fertility (AB-DLO). For the research in Mali, cooperation of the Institut d'Économie Rurale (IER) through the Projet Production Soudano Sahélienne (PSS-Niono) and the Équipe Systèmes de Production et Gestión des Ressources Naturelles (ESPGRN-Sikasso) is gratefully acknowledged.

## References

Berckmoes W M L, Jager E J, Koné Y (1990) L'intensification agricole au Mali-Sud: Souhait ou réalité? Royal Tropical Institute, Bulletin No. 318, Amsterdam, The Netherlands. 40 p.

Bevan D L, Bigsten A, Collier P, Gunning J W (1987) Peasant supply response in rationed economies. World Dev. 15:431–439.

Bosma R, Bengaly M P, Defoer T (1993) Pour un système durable de production: Plus de bétail, rôle des ruminants au Mali-Sud, dans le maintien du taux de matière organique des sols. Séminaire 'Elevage et cycle viable des éléments nutritifs dans les systèmes mixtes agriculture-élevage de l'Afrique sub-saharienne'. International Livestock Centre for Africa, Addis Ababa, Ethiopia. 14 p.

Breman H, De Ridder N (Eds.) (1991) Manuel sur les pâturages des pays sahéliens. Karthala, Paris, France. 485 p.

Brons J, Diarra S, Dembélé I, Bagayoko S, Djouara H (1994) Diversité de gestión de l'exploitation agricole: Etude sur les facteurs d'intensification agricole au Mali Sud. Direction Régionale des Systèmes de Production et de la Gestión des Ressources Naturelles. IER-DRSPR Document No. 94/33, Sikasso, Mali. 70 p.

Budd J W (1993) Changing food prices and rural welfare: A nonparametric examination of the Côte d'Ivoire. Econ. Dev. Cultural Change 41:587–603.

CMDT – Compagnie Malienne pour le Développement des Textiles (1993) Etude des prix céréaliers en zones cotonnières: Le cas de la région CMDT de Koutiala. Suivi Evaluation CMDT, Bamako, Mali.
CMDT (1994) Annuaire Statistique – Résultats de l'enquête agricole permanente 93/94. Suivi Evaluation CMDT, Bamako, Mali.
Deaton A S (1990) Savings in developing countries: Theory and review. World Bank Econ. Rev. 4:61–96.
De Janvry A, Fafchamps M, Sadoulet E (1991) Peasant household behaviour with missing markets: Some paradoxes explained. Econ. J. 101:1400–1417.
Delgado C, Hazell P, Hopkins J, Kelly V (1994) Promoting intersectoral growth linkages in rural Africa through agricultural technology and policy reform. Am. J. Agric. Econ. 76:1166–1171.
DNSI – Direction Nationale de la Statistique et de l'Information (1991a) Enquête budget consommation (1988–1989). DNSI/PADEM/DSA, Bamako, Mali. 96 p.
DNSI (1991b) Recensement général de la population et de l'habitat. DNSI/Bureau Central de Récensement, Bamako, Mali. 450 p.
DNSI (1993) Enquête agricole de conjuncture campagne 1992–1993: Résultats définitifs. DNSI/PNUD-PADEM/CILLS-DIAPER, Bamako, Mali. 101 p.
DRSPR – Direction Régionale des Systèmes de Production et de la Gestión des Ressources Naturelles (1992) Commision technique sur les systèmes de production rurale. Synthèse des résultats de la campagne 1991-1992. DRSPR, Sikasso, Mali. 199 p.
Fafchamps M (1993) Sequential labor decisions under uncertainty: An estimable household model of West-African farmers. Econometrica 61:1173–1197.
Goetz S J (1992) A selectivity model of household food marketing behaviour in Sub-Saharan Africa. Am. J. Agric. Econ. 64:444–452.
Hazell P B R, Norton R D (1986) Mathematical programming for economic analysis in agriculture. MacMillan, NY, USA. 400 p.
Hengsdijk H, Quak W, Bakker E J (1996) A land use system and technology generator for the Koutiala Region in South Mali. DLV Report No. 5, DLO-Research Institute for Agrobiology and Soil Fertility, Wageningen, The Netherlands. (in press).
Kruseman G, Ruben R, Hengsdijk H, Van Ittersum M K (1995) Farm household modelling for policy scenarios on sustainable land use. Neth. J. Agric. Sci. 43:111–123.
Kuyvenhoven A, Ruben R, Kruseman G (1995) Options for sustainable agricultural systems and policy instruments to reach them. Pages 187–221 in Bouma J, Kuyvenhoven A, Bouman B A M, Luyten J C, Zandstra H G (Eds.) Eco-regional approaches for sustainable land use and food production. Kluwer Academic Publishers, Dordrecht, The Netherlands.
Leesberg J, Kalé Sanogo M, Diallo O (1990) La recherche sur les systèmes de production à l'Office du Niger; Production et intensification dans les secteurs de Niono et de Kokry. Office du Niger/Programa ARPON, Niono, Mali. 38 p. + 15 annexes.
MAEE - Ministère de l'Agriculture, de l'Elevage et de l'Environnement (1992) Schema directeur du secteur développement rural. MAEE, Bamako, Mali. 45 p.
Ministère du Plan (1988) Plan quinquennal de développement économique et social 1987–1991. Direction Nationale de la Planification, Bamako, Mali. 121 p.
Reardon T (1993) Cereals demand in the Sahel and potential impacts of regional cereal protection. World Dev. 21:17–35.
Reardon T, Vosti S (1992) Issues in the analysis of the effects of policy on conservation and productivity at the household level in developing countries. Q. J. Int. Agric. 31:380–396.
Reardon T, Matlon P, Delgado C (1988) Coping with household-level food insecurity in drought-affected areas of Burkina Faso. World Dev. 16:1065–1074.
Reardon T, Crawford E, Kelly V (1994) Links between nonfarm income and farm investment in African households: Adding the capital market perspective. Am. J. Agric. Econ. 76:1172–1176.
Romero C (1993) A research on Andalusian farmers' objectives: Methodological aspects and policy implications. Pages 89–99 in Proceedings of the VIIth European Conference of Agriculture Economics Congress, Sresa, Italy. Dartmouth Publishing Company, Aldershot, UK.
Rosenzweig M R, Wolpin K I (1993) Credit market constraints, consumption smoothing, and the accumulation of durable production assets in low-income countries: Investments in bullocks in India. J. Pol. Econ. 101:223–244.
Ruben R, Kruseman G, Hengsdijk H (1994) Farm household modelling for estimating the effectiveness of price instruments on sustainable land use in the Atlantic Zone of Costa Rica. DLV Report No. 4, DLO-Research Institute for Agrobiology and Soil Fertility, Wageningen, The Netherlands. 44 p.
Singh I J, Squire L, Strauss S (Eds.) (1986) Agricultural household models: Extensions, applications and policy. The Johns Hopkins University Press, Baltimore, London. 335 p.

Toulmin C (1994) Staying together: Household responses to risk and market malfunctioning in Mali. Pages 115–140 in Haswell M, Hunt D (Eds.) Rural households in emerging societies. Berg Publishers, Oxford, UK.
Van Der Pol F (1992) Soil mining. An unseen contributor to farm income in Southern Mali. Royal Tropical Institute Bulletin No. 325, Amsterdam, The Netherlands. 47 p.
Van der Pol F (1993) Profits de l'intensification. Royal Tropical Institute, Amsterdam, The Netherlands. 38 p.
Van Keulen H, Breman H (1990) Agricultural development in the West African Sahelian Region: A cure against land hunger? Agric. Ecosys. Environ. 32:177–197.
Weber M T, Staatz J M, Holtzman J S, Crawford E W, Bernstein R H (1988) Informing food security decisions in Africa: Empirical analysis and policy dialogue. Am. J. Agric. Econ. 70:1044–1052.
World Bank (1994) Mali: Policy framework paper 1994-1996. World Bank, Washington DC, USA. 24 p.

# Characterization of rice pest constraints in Asia: an empirical approach

S. SAVARY[1], F.A. ELAZEGUI[2], H.O. PINNSCHMIDT[2] and P.S. TENG[2]

[1]*Institut Français de Recherche pour le Développement en Coopération (ORSTOM), International Rice Research Institute, P.O. Box 933, 1099 Manila, Philippines*

[2]*International Rice Research Institute, P.O. Box 933, 1099 Manila, Philippines*

**Key words:** risk analysis, attainable yield, actual yield, yield loss, risk probability, risk magnitude, correspondence analysis, cluster analysis, principal component analysis

**Abstract**

A methodology has been developed which allows the risk posed by pests to be addressed in the context of shifting intensification levels of rice production in Asia. The term 'pest' is used here to encompass any harmful agent, including pathogens, insects, and weeds. This methodology combines two separate approaches: assessing risk probability, and quantifying risk magnitude. Risk probability may be defined as the probability of a particular pest combination occurring in a given production situation. It can be derived from surveys in selected sites. Yield loss – the gap between attainable and actual yields – is taken as one measure of risk magnitude. It is assessed from a crop loss data base developed from a series of field experiments, where inputs to the rice crop and pests are manipulated independently. The analysis indicates that rice tungro virus disease, rice sheath blight, and weed infestation correspond to very high risk magnitudes. Yield losses due to pest injuries are often strongly influenced by changes in production situation, and, therefore, by variation in attainable yield. The pattern of this relationship strongly depends on the nature of the pest, however. Different pest scenarios from contrasting production situations in Uttar Pradesh, Central Luzon, eastern Thailand and the Mekong Delta can be used to illustrate the present situation, and forward hypotheses on the future need of research at the system level.

## Introduction

This paper presents methodologies currently used at IRRI (International Rice Research Institute) to characterize rice pests, i.e. to describe their association with rice environments and cropping practices, to quantify the damage they do to crops, and to assess their importance in the context of agricultural changes. In this paper, the term 'pest' encompasses any injurious biological agent, from viruses to weeds. Our emphasis is not on the biological nature of pests (and their diversity), but rather on their relationships with crop yield (and their commonalty in that respect), measured as injuries that are physiologically translated by the crop as a damage.

## Objectives of characterizing pest constraints

Characterization of pest constraints at the field level starts with the collection of large data sets. This paper does not specifically deal with the collection process itself, but rather with the mobilization of the data. In the case of rice pests (pathogens, insects, weeds), Elazegui et al. (1990) provide a detailed methodology.



*P.S. Teng* et al. *(eds.): Applications of Systems Approaches at the Farm and Regional Levels, 83–98.*

A major difficulty encountered in the analysis of such data sets is the combination of variables with differing characteristics: some variables are quantitative, while some are qualitative; some are time dependent, while some are time independent. Qualitative attributes of agricultural systems often are of prime importance, such as, in the case of studies at the field level, the nature of the previous crop, or the genotype of the present crop plant. The approach outlined here attempts to fully exploit and represent the variation among such systems that is accounted for by both their qualitative and quantitative attributes. De Wit (1982) coined the term 'production situation' to express the unique combination of physical, biological, and socio-economic attributes that characterize a particular agrosystem.

Consideration of time-dependent and time-independent variables reflects the need to incorporate the dynamics of some processes in the analysis. For example, the date of crop establishment is not considered a time-dependent variable when analysing yield variation in farmers' fields, as this date is set for a whole cropping season in a given field. However, the amount of injury inflicted by a pathogen is dynamic, often with an initial phase of slow progress, a rapid increase, and a final decline. Such dynamics need to be taken into account if their effects on yield variation are to be detected. Several alternatives have been offered (Teng 1983, 1987; Savary and Zadoks 1992), and this aspect is dealt with further by Pinnschmidt et al. (1996).

A very large fraction of the variation in production situations is reflected in qualitative attributes of agroecosystems. Most quantitative information pertains to pest injuries. A methodology to combine these groups of variables, which consists primarily in converting the quantitative data into categorized, ordinal information, is needed. The rationale of this approach and its application to plant protection is described in more detail in Savary et al. (1995). When variation of yield is considered, yield data also have to be categorized, e.g. from very low, to medium, to very high. Concerns may be raised on the loss of precision in doing so. These concerns are mitigated when considering the considerable error associated with yield estimation methods in farmers' fields (e.g. Poate 1988). It should further be emphasized that the objective of characterization is to generate a relevant, accurate representation of interacting factors at the field level, rather than produce precise statistical estimates of yields.

Characterization of pest constraints to crops has several objectives. In the context of the current changes world agriculture is undergoing, and especially rice production in Asia, the identification of pest injuries that are influenced by shifts in production situations is of prime importance. More specifically, it becomes crucial to identify injuries whose importance would be increased when agricultural intensification takes place. Another objective is to define combinations of injuries that affect a crop in the course of a cropping – injury profiles – that are specifically associated with some production situations. This allows a third objective to be addressed, the delineation of recommendation domains that are production situation-specific.

The approach (Savary et al. 1994) we have used involves four steps (Figure 1):

(1) Categorization of quantitative information, i.e. its translation into qualitative, ordinal data.

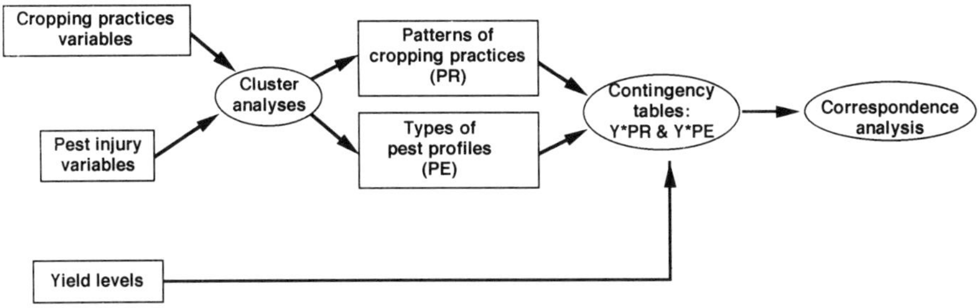


*Figure 1.* Steps for characterization of pest constraints.

(2) Cluster analyses of information pertaining to the cropping practices and environment, or to the pest constraints, using a chi-square distance (Benzécri 1973).
(3) Building contingency tables and hypothesis testing, such as independence of patterns of cropping practices from yield levels.
(4) Correspondence analysis.

Step 3 provides a means to assess the representability, or the relevance, of the categories defined in step 1, and of the clusters generated in step 2. Step 4, correspondence analysis (Benzécri 1973; Dervin 1988), consists of a non-parametric multivariate method allowing eigenvectors and axes to be generated in a way similar to principal component analysis. One of the main outputs of such analysis are graphs, where the variables (here, clusters) are mapped using a chi-square distance.

*An example of characterization: rice pests in Uttar Pradesh, India*

These techniques have been used to characterize pest injuries of rice in Uttar Pradesh, India (Singh et al. 1994). Figure 2 shows the main output of an analysis performed after the first year of survey, and is based on 90 fields only. The survey was later pursued, and we report results on three years and 251 fields later in this paper. The upper part of the figure is generated from correspondence analysis, and shows the relationships among three categories of attributes: patterns of cropping practices, injury profiles, and yield levels. The lower part of the figure is a summary of cluster analyses, where the main characteristics of the clusters that were generated are indicated.

In this preliminary analysis, four clusters of cropping practices were identified. Cluster A, for instance, represents rice production in the typical rice-wheat system, with high labour (transplanting) and financial (fertilizer) inputs. This production situation is predominantly associated with a particular injury profile, which is characterized by its diversity. It strongly contrasts, for instance, with the very poor injury profile that is predominantly associated with production situation C. Correspondence analysis provides a means to show the relationships among these clusters. The two axes involved in the upper part of Figure 2 account for a very large fraction

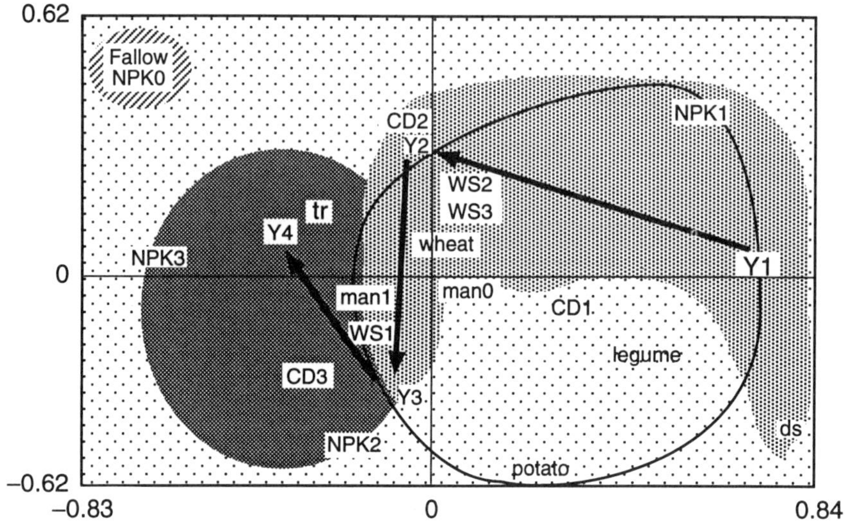


| domain | cropping practices | pest constraints |
|---|---|---|
| A | • high fertilizer input<br>• some manure<br>• previous crop: wheat predominant<br>• transplanted rice<br>• medium-high yield | • deadhearts<br>• whiteheads<br>• army worms<br>• rice bugs<br>• sheath blight<br>• sheath rot<br>• neck blast |
| B | • low fertilizer input<br>• some manure<br>• diverse previous crops<br>• water stress often high<br>• transplanted rice predominant<br>• short cycle<br>• low yield | • glume discoloration<br>• (leaf blast) |
| C | • low fertilizer input<br>• no manure<br>• previous crop: wheat only<br>• medium-low water stress<br>• transplanted or direct-seeded rice<br>• long cycle varieties | • weeds above the rice canopy<br>• glume discoloration |
| D | • no fertilizer<br>• fallow | • deadhearts<br>• whiteheads<br>• sheath rot |

*Figure 2.* A correspondence analysis of patterns of cropping practices, injury profiles and yield levels in Uttar Pradesh, India (Singh et al. 1994). The upper part of the figure shows the two first axes generated by the analysis, and a few attributes of the ecosystem – ds: direct seeding; tr: transplanting; legume, potato, wheat, fallow: nature of the crop preceding rice; CD1 to CD3: cultivars of increasing cycle duration; WS1 to WS3: increasing water stress; NPK0 to NPK2: increasing mineral fertilizer supply; man1 and man0: application of manure or not; Y1 to Y4: rice yield, low to high. The lower part shows the overall association between two groups of clusters independently generated from the data: patterns of cropping practices, and pest constraints (injury profiles).

(over 80%) of the information captured by the contingency tables. This diagram allows us to show the relationships between clusters and yield levels. Four yield levels (from very low, Y1, to high, Y4) are considered, and a path of increasing yields is delineated. This type of analysis provides a means to compact the information to a very high degree: clusters of cropping practices – production situations – and injury profiles are defined; their linkage may be tested; and their overall relationship with yield variation can be mapped on a summarized representation of a multivariate data set.

## Risk assessment: overall methodology

### *Background*

Analyses of surveys conducted in the Philippines (Savary et al. 1994), in Uttar Pradesh (Singh et al. 1994), and in Vietnam and Thailand (Pinnschmidt et al. 1995; Pinnschmidt et al. 1996), have shown that the approach illustrated in the example of Figure 1 is efficient in defining clusters of cropping practices. These clusters reflect specific combinations of production methods that are adaptations to environments and reflect production situations *sensu* De Wit (1982). To a large extent, qualitative rather than quantitative variables determine this clustering. In a similar way, pest injuries tend to be associated in specific patterns of combinations. The approach allows these injury profiles to be defined. There is usually a very strong linkage between production situations and injury profiles, which can be tested using a chi-square tcst (Savary et al. 1994). In fact, rejection of the null hypothesis of independence between production situations and injury profiles is the very basis of conducting correspondence analysis, which can be seen as a procedure to represent complex relationships in a set of contingency tables.

The approach also provides an easy means to describe the variation in yield levels with production situations. Generally, the linkage of yield levels with production situations is much stronger than with injury profiles (Savary et al. 1994). This may be attributed to (i) the very strong effect of production situations on attainable yields (the yield of a crop that would not have suffered any injury), and (ii) the comparatively smaller effect of injuries in reducing yields from their attainable to their actual levels.

When considering the risk imposed on crops by pests, the consequences of raising the attainable yield on the shape of damage functions (Zadoks 1985) should also be considered. Depending on the specific pest, and on the physiological nature of the damage mechanisms, an interaction between attainable yield and injury may occur. If so, damage becomes a function not only of injury, but also of attainable yield.

### *Strategy*

To address the risk imposed by pests on crop production, an overall methodology therefore is needed, where estimates of injury probability are derived from surveys in various production situations, and measurements of damage are made at varying attainable yields – which reflect changes in production situations.

The approach here to analyse the risk associated with pests has two levels. In the first level, the probability $P$ of the risk (Rowe 1980; Savary 1991) has to be determined. Given a sufficient sample size, the probability of occurrence of each injury (or specific combinations of injuries) can be determined in a series of subsamples, defined as the fields belonging to a specified production situation where this injury occurs. In a second level, the risk magnitude $M$ is determined as the amount of damage (yield reduction) caused by each injury in each production situation.

*Operational definition of risk*

Risk is defined as the product of risk probability $P$ and risk magnitude $M$, as a dimensionless index with % as the unit, that is:

$$R = P \times M$$

Experiments have been designed to measure $M$, expressed in relative terms (% yield lost to a particular pest). Simultaneously with the characterization of production situations, estimates for $P$ have been derived from surveys in farmers' fields. The risk corresponding to each individual pest is then defined as the product: $R = P \times M$. $R$ is a dimensionless index with % as unit.

## Surveys in farmers' fields: risk probability

*Characterization of production situations and injury profiles*

The characterization procedure outlined in Figure 1 was applied to survey data generated on 635 fields representative of four countries: the Philippines, India (eastern Uttar Pradesh), Thailand and Vietnam. Results pertaining to the Philippines (Central Luzon) are used here to further illustrate the approach. Figure 3 (Savary et al. 1994) provides an overview of these results, with seven production situations (Figure 3C), seven injury profiles (Figure 3D), and several yield levels (Figure 3B). Figure 3C shows the contrast between high-input, dry season crops (PR3, on the left hand side of the horizontal axis) and low-input, rainy season crops (PR4). Production situations such as PR5, i.e. rainy season crops with comparatively high financial (fertilizer and herbicides) and labour (transplanting and handweeding) inputs, are in a central position on the graph. Figure 3D shows contrasting injury profiles, some of them involving a number of pests, and where weeds often dominate (e.g. PE4: bacterial leaf streak, and weed infestation above and below the rice crop canopy), and some with only one dominant pest (PE6: sheath rot; PE1: sheath blight).

The corresponding path of increasing yield (Figure 3B) matches changes in production situations, with an overall movement from right (Y1 is close to PR1 and PR4) to left (Y6 is close to PR3). While the horizontal axis accounts for the largest variation in actual yield, and reflects the corresponding variation in production

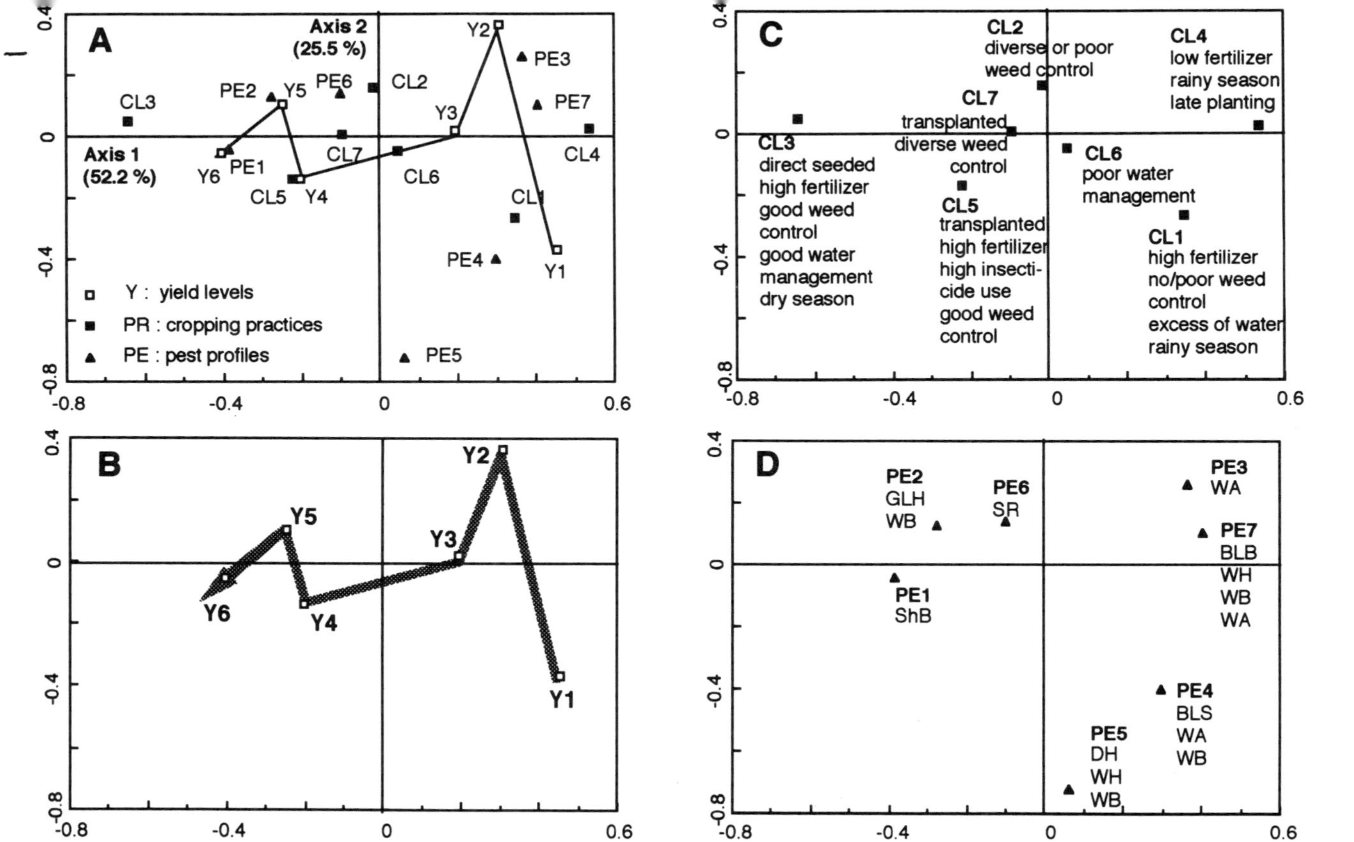


*Figure 3.* Analysis of correspondences between production situations (CL1 to CL7), pest injury profiles (PE1 to PE7), and yield levels (Y1 to Y6) (*Source:* Savary et al. 1994). (A) Plot of all the categories of each variable on the two first axes (the figures indicate chi-square distance along these axes). The successive yield levels have been linked. (B) Path of increasing yield levels. (C) Plot of the clusters representing production situations, and their main features. (D) Plot of the clusters representing injury profiles, and their main components (ShB: sheath blight; GLH: green leaf hopper; WB: weed below the rice crop canopy; SR: sheath rot; WA: weed above the rice crop canopy; BLB: bacterial leaf blight; WH: white heads; BLS: bacterial leaf streak; DH: dead hearts due to stem borers).

situations, departures along the vertical axis may be attributable partly to injury profiles. One interesting feature of this analysis is the close association of PE1 (high sheath blight) with high yields. Two interpretations may be forwarded: sheath blight injury is enhanced by increased inputs to the crop (especially nitrogen fertilizer), and/or this injury does not cause measurable yield reductions.

*Links between production situations and injury profiles*

This analysis again demonstrates a very strong link between injury profiles and production situations. When the null hypothesis of independence of distribution of fields among production situations and injury profiles was tested (Savary et al. 1994), a chi-square value of 55.2 was obtained, and the null hypothesis was rejected ($P < 0.005$, df=36).

*Risk probabilities in different production situations*

The proportion of fields affected by injuries was taken as a preliminary estimate of risk probability. These proportions were computed for nine injuries in each of the production situations identified in the four countries. Figure 4 shows these proportions for a selection of production situations in the Philippines (CL), Uttar Pradesh (IN), and Vietnam and Thailand (VT). Strong differences among production situations are apparent, some pests showing large variation in the proportion of affected fields, especially sheath blight, ShB, and brown spot, BS. Rice tungro disease (RTD) is detected in some production situations only at low proportions of affected fields. Insect injuries (dead hearts, DH, and white heads, WH) are omnipresent in high proportions, with the exception of whorl maggot (WM). Weed infestation, mostly at high levels, appears to be the most common constraint in all production situations.

## Field experiments: risk magnitude

*The experimental crop loss data base*

Over the past years, a series of experiments has been conducted to measure the yield-reducing effects of pest injuries, either alone or in combination, at varying attainable yields (Castilla et al. 1993). An outline of a typical experiment is shown in Figure 5. The experimental design relies on the concept of isolated microplots (Parlevliet and Van Ommeren 1975) in order to minimize inter-plot interference. The area of individual plots is $5.8\,m^2$, and an experiment is commonly established in a $3750\,m^2$ rice field, so as to provide large buffer areas between plots. Pests are artificially introduced in the assigned plots, and no pesticide is used in the plots, so as to avoid possible side-effects on the host physiology. The (unreplicated) design involves two levels, with main plots representing combinations of input factors, and subplots consisting of combinations of inoculations or infestations. In any input factor combination, a non-injured plot (Figure 5C) is included, which allows assessment

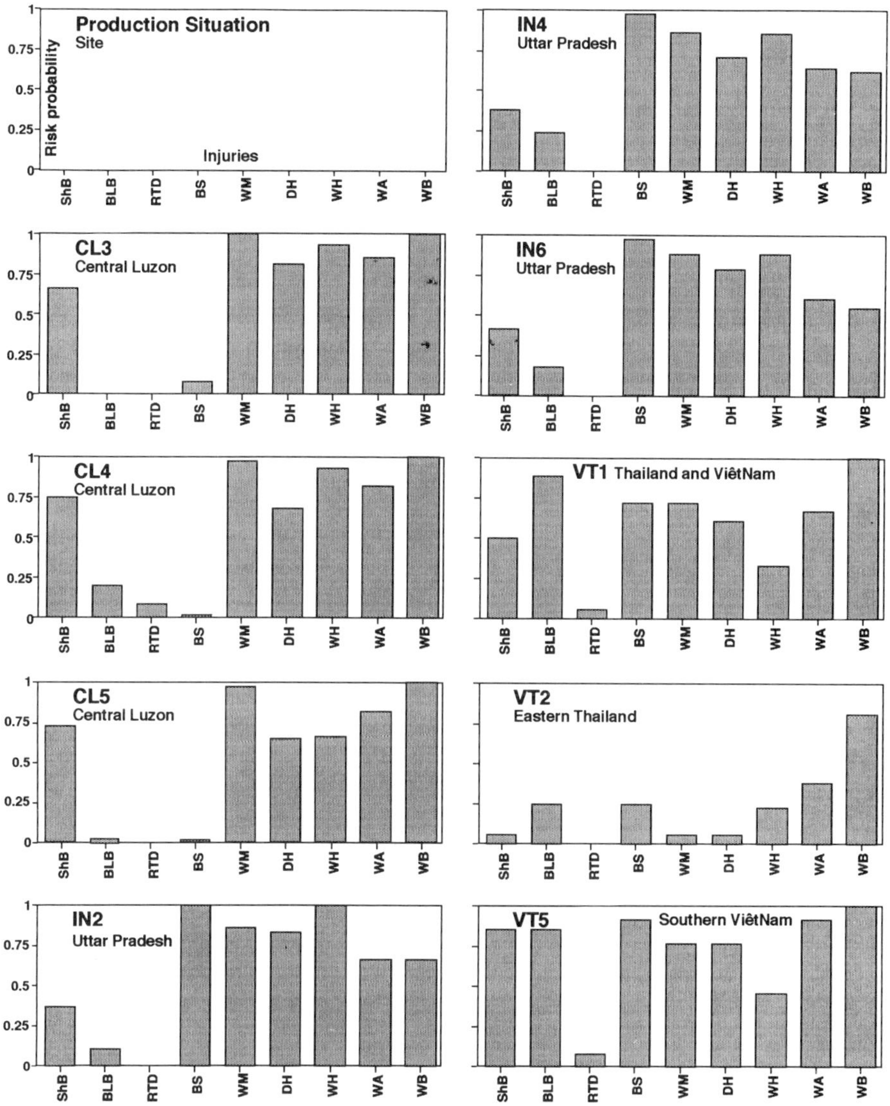


*Figure 4.* Proportion of fields affected by nine injuries in selected production situations in four countries. Seven production situations were determined by cluster analysis from each of the three survey data sets collected in the Philippines (CL1 to CL7), India (IN1 to IN7), and Vietnam and Thailand (VT1 to VT7). Only some production situations are represented. Each histogram represents a particular production situation in a given site or country (e.g. CL3, Central Luzon). The horizontal axis indicates injuries – ShB: sheath blight; BLB: bacterial leaf blight; RTD: rice tungro disease; BS: brown spot; WM: whorl maggot injury; DH: dead hearts; WH: white heads; WA; weed above the rice crop canopy; WB: weed below the rice canopy. The vertical axis indicates the proportion (0 to 1) of affected fields.

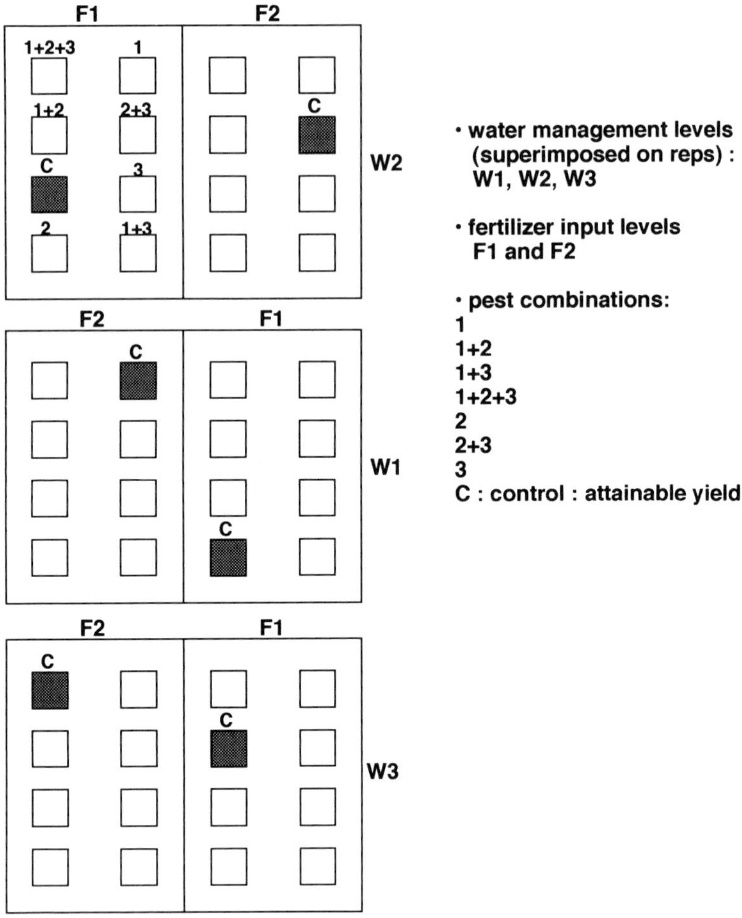


*Figure 5.* Experimental crop loss data base: design of a typical experiment.

of the attainable yield corresponding to each particular production situation. Ten such experiments have been conducted, either in farmers' fields in the neighbourhood of IRRI, Laguna, Philippines, or at IRRI's experimental farm, yielding a data base consisting of more than 400 individual plots.

Each experiment can be seen as a new building-block meant to explore the multidimensional surface of yield response (Teng and Gaunt 1980) to combinations of inputs, environmental factors and injuries. The data base currently addresses four input factors (potential yield of the rice cultivar, crop establishment method, water management and fertilizer supply) and nine injuries (bacterial blight, rice tungro disease, brown spot, sheath blight, whorl maggot injury, dead hearts, white heads, weed infestation above and below the rice crop canopy). It covers a range of attainable yields from 1 to 11 ton ha$^{-1}$.

*Analytical steps*

The steps involved in the analysis of the Experimental Crop Loss Data Base are outlined in Figure 5. First, principal component analysis is performed on the injury data to generate eigenvectors ($Fi$), i.e. independent linear combinations of the injury variables. Next, these eigenvectors are used as new variables together with attainable yield ($Ya$) and the $Ya \times Fi$ interactions in a step-wise, multiple regression procedure. The aim is to generate a regression equation that would provide an adequate description of the yield response surface at varying levels of attainable yield and injuries.

*Yield response surface and damage functions*

The analysis yielded the following equation:

$$Y = 0.280Ya - 0.203F2 - 0.208F3 + 0.326F6 - 0.016Ya \times F4 - 0.124Ya \times F6 - 0.020Ya \times F7 \quad (1)$$

This equation was tested for coefficient of determination ($R^2 = 0.974$, $n = 419$), aptness of residual plot, and $F$ test ($P < 0.0001$). One important feature of this equation is the significant contribution of $Ya \times Fi$ interactions, demonstrating that some injuries (those included in eigenvectors $F4$, $F6$, and $F7$) significantly interact with attainable yield in their yield-reducing effect.

The regression equation was then used to estimate yield values under a series of scenarios, where: (i) pre-set levels of attainable yield were chosen; and (ii) injuries were chosen one at a time and varied within a range of levels, thus allowing computation of the crop yield as a function of injury alone, at that specified attainable yield. Three sets of such scenarios are shown in Figure 6. The first set shows damage functions associated with increasing injury due to weed infestation above the rice crop canopy. With increasing attainable yield ($Ya = 1$ to 11 ton ha$^{-1}$), there is a steady, regular increase of the slope of the damage function when crop losses are expressed in absolute terms (ton ha$^{-1}$, Figure 6, upper part). As a result, when expressed in relative terms (% yield loss, lower part), the slope of the damage function is essentially the same at any attainable yield. In the case of sheath blight (Figure 6, centre), there is an increase in the slope of the (absolute) damage function, resulting in increasing (relative) damage when the attainable yields are increased. In the case of rice tungro disease (Figure 6, right), the situation is reversed: the slope of absolute damage tends to increase more slowly with increasing attainable yields and, as a result, the (relative) damage decreases with increasing attainable yield. These examples show that the rice crop responds differently to differing injuries with increasing attainable yield, i.e. improved production situations: the relative damage due to weeds remains essentially the same, while that due to sheath blight increases, and that due to tungro decreases. It must be emphasized, however, that damage due to tungro remains the highest of all nine injuries considered, with a range of 30% to 60% yield loss for severe infection.

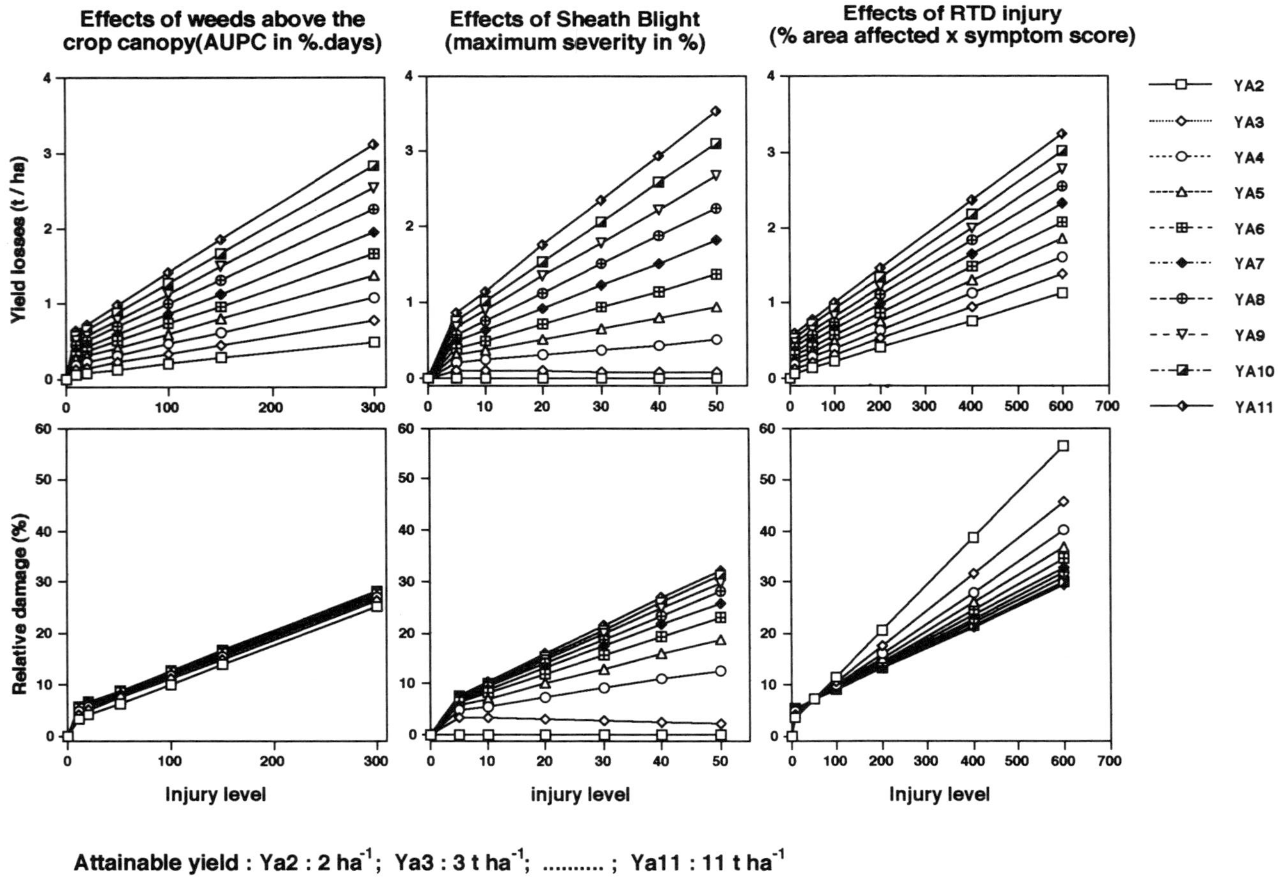


*Figure 6.* Damage functions for three injuries: weeds above the rice canopy, sheath blight injury, and rice tungro disease. Increasing injury levels are indicated on the horizontal axis. Increasing damage is indicated on the vertical axes, both in absolute (yield losses in ton $ha^{-1}$) and relative (relative damage in %) terms. Damage functions are shown for a range of attainable yields, from 2 ton $ha^{-1}$ (Ya2) to 11 ton $ha^{-1}$ (Ya11).

## Risk analysis across production situations

Equation (1) was also used to estimate the risk magnitude (% yield loss) corresponding to each of the production situations generated from survey data analyses. In each case, computations used the mean yield of each production situation as an indicator of attainable yield, and a set of two arbitrary injury levels considered to represent 'high' and 'low' injuries. The choice of these injury levels was based on field experience, so that they would remain within the range of possible injuries. The resulting estimates for risk magnitude were then multiplied by the estimates for risk probabilities in the corresponding production situations (Figure 4) to produce estimates for risk (Figure 7).

The variation in risk within production situations is very large, from 0% to 30%. This is, of course, attributable first to the fact that all injuries were not detected in all production situations, and, to a lesser extent, to variation in risk magnitude depending on the injury considered (e.g. Figure 6). The variation in risk patterns across production situations is even wider, as the effect of varying attainable yield is superimposed on the two first causes of variation. As a result, some production situations are affected by a number of different risks at high levels (e.g. VT5, southern Vietnam), while others seem to have fewer, and lower risks (e.g. VT2, eastern Thailand). Considerable variation among production situations can also be seen when a particular risk is considered, especially for sheath blight (e.g. CL3, Philippines vs VT1, Thailand and Vietnam), or brown spot (e.g. IN6, India vs CL4, Philippines). A striking result of this analysis is the comparatively low risk attached to tungro. This is first attributable to the choice of the sites and years where surveys have been conducted, which resulted in very small proportions of field affected. This, in turn, leads to extremely low risk probability estimates, cancelling the effect of comparatively very large risk magnitudes. Another conspicuous result is the comparatively stable and high risk attached to weeds. Weed infestation occurs to some degree in every production situation, and its risk magnitude can be very high.

## Conclusions

We have outlined methodologies that are currently being used to analyse survey data with two types of objectives: characterization of rice pests patterns, and crop loss assessment. Non-parametric, multivariate techniques appear to be robust and reliable tools for characterization purposes. They allow information that is diverse in nature to be mobilized, and they take due consideration of possible production situations. Typologies of production situations are strongly linked with typologies in injury profiles, and this link allows the risk associated with pests in these different production situations to be addressed.

Preliminary results on risk assessment have been presented here primarily to illustrate an empirical approach, based on a relatively large field sample size, but a very small number of sites. This is an ongoing effort, where additional information is needed to document both the probability and the magnitude components of risk.

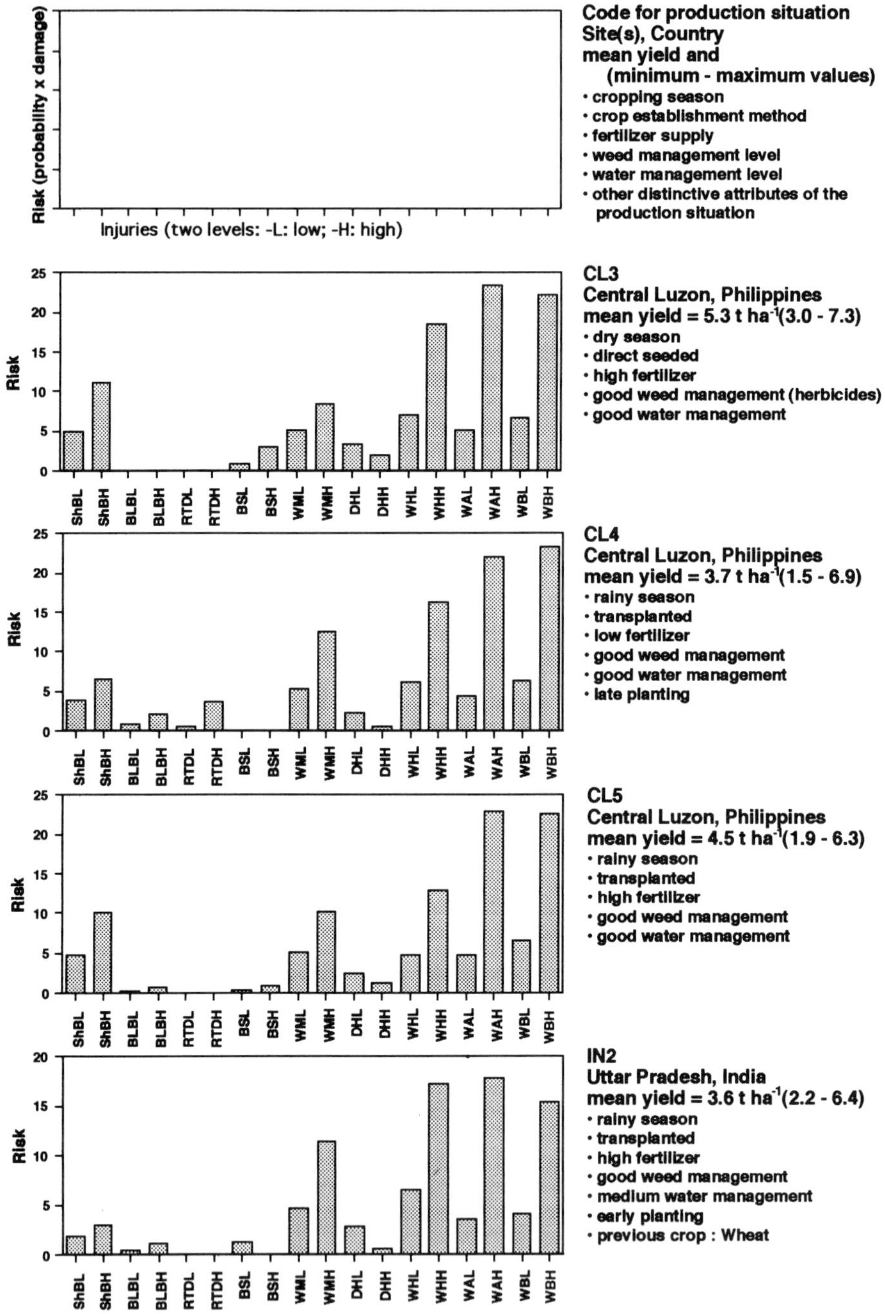


*Figure 7.* Estimates of risk magnitude in selected production situations in four countries. The horizontal axis indicates injuries due to ShB: sheath blight; BLB: bacterial leaf blight; RTD: rice tungro disease; BS: brown spot; WM: whorl maggot injury; DH: dead hearts; WH: white heads; WA: weed above the rice crop canopy; WB: weed below the rice crop canopy. For each injury, two levels, high (-H) or low (-L), are shown. These levels were defined as follows: ShB low: maximum severity = 10%; high = 40%; BLB low: AUSPC = 200% days; high = 2000% days; RTD low: area × intensity index = 50; high = 600; BS low:

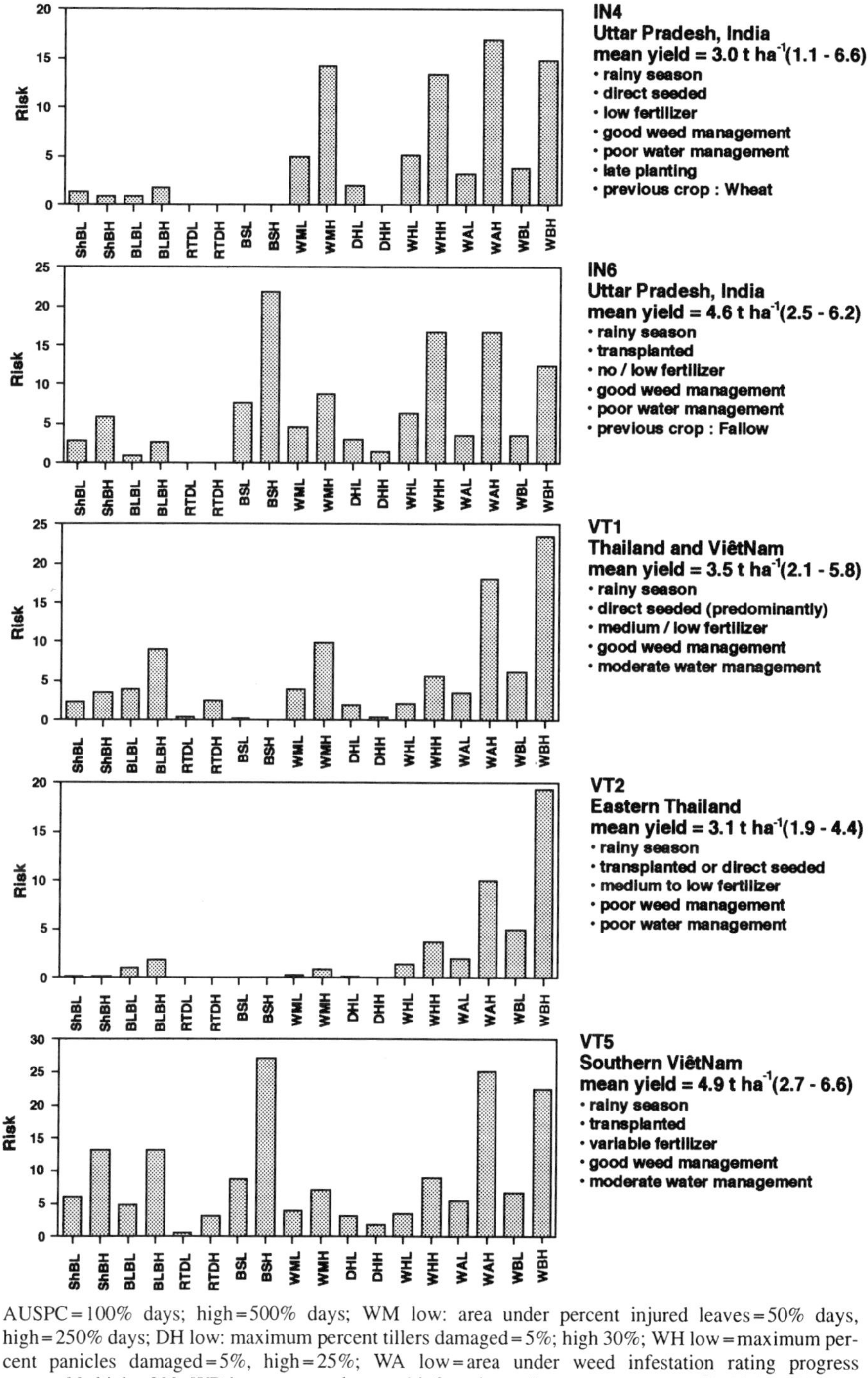


AUSPC = 100% days; high = 500% days; WM low: area under percent injured leaves = 50% days, high = 250% days; DH low: maximum percent tillers damaged = 5%; high 30%; WH low = maximum percent panicles damaged = 5%, high = 25%; WA low = area under weed infestation rating progress curve = 20; high = 300; WB low = area under weed infestation rating progress curve = 50; high = 400. The vertical axis indicates risk, i.e. the product of risk probability (Figure 4) and risk magnitude expressed as relative damage (see Figure 6 for three examples).

More information is needed to better quantify risk magnitude of some of the injuries that have been addressed here, such as brown spot, and also to include additional rice pests. There is a large variation among rice pests in their patterns of occurrence over time and space. The space and time dimensions of risk probability must, therefore, also be considered in the future. Additional survey work is under way in new sites where the likelihood of injuries differs from sites that have already been investigated.

## References


Benzécri J P (1973) L'Analyse des données. Tome 2. L'Analyse des correspondances. Dunod, Paris, France. 632 p.

Castilla N P, Elazegui F A, Mabbayad M O, Barrion A T, Savary S (1993) Combined effects of pests in farmers' fields: Methodological outlines of a yield loss data base in rice. Rice Res. Notes 18(3):41–42.

De Wit C T (1982) La productivité des pâturages Sahéliens. Pages 22–35 in Penning de Vries F W T, Djitèye M A (Eds.) La productivité des pâturages Sahéliens – une étude des sols, des végétations et de l'exploitation de cette ressource naturelle. Pudoc, Wageningen, The Netherlands.

Dervin C (1988) Comment interpréter les résultats d'une analyse factorielle des correspondances? ITCF, Paris, France. 75 p.

Elazegui F A, Soriano J, Bandong J, Estorninos L, Jonson I, Teng P S, Shepard B M, Litsinger J A, Moody K, Hibino H (1990) Methodology used in the IRRI integrated pest survey. Pages 241–271 in Crop loss assessment in rice. International Rice Research Institute, P.O. Box 933, Manila, Philippines.

Parlevliet J E, Van Ommeren A (1975) Partial resistance of barley to leaf rust: *Puccinia hordei*. II. Relationship between field trials microplot tests and latent period. Euphytica 24:293–303.

Pinnschmidt H O, Nguyen Dan Long, Tran Tan Viet, Le Dinh Don, Teng P S (1995) Characterization of pests, pest losses, and production patterns in rainfed lowland rice of the Mekong River Delta. Pages 223–242 in Denning G L, Vo-Tong Xuan (Eds.) Vietnam and IRRI: A partnership in rice research. International Rice Research Institute, P.O. Box 933, Manila, Philippines.

Pinnschmidt H O, Chamarerk V, Cabulan N, Dela Pena F, Long N D, Savary S, Klein-Gebbink H W, Teng P S (1996) Yield gap analysis of rainfed lowland systems to guide crop and pest management. In Kropff M J, Teng P S, Aggarwal P K, Bouma J, Bouman B A M, Jones J W, Van Laar H H (Eds.) Applications of systems approaches at the field level. Kluwer Academic Publishers, Dordrecht, The Netherlands (in press).

Poate D (1988) A review of methods for measuring crop production from small-holder producers. Exp. Agric. 24:1–14.

Rowe W D (1980) Risk assessment approaches and methods. Pages 3–29 in Conran G (Ed.) Society, technology, and risk assessment. Academic Press, New York, USA.

Savary S (Ed.) (1991) Approches de la pathologie des cultures tropicales. L'Exemple de l'arachide en Afrique de l'Ouest. Karthala-Institut Français de Recherche Scientifique pour le Développement en Coopération, Paris, France. 288 p.

Savary S, Zadoks J C (1992) Analysis of crop loss in the multiple pathosystem groundnut-rust-leaf spot. I. Six experiments. Crop Prot. 11:99–109.

Savary S, Elazegui F A, Moody K, Teng P S (1994) Characterization of rice cropping practices and multiple pest systems in the Philippines. Agric. Syst. 46:385–408.

Savary S, Madden L V, Zadoks J C, Klein-Gebbinck H W (1995) Use of categorical information and correspondence analysis in plant disease epidemiology. Advances in botanical research incorporating. Adv. Plant Pathol. 21:213–240.

Singh H M, Srivastava R K, Singh R K, Savary S (1994) Illustrating the concept of recommendation domain in IPM: An Indian case study. Int. Rice Res. Notes 19:28–30.

Teng P S (1983) Estimating and interpreting disease intensity and loss in commercial fields. Phytopathology 73:1587–1590.

Teng P S (Ed.) (1987) Crop loss assessment and pest management. APS Press, St. Paul, MN, USA. 270 p.

Teng P S, Gaunt R E (1980) Modelling systems of disease and yield loss in cereals. Agric. Syst. 6:131–154.

Zadoks J C (1985) On the conceptual basis of crop loss assessment: The threshold theory. Annu. Rev. Phytopathol. 23:455–473.

# Systems diagnoses at field, farm and watershed levels in diversifying upland agroecosystems: towards comprehensive solutions to farmers' problems

G. TRÉBUIL[1,2], S.P. KAM[2], F. TURKELBOOM[3] and B. SHINAWATRA[4]

[1]*Department of Annual Crops, Centre de Coopération Internationale en Recherche Agronomique pour le Développement, BP 5035, 34032 Montpellier Cedex 1, France*
[2]*International Rice Research Institute, P.O. Box 933, 1099 Manila, Philippines*
[3]*Soil Fertility Conservation Project, KU-Leuven, Maejo University, Chiang Mai, Thailand*
[4]*Multiple Cropping Center, Faculty of Agriculture, Chiang Mai, Thailand*

**Key words:** systems diagnosis, aggregation level, land-use dynamics, farmer typology, cropping system, north Thailand

**Abstract**

In mainland Southeast Asia, the increasing population pressure on montane agroecosystems, their growing integration into the market economies and the impact of national environmental protection policies are provoking rapid transformations in highland farmers' strategies and practices evolving from slash and burn to more intensive, diverse production systems. In such a context, most of the attempts at transferring standard recommended technologies are facing very low rates of adoption. These variable and heterogeneous environments provide opportunities for using holistic systems approaches to address the whole complexity of agricultural development issues. To boost the prospects of significant impact, however, the farmer must be put at the centre of a development-oriented research process. Diagnostic tools at field, farm and watershed levels are used to understand the rapidly increasing diversity of farmers' circumstances, the rationale of farmers' practices and strategies, and key dynamics at work. Such information must be integrated in the research agenda to design improved production systems preserving the resource base while increasing land and labour productivity. At farmers' field level, we use the on-farm agronomic experiment-survey procedure to inventory farmers' techniques, explain their practices and assess their impact on crop function and its environment. In actual farmers' conditions, limiting factors of yields are ranked, yield modelling is carried out, and hypotheses for new cropping systems are derived. At the farming system level, the diversity of farmers' objectives and strategies is analysed and their functioning summarized diagrammatically. Similar farming systems are grouped into a typology and trajectories of evolution displaying the process of accumulation/elimination on the farms along various pathways. Based on results obtained at the two previous scales, the analysis of land-use dynamics at the watershed level spatially distributes the diversity of situations and the extent of dominating trends, and points to the conflict areas. Risky practices regarding land degradation are mapped and the potential impact of improved cropping systems can also be generated at this scale. Illustrations from a case study of diversifying upland rice-based farming systems in montane upper northern Thailand are provided to demonstrate how these on-farm research tools can be articulated into an integrated systems research approach producing fine-tuned and well-targeted innovations. Their role in helping the needed institutional change in these less-favoured ecosystems is also underlined.

## Introduction

Apart from the extensive spatial heterogeneity and variability of soil and climatic conditions in upland mainland Southeast Asian agroecosystems, three kinds of con-



*P.S. Teng* et al. *(eds.): Applications of Systems Approaches at the Farm and Regional Levels, 99–114.*

temporary socioeconomic transformations are at the origin of the growing complexity of agronomic situations and rapidly increasing diversity of farmers' social demand for agricultural development: (i) the growing population pressure on resources, through natural growth but often due to sudden cross-border migration or population movement from densely populated lowlands, is rapidly bringing the agricultural systems to the limits of their carrying capacities; (ii) the rapid integration of highlanders in booming market economies which offer numerous opportunities for new activities, such as off-farm employment or cash cropping to meet the needs of changing food consumption patterns at home or export markets; and (iii) the impact of recent national environmental protection policies which, by restricting the traditional land-use system, accelerate the shortening of fallow periods and deepen the fertility maintenance crisis leading to lower soil productivity.

In a very short period of time, highland farmers have been forced to manage an agricultural revolution, a transition from a more or less typical self-subsistent slash-and-burn mode of exploitation of natural resources to market-oriented, intensive and diversified production systems. Because the Green Revolution has been ignoring these areas during the last three decades, the current farmer-led diversification of cultivation practices, cropping and farming systems is taking place without the support of traditional agricultural research and extension institutions. In these marginal areas, non-governmental organizations (NGOs) are more common partners of the development-oriented researcher. In the case of northern Thailand, the very limited success at introducing standard 'technological packages' for soil conservation farming, based on grass strips or hedgerows systems, illustrates the need for a new approach integrating a better management of the diversity of farmers' circumstances at field, farm and watershed, or village territory level. Such a multi-scale, on-farm systems approach to agricultural development is presented here to understand the complexity of local key issues necessary to respond efficiently to the increasingly varied social demand for development. Preliminary diagnostic analysis of the diversity of farmers' situations and practices leads to an understanding of their different adaptive mechanisms, the identification of key questions for research and leverage points to guide the conception and design of alternative adapted production systems preserving the resource base while increasing land and labour productivity. The methodological challenge consists in the interdisciplinary integration of knowledge acquired at complementary scales for the design of operational solutions to real farmers' problems. To manage the high level of complexity of such agroecosystems, complementary diagnostic tools at field, farm and watershed levels can be used to grasp the current dynamics at work, to understand the diversified farmers' strategies and practices, and evaluate their effects on crop productivity and the environment. Since the 1993 wet season, such an approach has been introduced to identify pathways for improving the sustainability of upland-rice based production systems in upper northern Thailand. Outputs from this case study are illustrated in this paper.

The on-farm research tools used have the following respective objectives:

- *Field level:* to understand the diversity of farmers' practices, itineraries of techniques, i.e. the logical and well-ordered succession of cultivation practices carried out on a field by the farmer to control the environment, and to achieve a given production objective (Sebillotte 1978), and cropping systems, i.e. the crops, their succession and the itineraries of techniques applied to each of them including the varietal choice (Sebillotte 1990a), to assess their effects on the environment and for *in situ* grading of limiting factors of yields.

- *Farm level:* to typify the main farming systems according to their differentiated objectives, strategies and related functioning, history and combination of productions to better document and focus the research agenda on key issues and target the future production of 'solutions' (Trébuil 1992; Capillon 1993; Trébuil and Dufumier 1993). Based on the historical analysis of the transformations of the local agricultural system, to build trajectories of evolution linking these farming systems and to elaborate a provisional evolution schema of the regional agriculture anticipating future changes.
- *Watershed level:* to analyse land-use dynamics and the spatial distribution of key issues and conflicts, such as the consequences of land-use changes on land degradation. This should be done by integrating knowledge obtained at the two previous scales.

## Cropping systems: yield variability, grading of limiting factors and effects on the environment

The on-farm agronomic survey (Crozat and Chitapong 1988; Meynard and David 1992; Leterme et al. 1994) is used to identify and prioritize limiting factors of the yield in farmers' fields. It is a prerequisite to the design or introduction of appropriate technical innovations (Jouve 1984; Sebillotte 1989; Byerlee et al. 1991). It is also used to inventory farmers' itineraries of techniques, to elucidate their rationale and to assess their impacts on the crop and its environment. On completion of the survey, limiting factors of farmers' yields are ranked, empirical yield modelling is carried out (Combes and Picard 1992) and hypotheses for the design of improved cropping systems are derived.

### *Yield variability analysis and ranking of yield limiting factors*

In transitional slash-and-burn production systems, there is sufficient variability in upland rice yields between farmers' fields, as well as within the fields, to conduct such a diagnostic analysis on the main staple crop (Table 1). The agronomic survey aims at explaining these variations in yields through the analysis of the observed diversity of field situations created by the shortening of fallow periods, its effects on the crop environmental conditions and the subsequent diversification of farmers' land preparation and weeding practices (Van Keer et al. 1995). Far from considering the diversity of fields as an obstacle, this kind of crop diagnostic analysis makes use of it by building an adapted experimental design to model the crop functioning under varying environmental conditions.

Data analysis is guided by an agrophysiological analysis of the yield build-up

*Table 1.* Variability of upland rice yields (early and late maturing varieties) in farmers' fields in upper northern Thailand, 1993–95 wet seasons (yields expressed in g dry matter $m^{-2}$).

| Year | 1993 | 1994 | 1995 |
|---|---|---|---|
| | | Field homogeneous zones | |
| Early maturing | 57–204 | 0–166 | 0–220 |
| Late maturing | 27–259 | 23–217 | 30–188 |
| | | Small monitoring squares | |
| Early maturing | 43–239 | 0–223 | 0–308 |
| Late maturing | 16–433 | 0–368 | 8–290 |

*Table 2.* Effect of soil erosion by concentrated flow on upland rice plant density (no. plants $m^{-2}$) in three farmers' fields in upper northern Thailand, 1994 and 1995 wet seasons.

| Landscape position | Field 1/94 | Field 2/94 | Field 2/95 | Average 3 fields |
|---|---|---|---|---|
| Top of slope | 118 | 76 | 86 | 93 |
| On the slope | 57 | 44 | 58 | 53 |
| % plant losses | 52 | 42 | 33 | 43 |

process to understand the functioning of the upland rice crop under varying environmental conditions and cultivation practices. Standard relationships between the levels of successive yield components or other key crop characteristics are established. In a first step of the analysis, these simple models are used to identify and date yield limitations. Crop functioning problems are diagnosed by comparing the actual values recorded for a given yield component to the potential ones given by the model.

In a second step, we look for the factor(s) or condition(s) which are at the origin of the yield limitations during each key phase of the crop cycle. To do so, the data sets on complementary crop characteristics and environmental conditions are linked to determine the environmental conditions and/or plant status responsible for the crop functioning problems diagnosed at a given stage of the cycle. In this step, it is useful to integrate upstream research results at the field level (Picard 1992). Finally, we look at the relationships between the favourable (no yield limitation) or unfavourable (limited yields) conditions created and the characteristics of the cropping systems performed to identify the kinds of cultivation practices to be promoted, as well as those to be avoided. At this stage, for the majority of sloping upland rice fields in Chiang Rai, erosion by concentrated flow was shown to lead to very significant plant losses (Table 2).

### *Influence of cropping systems on erosion risk*

In another complementary on-farm experiment-survey, the influence of slope characteristics (angle, length) and main types of annual cropping systems (upland rice, maize, soybean, vegetable-based) on the risk of erosion by concentrated flow was assessed. Observations were carried out during critical periods to identify at field level and to map at watershed level the risky situations as far as landforms and farmers' cultivation practices were concerned (Turkelboom et al. 1995). The extensive

*Table 3.* Effects of slope characteristics and soil surface states on the risk of erosion by concentrated flow in Chiang Rai Province, northern Thailand, 1994 and 1995 wet seasons.

| Surface states | | Slope angle (%) and length (m) | | | | | |
|---|---|---|---|---|---|---|---|
| Canopy cover | Roughness | 31–42% 11–25 m | 31–42% 27–50 m | 46–56% 11–25 m | 46–56% 27–50 m | 60–69% 11–25 m | 60–69% 27–50 m |
| 0–20% | Fine | **3/14*** | **5/25** | | **5/25** | **3/8** | **10/28** |
| | Medium | 1/3 | 0/2 | | **4/15** | 1/6 | **3/9** |
| | Rough | | 1/5 | 1/4 | | | 1/1 |
| 21–50% | Fine | 0/2 | 2/11 | 0/4 | | 4/22 | **3/11** |
| | Medium | 1/3 | 1/8 | | 0/6 | 1/5 | 1/12 |
| | Rough | | | | | 1/1 | 0/1 |
| >50% | Fine | 0/2 | 0/1 | 1/5 | | | 1/9 |
| | Medium | 0/1 | 1/4 | 1/8 | 2/15 | 0/4 | 2/15 |
| | Rough | 2/6 | | | | | 0/1 |

*Number of observations for which new erosion features were found and measured compared to the previous visit/total number of observations for this given slope and soil surface state situation. Fields with soil conservation measures and new clearings are not included. The most risky situations are displayed in bold characters when enough observations could be made.

range of soil surface states created by the diversity of farmers' cropping systems was used for making comparisons across the diversity of slope characteristics offered by the landforms. Taking the single rain period as time scale allowed the observers to make numerous observations on the effects of all different kinds of precipitation on soil surface degradation and to analyse them in terms of risk (Boiffin et al. 1988). The matrix presented in Table 3 displays, in bold characters, the most risky situations under which erosion is frequently generated. Based on this data set and field observations on farmers' practices, it was possible to grade existing cropping systems for their susceptibility to erosion. Upland rice and ginger were diagnosed as the crops generating the most erosion symptoms and soil losses (Healy et al. 1995). But at this stage, to better target interventions for soil conservation, and to take into account the strategies and socioeconomic conditions of the concerned groups of farmers, an understanding of the rapid diversification taking place among the local farming systems was necessary. This is because the type of farming system management affects the technical decisions made at field level (Sebillotte 1990b, 1994). This farm-level step of the analysis focuses on the interactions between biophysical, technological factors and conditions, and socioeconomic circumstances at the household-based farming system level.

## Farming systems differentiation: recent history, strategies, typology and trajectories of evolution

In montane upper northern Thailand, a rapid process of differentiation among the household-based farming systems has been taking place during the last two decades (Thong-Ngam et al. 1995). Because of the very composite nature of most of them (introduction of many cash crops, more crop-livestock interactions, articulation of on-farm/off-farm activities, etc.), a whole-farm approach is required to categorize these systems (Capillon 1986, 1993).

*What kind of farming systems classification?*

A pertinent farming systems classification for development purposes needs to provide understanding of the diversity of farmers' objectives and strategies that are feeding the farm diversification process (Sebillotte 1989). Their identification can start with the analysis of ecological and technological transformations in relation to socio-economic changes experienced by the local agricultural system in the recent past.

*An historical approach to farming systems diversity*

An historical approach helps to understand the nature, origins, causes, extent and consequences of key changes that led to the current diversity of farming systems. An historical profile of the local agriculture, identifying and dating how changes in the economic and social context influenced the transformations of ecological conditions and technical practices, was prepared from a review of secondary data and informal interviews with key informants. The format displayed in Figure 1 helped in understanding the origin of the current diversity of farmers' circumstances. Interpretation of such data leads to the identification of successive agrarian systems (Trébuil 1988) displaying coherent relationships between the prevailing ecological conditions, the dominating mode of exploitation of the environment, and the social conditions and needs during that period. In this case, a traditional agrarian system could be distinguished from a new one emerging after 1980. Such outputs are used to inventory the diversity of farming systems situations to be sampled, analysed and classified. This analysis of the pathways followed by the most dynamic and innovative farming systems was also used to identify evolution trajectories and anticipate future changes among the farming community.

*Analysis of farmers' strategies and farming systems functioning*

A diverse sample of farms is required to assess the objectives and strategies of the various types of farming systems. To analyse farmers' production systems is to elucidate the logical sequence of strategic and tactical decisions they are making, in relation to their families in a given set of constraints and potentials to achieve their objectives. It can be considered as part of the identification of the farmer's 'crop management action model' (Sebillotte 1994). The five aspects we analysed for this purpose were: (i) the description of the production combination; (ii) its relationships with the family and its objectives, the physical and socioeconomic environments, the labour force, the farm equipment and its history; (iii) identification of the determining factors and conditions of this combination; (iv) the presentation of the farmer's strategy; and finally (v) the presentation of the overall functioning of the farming system (Capillon and Manichon 1991). During the first two steps, the farmer's major

| Economic and social changes | Date | Agroecosystem transformations |
|---|---|---|
| **First settlement**: 42 Akha villagers migrate from the border because of insecurity | 1907 | Clearing part of the primary forest, **slash and burn agriculture** (upland rice, maize) |
| Migration from Ho Mae Cam<br>Total population: **37 families** (200 people) | 1954 | |
| Conflict concerning animal rearing: 20 families leave for Ban Tum Van (3 km) | 1965 | 1/3 of the area still covered by forest.<br>Rotation: 2 years of cultivation/7 year fallow |
| 10 families leave for Chiang Mai<br>Only **7 families left in Mae Salaep** | 1971 | |
| 6 families come from Humackam (Thailand) | 1973 | |
| Cattle disease → « bad spirit » on the village →10 families migrate from Mae Salaep Bon to Mae Salaep Lang (distance: 500 m)<br>**Construction of the road to Mae Salaep** | 1979 | **First subsistence problems**<br>Only small sacred forest left.<br>Rotation: 4 year cultivation/5 year fallow<br>UR yield: 50 kg harvested for 1 kg planted |
| **40 families** migrate from Praya Prai to Mae Salaep because of insecurity along border | 1981 | New settlers get progressively access to some land in the main watershed |
| Some villagers obtain Thai citizenship | 1982 | Maize becomes a cash crop |
| First agricultural and social development program held by TAHASD project | 1983 | First soil conservation measures (grass strips) and **paddy fields** introduced |
| | 1984 | Introduction of **groundnut as a cash crop** |
| Development of off-farm activities:<br>handicraft and hired labour in the lowlands | 1987 | Introduction of **ginger and cabbage as cash crops** |
| New agricultural and social development program | 1988 | Increase in the paddy area |
| | 1992 | Beginning of **fruit tree plantations** |
| **Intervention of public welfare department**<br>Electricity installed in the village<br>Some villages convert to Christianity | 1993 | **Reforestation project** by the Royal Forestry Department: concerns **50% of the total watershed area** |
| Increase in the number of villagers getting Thai citizenship<br>Total population:<br>Mae Salaep Lang: 72 families (250 people)<br>Mae Salaep Bon: 33 families (120 people) | 1994 | Beginning of reforestation.<br>High proportion of land under permanent cultivation.<br>UR yield: 20 kg harvested for 1 kg planted<br>35% of the land left under fallow. |

*Figure 1.* Historical profile of the Mae Salaep Lang agricultural system, Chiang Rai Province, upper northern Thailand.

strategic choices, the ones that cannot be modified at short notice and are leading to the selection of the farming system/subsystem production combination are identified. The determining factors of such a combination are understood through the characterization (constraints vs potentials) of the production system and its history, as well as biophysical and socioeconomic environments that either limit farmer choices or explain the occurrence of a given subsystem. The linkages between farmer's strategic choices and the farmer-family objectives and projects are displayed in the fourth step, leading to a diagram of the farming system functioning. This document is then used to propose a farmers' classification based on the different ways they produce.

## *Building a farming systems typology*

Based on the comparison and identification of similarities between the farming systems functioning diagrams, the main types of farming systems for the studied

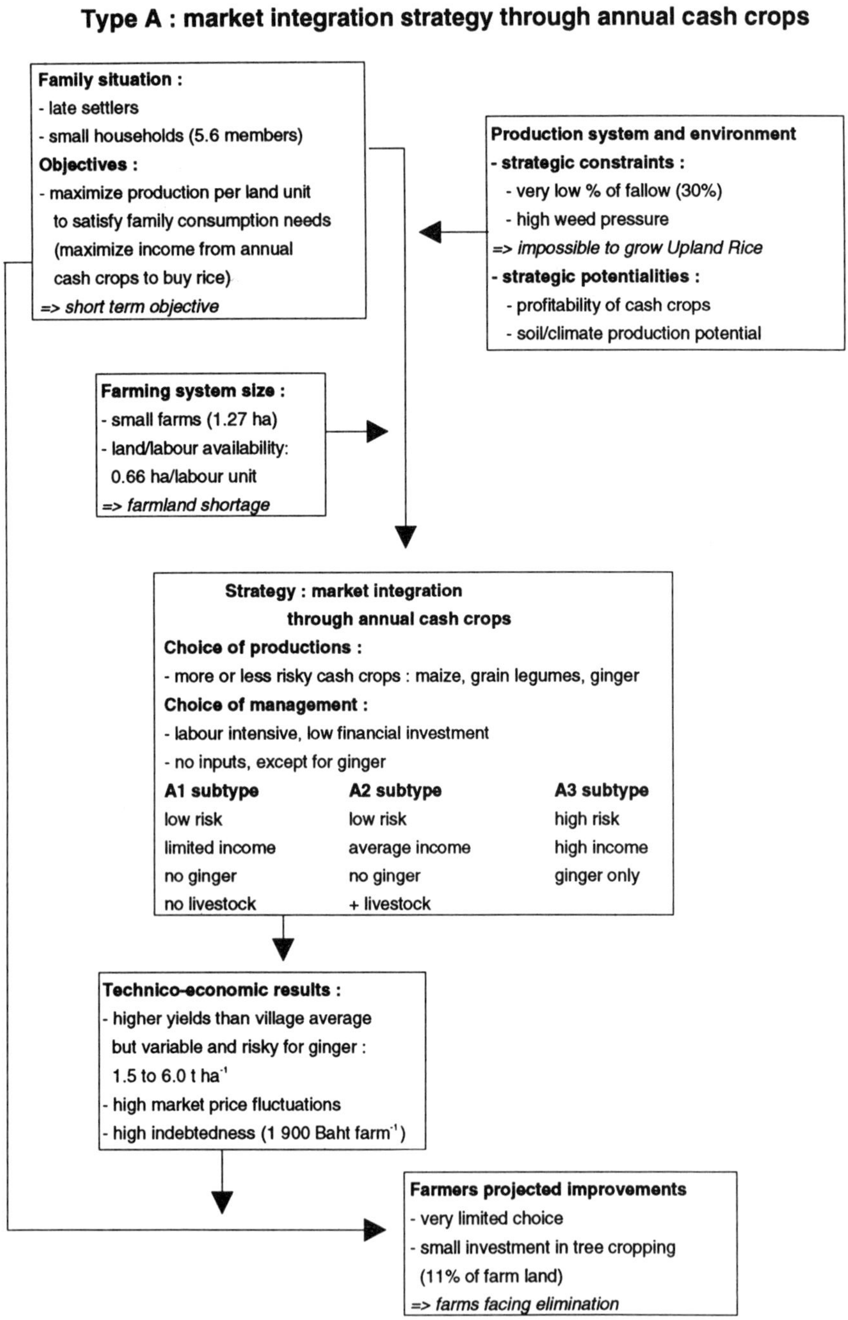


*Figure 2.* Diagrammatic presentation of the functioning of the three main types of farming systems in Mae Salaep Lang in Chiang Mai Province, upper northern Thailand, 1995.

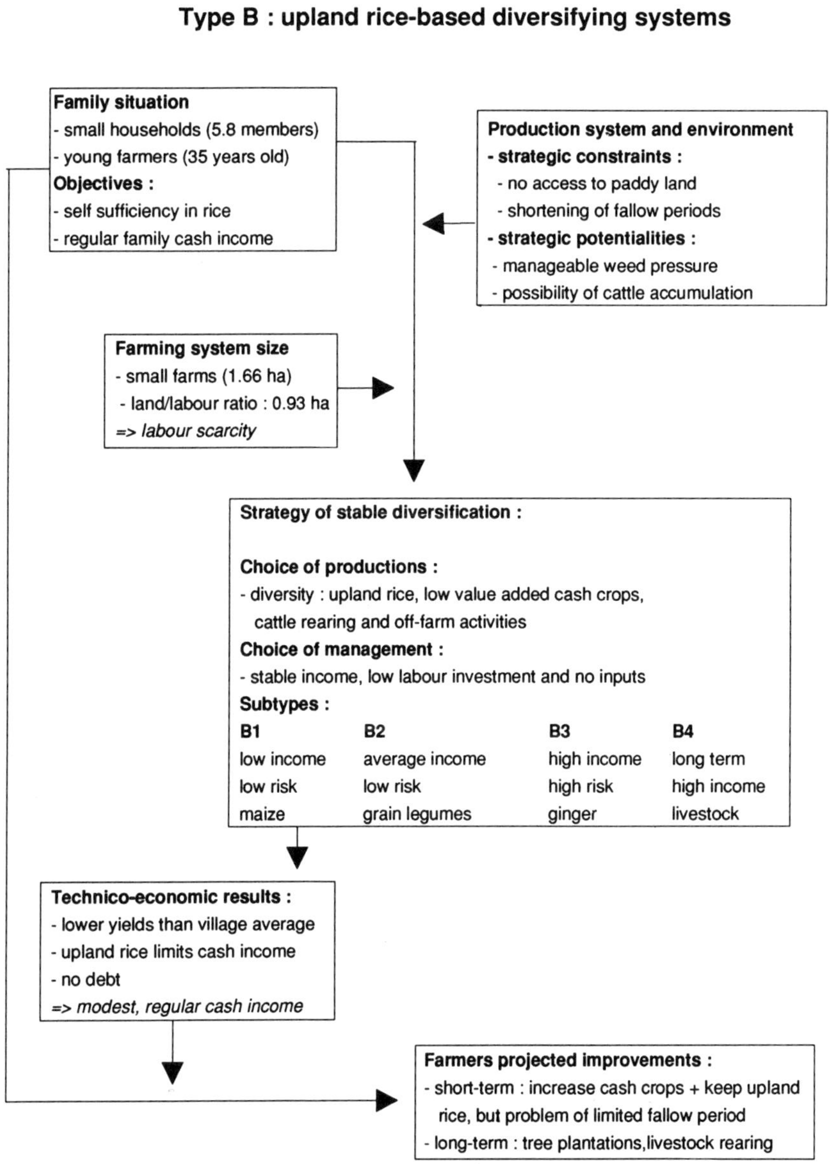


*Figure 2.* Continued.

**Type C : paddy rice-based investment and market oriented systems**

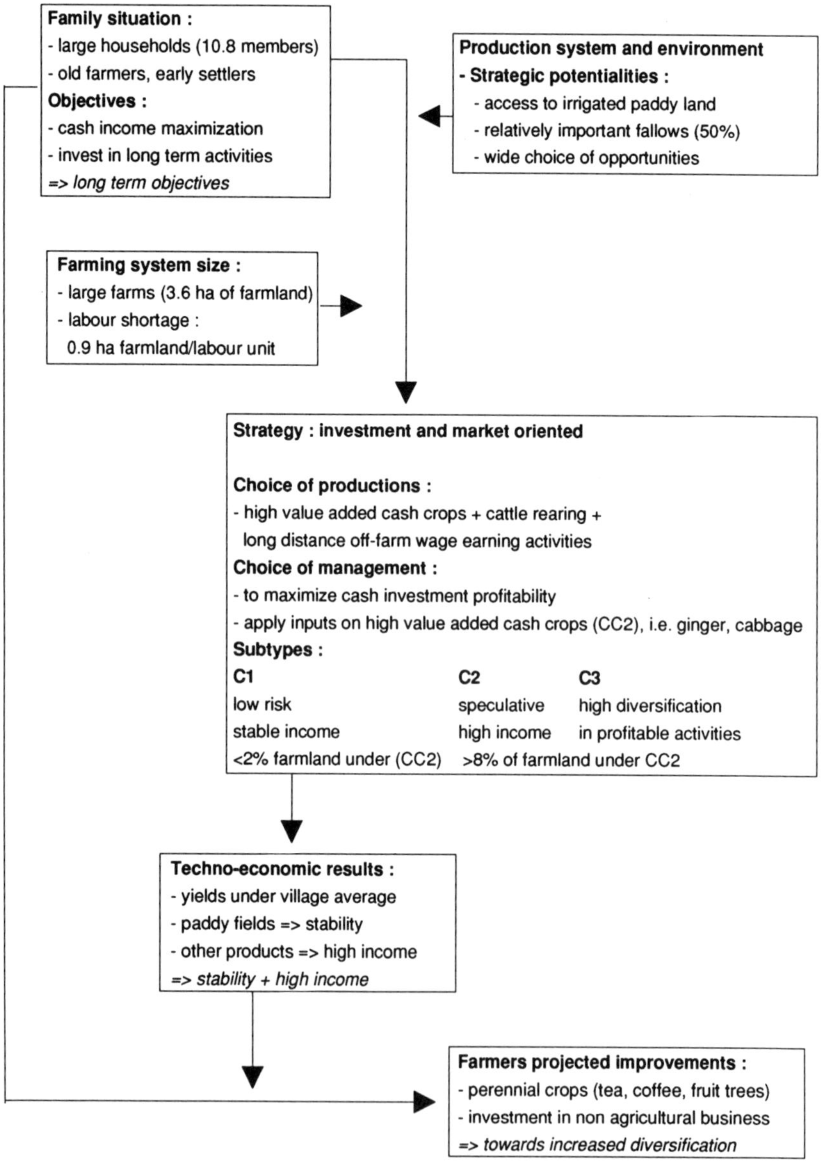


*Figure 2.* Continued.

*Table 4.* Assessment of the importance of soil erosion across the typology of farming systems (see also Figure 2) in Mae Salaep of Chiang Rai Province, upper northern Thailand, 1994.

| Sub-types | Cropping subsystems* | | | | | | Livestock | Off-farm | Risk of erosion[†] | No. of farms |
|---|---|---|---|---|---|---|---|---|---|---|
| | UR | FCC | HCC | IR | PCC | Fallow | | | | |
| A1 | | x | | | | | | | - | 2 |
| A2 | | x | x | | | | | | ++ | 2 |
| A3 | | x | x | | | | x | | + | 4 |
| B1–B2 | x | x | | | x | x | | x | + | 10 |
| B3 | x | x | x | | x | x | x | | ++ | 3 |
| B4 | x | | | | | | xx | x | - | 4 |
| C1 | | x | | x | xx | xx | | | -- | 5 |
| C2 | | x | x | x | xx | xx | | | + | 8 |
| C3 | | x | x | x | xx | xx | x | x | - | 2 |

*UR=upland rice, FCC=field cash crops, HCC=horticultural cash crops, IR=irrigated rice, PCC=perennial cash crops, Fallow=availability of fallow land; x=cropping subsystem is present and xx=important or key subsystem.
[†]Indicators --, -, +, ++: very low, low, limited and high risk, respectively.

area were grouped into a typology. Figure 2 shows the diagrams obtained for the three main types of farming systems in an Akha village of upper northern Thailand. A second level of variation in their strategies allowed the identification of several subtypes for each of these main categories according to farmers' different attitudes towards risks and their management. From these diagrams, key indicators were derived to assign rapidly, and with a reasonable level of accuracy, any given farming system to one of these main types and subtypes of farming systems.

Based on the knowledge of the susceptibility to erosion of the different cropping systems acquired at field level and the identification at farm level of the typical production combinations implemented by each main type of farming system, it was possible to rank them according to the importance of the soil degradation issue in their current functioning (Table 4). Subtypes A2 and B3, which represent 14% of the total number of farms in the study area, are the ones to be targeted for improving soil conservation. Their high level of susceptibility to soil erosion is here due to the importance of ginger (A2 and B3) and upland rice (B3) subsystems on the farms, as well as to the lack of other components, such as irrigated paddies, livestock production and off-farm activities, or relative abundance of fallow, that could ease the pressure on the sloping land. Another 60% of the farms (A3, B1, B2 and C2), while less affected by the soil erosion, could also benefit from such innovations. But for another 26% of the villagers, soil erosion is not a key issue, a statement we could hear from some farmers at the beginning of this research but that we can now explain.

During the construction of this kind of typology, linkages between the different categories of farming systems, particularly patterns of social relations of production, are also elucidated. Taking them into consideration when designing innovations helps to improve the effectiveness and impact of development-oriented systems research, particularly when dealing with resource-poor farmers having very little room to manoeuvre.

*Trajectories of evolution to target issues and forecast impact of interventions*

The construction of trajectories of evolution displaying linkages between farming system types is a step towards a regional model of the local agricultural system and its dynamics. Figure 3 shows such an output in the case of the study site in northern Thailand. These trajectories can be used to monitor changes across a limited network of reference farms, to assess the viability of these different categories of farms, to display dominant trends (like the contradictory process of accumulation-elimination among the local farming community), as well as to evaluate the impact of past interventions and to anticipate future possible transformations by looking at the choices of the most innovative producers.

## Synthesis at watershed level: land-use dynamics and spatial distribution of key issues

Trends as shown along the trajectories in Figure 3 can be used to help forecast the effects of the current rapid agricultural diversification on the soil erosion problem at the watershed level. The gradual disappearance of large upland rice fields, the smaller size (and so shorter slope lengths) of the horticultural cash crop fields and the increase in perennial plantation crops and paddies, as well as in off-farm employment opportunities helping to ease pressure on the land, are key changes indicating that diversification does not simply imply an increase in soil degradation problems in this watershed (Healy et al. 1995). The analysis of land-use dynamics conducted at the watershed level using GIS tools, while confirming these evolutions in land use, provided a spatial distribution and a quantification for most of these dominating trends (Table 5).

Following the integration of the results obtained from the diagnoses at field and farm levels, GIS tools were used to map the crucial areas and most risky situations regarding soil degradation at watershed level (Figure 4). They comprise upland rice and horticultural cash crop fields belonging to A2 and B3 types of farming systems and located on steep land (slope angle higher than 30% and for all slope lengths). Such a 'conflict map' can be used for assessing the potential impact of interventions, and to improve the inter-institutional dialogue between stakeholders to decide, for example in this case, what areas of the watershed should be reforested in agreement with the government policy.

## Conclusions

The proposed integrated approach for development-oriented research on rapidly diversifying highland cultivated ecosystems demonstrates the key role played by the initial diagnostic analysis of actual conditions of production to identify yield-limiting factors and conditions of environmental degradation (Picard 1995). The understanding of effects of the diversity of farmers' practices on the functioning of crop populations and the status of the resource base provides a platform for making choices and improving the dialogue between key institutional stakeholders concerning access to

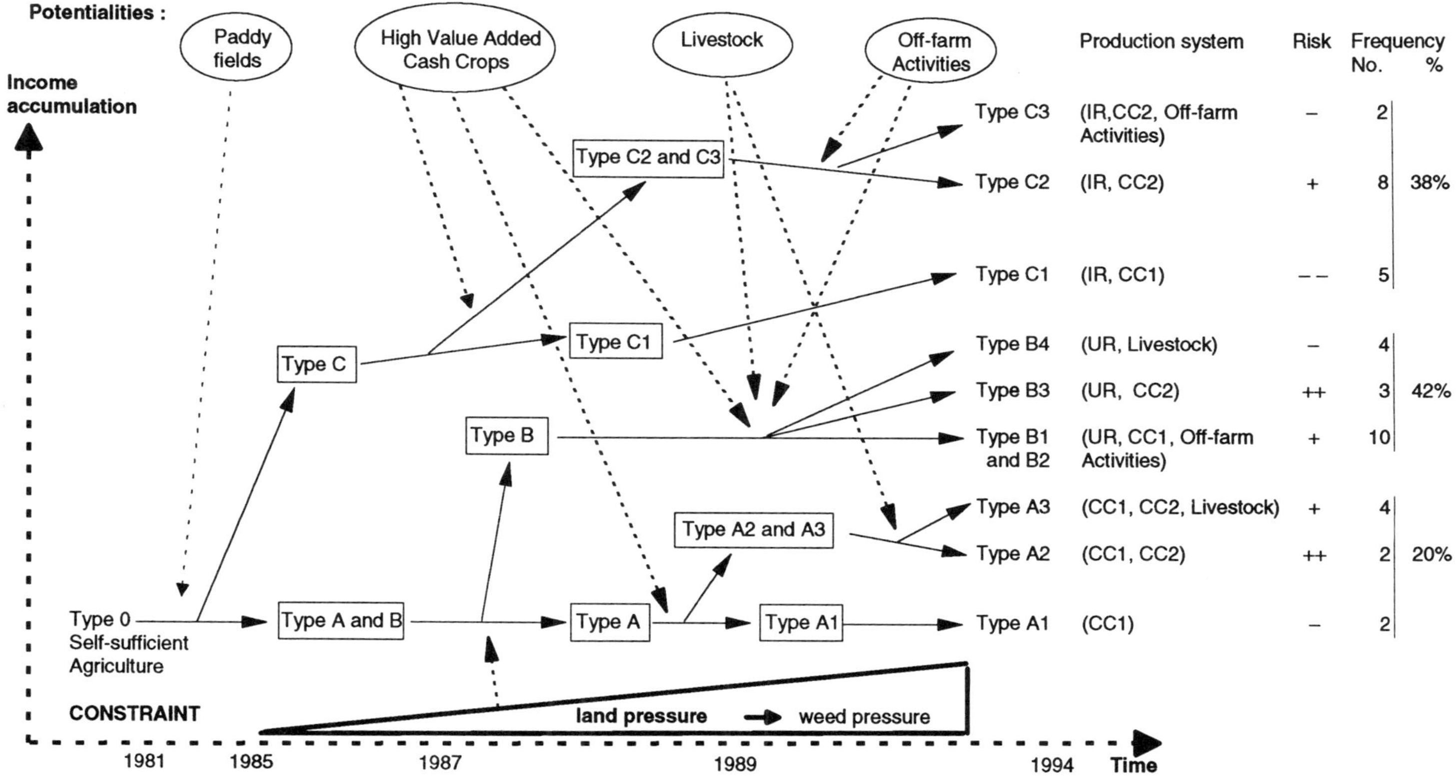


IR : Irrigated Rice
UR : Upland Rice

CC1 : Low Value Added Cash Crops
CC2 : High Value Added Cash Crops

*Figure 3.* Trajectories of evolution for farming systems in Mae Salaep Lang, Chiang Rai Province, upper northern Thailand.

*Table 5.* Changes (in hectares) in land-use types between 1990 and 1994 wet seasons in Mae Salaep Lang of Chiang Rai Province, upper northern Thailand.

| Land-use types | 1990 WS | 1994 WS | % change |
|---|---|---|---|
| Upland rice | 70.8 | 11.6 | −84 |
| Irrigated paddies | 14.0 | 16.5 | +18 |
| Field cash crops | 39.7 | 32.3 | −19 |
| Horticultural cash crops | 0 | 4.4 | n.a. |
| Perennial cash crops | 8.3 | 11.2 | +35 |

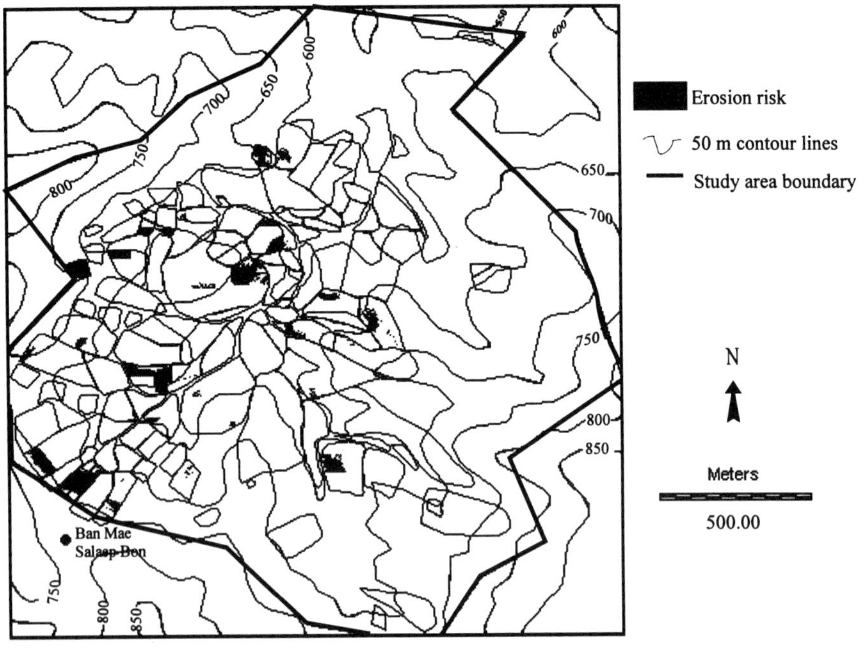


*Figure 4.* Farmers' practices and soil conservation conflict map in Mae Salaep Lang watershed, Chiang Rai Province, upper northern Thailand, 1994 wet season.

natural resources in agreement with the social demand. The diagnosis outputs can also be used to conceive and design adapted interventions that will be well integrated into the ecological-socioeconomic environment. In particular, interventions should take into account the strategies of target beneficiaries, as well as the specific constraints they are facing, in order to improve the linkages between the technical and socioeconomic dimensions of agricultural development issues. Based on the understanding of the functioning of the local agricultural system provided by the initial diagnosis, the potential impact of the adoption of new technological models in various niches can be assessed. This sound integrated approach and an emphasis on the management of agroecosystem diversity favours the emergence of 'double-green revolution' innovations (Griffon and Weber 1995), allowing an increase in productivity while preserving the natural resource base in those marginal lands.

Lastly, if complementary analyses of key commodity chains (listing all the key players to determine the linkages between their activities and the management of natural resources) and policies are added, the proposed approach leads to a functional model of the regional agricultural system. This would be a fundamental output towards an ecoregional approach to systems research for agricultural development (Manichon 1995).

## References

Boiffin J, Papy F, Eimberck M (1988) Influence des systèmes de culture sur les risques d'érosion par ruissellement concentré. I. Analyse des conditions de déclenchement de l'érosion. Agronomie 8(8): 663–673.

Byerlee D, Triomphe B, Sebillotte M (1991) Integrating agronomic and economic perspectives into the diagnostic stage of on-farm research. Exp. Agric. 27:95–114.

Capillon A (1986) A classification of farming systems, preliminary to an extension program: Methodology. Pages 219–235 in Farming systems research paper series 11. Kansas State University, Manhattan, KA, USA.

Capillon A (1993) Farming systems typologies and their contribution to regional development (in French, English summary). PhD thesis, National Agronomic Institute Paris-Grignon, France. 48 p.

Capillon A, Manichon H (1991) Guidelines for the study of farming systems for agronomists (in French). Adeprina/Apca, National Agronomic Institute Paris-Grignon, France. 65 p.

Combes L, Picard D (1992) Yield build-up of main annual crops (in French). INRA editions, Paris, France. 191 p.

Crozat Y, Chitapong P (1988) The on-farm agronomic survey, a tool for grading limiting factors of the yield. Pages 87–110 in Trébuil G (Ed.) Farming systems research and development in Thailand. GRET/Prince of Songkla University, Hat Yai, Thailand.

Griffon M, Weber J (1995) The doubly-green revolution: Economy and institutions. Paper presented at the International Seminar Towards a Doubly-Green Revolution, 8–9 November 1995, Poitiers, France. 17 p.

Healy S, Thong-Ngam C, Turkelboom F (1995) Analysis of the diversification of agricultural production systems in upper northern Thailand: Influence on land degradation (in French, English summary). ESAT-ENESAD dissertation, CNEARC, Montpellier, France. 89 p.

Jouve P (1984) The agronomic diagnosis as a preliminary step to development-oriented research projects (in French, English summary). Cahiers de la recherche-développement 3/4:67–75.

Leterme P, Manichon H, Roger-Estrade J (1994) Integrated analysis of wheat yields and their causes of variation in a network of farmers' fields in Thymerais (in French, English summary). Agronomie 14: 341–361.

Manichon H (1995) Ecoregionality in research for development: French research proposals. Paper presented at the International Seminar Towards a Doubly-Green Revolution, 8–9 November 1995, Poitiers, France. 7 p.

Meynard J M, David G (1992) Diagnosis on the yield elaboration of crops (in French, English summary). Cahiers Agric. 1:9–19.

Picard D (1992) The future of crop physiology. Paper presented at the 2nd Congress of the European Society for Agronomy, 24-28 August 1992, Warwick, UK. 4 p.

Picard D (1995) Future technology for the doubly-green revolution. Paper presented at the International Seminar Towards a Doubly-Green Revolution, 8–9 November 1995, Poitiers, France.

Sebillotte M (1978) Itinerary of techniques and evolution of the agronomic thought (in French, English summary). Comptes-rendus de l'Académie d'Agriculture Française 64:906–916.

Sebillotte M (1989) Approaches of the on-farm agronomist: Illustrated methodological considerations. Development-Oriented Research on Agrarian Systems Project, Kasetsart University, Bangkok, Thailand. 32 p.

Sebillotte M (1990a) Cropping system: An operational concept for agronomists (in French). Pages 165–196 in Combe L, Picard D (Eds.) Les systèmes de culture. INRA, Versailles, France.

Sebillotte M (1990b) Some concepts for analysing farming and cropping systems and for understanding their different effects. Paper presented at the Inaugural Congress of the European Society of Agronomy, 5–7 December 1990, Paris, France. 16 p.

Sebillotte M (1994) Systems research and action (in French). Keynote address to the International Symposium on Systems Research in Agriculture and Rural Development, 21–25 November 1995, Montpellier, France. 29 p.

Thong-Ngam C, Shinawatra B, Healy S, Trébuil G (1995) Farmers' resource management and decision-making in the context of changes in the Thai highlands. Pages 462–487 in Proceedings of the Regional Symposium on Montane Mainland Southeast Asia in Transition, 12–16 November 1995, Chiang Mai, Thailand.

Trébuil G (1988) Principles and steps of the method of diagnosis on agrarian systems: A case study from Sathing Phra in southern Thailand. Pages 27–62 in Trébuil G (Ed.) Farming systems research and development in Thailand. GRET/Prince of Songkla University, Hat Yai, Thailand.

Trébuil G (1992) Setting research priorities with farmers. Paper presented at the Second Asian Farming Systems Symposium, 2–5 November 1992, Colombo, Sri Lanka. 19 p.

Trébuil G, Dufumier M (1993) Diagnoses on regional agrarian systems and sustainability of agricultural production systems in Thailand. J. Asian Farming Syst. Assoc. 4:557–568.

Turkelboom F, Trébuil G, Vejpas C (1995) Starting from the farmers' fields: On-farm analysis and development of conservation strategies for steep land. Paper presented at the IBSRAM International Workshop on Soil Conservation Extension, 4–11 June 1995, Chiang Mai, Thailand. 13 p.

Van Keer K, Vejpas C, Trébuil G (1995) Effects of fallow type and farmers' practices on weed infestation in an upland rice based swidden system in northern Thailand. Paper presented at the 2nd Weed Science Conference, 26–27 May 1995, University of Peradenyia, Sri Lanka. 14 p.

# Managing dairy-crop production systems for zero-excess phosphorus: a multi-criteria optimization approach

E.C. ALOCILJA[1], R.D. VON BERNUTH[1] and D.K. BEEDE[2]

[1]*Department of Agricultural Engineering, Michigan State University, East Lansing, MI 48824, USA*

[2]*Department of Animal Science, Michigan State University, East Lansing, MI 48824, USA*

**Key words:** phosphorus, multi-criteria optimization, manure management, dairy operation

**Abstract**

Dairying in Michigan is a major industry, accounting for half of all livestock receipts and one-fifth of all farm cash receipts. However, industry expansion is constrained by the challenging task of disposing of phosphorus-rich manure without it becoming a nuisance and a liability. A multi-criteria optimization-based phosphorus balance model for a dairy-crop operation is described in this paper. A typical medium-sized dairy farm with sufficient land area for crop production to support dairy feed production was analysed. In the model, manure is used as a fertilizer nutrient source in crop production; animal weight gain, reproduction, and import and export of nutrients through fertilizer, purchased feed, milk, and animal sales are accounted for. Rations for animals leading to zero excess manure phosphorus at best economic return are identified by simultaneously minimizing excess manure phosphorus, feed cost and cropland requirement.

## Introduction

The business of food production is struggling to survive in a populous world demanding more food supply of superior quality and at low cost. Under these circumstances, many food producers try to satisfy the demand by methods which potentially could compromise environmental quality, the very source of future food supply. For food production to be sustainable in future generations, producers may have to re-engineer system structures and management operations to satisfy conflicting objectives between profit maximization and environmental protection under land, labour, energy, technology and biological constraints.

One food sector of concern in this paper is the dairy-crop production system and its use of the nutrient phosphorus (P). Plants and animals require P for healthy growth, high yield, and increased output. On the other hand, P is also a major element in animal manure contributing to environmental problems. Phosphorus-loaded farm soils may be eroded during heavy rainfall and washed to rivers, lakes and other bodies of water as sediments, resulting eventually in lake eutrofication. In a study between 1990 and 1992, contaminated sediments contributed greatly to the degraded water quality in the Saginaw River and Saginaw Bay of Michigan (CGLAS 1993). During this study, mean total P concentration in suspended solids was very high in the Saginaw River (4043 $\mu g\, g^{-1}$). At this time, there were extensive blue-green algal blooms occurring in the river. In 1987, loading of P in the Saginaw Bay was estimated to be 665 tons, 50% over the limit set by the Great Lakes Water Quality Agreement between the USA and Canada (MUCC 1993). A large source of non-point P comes



*P.S. Teng* et al. *(eds.): Applications of Systems Approaches at the Farm and Regional Levels, 115–128.*

from agricultural operations involving livestock manure and commercial fertilizer (MUCC 1993).

Dairy production is an important industry in Michigan. In 1991, dairying accounted for half of all livestock receipts and one-fifth of all farm cash receipts (MDA 1993). With annual production of milk close to 2.5 billion kilograms (MDA 1993) and with almost three times as much manure as milk being produced from the milking cows alone (MPS 1985), disposal of dairy manure is a major consideration. With projected increasing livestock densities into the twenty-first century (Lanyon 1992), environmental impact of excess nutrient P will become even more of an important issue.

The Michigan Right to Farm statute (P.A. 93, enacted in 1981) authorized the Michigan Commission of Agriculture to develop generally accepted agricultural and management practices to promote environmental stewardship and maintain a person's right to farm (MCA 1993). One major area of recommended practice is the application of manure to cropland as fertilizer source (MCA 1995). However, if the soil P test (Bray P1) reaches 150 lb (70 kg) $acre^{-1}$, manure application should be reduced to a rate where manure P added does not exceed the P removed by the harvested crop. If the Bray P1 test reaches 300 lb (140 kg) $acre^{-1}$ or higher, manure applications should be discontinued until nutrient harvest by crops reduces P test levels to less than 300 lb $acre^{-1}$.

Recycling of manure as a fertilizer nutrient source in crop production has been recommended widely to minimize the problem of nutrient loading in surface and groundwater. Henry et al. (1995) developed a linear programming model for dairy nutrient management that includes nutrient recycling. The model minimized total feed cost based on the ability of crops to take up manure nutrients and supply forages for cattle rations. Westphal et al. (1989) and Lanyon (1992) used a linear programming model to study the relationship of nutrient management strategies to optimal herd size and net farm return. However, most dairy farms operate under land constraints, so that there may not be enough crop production acreage for total manure nutrient use (Van Horn et al. 1994). Total manure nutrient use is a management strategy in which all nutrients are applied at rates equal to crop requirements.

## Objectives and methodology

The objective of this research was to develop a multi-criteria optimization-based phosphorus Nutrient Balance Model of a Dairy-crop operation (hereafter referred to as the NBD model) that can be used as a tool in identifying economically sound and environmentally friendly animal rations and cropping systems that allow for total manure use at lowest feed cost on minimum cropland.

### *Multi-criteria optimization*

Process network theory (Alocilja 1990a, b) provides the system structure to formulate the optimization. The NBD model of a dairy farm is partitioned into the following components: commercial fertilizer application, crop production, soil profile, surface run-off, feed supplement, feed ration formulation, dairy herd, manure handling, manure storage and manure spreading. The dairy herd includes several animal groups from calves to lactating cows. A generalized system structure of the model is illustrated in Figure 1. The components are interconnected through the flow of materials per unit time. Although the physical form of the materials undergoes transformation during its flow from fertilizer to manure, the nutrient P is tracked in its elemental form.

The NBD model is a generalized representation of a typical dairy production system where the producer grows most of the feeds to support a dairy herd for milking and herd replacements. Feed supplement is provided when necessary. Manure is spread to the crop fields as a source of nutrient P. Whenever the P supply in this form is insufficient, commercial fertilizer is added. Manure handling and storage are assumed to have no losses. Phosphorus exits the system as run-off, excess manure, milk, cull animals, dead animals, afterbirth and fluids, and animals sent to market. Applied P not taken up by the crop and not leaving the system as run-off is stored in the soil profile. A run-off factor is included to account for variability in soil type and slope of the landscape. When the run-off factor results in a large nutrient loss, the remaining applied P could be less than the P needs of the crop. In this situation, the crop is allowed to 'mine' the soil for its nutrient needs. It is assumed that no significant loss of biomass occurs during harvesting and processing of feeds. Each unit of feed is assigned a market value to represent production cost. Costs of collection, storage and handling are not part of the economic analysis. The dairy operation is analysed over a period of one year at steady-state. Crop and animal management used in the model follow recommended practices (MSU 1992a, b).

*Objective functions*

The NBD model provides a mechanism to identify the best feed rations and cropping systems such that three objectives can be satisfied simultaneously: minimize excess manure P ($z_1$), minimize feed cost ($z_2$) and minimize cropland ($z_3$), subject to constraints. The optimization technique uses compromise programming which seeks to find a solution as close as possible to the ideal point (Romero and Rehman 1989) – the ideal point being defined as the ideal minimum values for excess manure P ($z_1^*$), feed cost ($z_2^*$) and cropland ($z_3^*$). The objective functions are normalized by introducing the anti-ideal points $z_{1*}$, $z_{2*}$ and $z_{3*}$, which represent undesirable situations in terms of excess manure, feed cost and cropland, respectively. The optimization algorithm is set up in the following manner:

$$\text{Min}\, w_1 \frac{z_1^* - z_1(x)}{z_1^* - z_{1*}} + w_2 \frac{z_2^* - z_2(x)}{z_2^* - z_{2*}} + w_3 \frac{z_3^* - z_3(x)}{z_3^* - z_{3*}} \tag{1}$$

s.t.

$$\sum_{i=1}^{3} w_i = 1.0$$

$$x \geq 0$$

$$x \in F$$

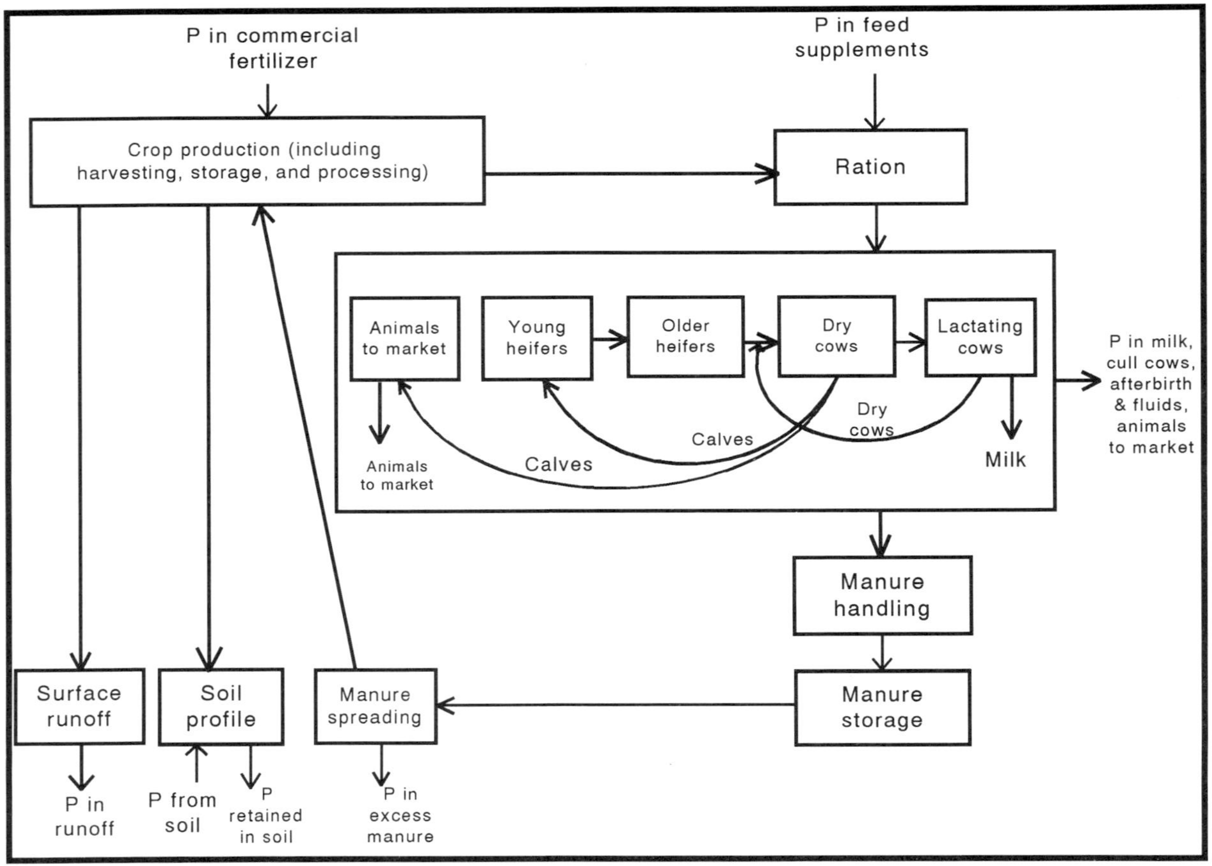


*Figure 1.* A generalized system structure of a typical dairy-crop operation.

where $x$ is a vector of control variables defined as the feedstuffs in the set of feasible solutions defined by $F$. The weights $w_1$, $w_2$ and $w_3$ assign the relative importance of one objective with the other. The three objective functions are expressed analytically as follows on a per animal basis:

1 Excess manure P, in kilogram (kg).

$$z_1(x) = \left(\sum_{i=1}^{n} P_i \times x_i - P_m \times my - P_a \times dg\right) - \left(\sum_{i=1}^{j} \frac{Pf_i}{Pr_i \times y_i} P_i \times x_i\right) \qquad (2)$$

2 Feed cost, in US dollar ($).

$$z_2(x) = \sum_{i=1}^{n} c_i \times x_i \qquad (3)$$

3 Cropland, in hectare (ha).

$$z_3(x) = \sum_{i=1}^{j} \frac{x_i}{y_i \times DM_i} \qquad (4)$$

The $P_i$ in equation (2) represents the P content of feed crop $i$, in kg P kg$^{-1}$ biomass, as a ratio of the P removal ($Pr_i$) and dry matter ($DM_i$) content of feed crop $i$, that is:

$$P_i = \frac{Pr_i}{DM_i} \qquad (5)$$

The $Pf_i$ in equation (2) represents the fertilizer P recommendation of feed crop $i$, in kg P ha$^{-1}$, as a function of crop yield goal ($y_i$), in kg ha$^{-1}$ biomass, and the soil P content of field $i$ ($Ps_i$), that is:

$$Pf_i = \alpha_i + \beta_i \times y_i - 1.25 \times Ps_i \qquad (6)$$

where $\alpha_i$ and $\beta_i$ are the regression coefficients in the fertilizer equation (equation (6)).

The other variables in equations (2) through (4) are $x_i$, which represents crop/feed $i$, $i = 1, 2, \ldots, j$ as farm-grown feed crops and $i = j+1, \ldots, n$ as purchased feed supplements; $P_m$, the P content of milk in kg P kg$^{-1}$ biomass; $my$, milk yield in kg d$^{-1}$; $P_a$, the P content of animal body in kg P kg$^{-1}$ body weight; $dg$, the daily weight gain in kg; and $c_i$, the cost of feed $i$ in \$ kg$^{-1}$ biomass. The following feed crops and supplements are labelled as follows: 1 = alfalfa silage; 2 = high-moisture corn; 3 = corn silage, 4 = soybeans; 5 = dry corn; 6 = orchardgrass hay; 7 = dicalcium phosphate; 8 = soybean meal; 9 = urea; 10 = blood meal; and 11 = calcium carbonate.

*Constraint functions*

Daily ration is formulated based on the dry matter intake (DMI) and nutrient requirements of each animal group according to the equations outlined in the Spartan Dairy Ration Evaluator/Balancer (MSU 1992b). The following nutrients are included in the constraint functions: net energy for lactation (NE1), net energy for maintenance

(NEm), net energy for gain (NEg), crude protein (CP), neutral detergent fibre (NDF), calcium (Ca) and phosphorus (P). The DMI and nutrient requirements of each animal group vary according to several factors, namely: body weight (*bw*), daily weight gain (*dg*), fat-corrected milk (*fm*), days in milk (*dm*), daily milk yield (*my*), percent milk fat (*mf*), days till freshening (*df*) and gestation factor (*gf*). Although the nutrient requirement also is a function of body weight gain (reserves), the net gain in reserves is assumed to be zero over a period of one year. The nutrient requirements of a dry cow and a pregnant heifer are similar to that of a lactating cow without milk yield, except that the pregnancy requirement, in the form of *gf*, is added to NE1 and CP. The *gf* is a function of *df*, that is:

$$gf = \frac{1.1-500}{284-df} + \frac{70000}{(284-df)^2} \tag{7}$$

The constraint functions are formulated as follows, where the actual calculated amount is on the left-hand side of the equation and the requirement or constraint is on the right-hand side of the equation.

1 Actual ration (left-hand side of the equation) is equal to the daily dry matter intake (DMI) requirement (right-hand side of the equation).

$$\sum_{i=1}^{n} x_i = a1(0.02 \times bw + 0.3 \times my(0.15 \times mf + 0.4) + dg) \tag{8}$$

where

$$\begin{bmatrix} a1 = 0.8 + \dfrac{0.2 \times dm}{80} & dm \leq 80 \text{ (for lactating cow)} \\ a1 = 0.75 + \dfrac{0.2 \times df}{60} & df \leq 60 \text{ (for dry cow)} \\ a1 = 1 & \text{otherwise} \end{bmatrix} \tag{9}$$

2 Actual NE1 content of ration is at least equal to or greater than the NE1 requirement.

$$\sum_{i=1}^{n} a_i \times x_i \geq 0.08 \times bw^{0.75} + my(0.3512 + 0.0962 \times mf) + 5.12 \times dg + \frac{0.77}{gf} \tag{10}$$

where $a_i$ is the NE1 content of feed $i$, in megacalorie (Mcal) $kg^{-1}$ biomass. For the dry cow and pregnant heifer, the right-hand side of equation (10) is not allowed to be less than 1.1×DMI or more than 1.7×DMI.

3 Actual NEm content of ration is at least equal to or greater than the NEm requirement.

$$\sum_{i=1}^{n} s_i \times x_i \geq 0.086 \times bw^{0.75} \tag{11}$$

where $s_i$ is the NEm content of feed $i$, in Mcal $kg^{-1}$ biomass.

4 Actual NEg content of ration is at least equal to or greater than the NEg requirement.

$$\sum_{i=1}^{n} t_i \times x_i \geq 0.035 \times bw^{0.75} \times dg^{1.119} + dg \tag{12}$$

where $t_i$ is the NEg content of feed $i$, in Mcal kg$^{-1}$ biomass.

5 Actual CP content of ration is at least equal to or greater than the CP requirement.

$$\sum_{i=1}^{n} g_i \times x_i \geq 0.00683 \times bw^{0.668} + my(1.9 + 0.4 \times mf)/39 + 0.65 \times dg + \frac{0.2}{gf} \tag{13}$$

where $g_i$ is the CP content of feed $i$ (%). For the dry cow and pregnant heifer, the right-hand side of equation (13) cannot be less than 13% or greater than 18% of DMI.

6 Actual NDF content of ration is at least equal to or greater than the NDF requirement.

$$\sum_{i=1}^{n} q_i \times x_i \geq \begin{bmatrix} 0.30\ \text{DMI}, & rfm < 30 \\ 0.29\ \text{DMI}, & 30 \leq rfm < 35 \\ 0.28\ \text{DMI}, & 35 \leq rfm < 40 \\ 0.27\ \text{DMI}, & rfm \geq 40 \end{bmatrix} \tag{14}$$

where $q_i$ is the NDF content of feed $i$ (%), and the relative fat-corrected milk ($rfm$) is:

$$rfm = \frac{my(0.15 \times mf + 0.4)}{(bw/600)^{0.75}} \tag{15}$$

7 Actual Ca content of ration is at least equal to or greater than the Ca requirement.

$$\sum_{i=1}^{n} r_i \times x_i \geq \frac{a6 \times 0.0154 \times bw + 1.34 \times my(0.15 \times mf + 0.4)}{0.38 \times 1000} \tag{16}$$

where $r_i$ is the Ca content of feed $i$ (%). However, actual Ca content cannot exceed 2% of DMI, that is:

$$\sum_{i=1}^{n} r_i \times x_i \leq 0.02 \times \text{DMI} \tag{17}$$

8 Actual P content of ration is at least equal to or greater than the P requirement.

$$\sum_{i=1}^{n} P_i \times x_i \geq \frac{a6 \times 0.0143 \times bw + 0.99 \times my(0.15 \times mf + 0.4)}{0.5 \times 1000} \tag{18}$$

where $P_i$ is the $P$ content of feed $i$ (%). However, actual $P$ content cannot exceed 1%

of DMI, that is:

$$\sum_{i=1}^{n} P_i \times x_i \leq 0.01 \times \text{DMI} \tag{19}$$

9 Constraints imposed on certain feeds.

$$\begin{aligned} 0.25 \times \text{DMI} \leq x_2 + x_5 \leq 0.4 \times \text{DMI} \\ x_3 \leq 0.5 \times \text{DMI} \\ \begin{bmatrix} x_6 \leq 0.2 \times \text{DMI} & \text{for dry cows, pregnant heifers} \\ x_6 \leq 0.1 \times \text{DMI} & \text{otherwise} \end{bmatrix} \\ x_9 \leq 0.005 \times \text{DMI} \\ x_{10} \leq 0.01 \times \text{DMI} \end{aligned} \tag{20}$$

10 Dry matter manure P (PM), in kg, is non-negative.

$$\text{PM} = \sum_{i=1}^{n} P_i \times x_i - P_m \times my - P_a \times dg \geq 0 \tag{21}$$

Rations are prepared for seven animal groups, namely: non-replacement animals (average body weight = 178 kg), young heifers (178 kg), old heifers (430 kg), pregnant heifers (568 kg), first lactating cows (589 kg), dry cows (658 kg) and mature lactating cows (627 kg). The non-replacement animals are fed for a year and then sold.

*System parameterization and assumptions*

The dairy-crop system in the model is formulated based on several assumptions, namely: the P content of milk is 0.1% (Van Horn et al. 1994); the P content of animal bodies is 0.194% and the P content of afterbirth and fluids is 0.47% (Van Saun 1991); a dry cow or pregnant heifer delivers one calf; new-born calves are 50% female and 50% male; 67% of DMI is digested, while 33% is excreted as manure. The P contents of the feed supplements are assigned according to their feed composition description by the NRC (1989). The amount of nutrient P that may be applied to a field cannot exceed the nutrient P requirement of the crop.

## NBD model validation

The NBD model is validated through the crop production and lactating cow subsystems. The integrated dairy-crop system is validated through a demonstration of nutrient P below.

The first validation of the model is on the P content of the feeds grown in the farm. Table 1 shows a comparison on the P content of the feeds between the simulated results and the values used by NRC. Except for orchardgrass hay, which has 24% and 37% variability from the NRC values for early and late bloom respectively, the other five feedstuffs have an average of 9% difference from NRC values.

*Table 1.* A comparison of the P content of feeds predicted by the NBD model and values used by the National Research Council (NRC 1989).

| Feed | P content of feed by NBD model (%) | P content of feed by NRC (%) | Variability of model from NRC values (%) |
|---|---|---|---|
| Alfalfa silage, early bloom (35% DM) | 0.20 | 0.22 | 9 |
| Corn silage, well-eared (33% DM) | 0.24 | 0.22 | 9 |
| High-moisture corn (77% DM) | 0.34 | 0.32 | 6 |
| Dry corn grain, cracked (89% DM) | 0.31 | 0.29 | 7 |
| Soybeans (92% DM) | 0.75 | 0.65 | 15 |
| Orchardgrass hay, sun-cured, early bloom (89% DM) | 0.42 | 0.34 | 24 |
| Orchardgrass hay, sun-cured, late bloom (91% DM) | 0.41 | 0.30 | 37 |

*Table 2.* A comparison between the model output and data from Van Horn et al. (1994) of a 635 kg Holstein cow producing 22.7 kg d$^{-1}$ milk and consuming 17.8 kg d$^{-1}$ DM (50% corn silage).

| Variable | NBD model output | Data from Van Horn et al. (1994) |
|---|---|---|
| Dry matter intake (DMI) (kg d$^{-1}$) | 19.2 | 17.8 |
| Manure produced per cow (kg DM d$^{-1}$) | 6.33 | 5.92 |
| P content of faeces (% DM) | 0.94 | 0.84 |
| Diet composition | 0.43% P<br>0.54% Ca<br>12.8% CP | 0.43% P<br>0.82% Ca<br>14.5% CP |

The second validation is on the lactating cow subsystem using a compiled data of a typical Holstein cow from the paper by Van Horn et al. (1994). A 635 kg Holstein cow, producing 22.7 kg d$^{-1}$ milk and consuming 17.8 kg d$^{-1}$ DM (0.43% P content; 50% corn silage), was used in the validation. The model was run for a 635 kg Holstein cow producing 22.7 kg d$^{-1}$ milk, where the ration was 50% corn silage and the P diet was 0.43%. In the absence of additional information from Van Horn et al. (1994), the following parameter values were supplied to run the model: the cow was in its second lactation, on average, 150 days in milk (to estimate feed requirement), 3.5% milk fat and daily gain of 0.14 kg. Table 2 shows a comparison of DMI, amount of manure produced, P content of manure and diet composition in terms of P, Ca and CP (crude protein) between the model output and the data from Van Horn et al. (1994).

## Results and discussion

The dairy-crop farm is analysed with the following scenarios: each field has a soil P of 34 kg ha$^{-1}$, run-off rate is 1%, 100 milking cows are to be maintained, average milk yield is 30 kg d$^{-1}$; milk fat is 3.5%; milking cows are 150 days in milk on average, dry cows and pregnant heifers are in a stage of 30 days till freshening; culling rate of mature lactating cows is 25%; and mortality is zero. Results of the model show that, in order to maintain a Holstein milking cow herd of 100, a total of 220 animals are in the farm, categorized as follows: 100 milking cows, 20 older heifers and 20 young heifers as replacement, and 80 young animals (30 heifers and 50 bulls) to be sold to the market. Twenty of the milking cows are in the first lactation, whereas 80 are in their second and greater lactation.

*Minimum cropland strategy*

Due to land constraints, many dairy operations have more cows per unit land area than desired for environmental sustainability. Less and less feeds are grown in the farm resulting in more and more excess manure. The model was run with the objective of finding the best ration for each of the seven animal groups, and corresponding cropping system, that lead to a minimum cropland requirement. This problem is equivalent to running the model with the third objective function (minimize cropland) only in effect. The weights assigned to the objective functions were $w_1=0$, $w_2=0$ and $w_3=1$. The results show that 78 ha are sufficient to maintain 220 animals. The formulated rations of some selected animal group (mature lactating cows, dry cows, older heifers and young heifers) are shown in Table 3. The total feed cost from this strategy is \$130 866 $yr^{-1}$. Feed supplements are 41% of the total ration. The cropping system, consisting of corn (silage), corn (grain) and orchardgrass, requires a total of 1480 kg of nutrient P for plant growth. A total of 4122 kg manure P are produced by the 220 animals and the average P content of the dry matter manure is 1.34%. This strategy results in an excess manure P of 2642 kg that could be a potential nuisance to the environment. If confined to the farm, this means a nutrient addition equivalent to 34 kg P $ha^{-1}$ for the 78 ha. This amount of P is about 59% of corn fertilizer P usage in 1992, which is about 58 kg $ha^{-1}$ (MDA 1993). A yearly accumulation of this magnitude will have serious implication in the long run due to soil P saturation. The P eventually will either move down through the soil profile to the groundwater or wash out through soil erosion into surface water. The obvious option is to export the excess manure P, which is equivalent to 197 164 kg dry matter manure.

*Minimum feed cost strategy*

Concern for profitability requires a dairy operation to minimize feed cost. The model was run with the objective of finding the best ration for each of the seven animal groups and cropping pattern that would yield the least feed cost. This problem is equivalent to implementing the second objective function only, to minimize feed cost. The assigned weights were: $w_1=0$, $w_2=1$ and $w_3=0$. The results show that the total feed cost is \$90 974 $yr^{-1}$ and 112 ha are required. The formulated rations for the mature lactating cows, dry cows, older heifers and young heifers are shown in Table 4. The cropping system, consisting of alfalfa, corn (silage), corn (grain) and orchard-grass, requires a total of 2220 kg of nutrient P for plant growth. A total of 2425 kg manure P are produced by the 220 animals and the average P content of the dry matter manure is 0.79%. This strategy results in 205 kg of excess manure P or equivalent to 25 949 kg of dry matter manure.

*Table 3.* Formulated rations of selected animal groups under the minimum cropland strategy.

| Feeds | Mature lactating cow | Dry cow | Older heifer | Young heifer |
|---|---|---|---|---|
| Corn silage (%) | 35.9 | 20.8 | 0 | 0 |
| Dry corn (%) | 25.0 | 25.0 | 25.0 | 25.0 |
| Orchardgrass hay (%) | 10.0 | 20.0 | 0 | 0 |
| Soybean meal (%) | 28.0 | 34.0 | 67.9 | 67.9 |
| Dicalcium phosphate (%) | 0 | 0 | 2.3 | 2.3 |
| Urea (%) | 0 | 0 | 0.5 | 0.5 |
| Blood meal (%) | 0 | 0 | 1.0 | 1.0 |
| Calcium carbonate (%) | 1.1 | 0.2 | 3.3 | 3.3 |
| Total dry matter (kg) | 21.0 | 11.8 | 9.2 | 4.3 |

### *Multi-criteria optimization strategy*

Neither the minimum area nor the minimum feed cost strategies gave an environmentally optimum solution. An ideal management strategy would be zero excess manure P while satisfying economic returns and reasonable cropland requirement. The model was run as multi-criteria optimization (MCO) with three objectives: minimize excess manure P, minimize feed cost and minimize cropland. The corresponding weights were $w_1=0.3$, $w_2=0.2$ and $w_3=0.5$. The results show that feed cost is \$92 155 $yr^{-1}$ and 112 ha are required to grow the feed crops. The formulated rations of some selected animal groups are shown in Table 5. A total of 2276 kg of manure P is produced from the 220 animals and the average P content of the dry matter manure is 0.74%. The cropping system, consisting of alfalfa, corn (silage), corn (grain) and orchardgrass, requires 2276 kg of nutrient P, resulting in zero excess manure.

Table 6 shows a comparison among the various optimization strategies in terms of total cropland requirement, cropping system, P demand by crops, P supply from manure and feed cost. The cropland requirement of the MCO strategy is comparable to the minimum feed cost strategy, but the MCO strategy has the most nutrient P demand by the crops, and the least P supply from manure, which allows for all manure P to be applied, resulting in zero excess. The cropping system of the MCO strategy has more land devoted to alfalfa and lesser land to corn silage, compared to the minimum feed cost strategy. However, the minimum feed cost strategy is cheaper by \$1181 than the MCO strategy. On the other hand, it has excess dry matter manure of 25 949 kg. If the savings from the minimum feed cost strategy are used to pay for the disposal of the excess manure, disposal cost should not exceed \$0.05 $kg^{-1}$ dry matter manure to be comparable to the MCO strategy in economic terms.

### *Mass balance in P for the dairy system*

Using the NBD model, a mass balance in P is demonstrated for the MCO strategy and shown in Figure 2. A total of 4352 kg P enter the 220 animal, 112 ha dairy-crop

*Table 4.* Formulated rations of selected animal groups under the minimum feed cost strategy.

| Feeds | Mature lactating cow | Dry cow | Older heifer | Young heifer |
|---|---|---|---|---|
| Alfalfa silage (%) | 3.3 | 0 | 3.1 | 0 |
| Corn silage (%) | 50.0 | 45.7 | 50.0 | 41.0 |
| Dry corn (%) | 25.0 | 25.0 | 26.9 | 25.0 |
| Orchardgrass hay (%) | 10.0 | 20.0 | 20.0 | 20.0 |
| Soybean meal (%) | 10.0 | 8.6 | 0 | 13.5 |
| Dicalcium phosphate (%) | 0.4 | 0 | 0 | 0 |
| Urea (%) | 0.5 | 0.5 | 0 | 0.5 |
| Calcium carbonate (%) | 0.8 | 0.2 | <0.1 | <0.1 |
| Total dry matter (kg) | 21.0 | 11.8 | 9.2 | 4.3 |

*Table 5.* Formulated rations of selected animal groups under the multi-criteria optimization strategy.

| Feeds | Mature lactating cow | Dry cow | Older heifer | Young heifer |
|---|---|---|---|---|
| Alfalfa silage (%) | 0.8 | 9.4 | 3.2 | 10.8 |
| Corn silage (%) | 50.0 | 38.5 | 50.0 | 32.5 |
| Dry corn (%) | 25.0 | 25.0 | 25.0 | 25.0 |
| Orchardgrass hay (%) | 10.0 | 20.0 | 20.0 | 20.0 |
| Soybean meal (%) | 13.0 | 6.5 | 0 | 11.2 |
| Dicalcium phosphate (%) | 0.2 | 0 | 0 | 0 |
| Urea (%) | 0.1 | 0.6 | 0 | 0.5 |
| Calcium carbonate (%) | 0.9 | 0 | 1.8 | 0 |
| Total dry matter (kg) | 21.0 | 11.8 | 9.2 | 4.3 |

*Table 6.* Comparison of various optimization strategies.

| Variables/optimization strategy | Minimize excess manure P, feed cost, and cropland | Minimize feed cost only | Minimize cropland only |
|---|---|---|---|
| Total cropland ($ha\,yr^{-1}$) | 12 | 12 | 78 |
| Cropping system: | | | |
| – alfalfa (ha) | 7 | 5 | 0 |
| – corn (silage) (ha) | 53 | 55 | 30 |
| – corn (grain) (ha) | 42 | 42 | 42 |
| – orchardgrass (ha) | 10 | 10 | 6 |
| Nutrient P demand for crop growth ($kg\,yr^{-1}$) | 2276 | 2220 | 1480 |
| Nutrient P supply for manure ($kg\,yr^{-1}$) | 2276 | 2425 | 4122 |
| Excess manure P ($kg\,yr^{-1}$) | 0 | 205 | 2642 |
| Total feed cost ($\$\,yr^{-1}$) | 92155 | 90974 | 130866 |

system. Of the total P input, 955 kg come from feed supplements and 3397 kg come from the soil. A total of 4352 kg P exit the system in the form of 915 kg in milk, 34 kg in cull cows and afterbirth and fluids, 49 kg in animals to market, 23 kg in surface run-off and 3331 kg retained in the soil profile. The nutrient P needs of 2276 kg by the crops are supplied fully from the manure, without excess.

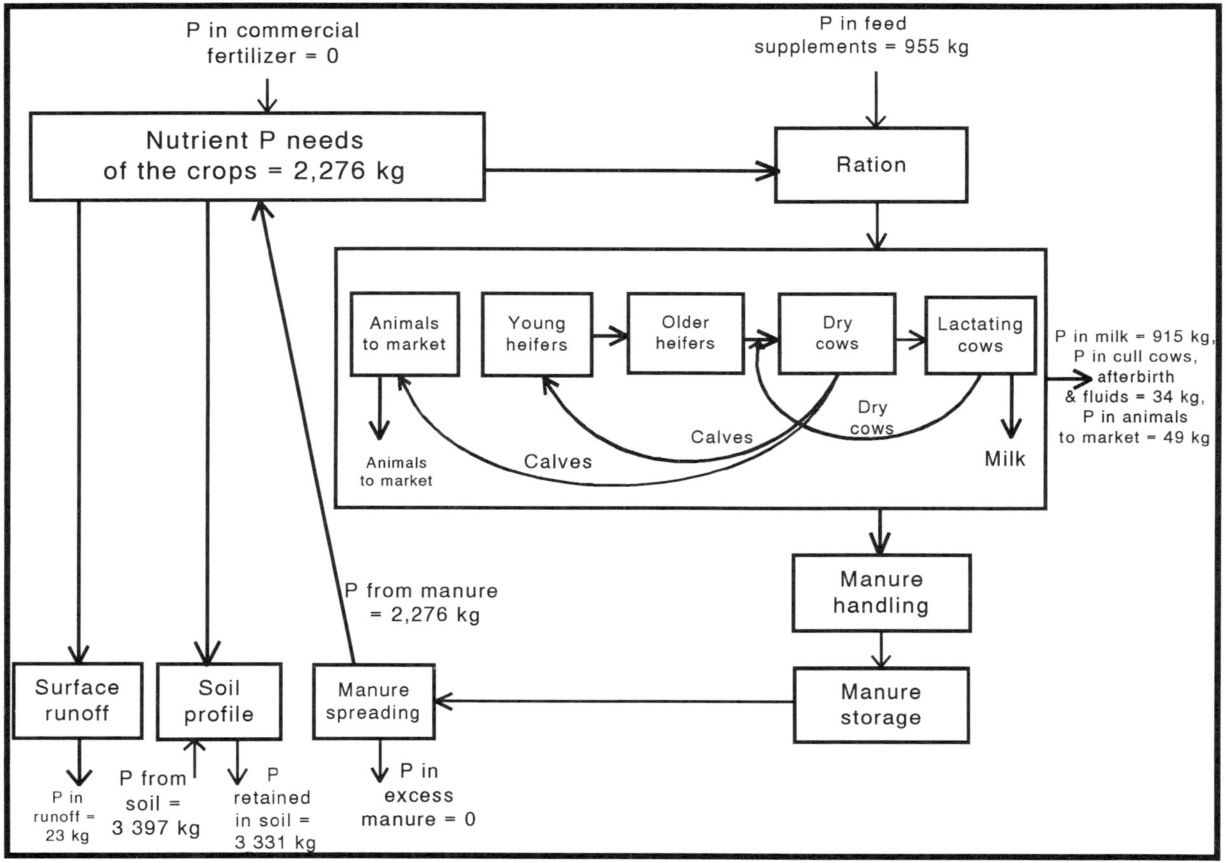


*Figure 2.* An illustration of a mass balance in nutrient P in a dairy-crop system.

## Conclusions

Environmental protection, profit considerations and land requirement may be conflicting objectives. However, the NBD model has shown that rations can be formulated for animals in various stages of growth, and cropping systems can be identified that result in zero excess manure P at a cost and cropland requirement comparable to the minimum feed cost management strategy. The MCO strategy is sustainable both in the short and long terms. The results show that environmental protection can be addressed simultaneously with profitability in a market-oriented economy. The NBD model is a useful tool in identifying environmentally and economically viable solutions.

## References

Alocilja E C (1990a) Process network theory. Pages 66–71 in Proceedings of the IEEE International Conference on Systems, Man, and Cybernetics, 4–7 November 1990, Los Angeles, CA, USA.

Alocilja E C (1990b) The application of process network theory in Agroecosystems. ASAE International Winter Meeting, Paper No. 907578. ASAE, St. Joseph, MI, USA. 13 p.

CGLAS – Center for Great Lakes and Aquatic Sciences (1993) Suspended solids and bedload transport of nutrients, heavy metals, and PCBs in the 16 major tributaries to Saginaw Bay, 1990–1992. University of Michigan, Ann Arbor, MI, USA. 160 p.

Henry G M, DeLorenzo M A, Beede D K, Van Horn H H, Moss C B, Boggess W G (1995) Determining optimal nutrient management strategies for dairy farms. J. Dairy Sci. 78:693–703.

Lanyon L E (1992) Implications of dairy herd size for farm material transport, plant nutrient management, and water quality. J. Dairy Sci. 75:334–344.

MCA – Michigan Commission of Agriculture (1993) Michigan's right to farm act. Lansing, MI, USA.

MCA (1995) Generally accepted agricultural and management practices for manure management and utilization. Lansing, MI, USA. 25 p.

MDA – Michigan Department of Agriculture (1993) Michigan agricultural statistics. Lansing, MI, USA. 92 p.

MPS – Midwest Plan Service (1985) Livestock waste facilities handbook. 2nd Edition. Iowa State University, Ames, IA, USA.

MSU – Michigan State University (1992a) Fertiliser recommendations for field crops in Michigan, Extension Bulletin E-550A, Cooperative Extension Service, East Lansing, MI, USA. 35 p.

MSU (1992b) Spartan dairy ration evaluator/balancer user's manual, version 2.01, East Lansing, MI, USA. 225 p.

MUCC – Michigan United Conservation Clubs (1993) Saginaw Bay watershed land use and zoning study. Lansing, MI, USA. 313 p.

NRC – National Research Council (1989) Nutrient requirements of dairy cattle. National Academy Press, Washington DC, USA. 158 p.

Romero C, Rehman T (1989) Multiple criteria analysis for agricultural decisions. Elsevier, New York, Amsterdam. 257 p.

Van Horn H H, Wilkie A C, Powers W J, Nordstedt R A (1994) Components of dairy manure management systems. J. Dairy Sci. 77:2008–2030.

Van Saun R J (1991) Dry cow nutrition: The key to improving fresh cow performance. Veterinary Clinics of North America. Food Anim. Pract. 7(2):599–620.

Westphal P J, Lanyon L E, Partenheimer E J (1989) Plant nutrient management strategy implications for optimal herd size and performance of a simulated dairy farm. Agric. Syst. 31:381–394.

# The role of systems research in grazing management: applications to sustainable cattle production in Latin America

M. HERRERO[1,2], R.H. FAWCETT[1], E. PÉREZ[2] and J.B. DENT[1]
[1]*Institute of Ecology and Resource Management, The University of Edinburgh, West Mains Road, Edinburgh EH9 3JH, UK*
[2]*School of Veterinary Medicine, National University, Heredia, Costa Rica*

**Key words:** decision-support systems, models, cattle production, farm management, grazing management, land use, Latin America

**Abstract**
Advances in pasture science over the past 25 years have considerably widened the scope for improvement in ruminant livestock production from grazing systems. The introduction of improved pastures and legumes to increase output per animal and/or per hectare is a classic example of such advances. To ensure the long-term success of these technologies, management of the grazing system has to be improved, and must deal with strategies integrating multiple choices over an extended planning horizon. For example, the control of stocking rates, paddock rotation lengths and fertilizer practices is important to maintain pastures in the long run. However, sustainable grazing cannot be achieved if management strategies like these are taken without understanding the whole farming system and the interactions between its components. This paper describes current efforts and the role of decision-support systems created by linking livestock information systems and whole-farm models to support management decisions in cattle farms of Latin America.

## Introduction

Grasslands are one of the major natural resources of Latin America. They comprise about 28% of the total land of the region and support approximately 25% of the world's cattle population (1.3 billion head) (FAO 1994). A large proportion of them are rangelands which support relatively low levels of animal production. Therefore, in the past 25 years, efforts to increase animal production have been directed towards technologies such as introduction of improved pastures, legumes and their associations (Lascano 1990).

Substantial evidence of the benefits of these introductions in terms of animal performance is available (see Skerman et al. 1988; or Humphreys 1991). However, a delicate balance between animal production per unit of area and per animal is required (Lascano 1990). The technologies to maintain this balance require an increased managerial capacity from the farm household and, in many cases, an increased interaction between the farmer and extension personnel. This suggests the need for decision-support tools to assist in the selection of appropriate grazing system designs and management practices. A requirement of these tools is that they provide a holistic view of the farming system in order to provide realistic management solutions (Dent et al. 1994; Herrero et al. 1996).



*P.S. Teng* et al. *(eds.): Applications of Systems Approaches at the Farm and Regional Levels, 129–136.*

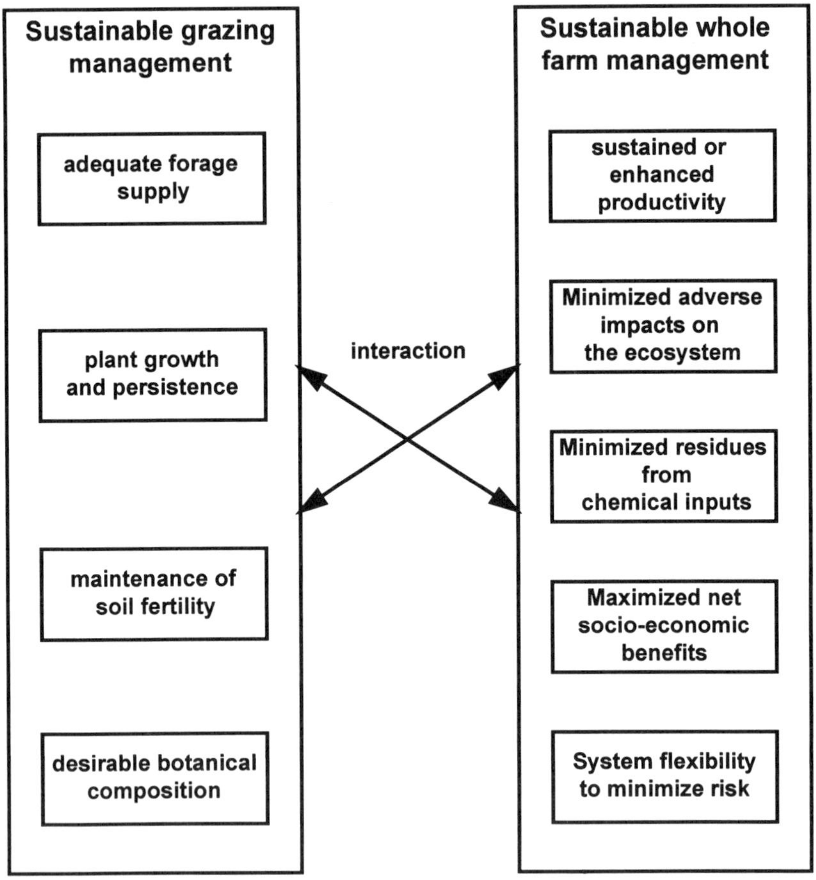


*Figure 1.* Attributes of sustainable grazing and sustainable farm management.

The objectives of the present paper are to show the importance of a systems approach for sustainable cattle production systems in Latin America and to describe tech-nologies that may help to ensure their long-term success.

## Sustainable cattle production systems: concepts and compromises

Many definitions of sustainability have been formulated depending on the context of its use. One of the most quoted definitions is that by the World Commission on Environment and Development (WCED 1987): '*...development that meets the needs of the present without compromising the ability of future generations to meet their own needs.*'

In a pastoral context, Figure 1 illustrates grazing management issues (left column) which, although often conflicting, lead to sustainability *sensu* WCED at the field level, when combined in a compromise manner. Figure 1 also shows how sustainability

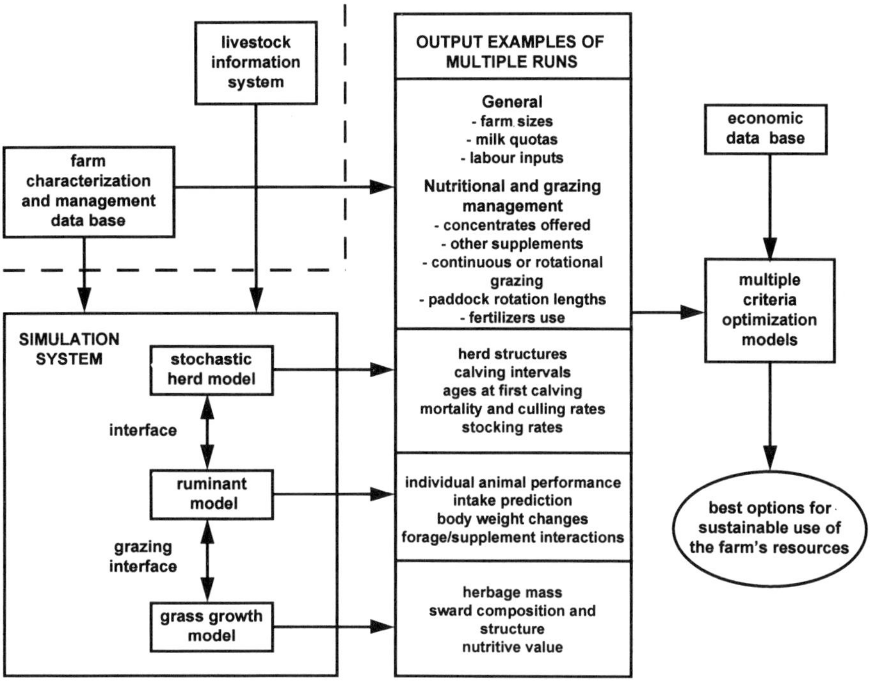


*Figure 2.* General pastoral systems modelling methodology. (*Source:* adapted from Herrero et al. 1996.)

measures at different levels within the farm demand compromise management to achieve overall sustainability. It is evident that grazing management is one of the key determinants of sustainability for the whole farm in the long run. However, because all the requirements (see Figure 1) have to be met simultaneously, a series of compromises arise which cannot be studied unless the whole system and associated objectives are represented. For example, in biological terms, the provision of an adequate year-round forage supply of a desired botanical composition and quality is central to animal performance. However, to obtain this while at the same time maintaining soil fertility or the legume component, keeping a flexible system to minimize risks due to climate and markets, and minimizing residues from fertilizer and chemical inputs, is a formidable task (Humphreys 1994). The problem is multiplied several fold when socioeconomic and external aspects which are fundamental to the decision-making process, are included (Dent et al. 1994).

## Decision support for pastoral systems

The complexity of pastoral systems justifies the need to build dynamic decision-support systems (DSS) based on simulation and optimization techniques to represent different components and the main interactions (i.e. Dent et al. 1994; Herrero et al. 1996). Figure 2 shows the general methodology for modelling pastoral production systems. Given the overall framework, it is obvious that grazing management decisions cannot be taken excluding the other components of the system. For example, grazing management strategies cannot be taken without consideration of herd or nutritional management, since herd dynamics or feed supplementation practices will determine the grazing intensity and use of the forage resource and subsequently animal performance. Thus, within the modelling framework (see Figure 2), the simulation system representing the biology of pastoral enterprises includes flexible mechanistic models representing: pasture growth, structure and quality; an individual ruminant digestion and metabolism model to test nutritional strategies; and a population model describing management practices at herd or flock level (i.e. stocking rates, sales of animals, mortality or replacement rates, calving intervals, etc.), which subsequently determine animal numbers and their age or physiological state (lactating vs pregnant cows, heifers, calves, etc.) classes.

Individual mechanistic models can be used to improve the understanding of a particular part of the system and the biological feasibility of some of the strategies. For example, Herrero (1995) used a pasture growth simulator, representing the growth and structure of tropical pasture swards under rotational or continuous grazing, with responses to N fertilizer applications, temperature and irradiance, to devise physiologically based management guidelines for rotational grazing systems in the highlands of Costa Rica.

Land-use options are central to management outcomes and require a mechanism to select between alternative farm resource allocation strategies. Due to the nature of the problem and the variety of objectives, multiple criteria decision models (MCDMs) are ideal structures to study these compromises (Romero and Rehman 1989; Fawcet 1995; Herrero et al. 1996). The outputs of the validated simulation system become the inputs for the MCDMs (i.e. Veloso et al. 1992; see Figure 2) and these explore the effects of different management options on the resource use strategies of the farm. Finally, the best resource use alternatives are selected according to the various objectives of the decision maker and are implemented at field level.

It is important to recognize the difference between systems analysis and decision support because modelling objectives and the variables under study can be very different. Systems analysis can be done at a variety of levels of aggregation but does not require a dynamic interaction with the decision makers, these being the farm household, the policy makers, extension service, etc. Decision support implies an analysis of the system, support to the management cycle and a constant interaction with the decision makers via performance monitoring.

The variables driving the system should represent those handled by the decision makers, according to the planned time horizons for the different goal achievements. The variables and the decision rules can be quantitative or qualitative depending on the information available (Röling 1994; Dent 1995). In quantitative systems, there is a trade-off between the level of aggregation, the accuracy required and the ease of validation and understanding. 'Hybrid' models based on a mechanistic platform but with empirical site-specific parameters obtained from local data may be the ideal for decision-support purposes.

## Livestock information systems

A key element of any DSS involves monitoring the performance of the system and analysing data for the decision maker. This is shown in the top left-hand corner of Figure 2.

In the context of tropical cattle production systems, this concept was pioneered by the Herd Health Project of the School of Veterinary Medicine of the National University of Costa Rica (Pérez et al. 1989). The VAMPP (Veterinary Automated Management and Production control Program), originally developed by Utrecht University, The Netherlands (Noordhuizen 1984), was introduced and adapted to the local farming systems and information requirements. The main objective of its introduction was to provide information to the farmer for day-to-day farm management while, at the same time, creating livestock data bases for research and teaching purposes. This function was developed as a farmer-oriented livestock information system (LIS). The system works by periodic monitoring of individual farms and the collection of production, reproduction and health data from individual animals. Data bases are created and analysed and immediate feedback to the farmer is given in the form of action lists (cows to inseminate, to calve, to dry-off, etc.) to support his operational management. Other longer-term performance indexes are also provided (calving intervals, ages at first calving, replacement rates, etc.) (Baaijen and Pérez 1995).

In 1988, the system was introduced to 23 farms in a small pilot project, through which more than 400 dairy, dual purpose and beef cattle farms in Costa Rica are now periodically monitored (every two weeks or once a month).

After the positive acceptance of the information system by the Costa Rican livestock farmers, links with other Latin American countries began in 1990 to establish similar systems. These were made through farmers' organizations, cooperatives or universities, and a regional network was created. Today, the livestock monitoring system is present in Nicaragua, Guatemala, Honduras, Panama, Colombia and Mexico and there are plans to introduce it in Bolivia, El Salvador and Venezuela.

The implementation of the LIS in the region has proved to be an efficient method for providing operational technical support to farmers while, at the same time, capturing information for research and teaching purposes (Baaijen and Pérez 1995). The decentralized nature of the system also allows data to be collated from individual farms to provide regional or group information to assist advisers, policy makers and researchers if required.

## The link between the modelling framework and the livestock information system

Drawing together the modelling framework (planning and resource use allocation) and the LIS (operational management support and monitoring) creates a DSS (Figure 3). The LIS supports operational decisions while the DSS creates the opportunity for strategic and tactical planning and target setting.

Once the DSS is sufficiently validated across a range of production systems, its

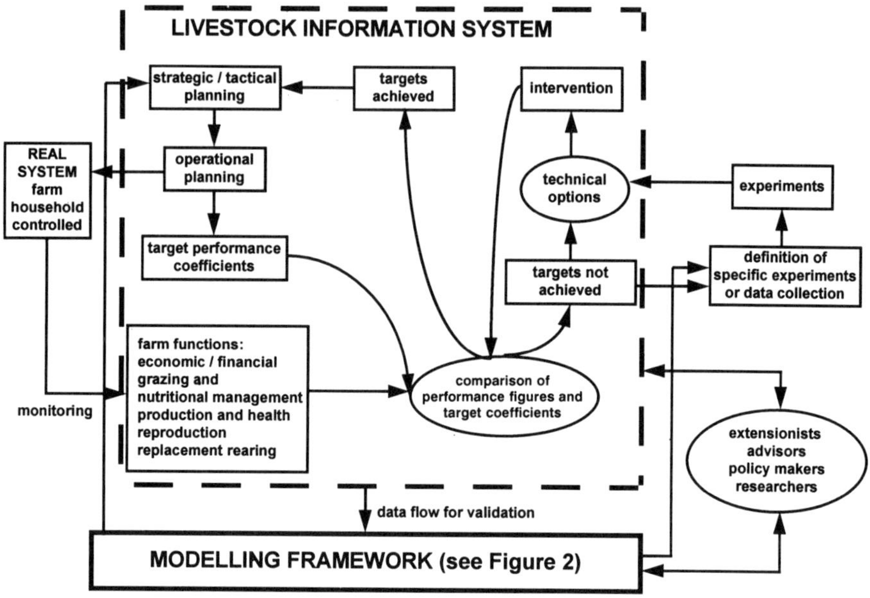


*Figure 3.* The decision-support system for pastoral enterprises.

role will be as a planning tool in conjunction with extension services to improve farm management and resource use. For individual farms it will:

- provide alternative strategies dealing with technical and land-use options;
- set target performance figures for the farm functions (i.e. grasslands use, replacement rates) for the preferred selected strategy;
- monitor the current performance of each farm on a routine basis;
- do periodical analyses to check if the targets are being achieved;
- suggest ways to adjust the current system if targets are not being met; and
- provide a basis for replanning if necessary.

At a more aggregated level it will be used as a policy-making and as a research/teaching tool. This use has the potential to create a link between the farmer, the extension agent and the policy maker while, at the same time, creating livestock performance data bases for research purposes. As the livestock information network is already working in several countries, a robust validation of the modelling framework would give the whole system the potential for regionalization according to the farming systems and farm characteristics of the countries of the region.

## Required research

In terms of the modelling framework, the agroecological diversity of Latin America and the distribution of pasture species mean that the models need to be adapted to deal with the variety of ruminant production systems (i.e. dual purpose, beef cattle) and pasture species with different growth habits and phenological attributes (i.e. *Brachiaria* spp., *Cynodon* spp.), grass/legume associations and rangelands.

Successful adaptation of the models is largely dependent on the availability of appropriate data for validation and parameter values. It is believed that, if sustainability is the key issue, more research is required on soil N dynamics in tropical grassland systems and on the physiology of tropical pastures, legumes and their associations. Although several efforts have been made (Humphreys 1994), there is still a lack of quantitative information on these processes for tropical Latin American grazing systems.

Studies on intake, diet selection and grazing behaviour of ruminants grazing rangelands and improved pasture/legume swards with different morphological characteristics are also necessary to provide understanding and to improve the prediction of intake of ruminants at pasture.

Sociological studies on the farmers' decision rules and behaviour are required in order to understand the justification of some of their actions. This is important to provide decision support in areas of real importance to the farming community.

## References

Baaijen M, Pérez E (1995) Information technology in the Costa Rican dairy sector: A key instrument in extension and on-farm research. Agric. Hum. Values (Spring 1995):45–51.

Dent J B (1995) Towards a general paradigm for decisionmaking. In Central America: Conservation and sustainable development. Proceedings of a Workshop on Sustainability of Livestock Production Systems, 8–11 August 1995, San José, Costa Rica. Cienc. Vet. 17:51–21.

Dent J B, McGregor M J, Edwards G (1994) Integrating livestock and socio-economic models into complex models. Pages 25–36 in Gibon A, Flamant J C (Eds.) The study of livestock farming systems in a research and development framework. EAAP Publication No. 63. Wageningen Press, The Netherlands.

FAO – Food and Agriculture Organization (1994) Pages Production Yearbook 1994. Volume No. 48. FAO of the United Nations. Rome, Italy. 243 p.

Fawcett R H (1995) Some practical approaches to the design and selection of optimal farming systems using MCDM techniques. In Central America: Conservation and sustainable development. Proceedings of a workshop on Sustainability of Livestock Production Systems, 8–11 August 1995, San José, Costa Rica. Cienc. Vet. 17:44–49.

Herrero M (1995) Grassland modelling: A decision-support tool. In Central America: Conservation and sustainable development. Proceedings of a workshop on Sustainability of Livestock Production Systems, 8–11 August 1995, San José, Costa Rica. Cienc. Vet. 17:72–79.

Herrero M, Fawcett R H, Dent J B (1996) Integrating simulation models to optimize nutrition and management for dairy farms: A methodology. In The Study of livestock farming systems in a research and development framework: Research, socio-economics and the land manager. European Association for Animal Production Publications (in press).

Humphreys L R (1991) Tropical pasture utilisation. Cambridge University Press, Cambridge, UK. 206 p.

Humphreys L R (1994) Tropical forages: Their role in sustainable agriculture. Longman Scientific & Technical, Essex, UK. 414 p.

Lascano C (1990) Managing the grazing resource for animal production in savannas of tropical America. Trop. Grassl. 25:66–72.

Noordhuizen J P T M (1984) Veterinary herd health and production control on dairy farms. PhD thesis, Utrecht University, Utrecht, The Netherlands.

Pérez E, Baaijen M T, Capella E, Barkema H (1989) Development of a livestock information system for Costa Rica. Pages 221–224 in Livestock production and diseases in the tropics. Proceedings of the 6th International Conference of Institutes for Tropical Veterinary Medicine, Utrecht, The Netherlands.

Röling E (1994) Interaction between extension services and farmer decisionmaking: New issues and sustainable farming. Pages 280–291 in Dent J B, McGregor M J (Eds.) Rural and farming systems analysis: European perspectives. CAB International, Wallingford, UK

Romero C, Rehman T (1989) Multiple criteria analysis for agricultural decisions. Elsevier Science Publications, Amsterdam, The Netherlands. 257 p.

Skerman P J, Cameron D G, Riveros F (1988) Tropical forage legumes. 2nd Edition. Food and Agriculture Plant Production and Protection Series No. 2. FAO, Rome, Italy. 692 p.

Veloso R F, McGregor M J, Dent J B, Thornton P K (1992) Farm planning for the Brazilian 'cerrado' region: Application of crop simulation models for yield forecasting. Pages 13–18 in Computers in agricultural extension programs. Proceedings of the 4th International Conference, Gainesville, Florida, USA.

WCED – World Commission on Environment and Development (1987) Our common future. WCED, Oxford University Press, Oxford, UK. 400 p.

# From decision-making process analysis to modelling and simulation of farm management: application to work organization in two rice-based farming systems

P.-Y. LE GAL

*Centre de coopération Internationale en Recherche Agronomique pour le Développement (CIRAD), BP 5035, 34032 Montpellier Cedex 1, France*

**Key words:** farm management, work organization, advice, modelling, rice

**Abstract**

In both industrialized and non-industrialized countries, the economic environment faced by farmers is undergoing profound changes due to uncertainty and economic liberalization. Decision-making processes related to farm management are more complex and diversified than ever. Normative models of management and research offering 'one best way' solution are no longer efficient in such a context. A new approach is presented here, based on (i) the concept of action model, (ii) a sectoral analysis of farm management, (iii) general frameworks of the decision-making processes adapted to complex problems like work organization and (iv) modelling and simulation to stimulate farmers' reasoning rather than giving them recommendations. This approach is illustrated by two cases dealing with work organization in rice-based farming systems. The first one is in Camargue (France) where rice establishment in spring is critical because of cold temperatures. We show how farmers organize their work from ploughing to seeding, how we can model and simulate this specific organization, and how the advice process is structured around this knowledge. The second case is in the Senegal River delta (Senegal), where farmers collectively manage irrigation schemes and try to implement a double cropping of rice using contractors for land preparation and harvest. This plurality of actors increases uncertainty and makes the coordination necessary between them more complex. In this case, advice is mainly an aid to formalize technical and organizational problems, as no modelling tool is yet available to represent these situations. The interests and limits of the approach, as well as further research perspectives, are discussed.

## Introduction

Farmers' economic environments have evolved recently, leaving them in a critical position to consider uncertainty and risk management. In Europe, common agricultural policy reform has led to unstable prices of farm products and reconsideration of production models. In developing countries, the turn-over of responsibilities from states to farmer organizations and the private sector has involved complex relationships between actors, especially in irrigated agriculture (Le Moigne and Barghouti 1990; Bélières et al. 1995).

A common question emerging from these new contexts is: how can advisers be more helpful to farmers in making their strategic as well as current management decisions? Normative models and research of a 'one best way' solution through mathematical programmes have become of limited value, as technical and economic conditions are liable to change quickly and in unknown ways (Attonaty and Soler



*P.S. Teng* et al. *(eds.): Applications of Systems Approaches at the Farm and Regional Levels, 137–152.*

1991). New approaches to advice have to be imagined, taking into account the farmers' specific knowledge and situation, progress in computer language and modelling of biophysical mechanisms (Nitsch 1991). This paper presents an approach designed by a team of French agronomists and management researchers, and its application in rice-based farming systems in the south of France (Camargue) and in the Senegal River delta (Senegal).

## Theoretical and methodological framework

### *From agricultural practices to decision-making process analysis*

Since the 1970s and the failure of 'top-down' organization of advice for farmers, some researchers have carried out studies to understand farmers' agricultural practices through holistic farm analysis (Chambers 1983; Landais and Deffontaines 1988; Byerlee et al. 1991). This knowledge was supposed to facilitate the production and extension of solutions to the specific constraints encountered by farmers in their activities (Capillon and Caneill 1987; Collinson 1987). This approach has been quite successful in extending simple technology like new varieties or fertilization techniques, based on on-farm trials (Tripp 1991), but shows significant shortcomings in the present economic situation of agriculture.

From a cognitive point of view, it does not provide reliable information about the process of choice followed by farmers before making their decisions, and about their ways of dealing with the future and uncertainty (Papy 1994). From an advisory point of view, it is not adapted to complex problems that cannot be solved with on-farm trials, like work organization or the choice of new crops. It is also seriously limited when assessing the climatic and economic risks taken by farmers adopting an innovation. A new approach has been designed to remove these difficulties, based on analysis of farmers' decision-making processes, and analysis of farm management from a sectoral point of view.

*Analysis of farmers' decision-making processes.* Farmers' decision-making processes can be thought of as the 'engine' of agricultural practices. Therefore, understanding these processes is necessary to improve the effectiveness of advice. This research has been guided by results taken from decision and management sciences. The decision-making process is dynamic, following several steps such as intelligence, design and choice activities (Simon 1960). It has a cognitive dimension, as choices are guided by mental representations which change as the shortfall between the expected and actual results of action are evaluated (Courbon 1982). It is predictable, as decision makers tend to design plans of action to structure their choices (Heiner 1983; Hoc 1991). This behaviour is particularly adapted to decisions made regularly, like work organization at the farm level. In this case, farmers have to choose during the cropping season, the specific arrangement of equipment and manpower necessary to achieve their objectives of time schedule and work quality, under diverse climatic conditions.

Thus, research conducted in arable farming systems in the Paris basin has shown that farmers make these choices through a permanent frame called the 'action model' (Sebillote and Soler 1990; Papy 1994). This includes: (i) a final goal to be reached, for example a sowing date not to be overstepped; (ii) some intermediate objectives where the work achieved is assessed using specific control indicators such as the acreage ploughed at a certain date; (iii) a set of rules under an 'If Conditions Then Decisions' format, which define the nature of decisions to be made when specific conditions are encountered, like climatic events; and (iv) a set of substitute solutions to be used when the expected goals cannot be achieved. Each farmer has a specific action model, which is designed from his or her own experience and knowledge, and which changes when new situations are met like technological innovations or evolution in market organization.

This concept seems valid for cyclic decisions made every year by experienced farmers. It has also been tested on forage systems management in France (Duru et al. 1994), but needs to be extended to equivalent tropical conditions. Furthermore, it is an efficient tool enabling researchers and advisers to carry out a diagnosis of complex problems like work organization, particularly from the farmer's point of view: did the farmer achieve what he or she wanted? What are the farmers' major constraints when running their model?

*Analysis of farm management from a sectoral point of view.* An holistic analysis is needed to classify the regional diversity of farms by designing typologies (Capillon 1993). But this approach is no longer useful to helping farmers make their management decisions for two main reasons: there can be a wide variety of decision-making processes within the same type of farm; and in some arable crop-based systems, farmers do not see their farm as a global system, but rather as a sum of management units between which they try to reduce the interactions (Hemidy et al. 1993). It seems more suitable to analyse farm management from a sectoral point of view, corresponding to the complex problems that farmers have to face like work organization, cropping and forage systems management or strategic management (investment decisions, choice of activities).

For each topic, researchers are aiming to develop a general framework of decision-making processes, which can be used for different farmers. One has been designed for work organization (Attonaty et al. 1990). It is structured around three management levels (Figure 1):

- Tasks include one or several operations performed at the same time. They use manpower and equipment resources and are characterized by specific performances which vary according to working conditions (soil, climate, field size).
- Sequences correspond to a series of tasks performed on a set of fields. These areas are defined by rules related to cropping systems, and the physical and economic environment. The succession of tasks is governed by specific rules. It can be sequential or simultaneous, according to the farm resources in equipment and human resources.
- Periods are defined by farmers according to their choice of priorities between

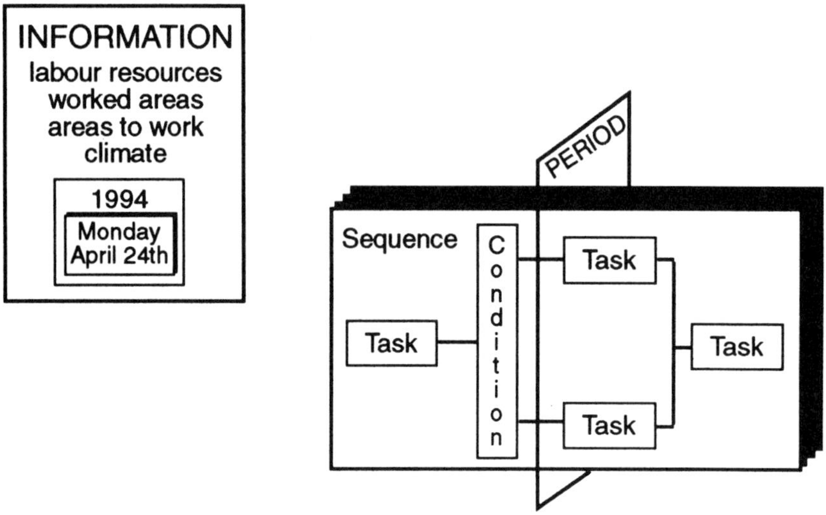


*Figure 1.* Structure of the daily work organization at three management levels: task, sequence and period (*Source:* Attonaty et al. 1990).

several sequences performed at the same time with limited human and equipment resources. Each period has starting and finishing dates, as well as rules of change in case of delay. These dates are used by farmers to assess their organization with a specific level of risk related to climatic variability.

The definition of tasks and the set of rules use daily information which includes indicators concerning the physical environment (soil moisture, climatic events), the statement of the work done, the calendar and available resources.

### *From decision-making process analysis to advice*

Giving farmers solutions and recommendations, even though they may be based on a correct diagnosis, will have little impact in solving the complex problems of management when the environment is uncertain. The relative positions of farmers and advisers have to be reconsidered in a way that stimulates farmers' learning processes through a better understanding of their own knowledge and decision-making processes, and a methodology able to assess the consequences of innovations on the risk level chosen by the farmer.

In view of this, advice has two main objectives: first, to help farmers design strategies to reach a specific goal; second, to help them perform a given strategy. The approach, already used in the north of France, is based on the following sequences, which structure the dialogue between farmers and advisers on an individual or collective basis (Attonaty et al. 1993):

- The farmer's actual action model is formulated by focusing on the problem periods, like bottlenecks in work organization.

- This knowledge is modelled with a specific software, for example OTELO (Organisation du Travail Et Langage à Objet, Work organization and object-oriented language), that has been designed for work organization. The general framework is described in Figure 1 (Attonaty et al. 1990). It mimics the decision-maker's behaviour when faced with climatic hazards, by using the principles of artificial intelligence. Rules, indicators and farm structure are programmed with a specific computer language which is close to the words and concepts that farmers use when organizing their work.
- The model incorporates various climatic scenarios, provided by a nearby meteorological station. For each day of the period studied, the simulator executes the tasks according to the model. It checks the indicators, launches the verified rules, does the work and saves the elementary information in a diary (date of execution, characteristics of the treated area, the worked acreage, staff and equipment used, etc.). The collection of these records is saved in a data base.
- The data base is asked to assess the efficiency of the farmer's actual organization over several years, according to his own objectives in term of calendar, work quality and risk level.
- If these objectives are not met, the effects of changes in the initial model are simulated. The modifications can be suggested by the farmer as well as the adviser. They can concern the organizational rules, introduction of new technology, variation in staff and equipment or revision of unrealistic objectives.

This interactive and iterative methodology helps farmers to formalize better their knowledge and decision-making processes on problems they are not used to formalizing. They can also evaluate with their own criteria the consequences of technical or organizational changes on their farm management. From advisers' and researchers' points of view, the methodology provides a powerful tool for carrying out a diagnosis and giving information adapted to each case, and for improving discussion with farmers on a dynamic and specific basis. Using modelling of decision-making processes also facilitates connections with the biophysical models designed for crop management.

We illustrate this process with two examples related to work organization in rice-based farming systems. The first is in Camargue, where rice is grown in large and highly mechanized farms, managed by a single decision maker usually assisted by workers. The second is in the Senegal River delta, where a double crop of rice is grown on irrigation schemes managed collectively by a large number of farmers assisted by contractors for some tasks.

## Application to work organization in two rice-based farming systems

### *The Camargue case*

Camargue is situated in the northern part of the rice growing area in Europe (43° 45′ N, 4° 30′ E). Low temperatures are the main climatic constraint for rice growing. They can fall below 12°C in April and May, considered as a critical limit for germi-

*Table 1.* Structural characteristics of the three farms surveyed in Camargue.

| | | Farm A | Farm B | Farm C |
|---|---|---|---|---|
| Cultivated area (ha) | | 174 | 227 | 304 |
| % under rice in 1994 | | 96 | 90 | 80 |
| Tractors | number | 3 | 5 | 6 |
| Workers | number | 2 | 3 | 6 |
| | permanent | 1 | 3 | 4 |
| | temporary | 1 | - | 2 |

nation and seedling emergence (Yoshida 1981), and below 2°C at the flowering stage in August. According to these constraints the optimal sowing period is short, from 15–20 April to 10–15 May. Moreover, farmers try to obtain a thin seedbed to improve germination. This objective is achieved by intensifying tillage operations from March to April, with the risk of delay in rainy years.

To understand better how farmers manage these constraints and to test the validity of the approach described above in this particular context, a survey of three farms was carried out (Kedda 1994). The methodology adopted included: (i) interviews prior to the cultural operations to find out how each farmer was planning his work from ploughing to sowing; (ii) data recording during the operational phase (area worked daily by operation) and interviews to understand the reasons for differences between the planned and real actions; and (iii) discussions carried out using modelling and simulation results from each farm with OTELO.

The three farms were quite similar in terms of structural characteristics and land use, mainly based on monocropping rice (Table 1). Farmers had also similar objectives of sowing date and seedbed quality. But we observed three specific ways of organizing work from ploughing to sowing, along the following lines:

- their own perceptions of constraints concerning climate, type of soil and irrigation structure;
- their choice of sowing and land preparation equipment, especially for the second tillage as a large range of tools is available from driven power take-off to trailed implements; and
- their combined resources of staff and tractors, which allow the possibility of performing simultaneously two operations.

We illustrate these specific organizational choices with the case of Mr. B. (Figure 2). After harvesting rice in October, he burns the straw and ploughs the stubble with a chisel. The mouldboard plough is only used when he needs to finish the ploughing quickly in case of delay during the winter. It is then attached to another tractor and the two operations are conducted simultaneously. The low lands are ploughed first, because of their poor accessibility when it rains. The farmer stops the operation if he estimates that daily rainfall has exceeded 20 mm or that rainfall over the last three days has exceeded 30 mm.

In March, a second tillage is carried out with two tractors, one with a cover-crop and the other one with a rototiller. Levelling with a scraper and decompaction with

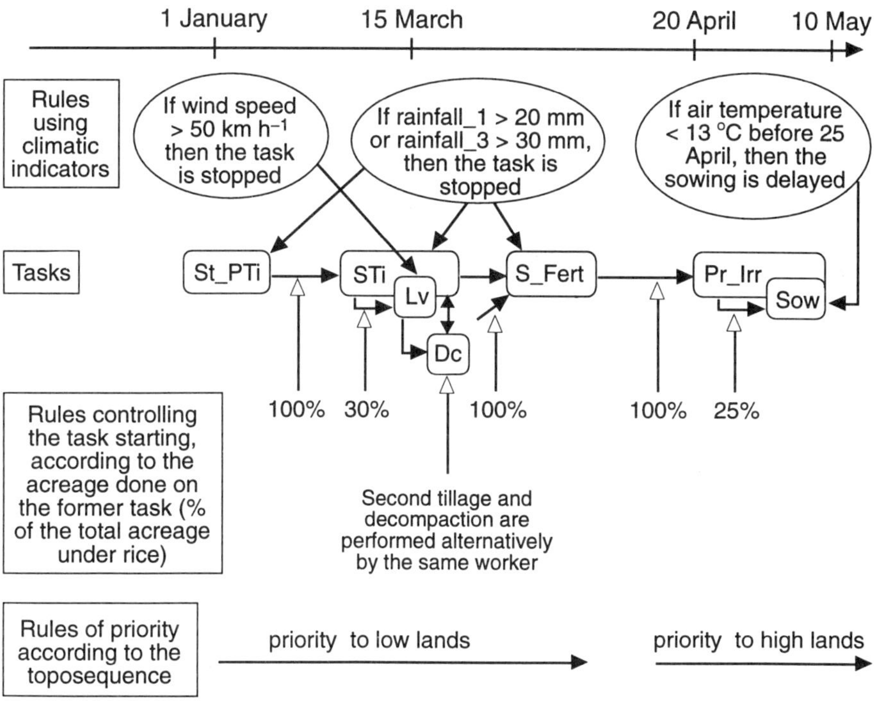


| Operation[a] | Tractors[b] | | | | | Agricultural implement[c] | | | | | | Workers[d] | | | | | Performance (ha d⁻¹) |
|---|---|---|---|---|---|---|---|---|---|---|---|---|---|---|---|---|---|
| | n | 60 | 90 | 110 | 150 | Pl | Ch | Rt | Sp | Cc | Sc | n | Fm | A | B | C | |
| St_PTi | 2 | | | x | x | x | x | | | | | 2 | | | x | x | 8–10 |
| STi | 2 | | | x | x | | | x | | x | | 2 | | | x | x | 25–30 |
| Lv | 1 | | | o | o | | | | | | x | 1 | | | o | o | 2–4 |
| DC | 1 | | | o | o | | x | | | | | 1 | | | x | | 12–16 |
| S_Fert | 2 | x | | | | | | | x | x | | 4 | x | x | x | x | 25–30 |
| Pr_Irr | | | - | | | | | | | | | 1 | | | | x | 10–12 |
| Sow | 1 | | x | | | | | | x | | | 4 | x | x | x | x | 25–30 |

[a]St_PTi: stubble ploughing + primary tillage; STi: second tillage; Lv: levelling; DC: decompaction; S_Fert: sealing fertilizers; Pr_Irr: pre-irrigation; Sow: sowing.
[b]Characterized by their engine power (cv).
[c]Pl: mouldboard plough; Ch: chisel; Rt: rototiller; Sp: spreader; Cc: cover-crop; Sc: scraper + levelling blade.
[d]Fm: farm manager; A, B, C: workers.
x: tractor, implement and worker specific to the operation.
o: tractor, implement and worker chosen.
rainfall_1: daily rainfall, rainfall_3: sum of rainfalls over the three days.

*Figure 2.* Example of a specific farmer's action model (farm B) for the rice seeding organization in Camargue (*Source:* Kedda 1994).

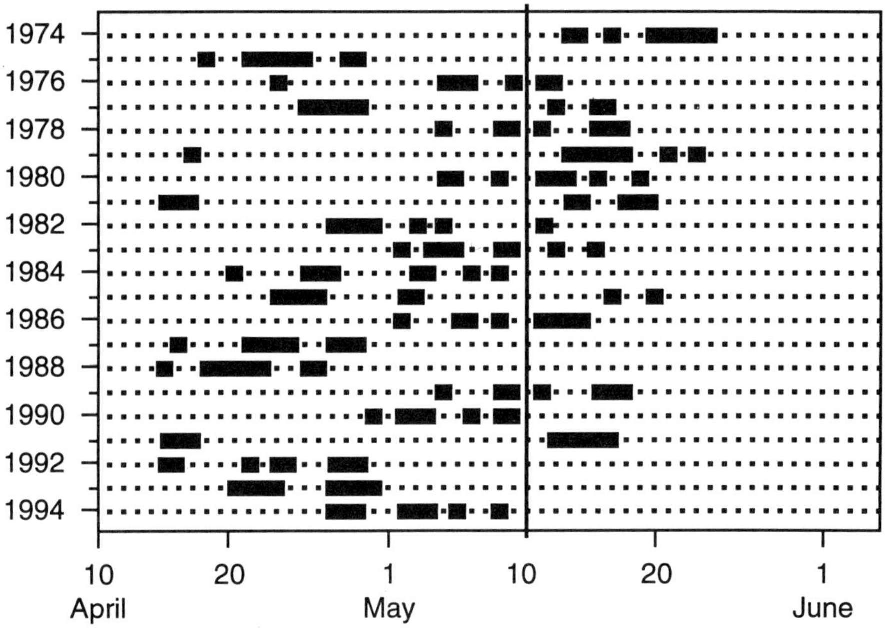


*Figure 3.* Simulated sowing dates for 21 climatic years for the Camargue, farm B. On the horizontal axis (the time axis) each dash represents one day. Each thick line corresponds to one or more sowing days; no work on the other days for different reasons (e.g. temperature, weekend). The vertical line corresponds to 10 May, the sowing date that the farmer tries to respect. The farmer's objective is not achieved 13 years out of 21.

the chisel are performed at the same time. As levelling is guided by laser, this operation is stopped when the wind blows over 50 km h$^{-1}$, which is regular in Camargue. Fertilizers are then spread and sealed in the soil with the cover crop. These two operations are performed simultaneously and need all the workers available on the farm to drive the two tractors and move fertilizer bags.

Pre-irrigation is conducted by a specialized worker and begins with the high lands in order to use water surplus or leakage on the low lands. The seedlings are then sown into the water, with the same spreader used for fertilizers. This operation also needs four workers, mainly for handling and marking. Its starting date is linked to the air temperature. Before 30 April, the sowing is delayed if the temperature is below 13°C. After this date Mr. B. sows in any case, as he prefers to avoid late seeding after 10 May because of the high risks of spikelet sterility related to cold temperature at the flowering stage.

Each task has a specific performance, the slowest being levelling (2–4 ha d$^{-1}$), the quickest second tillage and spreading (25–30 ha d$^{-1}$). Therefore, the farm manager has to define rules of succession between two sequential tasks according to his resources in staff and tractors. To simplify its organization Mr. B. tries to finish a task before starting the following one. There are two exceptions to this rule. First,

*Table 2.* Gap between simulated and observed sowing dates in 1994 for the three surveyed farms in Camargue (number of days).

| | Farm A | Farm B | Farm C |
|---|---|---|---|
| Start of sowing | −3 | −1 | −4 |
| End of sowing | +5 | 0 | +3 |

second tillage, levelling and decompaction are performed simultaneously to reduce delays with the slowest operation. Levelling starts when 30% of the acreage is ploughed and is limited to half the total acreage. Decompaction is performed alternatively with tillage by the same worker, as the farm manager usually does not participate in these operations. Second, sowing starts when 25% of the acreage is under water, so as not to be limited by the slower performance of pre-irrigation.

Execution of this programme depends on uncertain events like workers' illness or equipment breakdown, which are not taken into account from an advisory point of view, and random climatic events like rainfalls or cold temperature whose frequency can be estimated from meteorological records. The sensitivity of the farmer's action model to climatic constraints, according to his own calendar objectives, can be assessed using OTELO. Modelling requires some estimations or simplifications. For example, we translated the soil moisture indicators used by Mr. B. into rainfall amounts which are available in climatic data bases.

The models written with OTELO have been validated by comparing the simulated and observed sowing dates in 1994 for each farm (Table 2). The differences depend on the frequency of uncertain events during the observation period. They were very scarce in Mr. B.'s case. Validation based on observations obtained during earlier seasons is difficult because of (i) the lack of recorded data or (ii) the fact that the farmer's situation can have changed in terms of equipment, staff or cropping pattern.

Once validated, we simulated the models over the last 21 years, using climatic data from a nearby meteorological station. In Mr. B.'s example, we observed that his actual organization was quite risky, as he would achieve his objective of sowing 205 ha of rice before 10 May in only 8 years out of 21 (Figure 3). From subsequent discussion with him about the simulation results, it appears that this rice acreage was exceptional on his farm. It was due to the impossibility of establishing durum wheat in autumn 1993 because of heavy rainfalls. Our analysis confirmed his feeling that he should not increase his rice acreage beyond 150 ha, and new management ideas were discussed for the future.

These results were also presented to the three farmers altogether. They gave detailed and structured support to consider various problems linked with farm mechanization management, which they had not previously been used to discussing with one another. The approach has also been useful for our research team, enabling us to better formalize problems encountered by farmers at the establishment stage. We have designed new technical research programmes dealing with questions that are usually not raised; for example, the usefulness of basal nitrogen fertilization which appears difficult to fit in with pre-irrigation and sowing to reduce nitrogen losses.

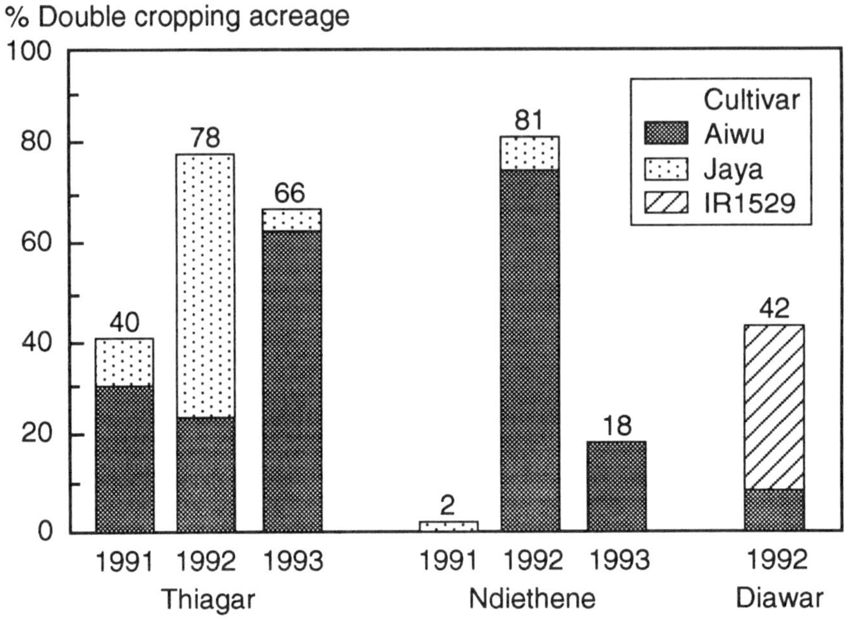


*Figure 4.* Evolution of double cropping of rice acreage sown after 15 August according to location, year and cultivar (Senegal River delta).

*Double cropping of irrigated rice in the Senegal River delta*

Double cropping, defined as growing two sequential crops the same year on the same field, is one way to intensify irrigated farming systems. In Sahelian countries (West Africa) its extension is promoted by the authorities to make the huge investments put into irrigation profitable. Although the physical conditions (annual water supply and temperatures) are favourable to its implementation, acreage under double cropping does not exceed 10% to 30% of the available irrigated acreage, Niger excepted (Le Gal 1995a).

In the Senegal River delta (16° 15′ N, 16° 00′ W), introduction of double cropping has been fairly recent (1988). It consists of two crops of rice, one in the dry season sown in February–March and harvested in July, followed by one sown in July–August and harvested in December–January. Farmers' success in managing this cropping system on a scheme covering from 500 to 1000 ha with electric pumping and surface irrigation, is quite irregular. We have assessed it with a simple indicator, used by farmers themselves, and defined as the acreage of rainy season rice sown after 15 August, which is the critical date for the medium-duration cultivar preferred by local people (Figure 4).

The complexity of problems faced by farmers in managing double cropping explains the diversity of results. Technical problems are related to cultivar, cultural operations and equipment choices. Available varieties are not really adapted to the dry season

*Table 3.* Mean and maximum daily performances ($ha\,d^{-1}$) of combine harvesters and tractors at three locations in the Senegal River delta (1991–93); n is number of observations to calculate the mean, maximum and coefficient of variation (cv) (*Source:* Le Gal 1995b).

| | | Thiagar | Ndiethene | Diawar | All |
|---|---|---|---|---|---|
| Combine harvester | n | 164 | 198 | 179 | 541 |
| | mean | 3.8 | 3.4 | 3.1 | 3.4 |
| | max. | 8.5 | 8.9 | 6.6 | 8.9 |
| | cv | 37 | 32 | 39 | 39 |
| Tractor + cover crop | n | 72 | 55 | 53 | 180 |
| | mean | 5.5 | 8.7 | 6.8 | 6.9 |
| | max. | 12.8 | 15.6 | 14.9 | 15.6 |
| | cv | 45 | 39 | 49 | 48 |

because of their sensitivity to cold temperatures at the establishment and the tillering stages (Dingkuhn et al. 1995a), which lengthen from 20 days in the rainy season to 30 days in the rainy season. Moreover their sensitivity to cold temperatures at the flowering stage means sowing the rainy season crop before cultivar-specific dates.

Mechanization of land preparation with heavy tractors and cover crops, and of harvest with large combine harvesters leads to accessibility constraints before harvest or after a rainfall. Daily performances of tasks are quite low and very irregular, as observed at three locations (Thiagar, Ndiethene and Diawar) where we conducted a research programme on collective management of double cropping from 1991 to 1993 (Table 3). These variations are related to several factors like the field characteristics (soil moisture, weed infestation, lodging, paddy yield, field size) and the age of equipment and its maintenance (frequency and duration of breakdowns).

But the main problems encountered by farmers are organizational. The management of the different tasks (land preparation, pre-irrigation, sowing and harvest) involves many actors:

- the individual farmers on their plots;
- the various farmer organizations in charge of water management and economic aspects;
- the private contractors who own tractors and combine-harvesters;
- the credit bank which loans the funds necessary to start the culture season; and
- the rice factories which buy paddy from farmers.

For example, the Thiagar scheme, spread over 800 ha, involves 850 farmers, 43 farmer organizations and 6 villages in its management.

In comparison with the Camargue case, management of this cropping system means not only coordination of several culture operations in a given time and space with limited resources, but also coordination of actors as in an industrial company. Using the methodology described above, we have tried in the first instance to understand better how farmer organizations deal with this double coordination. We have shown that, from a collective point of view, individual farmers and contractors provide

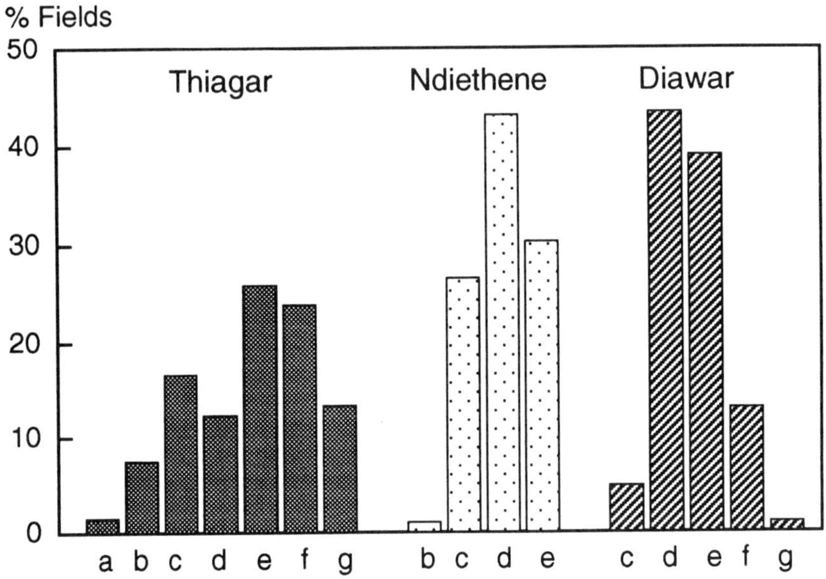


*Figure 5.* Field variation of the duration (days) between the flowering stage and the drainage date in the Senegal River delta for three locations during dry season 1992 (a: <10 d; b: 10–15 d; c: 16–20 d; d: 21–25 d; e: 26–30 d; f: 31–35 d; and g: >35 d).

uncertainty (Le Gal 1995b). For example, individual farmers choose their drainage date mostly on their own. They tend to drain their fields when the paddy is almost mature because they think it will increase their yield. But this behaviour is quite variable from one field to another (Figure 5). So farmer organizations, which choose the starting date of mechanized harvesting, have great difficulty assessing the situation at any time.

Contractors' behaviour is also quite uncertain as they do not really control their equipment management. Maintenance is usually poor because of lack of training and lack of cash flow when spare parts have to be bought. The high frequency of breakdowns is a major constraint for farmer organizations when planning machinery arrivals in the fields and task duration. To ensure actors' coordination in this uncertain environment, they have developed three types of behaviour whose relative weight varies from one location to another:

- Contractualization of business relationships, mainly by giving a monopoly to the contractors settled in their own village. By doing so, farmer organizations hope to improve the technical components of tasks by securing a minimum number of machines able to work in their fields, by getting priority treatment in the contractor's work plan and by forcing him to work in every plot, even if the characteristics (yield, soil moisture, size) are not profitable. But the monopoly is not ratified by a written contract, as the different members prefer to keep flexibility in case of problems like breakdowns.

- Simplification of collective decision-making processes, by designing collective management rules which help actors' coordination or give more flexibility to individual farmers when the scheme structure allows it. For example, some farmer organizations decide to standardize the harvest starting date in order to shorten the task duration. To deal with the large diversity of soil moisture and rice maturity at plot level, they choose this date with a simple visual indicator, namely when 80% of panicles are mature. This choice involves delays between maturity and harvest, but reduces the sinking risks for combine harvesters.
- Adaptation to the events met during the tasks. For example, farmer organizations seek extra combine harvesters when they consider that their current number is too low to complete harvesting in time. As it is the simplest way of dealing with uncertainty, such behaviour mainly structures their management processes. But its efficiency is low as it depends on the capacity of other members to react quickly to the farmer organizations' requests. Very often, contractors are not available as they are working on another area, or their equipment is not operational because of maintenance delays.

The excessive emphasis on adaptation versus contractualization and simplification in farmers' collective behaviour means that the action programmes they formulate before starting work are quite simplistic. The farmers' lack of experience of double cropping, the degree of uncertainty governing individual behaviour and the difficulty of actors' coordination explain this difficulty in structuring a collective action model. Nevertheless, the concept itself and the general framework of work organization have been useful to analyse the decision-making processes, carry out a diagnosis and help farmers to formalize their management choices. We also used OTELO to model and simulate various scenarios of collective management. We coupled it with a rice development model, RIDEV, designed by WARDA, to help farmers in their choices of variety and sowing date according to an assessment of climatic risks (Dingkuhn et al. 1995b). We had to simplify greatly the real situation. OTELO is a simulator that cannot take into account the plurality of actors and points of view, typical of collective management. Further research has to be done to provide decision-support systems adapted to these frequent situations.

## Conclusions: interests and limits of the approach

This approach has been in use for only a few years and its efficiency needs more validation using different cases. First results in France and our own experience in Senegal allow us to point out some major potentials and limitations from an advisory as well as research point of view.

### *Potentials*

By focusing on the farmers' learning processes rather than recommending technical or organizational solutions, this approach deeply modifies the relationships between farmers and advisers. The advice process is then defined as an exchange of knowledge between two experts, which supposes the adviser to be able to understand the farmer's decision-making processes, or at least to design a common view. To do so, the action model and the general framework of work organization are useful tools of analysis and diagnosis.

This formalization of farmer's behaviour is also the basis of modelling and simulation using artificial intelligence. This second component of the approach is important to explore the future and to help farmers design their own solutions according to their personal level of risk. It is also an efficient tool to deal with complex management problems that are difficult or impossible to experiment upon or to represent with mathematical programming.

Moreover modelling farmers' decision-making rules and indicators enables us to couple this knowledge with biophysical models. Thus, it is possible to assess the consequences of management choices on crops and the natural environment, or to design a suitable strategy to reach a certain productivity or product quality, on a simulation basis (Aubry et al. 1992). By treating quantitative as well as qualitative information, modelling becomes a powerful methodology to facilitate a multidisciplinary approach between the biophysical and social sciences.

*Limitations*

In its present form this approach has several limitations which need further research. First, the knowledge produced about farmers' behaviour and the efficiency of advice from the models are not easy to validate. This is a global problem for management science because of: (i) the relationship between the object studied, in this case farmers, and the external observer who wants to participate in their evolution; (ii) the difficulties in defining the object himself and his reality; and (iii) his irregularity and instability (Martinet 1990). Validating action models from their predictive capacity is not correct, as the approach aims to stimulate the farmers' reasons by prospecting possible futures. To solve these problems, we first try to confirm the farmers' declarations about their action plans with their observed practices. The validation of knowledge and of the advice process is then made by the farmers themselves, from their capacity to account for their perception of their management processes and to influence their decision making.

To implement such a validation process we are planning to use an adapted methodology, called research-action (Girin 1990). At first, research will be conducted on request from farmers. Then the research process will be controlled by two committees: the first, including farmers' and researchers' representatives, will periodically assess the correct running of the research work and will solve any possible conflicts between actors; the second will be a scientific peer review.

On the modelling side, tools adapted to collective management processes have to be found. The objective of modelling and simulation is then to give a correct representation of multiple actors' behaviour, and to facilitate their negotiation processes by exploring different types of relationships. Progress in artificial intelligence and multi-agent simulators will provide us with new opportunities (Bousquet et al. 1993). We are planning to use them in different situations, such as the management of the sugar-cane harvest among smallholders, contractors and sugar-cane factories in La Reunion, or water management in large irrigation schemes.

On the advisory side, this approach needs well-trained advisers, able to deal with the various and complex problems of farm management, to exchange knowledge with farmers and sometimes to use computers. This job description is quite different from that of the normal extension agent in tropical countries. But we think that the general trend of uncertainty and technological progress in agriculture, linked with farming diversity, needs a drastic evolution in advice methodology and institutions. In the present difficult financial context, some research has to be conducted to design new methods which are better adapted to farmers' needs but not too costly to run. The approach presented here gives some directions.

## References


Attonaty J-M, Soler L-G (1991) Renewing strategic decisionmaking aids. Eur. R. Agr. Eco. 18: 421–440.

Attonaty J-M, Chatelin M-H, Poussin J-C, Soler L-G (1990) Un simulateur à base de conaissance pour raisonner équipement et organisation du travail en agriculture. Pages 301–314 in Matarasso P (Ed.) Représentation, modélisation, développement. Agropolis, Montpellier, France.

Attonaty J-M, Chatelin M-H, Mousset J (1993) A decision support system based on farmer's knowledge to assess him in decision making about work organization and long term evolution. Paper presented at the International Seminar of CIGR on Models, Computer Programs, and Expert System for Agricultural Mechanization, 1–2 October 1993, Florence, Italy.

Aubry C, Chatelin M-H, Poussin J-C, Attonaty J-M, Masse J, Meynard J-M, Gérard C, Robert D (1992) DECIBLE: A decision support system for wheat management. Paper presented at the 4th International Congress for Computer Technology in Agriculture on Farm Computer Technology in Search of Users? 1–2 June 1992, Versailles, France.

Bélières J-F, Havard M, Le Gal P-Y (1995) Désengagement de l'état et dynamiques d'évolution de la riziculture irriguée dans le delta du fleuve Sénégal. Paper presented at the CNRS-CIRAD workshop on Quel avenir pour les rizicultures de l'Afrique de l'Ouest? 4–7 April 1995, CNRS-Centre de Coopération International en Recherche Agronomique pour le Développement (CIRAD), Bordeaux, France.

Bousquet F, Cambier C, Mullon C (1993) Simulating the interaction between a society and a renewable resource. J. Biol. Syst. 1(2):199–214.

Byerlee D, Triomphe B, Sebillotte M (1991) Integrating agronomic and economic perspectives into the diagnosis stage of on-farm research. Exp. Agric. 27(2):95–114.

Capillon A (1993) Typologie des exploitations agricoles, contribution à l'étude régionale des problèmes techniques. Thèse Doctoral, Institut National Agronomique Paris-Grignon, Paris, France. 48 p.

Capillon A, Caneill J (1987) Du champ cultivé aux unités de production: Un itinéraire obligé pour l'agronome. Cah. Institut Français de Recherche Scientifique pour le Développement en Coopération (ORSTOM), Sér. Sci. hum. 23(3/4):409–420.

Chambers R (1983) Rural development: Putting the last first. Longman, London, UK. 246 p.

Collinson M P (1987) Farming systems research: Procedures for technology development. Expl. Agric. 23:365–386.

Courbon J-C (1982) Processus de décision et aide à la décision. Economie et Sociétés. Série Sciences de Gestion 3, Tome XVI(12):1466–1476.

Dingkuhn M, Sow A, Samb A, Diack S, Asch F (1995a) Climatic determinants of irrigated rice performance in the Sahel. I. Photothermal and microclimatic responses of flowering. Agric. Syst. 48:385–410.

Dingkuhn M, Le Gal P-Y, Poussin J-C (1995b) RIDEV: Un modèle de développement du riz pour le choix des variétés et calendriers culturaux. Pages 205–222 in Boivin P, Dia I, Lericollais A, Poussin

J-C, Santoir C, Seck S M (Eds.) Nianga: Laboratoire de la culture irriguée. Institut Français de Recherche Scientifique pour le Développement en Coopération (ORSTOM), Paris, France.

Duru M, Bellon S, Chatelin M-H, Fiorelli J-L, Gibon A, Havet A, Mathieu A, Osty P-L (1994) Propositions pour l'aide à la gestion des ressources fourragères: Une approche système articulant enquête, expérimentation et simulation. Pages 104–109 in Proceedings of the International Symposium on Systems-oriented research in agriculture and rural development, 21–25 November 1994, Montpellier, France.

Girin J (1990) Analyse empirique des situations de gestion: éléments de théorie et de méthode. Pages 141–182 in Martinet A-Ch (Ed.) Epistémologies et sciences de gestion. Economica, Paris, France.

Heiner R A (1983) The origin of predictable behavior. Am. Econ. Rev. 73:560–595.

Hémidy L, Maxime F, Soler L-G (1993) Instrumentation et pilotage stratégique dans l'entreprise agricole. Pages 91–118 in Soler L G (Ed.) Instrumentation de gestion et conduite de l'entreprise, Cahiers Economie et Sociologie Rurales, Tome 28. INRA, Montpellier, France.

Hoc J-M (1991) La planification de l'activité. Pages 223–230 in Traité de psychologie cognitive. Tome 2. Dunod, Paris, France.

Kedda B (1994) Aide à la décision en riziculture inondée en Camargue: Modélisation des calendriers de travail des agriculteurs pendant la phase d'installation de la culture. Mémoire de fin d'études. CNEARC, Centre de Coopération International en Recherche Agronomique pour le Développement, INRA, Montpellier, France. 74 p.

Landais E, Deffontaines J-P (1988) Les pratiques des agriculteurs. Point de vue sur un courant de la recherche agronomique. Etud. Rurales 10:125–158.

Le Gal P-Y (1995a) Le développement de la double culture irriguée au Sahel: Contraintes et perspectives pour les agriculteurs. Paper presented to the WARDA workshop La riziculture irriguée au Sahel: Perspectives pour un développement durable. 27–31 March 1995, Ndiaye, Senegal.

Le Gal P-Y (1995b) Gestion collective des systèmes de culture en situation d'incertitude: Cas de l'organisation du travail en double culture dans le delta du fleuve Sénégal. Thèse Dr. Institut National Agronomique Paris-Grignon, Paris, France. 215 p.

Le Moigne G, Barghouti S (1990) Irrigation in sub-Saharan Africa: The development of public and private systems. World Bank Technical Paper No. 123, Washington DC, USA. 99 p.

Martinet A-Ch (1990) Grandes questions épistémologiques et sciences de gestion. Pages 9–30 in Martinet A-Ch (Ed.) Epistémologies et sciences de gestion. Economica, Paris, France.

Nitsch U (1991) Computers and the nature of farm management. Pages 99–106 in Kuiper D, Röling N G (Eds.) Knowledge management and information technology. Agricultural University, Wageningen, The Netherlands.

Papy F (1994) Working knowledge concerning technical systems and decision support. Pages 222–235 in Dent J B, McGregor M J (Eds.) Rural farming systems analysis: European perspectives. CAB International, Wallingford, UK.

Sebillotte M, Soler L-G (1990) Les processus de décision des agriculteurs. Première partie: Acquis et questions vives. Pages 93–101 in Brossier J, Vissac B, Le Moigne J L (Eds.) Modélisation systémique et système agraire: Décision et organisation. INRA, Paris, France.

Simon H A (1960) The new science of management decision. Harper and Row, New York, USA. 50 p.

Tripp R (1991) An overview of the cases of on-farm research. Pages 17–36 in Tripp R (Ed.) Planned change in farming systems: Progress in on-farm research. John Wiley and Sons, Chichester, UK.

Yoshida S (1981) Fundamentals of rice crop science. International Rice Research Institute, P.O. Box 933, Manila, Philippines. 269 p.

# Modelling socioeconomic influences on agroforestry adoption using a rule-based decision support system

M.P. ROBOTHAM

*Department of Agronomy and Soil Science, University of Hawaii at Manoa, 1910 East-West Rd, Honolulu, HI 96822, USA*

**Key words:** agroforestry, adoption of innovations, decision tree models, rule-based expert systems

**Abstract**

This paper describes AGFADOPT, a rule-based decision-support system designed to model some of the social and economic factors that influence small farmer adoption of tree-based innovations. The system is designed to help technically trained personnel incorporate social and economic criteria into project planning and implementation. To create AGFADOPT, information obtained from the literature was used to develop a set of general rules that simulate farmer decision making regarding the adoption of two tree-based technologies: fruit tree orchards and contour hedgerows. These rules were combined in the expert system using a yes/no decision tree framework. Preliminary verification of the system prototype using data from the Dominican Republic and evaluation of the system by potential users suggest that it is logically consistent, reasonably accurate (83% for tree orchard; 68% for contour hedgerows) and potentially useful.

## Introduction

Management problems of uplands and marginal areas in Southeast Asia have been the focus of recent analyses (Blair and Lefoy 1991; Allen 1993; Garrity 1993). This research and other related studies indicate that incorporating trees in some form into existing land management systems can be beneficial. Tree uses can include alley cropping and hedgerow cropping using nitrogen-fixing trees, expanded cultivation of trees for marketable fruit or nuts, and fuelwood and small timber plantations (Juo 1989; MacDicken and Vergara 1990; Watson 1990; Garrity 1993). However, all of these tree-based technologies, like any introduced technology, can provide benefits only if farmers are willing to adopt them into existing farming systems.

Farmer adoption of introduced technologies involves the interaction of a number of environmental and socioeconomic factors (Rogers 1983; Feder et al. 1985). However, government personnel at all levels from field technician to chief planning officer are usually trained in technical fields such as agriculture or forestry (Dove 1992). They often have limited knowledge of the social and economic factors which strongly influence project success. Most government agencies also lack the resources necessary to hire additional staff with social science expertise, or to extensively retrain current staff. Therefore, a system that would provide easy access to knowledge concerning the social and economic factors that affect technology adoption could prove useful.

Although a large number of factors both within and outside the farming system



*P.S. Teng* et al. *(eds.): Applications of Systems Approaches at the Farm and Regional Levels, 153–166.*

are believed to influence farmer decision making regarding introduced technologies, a relatively limited set of important factors appears repeatedly in a wide variety of different studies (Rogers 1983; Feder et al. 1985). AGFADOPT, the rule-based expert system presented in this analysis, is based on a simplified model of the decision-making process using general factors drawn from the existing scientific knowledge base. As such, it does not presume to include all the factors that can affect farmer decision making or to exactly replicate the decision-making process. Instead, this model relies on general principles to provide a first approximation of probable farmer behaviour in a wide variety of different circumstances.

## Existing knowledge base

There is a small but growing amount of literature examining the factors which influence the adoption of tree-based innovations such as alley cropping and contour hedgerows (Raintree 1983; Repollo and Castillo 1989; Fujisaka 1994), fuelwood plantings (Kummer, pers. commun.) and tree crops such as fruits, coffee, cacao and rubber (Watson 1990). Inference can also be made from the studies of adoption behaviour for soil conservation practices (Napier 1991; Napier et al. 1991) and from reviews of adoption behaviour in annual-crop based agricultural systems (Rogers 1983; Feder et al. 1985).

Taken as a group, these analyses identify several socioeconomic factors that are believed to influence the adoption of tree-based innovations. Some of the socio-economic factors are related to local perceptions about the innovation itself: the innovation must be designed to address a perceived problem and local residents must believe that the innovation will help to ameliorate the problem (Fujisaka 1994). These two factors are often collectively referred to as farmer awareness (Napier 1991; Napier et al. 1991). Other factors are related to the context into which the innovation is introduced. These include, but are not limited to: farm size, land tenure, human capital, labour availability, credit availability, farmer attitudes towards risk and uncertainty, and market availability (Raintree 1983; Feder et al. 1985; Repollo and Castillo 1989; Mercer 1992).

### *Local perception of a problem*

The first important concept from the adoption literature is that the proposed innovation must address a locally perceived problem (Raintree 1983; Fujisaka 1994). For example, farmers on deep fertile soils with little noticeable erosion in both Java, Indonesia and South Cotabato, Philippines, did not adopt contour hedgerows since they believed that erosion and soil fertility were not significant problems in their systems (Fujisaka 1994). Conversely, farmers in Misamis Oriental, Philippines, perceived soil erosion to be a problem and this perception contributed to their adoption of contour hedgerows (Fujisaka et al. 1994). Likewise, contour hedgerows were widely adopted in Cebu, Philippines, in part because they addressed farmer-perceived problems of soil erosion and declining soil fertility (Fujisaka 1989).

*Belief that innovation will solve the problem*

In addition to awareness of a problem that might be addressed by tree-based technologies, local farmers must believe that these technologies can solve the problem (Raintree 1983). Although evidence specifically related to tree-based technologies is scarce, corroborating evidence comes from numerous adoption studies in agriculture, focused on adoption of new crop varieties (Rogers 1983; Feder et al. 1985), as well as studies of soil and water conservation technology adoption (Napier 1991; Napier et al. 1991).

*Farm size*

One important contextual factor is farm size. Researchers (Fortmann 1990; Vergara and MacDicken 1990; Raintree 1991) suggest that in order to adopt any tree-based technology, farmers must have sufficient land for the trees while growing enough food crops to satisfy subsistence needs. Much of the evidence for this supposition is anecdotal or inferred. However, it is indirectly supported by Watson and Laquihon's (1987) observations that some farmers replace the double-hedgerow system recommended by their project with single hedgerows because they believe that double rows take up too much space. This indicates that farmers take space considerations very seriously when considering new technologies.

Farm size may not be important in all situations. In a different study from the one above, Laquihon (1989) found that land size did not have a significant effect on farmer adoption of contour hedgerows in the Philippines. Other results in the Dominican Republic (Hansen et al. 1987) support this view. In another study in the Philippines, farm size appeared to be inversely related to adoption of technologies (Segura de los Angeles 1986).

*Land tenure*

Another factor, often closely related to farm size, is land tenure. Land tenure is believed to be one of the keys to successful adoption of agroforestry projects (Granert and Sabueto 1987; Napier et al. 1991). In addition to clear rights to the land, a farmer must have control over the various rights which comprise tenure including the right to plant the tree, the right to own or inherit the tree, the right to use tree products, and the right to dispose of the tree (Fortmann 1987, 1990).

Conventional wisdom says that people only plant trees when they have secure land tenure. However, this may not be correct because it overlooks the relationship between trees and land (Fortmann 1990). Although the specific relationships between land and tree tenure are different in different parts of the world, it is generally suggested that secure land tenure is positively related to adoption of agroforestry technologies. In some places, it may be a necessary prerequisite (Riddell 1987). This contention is supported by results on alley farming adoption from Nigeria (Francis 1987), and the lack of agroforestry adoption among tenant farmers in the Philippines (Sajise 1987).

However, a clear relationship between land tenure and agroforestry adoption does not prevail in all cases. Laquihon (1989) did not find a significant relationship between land tenure and adoption of agroforestry practices in separate studies in the Philippines. In the Dominican Republic, deCaera (1987) also reported that farmers who occupied state forest land (officially as squatters) were as likely to adopt agroforestry practices as farmers who owned their land. In the Dominican case, it is likely that the perception of long-term use rights was more important than actual land tenure (deCaera 1987).

*Human capital*

Human capital, defined as educational level and experience of the farm household, may also influence adoption behaviour (Napier 1991; Mercer 1992). Laquihon (1989) found that formal education showed a strong, positive correlation with contour hedgerow adoption in the Philippines. However, Hansen et al. (1987) found no significant correlation between literacy and adoption of conservation practices in the Dominican Republic and Laquihon (1989) found no significant relationship between agroforestry adoption and farmer experience.

*Labour availability*

Closely tied to the concept of human capital is the availability and the distribution of labour, spatially, temporally, and between family members. Technologies with significant labour requirements, especially requirements which fall at the same time as other, potentially more important, farm tasks, have generally low levels of adoption (Raintree 1983; Vergara and MacDicken 1990). Lack of available labour has been shown to be a limiting factor in the adoption of agroforestry technologies in the Philippines (Segura de los Angeles 1986; Granert and Sabueto 1987). Labour availability also may be strongly affected by gender-based divisions of labour (Rocheleau 1987).

*Credit (financial capital)*

Since many agroforestry innovations have high up-front costs which may or may not be accompanied by deferred benefits, access to credit or other financial capital resources is believed to affect technology adoption (Edwards 1991). Hansen et al. (1987) found a significant correlation between use of credit and technology adoption in the Dominican Republic. However, Laquihon (1989) found that credit had only a minor effect on SALT adoption in the Philippines.

Providing credit to many small farmers may have negative consequences. Based on work in the Dominican Republic, Carrasco (1991) asserts that the provision of credit, while promoting adoption in the short run, can hinder long-term sustainable adoption of technologies.

*Risk and uncertainty*

Farmers' perceptions of risk and uncertainty are another major class of factors influencing adoption behaviour. Uncertainty is directly related to the amount of information available to the farmer about the technology. Risk, however, is more difficult to assess.

*Risk.* The farmer's situation can greatly affect his or her attitude towards risk. Subsistence farmers are assumed to want to minimize risk in adoption of new technologies. Producing sufficient staple crops for family consumption is the first priority of most small farmers. This is often referred to as the 'safety first' or the 'food first' strategy (Kunreuther and Wright 1979; Ortiz 1979; Rogers 1983).

One major way in which farmers reduce risk is by increasing diversity either within the farming system (Roumasset 1979) or outside the farming system. Farmers who are not completely dependent on farming operations for their livelihood usually are more risk taking than farmers who rely solely on farm incomes (Rogers 1983). In addition, farmers often accept higher levels of risk if the perceived benefits of the new technology are higher than perceived costs (Napier 1991).

All of these risk reduction strategies can influence agroforestry adoption. The desire to minimize risk through diversity tends to encourage the adoption of agroforestry technologies since they will increase system diversity. Other potential benefits to agroforestry systems in sloping upland areas include: (i) reduction of soil erosion; (ii) improvement of soil fertility status by providing green manure; (iii) improvement of the production of other crops (e.g. shade for coffee); (iv) provision of household needs (e.g. food, animal fodder, fuelwood); and (v) provision of products for sale off-farm (e.g. fuelwood, fruit) (MacDicken and Vergara 1990).

There are also risks associated with the adoption of agroforestry technologies and which may affect farmer adoption behaviour. These include reduced crop yields because of space occupied by trees, and further potential reductions in crop yield due to competition with trees for water and nutrients (MacDicken and Vergara 1990).

*Uncertainty.* The concept of uncertainty is closely related to risk. In general, the more information a farmer has about an introduced innovation, the smaller the role played by uncertainty in adoption. The farmer can then judge the technology on its own merits, not on its presumed attributes (Napier 1991). Farmer knowledge of an innovation can come from a variety of sources including: village meetings, written information, conversations with other farmers, on-farm extension agent visits, off-farm training sessions, and demonstration projects (Rogers 1983).

Research in the Philippines (Segura de los Angeles 1986; Laquihon 1989) has shown that farmer knowledge of technologies and contact with extension agents are positively related to adoption. This is supported by results from the Dominican Republic (Hansen et al. 1987; Robotham 1993; Witter et al. 1996).

*Market availability*

Another factor affecting farmer adoption of tree-based technologies is the availability of markets for tree products such as fruit, nuts, and fuelwood. Padoch and De Jong (1987) showed that the availability of markets influenced farmer adoption of tree orchards in the Peruvian Amazon. The presence of a market was also integral to the success of agroforestry projects in Kenya (Hosier 1989) and in Cebu, Philippines (Kummer, pers. commun.).

## Organizing the knowledge base

One of the ways to organize knowledge regarding adoption of tree-based technologies is to model how these factors influence farmer decision making. One of the simplest models is the decision-tree. Decision-trees are based on the psychological theory of elimination, by aspects first discussed by Tversky (1972) and elaborated on by Gladwin and Murtaugh (1980) and Gladwin (1983 and 1989). This theory states that a complex decision can be divided into discrete aspects each containing alternatives. Complex decisions are made on the basis of progressive elimination of alternatives in the decision. The decision criteria can be either orderings on alternatives on some aspect, dimension, or factor of the alternatives or they can be constraints that must be satisfied (Gladwin 1983).

In Gladwin's examples (Gladwin 1983, 1989), decision-tree models are constructed through directed interviews with farmers regarding the factors affecting their decision-making process. The information from these interviews is then synthesized by the researcher into a general model. In contrast, this analysis uses general information taken from the knowledge base regarding the adoption of tree-based innovations as the basis for model development.

The major advantages of decision-tree models are simplicity and transparency. The reported accuracy of the site-specific models developed by Gladwin (1983 and 1989) is also very high (80–95%). In addition, a decision-tree model can be easily transferred into a rule-based expert system. Since the primary goal of this analysis was the development of an expert system based on a simple and easily understandable model, a decision-tree format was adopted.

## Accessing and implementing the knowledge base

Even if a knowledge base is organized around an appropriate model, this does not guarantee that potential users of the knowledge will be able to access and implement the knowledge. With the increased availability of personal computers and advances in computer technology, computer programs that allow users to access a knowledge base in a convenient and understandable way have been developed. Expert systems have been successfully developed to capture and disseminate knowledge regarding the biophysical and environmental aspects of tree-based innovations (Warkentin et al. 1990; R.S. Yost, pers. commun.). Expert systems have also been developed for

*Table 1.* Categories of questions used in the model.

| Question category | No. of questions |
|---|---|
| Perception of the problem | 4 |
| Beliefs about tree-based technologies | 4 |
| Desires for the future | 3 |
| Market accessibility | 3 |
| Constraints | |
| Land | 5 |
| Knowledge | 2 |
| Labour | 2 |
| 6 | |
| Other | 2 |
| Perceived future benefits | 2 |
| Total | 33 |

farming systems in Indonesia (Colfer et al. 1989), and for a variety of management problems in agriculture (Yost et al. 1988; Mohan and Arumugam 1994) and forestry (Loh et al. 1994).

Expert systems are most appropriate when the knowledge to be used is primarily declarative; the knowledge base is hierarchical; and there is a knowledge bottleneck, a situation where the knowledge is needed by many different people to help them solve a relatively common problem (Plant and Stone 1991). In farmer adoption of tree-based technologies, the knowledge is primarily declarative and there is a widely recognized knowledge bottleneck. However, the knowledge base forms a very shallow hierarchy.

## Agroforestry decision-tree

The decision-tree model developed for this analysis uses several assumptions about farmer behaviour. First, it assumes that farmers adopt a risk averse or 'safety first' strategy. Under this scenario, a farmer must be reasonably assured of meeting subsistence production needs before taking the risk to incorporate trees into their system. This analysis also assumes that the farmer has the following possible courses of action: (i) plant traditional annual crops without changes in the existing system; (ii) plant a fruit or nut tree orchard on all or part of the land; or (iii) use an agroforestry technology based on contour planting of nitrogen-fixing tree hedgerows on their annual crop land.

The preconditions, opportunities, and constraints identified from the literature as well as the assumptions above were then used to formulate a set of 33 questions divided into six general categories (Table 1). These questions were combined to form the decision-tree that served as the basis for AGFADOPT, the expert system.

## Mapping the model into AGFADOPT

Mapping the rules from the decision-tree model into the expert system shell appeared to be reasonably straightforward. Rule-based expert systems, such as AGFADOPT

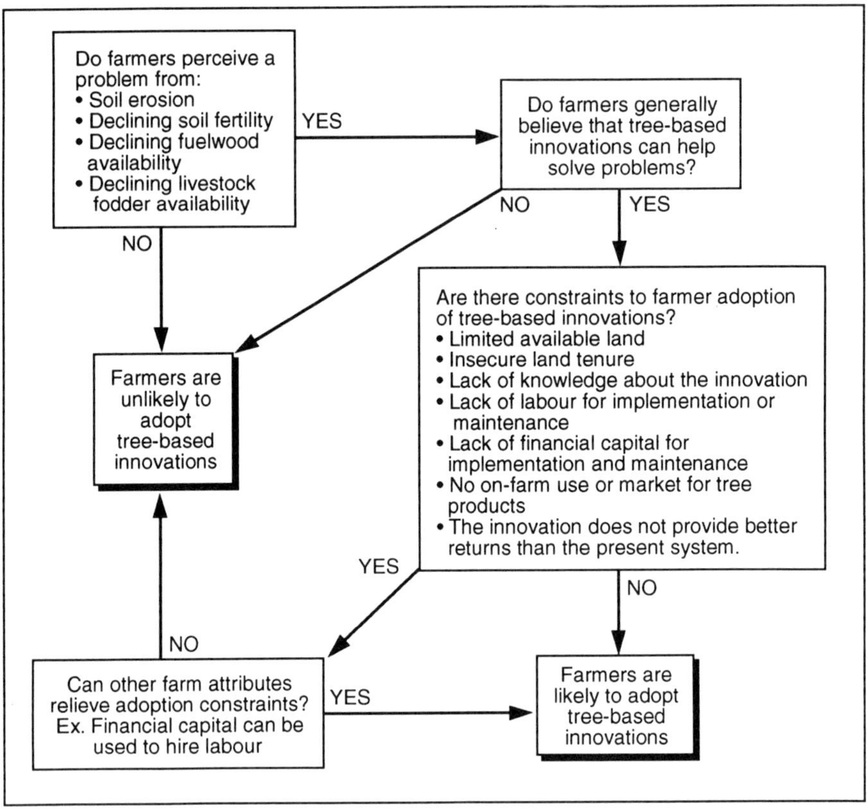


*Figure 1.* AGFADOPT decision logic.

contain several components: choices, the potential system outcomes; rules, the if-then-else statements representing the knowledge; and qualifiers, the individual statements that are combined to make rules. Rules can be further divided into three types (Clancey 1983): strategy rules, used to represent the plan or order of questioning; structural rules, used to reference a specific set of rules for a given conclusion; and support rules, that add conditions to the main conclusions. The THEN or ELSE conditions from one can serve as the IF conditions for subsequent rules. Through this process, rules are linked in a logical sequence that simulates reasoning. The questions in the decision-tree model could readily be translated into qualifiers in the expert system. For example, question 1 in the decision tree, 'Do farmers generally perceive that soil erosion is a problem?' became qualifier 1 in the expert system, 'Farmers generally perceive soil erosion to be <a problem, not a problem>.' These qualifiers were then combined to form a sequence of rules that reproduce the logic of the decision-tree (Figure 1).

Given the complexity of the decision-tree, rule development and organization proved difficult. Although the majority of the rules used in the system were structural rules, the complexity of the decision-tree required several levels of strategy rules. In addition, a number of support rules were added to the system to identify missing data and to provide the user with an indication of the likely effect of missing data on model predictions. After these additions, the prototype version of AGFADOPT increased from 33 model questions to 87 qualifiers used in 72 different rules.

## Evaluation of AGFADOPT

An expert system such as AGFADOPT can be evaluated on a number of criteria. The first and most simple of these is verification of the model syntax and confirmation that the rule-based system is free of error. The second type of evaluation involves verification of the model predictions, also called validation. Does the model correctly recreate the decisions of a human expert? In order to evaluate this question, the model must be tested against field data. The third type of evaluation involves an assessment of performance criteria. These include speed, reasonableness of the interaction and completeness (Plant and Stone 1991). AGFADOPT was assessed on all of these criteria.

The first type of verification, that of the model syntax, proved time-consuming but ultimately straightforward. Systematically developed sets of model inputs were used to test a large number of potential model situations. These included numerous unusual situations (such as large amounts of missing data) where the model was most likely to fail. After several modifications, the current expert system appears to be consistent and free of syntax errors.

Evaluating the accuracy of the model has proven more difficult. Because of the large amount of information used in the model, I have been unable to obtain the completely independent data set necessary to validate the model. However, the model was tested for consistency using data from a 1993 survey of 161 farmers in the Cordillera Central mountain range in the Dominican Republic. The survey was designed to collect baseline information on farming practices and farmer attitudes for use in the planning of future upland development efforts (Robotham 1993; Witter et al. 1996). It contained questions that corresponded to all of the questions used in the expert system. Although this survey data were not used extensively in the development of AGFADOPT, it was considered and, therefore, does not provide the completely independent data set necessary to validate the model. However, it was useful for model verification – to answer the question: 'Is the model able to reproduce the data that were used in model development?'

Results of the verification analysis for the adoption of tree orchard technologies differ significantly from the results for the adoption of contour hedgerows. For tree orchards, the expert system accurately reproduced farmer decisions for 83% (134 of 161) of the farmers surveyed (Table 2). However, for contour hedgerows the expert system accurately reproduced farmer decisions for only 68% (110 of 161) of the farmers surveyed (Table 3).

The third set of evaluation criteria focuses on the user interface and the overall utility of the model. In order to assess this for AGFADOPT, prototype copies were

*Table 2.* Model prediction vs observed farmer behaviour for tree orchard adoption.

| Model prediction | Observed behaviour | | |
|---|---|---|---|
| | Yes | No | Total |
| Yes | 83 | 15 | 98 |
| No | 12 | 51 | 63 |
| Total | 95 | 66 | 161 |

*Table 3.* Model prediction vs observed farmer behaviour for contour hedgerow adoption.

| Model prediction | Observed behaviour | | |
|---|---|---|---|
| | Yes | No | Total |
| Yes | 29 | 29 | 58 |
| No | 22 | 81 | 103 |
| Total | 51 | 110 | 161 |

distributed to colleagues in the USA and in other countries. These colleagues were asked for an evaluation of both system accuracy and performance. The initial responses regarding the AGFADOPT user interface have been positive. However, they represent only limited feedback and do not provide sufficient information for definitive conclusions.

## Limitations of the expert system

The accuracy of AGFADOPT for tree orchard technologies (83%) is similar to the accuracy (80–95%) presented for other decision-tree models (Gladwin 1989). However, the accuracy of AGFADOPT for contour hedgerows (68%) is much lower. There are a number of potential reasons for the discrepancy between the results for trees, which are similar to results of other models, and for contour hedgerows.

First of all, the model is based on community-level generalizations from the literature, while the data used in the model verification were from individual farmers. Scale may be an important issue in this type of analysis and may have different effects depending on the technology being considered. Since many of the benefits accrued from tree orchards result from the sale of products off-farm, tree orchard adoption may be more strongly affected by general, community-wide factors. In contrast, since the benefits from contour hedgerows are mainly confined to the farm itself (e.g. green manure, animal fodder, erosion control), site- and farmer-specific attributes may play a larger role in farmer decision making.

Second, the specific situation in the survey area where verification data were collected brought out other potential weaknesses of the expert system. One of the inaccuracies evident in Table 3 relates to a limitation of the model. The model does not incorporate the use of other contour barrier technologies such as the use of grasses or dead plant materials. Fifteen of the 29 farmers who did not adopt contour hedgerows but were predicted to do so by the model used an alternative barrier technology.

The model also does not differentiate between farmers who use the practices on all of their land and those who use the practices on only a small area. Both circumstances are classified as adoption. By equating partial and complete adoption of technologies, this model may be masking important variations in farmer behaviour that could have significant implications for land management.

## Evaluation of the approach

One goal of this analysis was to evaluate the utility and feasibility of using an expert system to organize and access a knowledge base of information related to farmer adoption of tree-based technologies. The development of AGFADOPT illustrates some difficulties but also some potential insights that come from incorporating even a relatively simple model into a deterministic, rule-based expert system. For example, comparison between system predictions and actual farmer behaviour helped identify the importance of alternative soil erosion technologies in the survey area. This insight provides a potential area for further study.

## Conclusions

The results indicate that even a simple model such as AGFADOPT, based on general principles, may provide a useful way to represent knowledge regarding how a variety of factors influence farmer adoption of tree-based technologies. However, the system still requires significant additional evaluation, regarding both its applicability to other areas and its utility to potential users.

In addition to potentially helping technically trained planners and field personnel make decisions on specific projects or programs, evaluators of the prototype system have suggested that AGFADOPT has potential use as an interactive teaching tool. As such, it might provide a way to introduce students in technical fields to socio-economic factors that affect adoption.

Expert systems such as AGFADOPT can also serve as the basis for much-needed research on those factors which affect agroforestry technology adoption under the wide variety of environmental, social, and political conditions where agroforestry is a potential land use. Comparisons between model predictions and field observations can help identify gaps in the knowledge base and facilitate hypothesis development. Additional research into areas where knowledge gaps exist can expand and improve the existing knowledge base. Effective use of this growing body of knowledge by planners, policy makers, and extension personnel is a prerequisite to the formulation and execution of effective upland development activities benefiting small farmers.

## Acknowledgements

Collection of the survey data used in the model verification was supported by a Title 12 USAID Institutional Support Grant to Dr. Scott Witter, Michigan State University.

The author's ongoing doctoral studies, including the development of AGFADOPT, are supported by a very generous grant from the East-West Center.

## References

Allen B J (1993) The problems of upland land management. Pages 225-241 in Brookfield H, Byron Y (Eds.) Southeast Asia's environmental future: The search for sustainability. United Nations University Press, Tokyo, Japan.
Blair G, Lefoy R (1991) Technologies for sustainable agriculture on marginal uplands in Southeast Asia. Proceedings of a seminar held at Ternate, Cavite, Philippines, 10–14 December 1990, ACIAR Proceedings No. 33. Australian Centre for International Agricultural Research, Canberra, Australia. 128 p.
Carrasco D (1991) The role of hillside farmers in achieving sustainable watershed management: A Dominican Republic example. PhD dissertation, Michigan State University, East Lansing, MI, USA.
de Ceara I A (1987) Land tenure and agroforestry in the Dominican Republic. Pages 301–322 in Raintree J B (Ed.) Land, trees, and tenure. Proceedings of an International Workshop on Tenure Issues in Agroforestry, 27–31 May 1985. International Centre for Research in Agroforestry, Nairobi, Kenya.
Clancey W J (1983) The epistemology of a rule-based expert system: A framework for explanation. Artif. Intel. 20:215–251.
Colfer C J P, Yost RS, Agus F, Evensen S (1989) Expert systems: A possible link from field work to policy in farming systems. AI Applications. Nat. Res. Manage. 3(2):31–40.
Dove M R (1992) Foresters' beliefs about farmers: A priority for social science research in social forestry. Agrofor. Syst. 17:31–41.
Edwards G (1991) Balancing cost-benefit analysis and ecological considerations in developing priorities in r and d in upland agriculture. Pages 15–24 in Technologies for sustainable agriculture on marginal uplands in Southeast Asia. Proceedings of a seminar held at Ternate, Cavite, Philippines, 10–14 December 1990. ACIAR Proceedings No. 33. Australian Centre for International Agricultural Research, Canberra, Australia.
Feder G, Just R E, Zilberman D (1985) Adoption of agricultural innovations in developing countries: A survey. Econ. Dev. Cultural Change 33(2):255–298.
Fortmann L (1987) Tree tenure: An analytical framework for agroforestry projects. Pages 17–33 in Raintree J B (Ed.) Land, trees, and tenure. Proceedings of an International Workshop on Tenure Issues in Agroforestry, 27–31 May 1985. International Centre for Research in Agroforestry, Nairobi, Kenya.
Fortmann L (1990) The view from the farmer: Social dimensions of agroforestry. Pages 63–73 in Agroforestry land-use systems. Proceedings of a Special Session on Agroforestry Land-use Systems in International Agronomy, American Society of Agronomy Annual Meeting, 28–29 November 1988. Nitrogen Fixing Tree Association Special Publication 90-02, Nitrogen Fixing Tree Association, Waimanalo, HI, USA.
Francis P (1987) Land tenure systems and the adoption of alley farming in southern Nigeria. Pages 175–180 in Raintree J B (Ed.) Land, trees, and tenure. Proceedings of an International Workshop on Tenure Issues in Agroforestry, 27–31 May 1985. International Centre for Research in Agroforestry, Nairobi, Kenya.
Fujisaka S (1989) The need to build upon farmer practice and knowledge: Reminders from selected upland conservation projects. Agrofor. Syst. 9:141–153.
Fujisaka S (1994) Learning from six reasons why farmers do not adopt innovations intended to improve sustainability in upland agriculture. Agric. Syst. 46:409–425.
Fujisaka S, Jayson E, Dapusala A (1994) Trees, grasses, and weeds: Species choices in farmer-developed contour hedgerows. Agrofor. Syst. 25:13–22.
Garrity D P (1993) Sustainable land-use systems for sloping uplands of Southeast Asia. Pages 41–66 in Technologies for sustainable agriculture in the tropics. ASA Special Publication No. 56. American Society of Agronomy, Madison, WI, USA.
Gladwin C H (1983) Contributions of decision-tree methodology to a farming systems program. Hum. Organ. 42(2):146–157.
Gladwin C H (1989) Ethnographic decision tree modeling. Sage University Publication Inc., Beverly Hills, CA, USA. 95 p.
Gladwin C H, Murtaugh M (1980) The attentive-preattentive distinction in agricultural decision making. Pages 115–134 in Barlett P F (Ed.) Agricultural decisionmaking: Anthropological contributions to rural development. Academic Press, New York, NY, USA.

Granert W G, Sabueto W Q (1987) Farmers' involvement and use of simple methods; Agroforestry strategies for watershed protection and management. Pages 227–235 in Vergara N T, Briones N B (Eds.) Agroforestry in the humid tropics. Southeast Asian Regional Center for Graduate Study and Research in Agriculture, College, Laguna, Philippines.

Hansen D O, Erbaugh J M, Napier T L (1987) Factors related to adoption of soil conservation practices in the Dominican Republic. J. Soil Water Conser. 42(5):367–369.

Hosier R H (1989) The economics of smallholder agroforestry: Two case studies. World Dev. 17(11): 1827–1839.

Juo A S R (1989) New farming systems development for the wetter tropics. Exp. Agric. 25:145–163.

Kunreuther H, Wright G (1979) Safety-first, gambling, and the subsistence farmer. Pages 213–230 in Roumasset J A, Boussard M, Singh I (Eds.) Risk, uncertainty, and agricultural development. Southeast Asian Regional Center for Graduate Study and Research in Agriculture, College, Laguna, Philippines.

Laquihon W A (1989) Some key determinants of SALT adoption in the Philippines: Viewpoints of farmer-cooperators. Pages 79–116 in Vergara N T, Fernandez R A (Eds.) Social forestry in Asia. Southeast Asian Regional Center for Graduate Study and Research in Agriculture, College, Laguna, Philippines.

Loh D K, Hsieh Y-T C, Choo Y K, Holtfrerich D R (1994) Integration of a rule-based expert system with GIS through a relational database management system for forest resource management. Comput. Electron. Agric. 11:215–228.

MacDicken K G, Vergara N T (1990) Introduction to agroforestry. Pages 1–30 in Agroforestry: Classification and management. John Wiley and Sons, New York, USA.

Mercer D E (1992) The economics of agroforestry. Pages 111–143 in Burch W R Jr, Parker J K (Eds.) Social science applications in Asian Agroforestry. Winrock International, Petit Jean Mountain, AR, USA.

Mohan S, Arumugam N (1994) CROPES: A rule-based expert system for crop selection in India. Trans. ASAE 37(3):1355–1363.

Napier T L (1991) Factors affecting acceptance and continued use of soil conservation practices in developing societies: A diffusion perspective. Agric. Ecosyst. Environ. 36:127–140.

Napier T L, Napier A S, Tucker M A (1991) The social, economic, and institutional factors affecting adoption of soil conservation practices: The Asian experience. Soil Tillage Res. 20:365–382.

Ortiz S (1979) The effect of risk aversion strategies on subsistence and cash crop decisions. Pages 231–246 in Roumasset J A, Boussard M, Singh I (Eds.) Risk, uncertainty, and agricultural development. Southeast Asian Regional Center for Graduate Study and Research in Agriculture, College, Laguna, Philippines.

Padoch C, De Jong W (1987) Traditional agroforestry practices of native and ribereno farmers in the lowland Peruvian Amazon. Pages 179–194 in Gholz H L (Ed.) Agroforestry: Realities, possibilities, and potentials. Martinus Nijhoff Publishers (Kluwer), Boston, MA, USA.

Plant R E, Stone N D (1991) Knowledge-based systems in agriculture. McGraw-Hill, New York, USA. 354 p.

Raintree J B (1983) Strategies for enhancing the adoptability of agroforestry innovations. Agrofor. Syst. 1:173–187.

Raintree J B (1991) Socio-economic attributes of trees and tree planting practices. Community Forestry Note No. 9. Food and Agriculture Organization, Rome, Italy. 115 p.

Repollo Jr. A Q, Castillo E R (1989) Agroforestry technology in hilly land households: Factors influencing its adoption. Pages 117–132 in Vergara N T, Fernandez R A (Eds.) Social forestry in Asia. Southeast Asian Regional Center for Graduate Study and Research in Agriculture, College, Laguna, Philippines.

Riddell J C (1987) Land tenure and agroforestry: A regional overview. Pages 1–16 in Raintree J B (Ed.) Land, trees, and tenure. Proceedings of an International Workshop on Tenure Issues in Agroforestry, 27–31 May 1985. International Centre for Research in Agroforestry, Nairobi, Kenya.

Robotham M (1993) The relationship between Plan Sierra outreach activities and the adoption and continued use of soil and water conservation technologies by upland farmers. MS thesis, Michigan State University, East Lansing, MI, USA. 150 p.

Rocheleau D E (1987) The user perspective and the agroforestry research and action agenda. Pages 59–87 in Gholz H L (Ed.) Agroforestry: Realities, possibilities, and potentials. Martinus Nijhoff Publishers (Kluwer), Boston, MA, USA.

Rogers E (1983) Diffusion of innovations. 3rd Edition. The Free Press, New York, USA.

Roumasset J A (1979) Introduction and state of the arts. Pages 3–22 in Roumasset J A, Boussard M, Singh I (Eds.) Risk, uncertainty and agricultural development. Southeast Asian Regional Center for Graduate Study and Research in Agriculture, College, Laguna, Philippines.

Sajise P E (1987) Agroforestry and land tenure issues in the Philippines. Pages 273–276 in Raintree J B (Ed.) Land, trees, and tenure. Proceedings of an International Workshop on Tenure Issues in Agroforestry, 27–31 May 1985. International Centre for Research in Agroforestry, Nairobi, Kenya.

Segura de los Angeles M (1986) Upland economics and economic impact analysis. Pages 169–188 in Fujisaka S, Sajise P E, del Castillo R A (Eds.) Man, agriculture, and the tropical forest. Winrock International, Bangkok, Thailand.

Tversky A (1972) Elimination by aspects: A theory of choice. Psychol. Rev. 79(4):281–299.

Vergara N T, MacDicken K G (1990) Extension and agroforestry technology delivery to farmers. Pages 354–373 in Agroforestry: Classification and management. John Wiley and Sons, New York, USA.

Warkentin M E, Nair P K R, Ruth S R, Sprague K (1990) A knowledge-based expert system for planning and design of agroforestry systems. Agrofor. Syst. 11:71–83.

Watson G A (1990) Tree crops and farming systems development in the humid tropics. Exp. Agric. 26: 143–159.

Watson H R, Laquihon W A (1987) Sloping agricultural land technology: An agroforestry model for soil conservation. Pages 209–226 in Vergara N T, Briones N D (Eds.) Agroforestry in the humid tropics. Southeast Asian Regional Center for Graduate Study and Research in Agriculture, College, Laguna, Philippines.

Witter S G, Robotham M P, Carrasco DA (1996) Plan Sierra, outreach programming, and sustainable adoption of soil and water conservation technologies by upland farmers. J. Soil Water Conserv. (in press).

Yost R S, Uehara G, Wade M, Sudjadi M, Widjaja-adhi I P G, Li Z C (1988) Expert systems in agriculture: Determining lime recommendations for soils of the humid tropics, University of Hawaii CTAHR Research Extension Series Publication No. 089. Honolulu, HI, USA.

# Systems thinking and practice for alley cropping research

S. SEITER and R.D. WILLIAM
*Department of Horticulture, Oregon State University, Corvallis, OR 97331, USA*

**Key words:** alley cropping, systems thinking, research methodology

**Abstract**
Systemic and reductionist methodologies were applied to explore alley cropping systems. Alder (*Alnus rubra*) and black locust (*Robina pseudoacacia*) are grown as green manure in sweet corn production systems. Our research includes the study of biological, ecological, and social interactions. The overarching approach follows the Hawkesbury systems agriculture model in which interactions are explored on four levels of research. On a basic science level, we study the soil food web composition and nitrogen cycling in the plant-soil system by using a new $^{15}N$ method that involves direct injection of the tracer into xylem of the trees. Applied field research involves the response of trees and crops to various planting patterns and N equivalence trials. Optimized systems inquiry is a third level of inquiry involving modelling of alley cropping systems limited by light and nitrogen. Field data are compared with predicted tree and crop growth. Finally, human action inquiry investigates the social components of this cropping system. We use participatory appraisal techniques such as farmer-scientist focus sessions and participatory mapping to improve the farmers' ability to deal with complexity in their farming system and to learn about farmers' decision making.

## Introduction

Agroforestry is a complex and dynamic land-use system for integrated food and fibre production and ecosystem protection. The practice requires an intimate knowledge of trees, crops, soil, and animals and an understanding of the interactions among these components. The ecology of an agroforestry system also involves human activities, decisions, and values that play a major role in its implementation.

Agroforesters have recognized that a systems approach would benefit research in complex cropping systems. For example, Kang and Van Heide (1985) employ a systems approach to nitrogen management in tropical agroforestry. A systems approach to study the interacting soil, plant and management processes was suggested by Nair (1984). Systems approaches in agroforestry have been influenced especially by the farming systems research and extension (FSR/E) movement and have considered also the social constraints on agroforestry systems. For example, Jones and Price (1985) describe how a FSR/E approach can be used in the analysis of agroforestry systems and suggest that farmer motivation and priorities play a crucial role. These examples demonstrate the diversity of systems approaches in agroforestry research. Common to all is the use of general systems-thinking ideas and practices.

Major concepts in systems thinking include: holism (i.e. a system is comprised of interacting components that function as a whole), communication and regulatory control between components (i.e. feedback loops), hierarchy (i.e. subsystems of lower complexity are nested in larger systems), and emergent properties on each level of



*P.S. Teng* et al. *(eds.): Applications of Systems Approaches at the Farm and Regional Levels, 167–176.*

complexity (Kauffman 1980; Checkland 1981). These concepts are used to explain and investigate agroforestry systems. For example, a feedback loop in an alley cropping system could consist of trees shading the crop, thereby reducing crop growth and residue which increases weediness, leading to reduced tree and crop yield. When these plant interactions cause farmers to change practices such as the coppicing regime, interactions involve the human management and the plant biology sub-systems, which illustrates the hierarchy concept.

## Systems practice

In agricultural and natural resource research, the scientific reductionist approach has been extremely effective to learn about basic biological and ecological processes. However, in situations that require analysis of complex interactions between system components and the synthesis of knowledge from multiple perspectives, traditional reductionist research approaches often were not very successful. Furthermore, traditional science has not been able to address people's concerns and values adequately. The ongoing world-wide debate about natural resource use clearly demonstrates that solving technical problems is only one part of the issue.

New methodologies such as Agroecosystems Analysis (Conway 1985) or various FSR/E approaches (Caldwell 1987) have been developed in recent decades to address complex agricultural and natural resource problems. Checkland and Scholes (1990) employ the soft systems methodology to resolve issues in human activity systems involving human values and decisions. This type of inquiry focuses on human management of agricultural resources and the expected consequences and interactions (Wilson and Morren 1990). The methodology may, therefore, be suited for alley cropping research.

Soft systems inquiry is one of four major approaches in agricultural science (Wilson and Morren 1990). The others are: hard systems inquiry, applied science, and basic science. Each approach handles a varying amount of complexity. Soft systems inquiry on one end and basic science on the other can be thought of as being the poles of an axis that stretches from holistic integration to reductionist separation. The four research approaches form the basis for 'Systems Agriculture', an over-arching methodology developed by faculty of the Hawkesburry College in Australia. Their vision is to train agricultural scientists who are familiar with each of these approaches and depending on the situation, can choose a systemic, a reductionist, or an integrated approach (Bawden 1991).

Regardless of the approach selected, a complete research or learning process involves at least four phases. Each phase requires different abilities from the learner (Kolb 1987). In the first phase, the learner observes and tries to obtain a complete picture of the situation (ability to diverge). Then the gathered information is assimilated and conceptualized to give the observations some meaning (ability to assimilate). This is followed by developing a plan of action (ability to converge) and, finally, the learner takes action and accommodates the outcome with the reality of the experience (ability to accommodate).

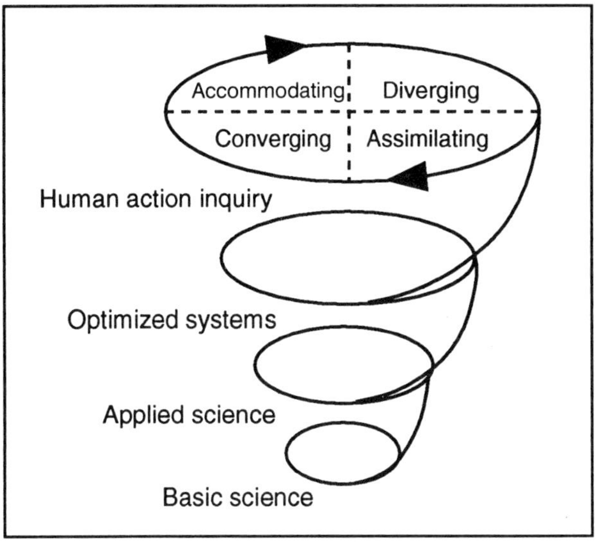


*Figure 1.* Spiral of interrelated research methodologies showing the cyclic learning process (*Source:* adapted from Bawden 1991).

The Hawkesburry group connected its research methodology with Kolb's learning theory. They visualized this connection as a spiral in which the levels of the spiral symbolize the different research methodologies (Figure 1). To complete a research process, the learner has to cycle through all learning phases. However, depending on the level of complexity of the problem, the learner can spiral up and down (i.e. select a certain method of inquiry). The spiral also symbolizes close linkages in the hierarchy of interrelated methodologies. As Srikandarajah et al. (1991) stated: 'Each level of inquiry provides a perspective and a clearer focus on intent for the subsequent level and each lower level provides insights for the higher levels'.

## The use of systems ideas in alley cropping

We used the Hawkesburry systems agriculture model with some modifications to guide our alley cropping research program. Most agricultural researchers are familiar with the terms and concepts of basic science and applied science. However, the terms 'hard systems' and 'soft systems' continually cause misunderstandings. The soft systems approach also has been criticized to be exclusive and magnifying existing power differences between people (Flood and Jackson 1991). Therefore, we changed the terms of our systems approaches. Instead of hard systems we use optimized systems reflecting the goal of this approach. 'Soft systems' is replaced by human action research. Ison (1995) defined action research as the linking of systems thinking to practice in a given context. Philosophically, our approach is close to participatory action research employing techniques such as participatory rural appraisal (Chambers

1992). Thus, human action research reflects our goal to move towards action and changes in resource management while integrating diverse perspectives.

Our alley cropping research is part of a long-term program exploring various vegetation management options that enhance profitable crop yield while protecting ecosystem integrity. Rows of the nitrogen fixing tree species *Alnus rubra, Robina pseudoacacia,* and *Alnus sinuta* were established in 1991 and intercropped with sweet corn since 1992. Corn and trees were planted in a variety of planting patterns to explore plant interactions. What follows is a summary of the key steps in each approach and a general description of the type of alley cropping research we initiated.

*Basic science inquiry*

In this methodology, the problem is reduced to small puzzles based on facts that are thought to be relevant. Hypotheses are proposed that may explain the perceived puzzles. These hypotheses then are experimentally tested and, after several replications, either confirmed or invalidated. In agricultural research, this often involves highly controlled experiments in the laboratory or small field plots to ascertain or quantify cause and effect relationships.

Many of the basic science questions remain to be answered in alley cropping research. In particular, basic plant physiological and soil ecological processes need to be explored to improve the cropping system. These processes include: quantification of biologically fixed $N_2$ by various woody species, role of mycorrhiza in nutrient uptake, and effect of trees on soil biota (Kang et al. 1990). In recent years, the number of alley cropping experiments conducted under highly controlled experimental conditions has increased, in particular studies that have used stable isotopes to investigate nutrient fluxes (Sanginga et al. 1990; Haggar et al. 1993; Xu et al. 1993).

Our research also focuses on nutrient fluxes. While nitrogen may be effectively recycled in alley cropping systems, precise quantification of nitrogen fluxes through the soil and plant pools and particularly nitrogen transfer between trees and crops is sparse. Our experiments investigate these N-fluxes involving a novel method in which $^{15}N$ is directly injected into the tree xylem (Horwath et al. 1992). With a single $^{15}N$ application, we are able to follow nitrogen movement into the various plant and soil pools over several years. The experimental design allows us to differentiate between nitrogen entering the system from tree roots and from tree prunings. In this study, one of our hypotheses was that nitrogen released from trees by fine root turnover is transferred quickly to the corn and slower from soil incorporated tree prunings. Preliminary results validate this. A substantial amount of nitrogen was transferred from tree roots to the annual crop while nitrogen from tree prunings remained immobilized in the undecomposed plant tissue and the soil organic matter.

Alley cropping may improve soil quality through the addition of different forms of organic matter. One important indicator of soil quality is soil microbial activity (Visser and Parkinson 1992). We are interested in quantitative and also qualitative changes in soil microbial activity resulting from alley cropping. For this reason, we are studying multiple trophic levels of the soil foodweb including soil microbes and their predators using direct microscopy methods. We hypothesized that the addition of green manure will generally increase the microbial biomass and change the soil foodweb composition compared to monocropped corn. Because of tree root activity, we expected to find differences in the microbial biomass depending on sample location in the alley. We found that microbial activity was much higher in the tree rows. As the growing season progressed, we observed more microbial biomass in corn rows next to trees compared to corn rows in the middle of the alley.

In systems-oriented research programs, linkages exist horizontally (i.e. same level of inquiry) and vertically (i.e. different levels of inquiry). The two studies described are closely linked horizontally since microbial biomass and type of soil microbes establish the balance between nitrogen mineralization and immobilization and thus determine the amount of nitrogen transferred from the trees to the other plant and soil pools. For us, the appropriate vertical linkage between basic science and applied science inquiry exists when the applied research seeks to demonstrate how relevant the basic processes studied are to production agriculture.

*Applied science inquiry*

In an applied science inquiry the problem is reduced to smaller units during the definition phase similar to the basic science approach. While basic science inquiry seeks to explain why something occurs, applied scientists ask what and how can it be done. Since most applied research is conducted in field plots where variable environmental conditions have to be integrated, applied science inquiry deals with a higher degree of complexity.

A review of research goals and needs shows the applied science focus in alley cropping research. The goal has been to contribute to the design of a cropping system that provides high crop yields while improving or maintaining soil resources. Reported research needs deal with what has to be done to achieve this goal and to gather information from various fields to obtain a greater understanding of what is happening in applied agroforestry situations (Huxley 1986). Information sought is concerned with finding suitable tree species, optimum plant densities and field management practices (Seskabembe 1985).

Our research on this level of the spiral involves nitrogen equivalence field trials. We hypothesized that, if green manure is decomposed fast enough, mineral nitrogen fertilizer could be substituted through alley cropping. To test this hypothesis, large field plots were planted with either monocropped or alley cropped sweet corn and fertilized at several nitrogen rates in a replicated design. We found that the green manure produced in the alley cropping system substituted some nitrogen but mineral nitrogen fertilizer always had to be added to obtain acceptable corn yield.

In another replicated field trial, trees and crops were planted in various densities. We compared potential tree species in several planting patterns. Under irrigated conditions, corn in rows next to the coppiced trees seemed to take advantage of higher levels of incoming radiation compared to corn in the middle of the alley. The increased corn yield almost compensated for the reduced corn growing area in the alley cropping systems compared to the monocropped corn. Very close spacing between trees and crops, however, significantly reduced corn yield.

The planting patterns that we compared were feasible for farmer implementation. Using farmer feasibility as a design criterion demonstrates the close linkage between applied science and human action inquiry. Participatory on-farm research demonstrates how farmers in particular, but also other players in a human activity system such as extensionists, food processors, or conservation groups, are easily integrated in design, implementation, and evaluation of an applied research program (Seiter et al. 1994). Another vertical linkage exists by using data from applied research experiments to validate predicted results of modelling activities during the optimized systems inquiry.

*Optimized systems inquiry*

In optimized systems inquiry, one attempts to deal with the system as if it is a whole with interactive parts. Research activities lead towards an agreed upon output or product (i.e. the optimization of a system). First, the problematic situation is described and the purpose of the inquiry defined. Synthesizing information from various perspectives, a systems model is designed next to study the interactions between system components. The model allows the systems analyst to test alternative strategies, from which one is selected and validated in relation to the original problem. Model types range from diagrammatic representation of component interactions to more elaborate models involving mathematical relationships to predict system behaviour under certain environmental conditions.

Hard system models now exist for the biophysical and socioeconomic component interactions in alley cropping systems. For example, Young et al. (1987) developed a computerized program that predicts changes in soil properties under specified agroforestry systems within various environments. Other models deal with radiation and nitrogen supply in alley cropping systems (Nygren and Jimenez 1993). They use geographic and climatic parameters of one location and growth data from published values to predict optimum tree arrangement. Swift et al. (1991) compared changes in soil organic matter predicted by a simulation model with experimental data for maize monocropping and alley cropping. Mendora (1987) proposed a mathematical allocation model that generates feasible land-use alternatives in Nigeria, including alley cropping.

We are currently developing a model of tree-crop competition for light and nitrogen (Figure 2). Observations from the applied and basic research experiments will be used to test the accuracy of the model. We used conceptual models at all stages and at all levels to clarify and communicate relationships between biotic and/or abiotic components of the cropping system. Frequently, these representations of agricultural systems or agroecosystem processes are diagrammed collaboratively by farmers, researchers, and extensionists and used in the exploration of human activity systems.

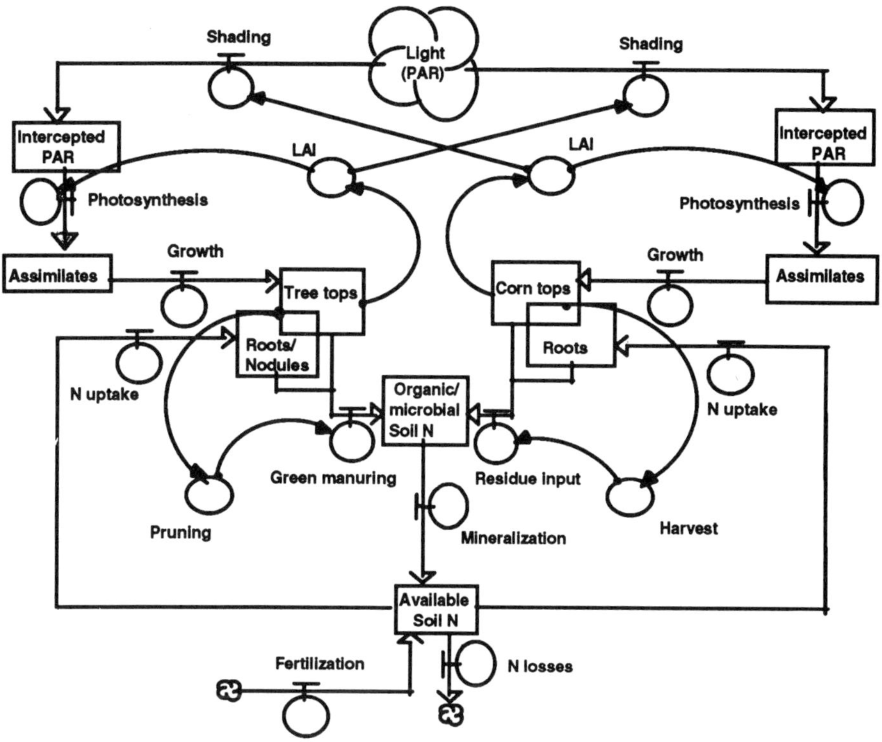


*Figure 2.* Model of an alley cropping system limited by light and nitrogen.

## *Human action inquiry*

Human action research is the exploration of views, values, and perceptions of various stakeholder groups. The research process begins by assembling a 'rich picture' of problematic situations or opportunities (Checkland 1981). This includes not only a diverse collection of factual information, but also an inquiry into the assumptions and worldviews or *Weltanschauungen* of the people who solicit, own, or control the information. Next, common themes among the many problem descriptions are identified and action statements (transformations) that focus on an improved future state are defined. Before changes are implemented, pictures of the future are compared to reality; then feasibility and desirability of the proposed changes are debated from multiple perspectives. Postponing judgement until this final stage is a key to success of this approach (Flood and Jackson 1991). Human action research is participatory and interdisciplinary. Participants agree to become co-learners who explore alternatives and discover the changes that would have to be made in the system to reach a common goal.

Participatory research has been an integral part of alley cropping research. The Diagnosis and Design methodology of the International Center for Research in

Agroforestry (ICRAF), for example, allows farmers and extension agents to be actively involved in the research process (Scherr 1989). Sumberg and Okali (1988) note that in complex technologies such as alley cropping, farmers' role in the technology development is critical and cost-effective. Many alley cropping studies now employ FSR/E approaches to take advantage of farmers' knowledge and to develop a better understanding of the production system (Heijer and Hombergh 1990; Dvorak 1991; Reynolds et al. 1991). There is, however, a critical difference between the systemic inquiry into human activity systems and the FSR/E perspective. FSR/E approaches are focused on the object of agricultural systems that exist in the world. Systems inquiry instead is focused on the process of people's perception of reality with the goal to create a learning system in which collaboration leads to desirable and feasible change (Bawden 1991).

Our efforts on the action inquiry level investigate human components of local farming systems. Agricultural production in western Oregon (USA) is dominated by high-value vegetable crops. Traditionally, these are planted as large monocrops with high input of pesticides and chemical fertilizer. Many growers, vegetable processors, government agencies, and the public are expressing concerns over potential detrimental effects to the environment caused by these practices. With them, social and agricultural science researchers explore the potential of alternative cropping systems, including alley cropping. To improve farmers' ability to deal with complexity in their farming system and to learn about farmers' decision making, we use participatory appraisal techniques such as participatory mapping, matrix scoring (Chambers 1992), and farmer-scientist focus sessions (Lev et al. 1993). Farmer-scientist focus sessions are specifically designed to share learning and improvements among the farmers, to investigate complex cropping system issues that require interdisciplinary examination and to explore diverse views, values and beliefs. The goal of this type of systemic action research is to give ownership of both process and results to participating farmers.

## Learning from systems practice

As we explore alley cropping and other alternatives that enhance profitable crop production with environmental stewardship, we have found the Hawkesburry model a powerful and challenging tool. For us, the spiral provides the path along which we keep our research flexible, truly interdisciplinary, and open to the changing demands and insights. Ideally, a research process begins with inquiry on the human-activity system level to learn about the opportunities and problems. With the participation of all players involved, broad research goals and specific objectives are formulated and the appropriate level of inquiry to investigate the situation is identified.

In this paper, the systems agriculture approach has been described as an overarching methodology for our research program. Initially, however, the approach had been developed as a way to improve student education (Bawden et al. 1984). By employing this approach, education shifts from a teaching to a learning paradigm. A learning environment is created where the boundaries between teachers and students blur. All participants in systems education become co-learners. If an institutional change goes hand in hand with the changes in the research and education programs, then conditions are provided in which learning can take place through experience, open and equal interactions, and personal exploration and experimentation (Pretty and Chambers 1994).

The nature of problems facing agricultural science is changing. The public asks farmers to produce food and fibre without harming the environment. Farmers in Oregon are looking for ways to meet this demand and to sustain agricultural productivity for future generations (Hatfield et al. 1994). The questions they raise often transcend disciplinary boundaries. To answer these questions, Jiggins (1994) suggests that future graduates in agricultural science will be required to integrate information from diverse perspectives and disciplines and to have a strong conceptual, methodological, and analytic understanding of how to introduce desired emergent properties in dynamic systems. By training scientists who are comfortable moving along the spiral through all phases of the learning cycle, we believe that we are able to better meet these challenges. At the same time, we feel that our action research approach contributes to collaborative learning among farmers and researchers and strengthens farmers' capability to experiment and manage the resource of their production system in harmony with their physical and social environment.

## References


Bawden R J (1991) Systems thinking and practice in agriculture. J. Dairy Sci. 74:2362–2373.

Bawden R J, Macadam R D, Packham R J, Valentine I (1984) Systems thinking and practice in the education of agriculturalists. Agric. Syst. 13:205–225.

Caldwell J (1987) An overview of farming systems research and development: Origins, applications, and issues. Pages 165–178 in Gittinger J, Leslie J, Hoisington C (Eds.) Food policy. Johns Hopkins Press, Baltimore, MD, USA.

Chambers R (1992) Rural appraisal: Rapid, relaxed, and participatory. Discussion Paper No. 311, Institute of Development Studies, University of Sussex, Brighton, UK. 90 p.

Checkland P (1981) Systems thinking, systems practice. John Wiley and Sons, New York, USA. 319 p.

Checkland P, Scholes J (1990) Soft systems methodology in action. John Wiley and Sons, New York, USA. 329 p.

Conway G (1985) Agroecosystems analysis. Agric. Administration. 20:31.

Dvorak K A (1991) Methods of on-farm, diagnostic research on adoption of alley cropping. Agrofor. Syst. 15(2–3):167–181.

Flood R L, Jackson M C (1991) Creative problem solving: Total systems intervention. John Wiley and Sons, New York, USA. 347 p.

Haggar J P, Tanner E V J, Beer J W, Kass D C L (1993) Nitrogen dynamics of tropical agroforestry and annual cropping systems. Soil. Biol. Biochem. 25(10):1363–1378.

Hatfield C, Hatfield D, Boop P, Boop W, William R D (1994) Communicating between farmers and scientists: A story about stories. Am. J. Alt. Agric. 9(4):186–187.

Heijer R P F, Hombergh H G M (1990) A model based approach to farming systems research: The construction of a theoretical framework applied to the *Leuceana* base farming systems in two villages of Sumbawa, Indonesia. FONC project communication. Yogyakarta, Indonesia.

Horwath W R, Paul E A, Pregitzer K S (1992) Injection of $^{15}N$ into trees to study nitrogen cycling in soil. Soil Sci. Soc. Am. J. 56(1):316–319.

Huxley P A (1986) Rationalizing research in hedgerow intercropping: An overview. International Council for Research in Agroforestry, Nairobi, Kenya. 146 p.

Ison R L (1995) Designing learning systems: How can systems approaches be applied in the training of research workers and development actors? In Proceedings of the International Symposium on Systems-

oriented Research in Agriculture and Rural Development, November 1994, Montpellier, France (in press).
Jiggins J (1994) Systems approaches: The challenge to university teaching. Pages 791–795 in Proceedings of the International Symposium on Systems-oriented Research in Agriculture and Rural Development, November 1994, Montpellier, France.
Jones J R, Price N (1985) Agroforestry: An application of farming systems approach to forestry. Human Org. 44(4):322–331.
Kang B T, Reynolds L, Atta-Krah A N (1990) Alley farming. Adv. Agron. 43:315–359.
Kang B T, Van Heide J (1985) Nitrogen management in farming systems in the humid and subhumid tropics. Institute for Soil Fertility, Haren, The Netherlands. 362 p.
Kauffman D L Jr. (1980) An introduction to systems thinking. Pegasus Communications, Cambridge, MA, USA. 40 p.
Kolb D (1987) Experiential learning: Experience as the source of learning and development. Prentice-Hall Inc., NJ, USA. 256 p.
Lev L D, McGrath D, Murrary H, William R D (1993) Organising and conducting farmer-scientist focus sessions. J. Nat. Res. Life Sci. Edu. 22(2):148–152.
Mendora G A (1987) A mathematical model for generating land-use allocation alternatives for agroforestry systems. Agrofor. Syst. 5(4):443–453.
Nair P K R (1984) Soil productivity aspects of agroforestry. International Council for Research in Agroforestry, Nairobi, Kenya. 85 p.
Nygren P, Jimenez J M (1993) Radiation and nitrogen supply in modeled alley cropping systems of *Erythrina poeppigiana* with sequential maize-bean cultivation. Agrofor. Syst. 21(3):271–285.
Pretty J N, Chambers R (1994) Towards a learning paradigm: New professionalism and institutions for sustainable agriculture. Pages 182–202 in Scoones I, Thompson J (Eds.) Beyond farmers first. Intermediate Technology Publications, London, UK.
Reynolds L, Di Domenico C, Atta-Krah A N, Cobbina J (1991) Alley farming in south-western Nigeria: The role of farming systems research in technology development. Pages 85–108 in Tripp R (Ed.) Planned change in farming systems: Progress in on-farm research. John Wiley and Sons, Chichester, UK.
Sanginga N, Zapata F, Danso S K A, Bowen G D (1990) Effect of successive cutting on uptake and partitioning of $^{15}N$ among plant parts of *Leuceana leucocephala*. Biol. Fertil. Soils. 9:37–42
Scherr S J (1989) The diagnosis and design approach to agroforestry planning and implementation: Examples from western Kenya. Paper presented at the International Symposium on Planning for agroforestry, April 1989, Pullman, WA, USA.
Seiter S, William R D, Luna J, McGrath D (1994) Mutual learning in participatory on-farm research project in Oregon. Pages 863–864 in Proceedings of the International Symposium on Systems-oriented research in agriculture and rural development, November 1994, Montpellier, France.
Seskabembe C K (1985) Perspectives in hedgerow intercropping. Agrofor. Syst. 3(4):339–356.
Srikandarajah N, Bawden R J, Packham R G (1991) Systems agriculture: A paradigm for sustainability. Association of Farming Systems Research and Extension (AFSRE) newsletter Vol. 2(3).
Sumberg J, Okali C (1988) Farmers, on-farm research and the development of new technology. Exp. Agric. 24:333–342.
Swift M J, Kang B T, Mulongoly K, Woomer P (1991) Organic matter management for sustainable soil fertility in tropical cropping systems. Pages 307–326 in Proceedings of the Evaluation for Sustainable Land Management in the Developing World, Volume 12. International Board for Soil Research and Management, Bangkok, Thailand.
Visser S, Parkinson D (1992) Soil biological criteria as indicators of soil quality: Soil microorganisms. Am. J. Alt. Ag. 7(1):33–37.
Wilson K, Morren G E B Jr. (1990) Systems approaches for improvement in agriculture and resource management. Macmillan, New York, USA. 361 p.
Young A, Cheatle R J, Muraya P (1987) The potential of agroforestry for soil conservation, III. Soil changes under agroforestry (SCUAF): A predictive model. Working Paper No. 44, International Council for Research in Agroforestry, Nairobi, Kenya.
Xu Z H, Meyer R J K, Saffigna P G, Chapman A L (1993) Nitrogen fertilizer in *Leuceana* alley cropping. Fertil. Res. 34(1):1–8.

# Economic and financial performance evaluation of a farm in the Brazilian savannas

R.F. VELOSO[1], E.R.O. CARVALHO[2] and A.M. GOULART[2]

[1]*Empresa Brasileira de Pesquisa Agropecuária, Centro de Pesquisa Agropecuária dos Cerrados (EMBRAPA/CPAC), Caixa Postal 08 223, CEP 73301-70, Planaltina DF, Brazil*

[2]*Cooperative Coopertinga, Rua Olimpio Jacinto 538, CEP 73800-00, Formosa, Goiás State, Brazil*

**Key words:** farm management, agricultural project, cerrados agricultural development

**Abstract**

The Brazilian savannas (cerrados) represent one of the last major agricultural frontiers in the southern hemisphere. The Japanese government has been interested in promoting the rational development of this region through pilot projects with full support of the Brazilian government and selected agricultural producers organized in cooperatives. Some farm management and system research tools have been proposed to provide new instruments for planning sustainable development in environmental, economic and social terms for the cerrados. Before formulating a mathematical model for the agricultural production processes of the cerrados, under given environmental, social, economic and political constraints, a data base containing current financial and economic performance parameters of an established farming system, together with some qualitative information on job creation and essential infrastructure, is being developed. The major objective is to develop a framework for *ex ante* evaluation of integrated regional planning schemes based on evaluation of real-world data. This evaluation effort is in progress and is generating farm management information that will be used to develop a comprehensive farming system model. Other institutions in Latin America such as CIAT in Colombia and the 'Universidade de Uberlândia – Minas Gerais State' have shown interest in this modelling approach, and a cooperative effort could be established in the near future to develop new instruments to support policy makers in agricultural and rural sectors.

## Introduction

The purpose of this paper is to present a cerrados farm case study which has been carried out to support (*ex ante*) resource allocation decisions at the farm and integrated regional planning levels.

Initially, a crop farm development project is described. This project incorporates 23 000 ha of cultivated land and 15 000 ha as ecological reserve in 53 farms located 260 km from Brasília in the northwest of Minas Gerais state in Brazil. The farmers involved in this project have organized a cooperative with a US $5 million asset value and US $8.6 million annual business turn-over. A description of a specific farm-level business and its activities follows. Finally, the cerrados crop farm outlook in the near future and the importance of the study in integrated regional planning schemes are discussed.

The Coopertinga cooperative was selected due to its location, previous work leading to the generation of technical data and the fact that it did not significantly differ from other cooperatives participating in the cerrados development program (PRODECER). The farm for the case study was selected by the cooperative's



*P.S. Teng* et al. *(eds.): Applications of Systems Approaches at the Farm and Regional Levels, 177–190.*

technical director as representative of other farms on the basis of its size, nature of soil, crop activities and the volume of business. Also, the selection was influenced by the availability of agronomic data.

## Problem definition

In the past twenty years, Brazilian agriculture has changed significantly as a consequence of the incorporation of new areas in the Brazilian midwest. Due to agricultural policies aiming to establish a basic infrastructure (such as storage and processing facilities) and research and development investments, the development (with subsidized credit) of highly capital-intensive annual crop production systems in the plateau areas of the cerrados was made possible (Mueller et al. 1992). In 1994, the Brazilian cerrados supplied around 8.75 million tons of soybean and supported 50 million head of cattle. However, this has been achieved at the cost of considerable environmental degradation and in spite of obstacles such as distance to markets, transportation problems, energy supply costs, location of the processing centres and the existence of many farmers with poor farm management skills.

Agricultural production in 1994, was around 11% in the US $470 billion total value of GDP (gross domestic product), but the contribution of the entire Brazilian agribusiness was approximately 40% of GDP. The level of modern input used by Brazilian agriculture is relatively low compared with that of developed countries. In 1993, the fertilizer (NPK) consumption was around 3 500 000 metric tons, the chemical (active ingredient basis) was used around 50 000 metric tons and the land use was estimated at 59 million ha cultivated (including around 2.95 million ha under irrigation) with annual and perennial crops, 85 million ha occupied pasture and grazing and 346 million ha were in forest lands (USDA 1994).

The government's present three-phased fiscal and monetary adjustments program (including abnormally high interest rates and trade liberalization) started in December 1993. Inflation has been reduced from 3000% to 35% per year. At the same time, agricultural input prices have increased, commodity prices are steady, and farmers' debt has increased significantly in US$ terms. Because of the strengthening of the 'Real' against the US$, agricultural and related imports have increased substantially. Normally the Brazilian trade balance is positive but in the first seven months of 1995, there was a trade deficit of nearly US $4.3 billion. Consequently, farm business performance has been drastically reduced and capital-intensive annual crop production is unlikely to spread to new remote frontiers in the cerrados in the near future. Part of the problem is the increased costs of credit and inputs for farm management.

Brazil's agricultural outlook for the next two years will depend not only on weather conditions but also on the future of the current economic stabilization program. In 1995, the Brazilian economy stabilized, but major political changes involving constitutional reforms are under way and will play a major role in terms of the economic outlook. The Brazilian trade balance is expected to be negative due to rising imports. This will probably result in some adjustments in exchange rates and duties on imports of both non-agricultural and agricultural products.

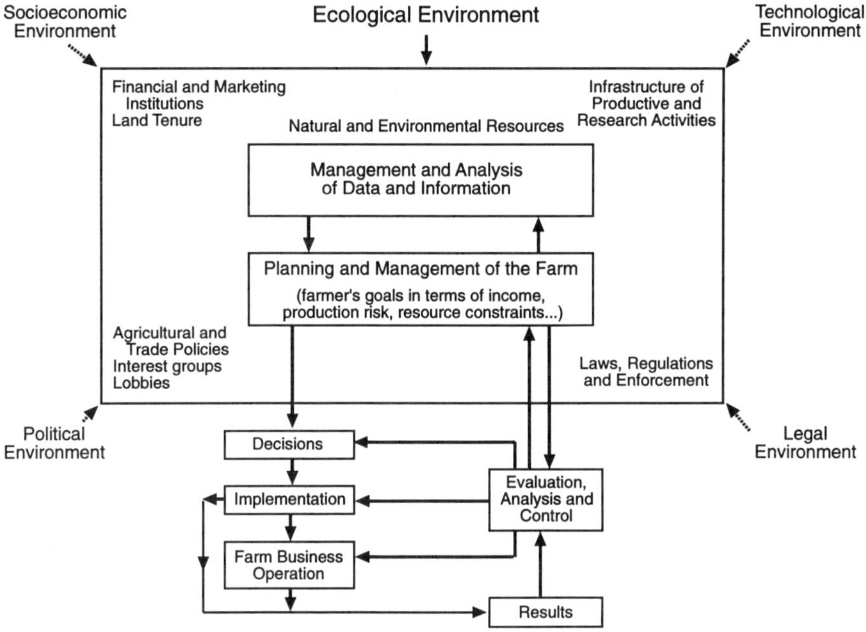


*Figure 1.* Farm planning and management concept.

In the Brazilian cerrados, agricultural production risks are associated with weather, soil fertility management and farm management generally in a very unstable economic environment. In the case study described below, farm development investments were financed in 1988 at 9% interest per year. Rules concerning debt servicing changed after the monetary adjustments initiated in December 1993.

In 1995, the real interest rate for agricultural credit was over 40% per year and, although production results in the past six years have been as expected, the farmers are facing a very serious financial problem. In addition, the terms of trade have been turned against Brazilian agricultural exports because the exchange rate of the 'Real' was held stable throughout a year that had 23.17% inflation. Even though Brazil had the largest harvest in its history, the net farm income fell by 20–25% due to low local and export prices.

Farmers in the cerrados with capital-intensive farms established through credit are currently in serious financial difficulties and are certainly favourably inclined towards less capital-intensive alternatives.

## The farm case study

The central idea of this case study is the understanding of the processes of economic changes in which a crop farm in the Brazilian cerrados has been involved. Farmers face various problems, some are quite perplexing and serious. Figure 1 presents a

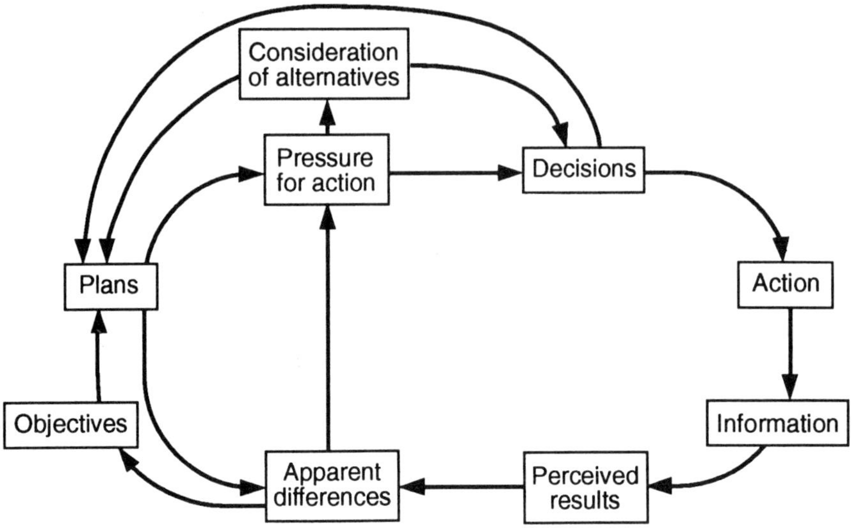


*Figure 2.* Decision-making process (*Source:* Veit 1978).

farm planning and management concept illustrating the broad farm environment. As in any business the key steps in the farm decision-making process can be described in the following sequence:

- define objectives;
- consider plans;
- plans lead to action;
- make some immediate decisions, consider alternatives;
- the consideration of alternatives leads to other decisions, which are fed back into the planning process;
- decisions are followed by action;
- action leads to results;
- results generate information;
- information generates perceived results;
- perceived results affect future objectives;
- perceived results, when compared with plans, identify differences; and
- these differences lead to pressure for action independent of the ongoing planning process as presented in Figure 2 (Veit 1978).

The dynamic (time-varying) behaviour of a farm system is caused both by the formal planning process and by on-going events.

The farmers involved in the Piratinga agricultural project (described below) are at a turning-point because, as a consequence of the low economic performance of their farms, they are under pressure and are looking for alternative agricultural activities that do not require credit from financial institutions. They have new plans related to an irrigated coffee crop, fruit crops and other technologies such as mini-

*Table 1.* Investment in farm machinery (numbers) by the Piratinga agricultural project (*Source:* Cooperative Coopertinga files, 1995).

| Description | Initially in 1988 | At the end of 1994 |
|---|---|---|
| Tractor | 105 | 130 |
| Planting equipment | 53 | 101 |
| Combine | 45 | 45 |
| Bean thrasher | 0 | 5 |
| Stalk chopper | 0 | 8 |
| Chisel plough | 0 | 30 |
| Lorry | 0 | 16 |
| Off-set disc harrow | 53 | 75 |
| Disc harrow | 53 | 93 |
| Sprayer (2000 l capacity) | 0 | 15 |
| Grain cleaning machine | 0 | 13 |
| Dry bean removal plant | 0 | 6 |
| Farm wagon | 53 | 72 |
| Generator set | 0 | 38 |
| Chopper | 0 | 19 |

mum tillage and a cheaper electricity supply, but they do not have the access to capital.

## Piratinga agricultural project

The Piratinga agricultural project has been financed at 9% interest per year (after inflation has been corrected in accordance with the official index). The funds provided by the Brazilian and Japanese governments to this project were lent to Banco do Brasil S.A. which followed a cerrados development program called PRODECER. Agricultural credit rules were defined by the Brazilian Central Bank.

The project is located in Formoso (Minas Gerais state), 260 km from Brasília (Federal District), near the border with Bahia and Goiás states. The project has incorporated 23 000 ha of cerrados soils distributed on 53 farms. This project, together with many others developed in Brazil in the last 20 years, has followed the Green Revolution approach. Initially it was designed to develop soybean, maize, orange and coffee crops without irrigation. Its implementation started in 1988 with young farmers who had some experience with soybean and maize crops. However, most of the farmers had no experience with perennial crops (such as coffee and orange). Moreover, they had little knowledge of local rainfall distribution and other weather information. After three years of local experience, most of the farmers decided to invest in irrigation systems. By the end of 1993, 40 central pivots were implanted on a total area of 1998 ha.

The farmers paid less than US $100 per hectare of land. It was then necessary to invest large sums of money in machinery, soil correction and preparation, and facilities such as houses, storage, roads, and electricity and water supplies. The large number of tractors, harvesters, trucks and implements described in Table 1 shows some of the investments realized by Piratinga project. Some US $2000 was invested per ha of cultivated soil.

*Table 2.* Piratinga agricultural project performance (*Source:* Coopertinga Cooperative files, Formosa-GO, 1995).

| Crop | Agricultural year 1992–93 | | | Agricultural year 1993–94 | | | Agricultural year 1994–95 | | |
|---|---|---|---|---|---|---|---|---|---|
| | Project | | Brazil | Project | | Brazil | Project | | Brazil |
| | Area (ha) | Average yield (ton $ha^{-1}$) | | Area (ha) | Average yield (ton $ha^{-1}$) | | Area (ha) | Average yield (ton $ha^{-1}$) | |
| Irrigated maize | 1473 | 6.00 | | 1990 | 7.20 | | 1876 | 8.40 | |
| Rain season maize | 1053 | 3.84 | 2.53 | 1306 | 4.80 | 2.44 | 1914 | 4.50 | 2.59 |
| Soybean | 8789 | 1.44 | 2.13 | 8458 | 2.58 | 2.16 | 9782 | 2.40 | 2.20 |
| Irrigated dry bean | 1840 | 2.58 | 0.63 | 1890 | 2.58 | 0.60 | 1998 | 2.70 | 0.59 |

*Table 3.* Agricultural products by the Coopertinga cooperative in 1992, 1993 and 1994 (*Source:* Cooperative Coopertinga files, 1995).

| Crop | Production (in metric tons) | | | Total revenue (in 1000 US$) | | |
|---|---|---|---|---|---|---|
| | 1992 | 1993 | 1994 | 1992 | 1993 | 1994 |
| Soybean | 21876 | 17196 | 25109 | 4193 | 3296 | 4812 |
| Maize | 13469 | 15263 | 22082 | 1459 | 1653 | 2398 |
| Dry bean | 1602 | 4031 | 2848 | 801 | 2015 | 1424 |
| Total | 36948 | 36492 | 50095 | 6453 | 6965 | 8635 |

Although the Piratinga project had been designed to settle new farmers and their families in new cerrados areas, most of them are living in Brasília and its satellite towns (in the Federal District). A reason for this is the lack of an integrated rural development program with the provision of health, education and other infrastructural facilities required by the local community. Consequently, family maintenance costs aggravate the farmer's financial situation because their businesses are capital-intensive in terms of machinery and equipment but their cultivated areas (on average, less than 300 ha) do not allow for needed economies of scale or size.

### *The Coopertinga cooperative*

The Coopertinga cooperative has incorporated all farmers involved in the Piratinga project. It was effectively established in 1990 and has been important for the farmers' commercial operations, including acquisition of agricultural inputs and marketing of agricultural products. It has presented above national average production results (Tables 2 and 3), and its tax payments in 1994 reached US $719993.

Strategically, the Coopertinga cooperative has planned a hydroelectric power unit estimated to cost US $7 million. The unit is very important to reduce the irrigated crop production costs and cooperative service costs (such as processing and classification of grain and cereals).

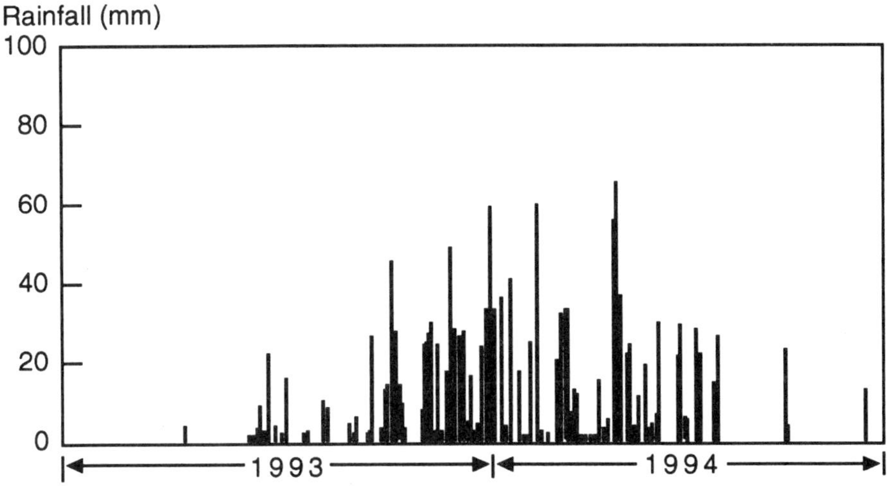


*Figure 3.* Precipitation at the Piratinga Project weather station (1 July 1993–30 June 1994).

*Table 4.* Evaluation of the São Francisco farm investments in 1993 (*Source:* PRODECER Program and Cooperative Coopertinga files, 1995).

| Description | Value in June 1993 (US$) |
|---|---|
| Land acquisition | 25 819 |
| Land valourization | 26 579 |
| Farm buildings | 67 175 |
| Investments in soil correction and preparation | 109 738 |
| Machinery and equipment | 151 099 |
| (initial cost US $361 817) | |
| (including 1 central pivot, 2 tractors, 1 combine and much equipment) | |
| Total | 380 412 |

## The São Francisco farm

The São Francisco farm is located on a plateau on cerrados soils at 840 m above sea level. An example of the local rainfall distribution is shown in Figure 3. This farm was evaluated in 1993 as shown in Table 4. Table 5 presents the production revenue and yield results in the first five years of the São Francisco farm.

The capital requirements to establish the São Francisco farm were estimated in June 1993 at about US $665 950 for investments as shown in the aggregated cash flow in Table 6. A 50 ha central pivot was the most expensive investment. The sum of US $578 122 was borrowed from the 'Banco do Brasil'. Up to 9 May 1993, the farmer had paid US $78 667 as debt service and the amount of US $476 589 is due over the next seven years. After the 'Real' stabilization plan the farm debt was corrected by a referential rate or index established by the Brazilian financial system. In November 1994, the farmer was obliged to sign a new contract, involving his debt, with the Banco do Brasil. This was part of an effort to normalize his credit. In

*Table 5.* São Francisco farm results (yields, products prices and revenues) (*Source:* Cooperative Coopertinga files, 1995).

| Description | Area (ha) | Yield[1] (kg ha$^{-1}$) | (bag ha$^{-1}$) | US\$[2]/bag (60 kg) | Value (US\$) | Subtotal (US\$) |
|---|---|---|---|---|---|---|
| Agricultural year (88–89) | 276 | | | | | |
| 1687 bags (60 kg) of soybean grain | | 366 | 6.11 | 8.50 | 14339 | |
| Subtotal | | | | | | 14339 |
| Agricultural year (89–90) | 246 | | | | | |
| 1630 bags (60 kg) of soybean grain | | 495 | 8.25 | 8.50 | 13855 | |
| 400 bags (60 kg) of soybean seed | | 495 | 8.25 | 8.93 | 3572 | |
| Subtotal | | | | | | 17427 |
| Agricultural year (90–91) | | | | | | |
| 256929 kg of soybean grain | 166 | 2687 | 44.79 | 7.56 | 32354 | |
| 189149 kg of soybean seed | | 2687 | 44.79 | 8.05 | 25378 | |
| 303036 kg of maize grain | 80 | 3787 | 63.13 | 4.76 | 24018 | |
| 5800 kg of coffee grain | 30 | 193 | 3.22 | 40.12 | 3877 | |
| Subtotal | | | | | | 85629 |
| Agricultural year (91–92) | | | | | | |
| 381290 kg of soybean grain | 186 | 2675 | 44.58 | 8.23 | 52294 | |
| 116267 kg of soybean seed | | 2675 | 44.58 | 11.07 | 21441 | |
| 156632 kg of maize grain | 50 | 3132 | 52.21 | 6.85 | 17890 | |
| 5520 kg of coffee grain | 30 | 184 | 3.07 | 25.00 | 2300 | |
| Subtotal | | | | | | 93927 |
| Agricultural year (92–93) | | | | | | |
| 156837 kg of soybean grain | 190 | 1291 | 21.52 | 7.05 | 18433 | |
| 88524 kg of soybean seed | | 1291 | 21.52 | 7.33 | 10810 | |
| 278052 kg of maize grain | 50 | 5561 | 92.68 | 5.15 | 23878 | |
| 92148 kg of dry bean | 50 | 1842 | 30.72 | 24.16 | 37098 | |
| 6217 kg of coffee grain | 30 | 207 | 3.45 | 24.54 | 2543 | |
| Pasture | 6 | 0 | 0.00 | 0.00 | 0.00 | |
| Subtotal | | | | | | 92763 |
| Total | | | | | | 304087 |

[1]In 1989–90, the dry period 'veranico' in the rainy season was 44 days long.
[2]After discounting the processing product costs (1.5%), a cooperative rate (2.0%), a marketing rate (2.5–4.0%) and a tax rate (19.2–19.5%).

consequence, his debt was corrected to US $742200 and his financial situation became very difficult. This was the situation of all Coopertinga's farmers involved. The lowest debt was US $630000. Figure 4 illustrates how the referential rate, inflation indexes (IGP and IPCr), and US$ in R$ terms, have varied in a period of 12 months (July 94 to July 95).

In 1995, the farmer was 40 years old. He lived on the farm from 1988 until 1993. However, in 1994, he moved with his family to a town close to Brasília in search of better education for his three children.

The original Piratinga project was designed to have about 10% of the crop land in coffee and citrus. Initially, the project master plan assumed that a 30 ha coffee crop could produce around US $70000 of revenue per year. In São Francisco farm, a 30 ha coffee plantation was established. After three years of experience with the coffee crop, the farmer felt that it was a bad investment and abandoned it because

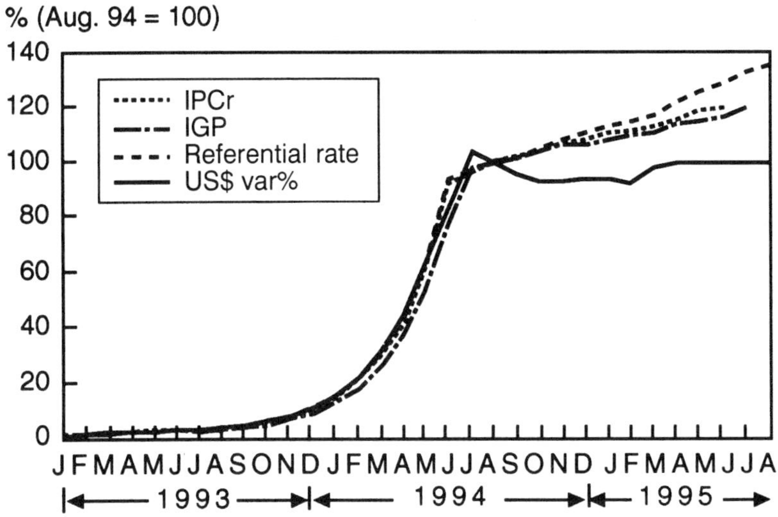


*Figure 4.* Brasilian economic indices (inflation indices: *IPCr*=Indice de Preços ao Consumidor; IGP=Indice Geral de Preços; see also text).

the dry season is too severe for coffee cultivation without irrigation. As a consequence of the low coffee yields, the farmer's financial situation was further aggravated.

The other crops developed on the farm were soybean, maize and dry beans. The soybean crop has been the main activity. The irrigated dry bean crop presents the most expensive production cost (US $960) per hectare and is risky in terms of market price, diseases and insect attacks.

## A farm scenario

Using expected crop yields and average product prices a cash flow was projected for the agricultural years from 1994–1995 to 1999–2000.

According to the assumptions (based on expected yields under similar conditions and past prices received by farmers in the region) incorporated in the projected cash flow present in Table 7, no more than US $15 647 per year could be used to pay off investment debt. This is less than 2.5% of the US $665 950 of investments used to set up São Francisco farm. In other words, under the current economic situation of input and output prices, the project cannot be amortized and can only pay a nominal interest rate of 2.5%. A farmer family maintenance cost of US $12 000 has been included as an operational cost (Tables 6 and 7).

Low-value crops like soybean and maize do not economically justify heavy investment in machinery and equipment under both dry and irrigated conditions. At present prices, only some perennial high-value crops such as irrigated coffee and fruits justify the investment for opening new cerrados frontiers.

*Table 6.* Aggregated cash flow of the São Francisco farm.

| | Year | | | | | | |
|---|---|---|---|---|---|---|---|
| | 1988–89 | 1989–90 | 1990–91 | 1991–92 (in US$) | 1992–93 | 1993–94 | Total invest |
| **Investments:** | | | | | | | |
| Land purchase | 25 819 | | | | | | |
| Land clearing | 23 212 | | | | | | |
| Lime and phosphate application | 75 701 | | 10 824 | | | | |
| Farm buildings and water supply | 26 870 | 40 305 | | | | | |
| Planting of coffee | | 20 560 | 27 064 | | | | |
| Farm machinery and equipment | 116 968 | 61 277 | | 8410 | 2400 | | |
| Central pivot | | | | | 172 761 | | |
| Miscellaneous | 53 773 | | | | | | |
| Subtotal (investments) | 322 343 | 122 142 | 37 888 | 8410 | 175 161 | | 665 944 |
| **Operational costs:** | | | | | | | |
| 246 ha of soybean crop (rainy season), 1988–89 | 60 570 | | | | | | |
| 246 ha of soybean crop (rainy season), 1989–90 | | 63 894 | | | | | |
| 246 ha of soybean crop (rainy season), 1990–91 | | | 49 916 | | | | |
| 80 ha of maize crop (rainy season), 1990–91 | | | 26 933 | | | | |
| 186 ha of soybean crop (rainy season), 1991–92 | | | | 56 404 | | | |
| 50 ha of maize crop (rainy season), 1991–92 | | | | 16 146 | | | |
| 50 ha of irrigated dry bean crop, 1992 | | | | | 40 081 | | |
| 190 ha of soybean crop (rainy season), 1992–93 | | | | | 58 457 | | |
| 50 ha of irrigated dry bean crop, 1992–93 | | | | | 26 061 | | |
| 28 ha of dry bean crop, 1993 | | | | | | 27 419 | |
| 17 ha of irrigated dry bean, 1993 | | | | | | 16 331 | |
| 5 ha of irrigated dry bean, 1993 | | | | | | 4778 | |
| 170 ha of soybean crop (rainy season), 1993–94 | | | | | | 51 658 | |
| 50 ha of irrigated maize crop, 1993–94 | | | | | | 26 061 | |
| 20 ha of maize crop (rainy season), 1993–94 | | | | | | 6596 | |
| Central pivot depreciation | | | | | | 96 787 | |
| Subtotal (operational costs): | 60 570 | 63 894 | 76 849 | 72 550 | 124 600 | 229 630 | |
| Total of investments + operational costs: | 382 913 | 186 036 | 114 737 | 80 960 | 299 761 | 229 630 | |

*Table 6.* continued

*Table 6.* continued

| | | | | | | |
|---|---|---|---|---|---|---|
| **Net sale revenues:** | | | | | | |
| 1687 bags of soybean (grain), 1988–89 | 14339 | | | | | |
| 1630 bags of soybean (grain), 1989–90 | | 13855 | | | | |
| 400 bags of soybean (seed), 1989–90 | | 3572 | | | | |
| 256929 kg of soybean (grain), 1990–91 | | | 32354 | | | |
| 189149 kg of soybean (seed), 1990–91 | | | 25378 | | | |
| 303036 kg of maize (grain), 1990–91 | | | 24018 | | | |
| 5800 kg of coffee (grain), 1990–91 | | | 3877 | | | |
| 381290 kg of soybean (grain), 1991–92 | | | | 52294 | | |
| 116267 kg of soybean (seed), 1991–92 | | | | 21441 | | |
| 156632 kg of maize (grain), 1991–92 | | | | 17890 | | |
| 5520 kg of coffee (grain), 1991–92 | | | | 2300 | | |
| 92148 kg of dry bean (grain), 1992 | | | | | 37098 | |
| 156837 kg of soybean (grain), 1992–93 | | | | | 18433 | |
| 88524 kg of soybean (seed), 1992–93 | | | | | 10810 | |
| 278052 kg of maize (grain), 1992–93 | | | | | 23878 | |
| 6217 kg of coffee (grain), 1992–93 | | | | | 2543 | |
| 6 ha of pasture, 1992–93 | | | | | 360 | |
| 144000 kg of dry bean (grain), 1993 | | | | | | 64377 |
| $x$ kg of maize (grain), 1993–94 | | | | | | 27000 |
| $y$ kg of maize (grain), 1993–94 | | | | | | 8100 |
| $z$ kg of soybean (grain), 1993–94 | | | | | | 65025 |
| Subtotal (net sale revenue): | 14339 | 17427 | 85627 | 93925 | 93122 | 164502 |
| **Crop insurance:** | | 3360 | 26673 | 152 | 1958 | |
| Total of (net sale revenue + crop insurance) | 14339 | 20787 | 112300 | 94077 | 95080 | 164502 |
| **Annual cash flow result:** | (368574) | (165249) | (2437) | 13117 | (204681) | (65128) |
| = (investments + oper. costs) – (net sale revenue + crop insurance) | | | | | | |
| Depreciation allowance (machinery and equipment): | | | | | | 113930 |
| Depreciation allowance (central pivot): | | | | | | 96787 |
| **Aggregated cash flow:** | | (533827) | (536262) | (523145) | (727824) | (582237) |

*Notes:*
(i) Average soybean production cost (rainy season crop) per ha $303.87.
(ii) Average maize production cost (irrigated crop) per ha $521.24.
(iii) Average maize production cost (rainy season crop) per ha $329.80.
(iv) Average dry bean production cost (irrigated crop) per ha $886.11.

*Table 7.* Aggregated cash flow of the São Francisco farm, based on expected yield and price values.

| | Unit | 1994–95 (US$) | Unit | 1995–96 (US$) | Unit | 1996–97 (US$) | Unit | 1997–98 (US$) | Unit | 1998–99 (US$) | Unit | 1999–00 (US$) |
|---|---|---|---|---|---|---|---|---|---|---|---|---|
| **Investment costs to be paid 1993–94:** | | **(582 237)** | | | | | | | | | | |
| **Investments:** | | | | | | | | | | | | |
| Lime and phosphate application | | 0 | | 0 | | 0 | | 0 | | 0 | | 0 |
| Farm buildings | | 0 | | 0 | | 0 | | 0 | | 0 | | 0 |
| Farm machinery and equipments | | 20 000 | | 0 | | 0 | | 0 | | 0 | | 0 |
| **Subtotal (investments):** | | **20 000** | | **0** | | **0** | | **0** | | **0** | | **0** |
| **Operational costs (estimated):** | | | | | | | | | | | | |
| Irrigated dry bean crop (ha) | 50 | 44 305 | 50 | 44 305 | 50 | 44 305 | 50 | 44 305 | 50 | 44 305 | 50 | 44 305 |
| Rainy season soybean crop (ha) | 180 | 54 697 | 200 | 60 774 | 200 | 60 774 | 200 | 60 774 | 200 | 60 774 | 200 | 60 774 |
| Irrigated maize crop (ha) | 50 | 26 061 | 50 | 26 061 | 50 | 26 061 | 50 | 26 061 | 50 | 26 061 | 50 | 26 061 |
| Rainy season maize crop (ha) | 40 | 13 192 | 20 | 6596 | 20 | 6596 | 20 | 6596 | 20 | 6596 | 20 | 6596 |
| Coffee crop (ha) (abandoned) | 0 | 0 | 0 | 0 | 0 | 0 | 0 | 0 | 0 | 0 | 0 | 0 |
| Pasture | 6 | 120 | 6 | 120 | 6 | 120 | 6 | 120 | 6 | 120 | 6 | 120 |
| Depreciation cost (5% of the central pivot value) | | 8638 | | 8638 | | 8638 | | 8638 | | 8638 | | 8638 |
| **Subtotal (operational costs):** | | **147 015** | | **146 496** | | **146 496** | | **146 496** | | **146 496** | | **146 496** |
| **Total (investments + operational costs):** | | **167 015** | | **146 496** | | **146 496** | | **146 496** | | **146 496** | | **146 496** |
| **Net sale revenue:** | | | | | | | | | | | | |
| Dry bean grain (irrig. crop) (60 kg bag) | 2150 | 51 944 | 2150 | 51 944 | 2150 | 51 944 | 2150 | 51 944 | 2150 | 51 944 | 2150 | 51 944 |
| Soybean seed (60 kg bag) | 0 | | 0 | | 0 | | 0 | | 0 | | 0 | |
| Soybean grain (60 kg bag) | 7740 | 60 642 | 8600 | 67 381 | 8600 | 67 381 | 8600 | 67 381 | 8600 | 67 381 | 8600 | 67 381 |
| Maize grain (irrig. crop) (60 kg bag) | 6000 | 33 520 | 6000 | 33 520 | 6000 | 33 520 | 6000 | 33 520 | 6000 | 33 520 | 6000 | 33 520 |
| Maize grain (60 kg bag) | 3200 | 17 877 | 1600 | 8938 | 1600 | 8938 | 1600 | 8938 | 1600 | 8938 | 1600 | 8938 |
| Coffee grain (60 kg bag) | 0 | | 0 | | 0 | | 0 | | 0 | | 0 | |
| Pasture (ha) | 6 | 360 | 6 | 360 | 6 | 360 | 6 | 360 | 6 | 360 | 6 | 360 |
| **Subtotal (net sale revenue):** | | **164 344** | | **162 143** | | **162 143** | | **162 143** | | **162 143** | | **162 143** |
| **Insurance revenue:** | | **0** | | **0** | | **0** | | **0** | | **0** | | **0** |

*Table 7.* Continued

*Table 7.* Continued

| | | | | | | |
|---|---|---|---|---|---|---|
| **Total (net sale revenue + insurance revenue)** | **164 344** | **162 143** | **162 143** | **162 143** | **162 143** | **162 143** |
| **Annual cash flow:** (total investments + total operating costs) – (net sale revenue + insurance revenue): | **(2670)** | **15 647** | **15 647** | **15 647** | **15 647** | **15 647** |
| **Aggregated cash flow:** | **(584 908)** | **(546 685)** | **(508 463)** | **(470 241)** | **(432 019)** | **(393 796)** |
| **Depreciation allowance (machinery and equipment) 1/15:** | **13 937** | **13 937** | **13 937** | **13 937** | **13 937** | **13 937** |
| **Depreciation allowance (central pivot) 5%:** | **8638** | **8638** | **8638** | **8638** | **8638** | **8638** |
| **Capital to be paid:** | **(562 333)** | **(524 110)** | **(485 888)** | **(447 666)** | **(409 443)** | **(371 221)** |

*Note:* It is assumed that the farmer will invest 20 000 US dollars in planting equipment specific for minimum tillage.

| | | | | | |
|---|---|---|---|---|---|
| | | | | | **Expected** |
| (i) Av. soybean production cost (rainy season crop) per ha | $303.87 | expected soybean yield (rainy season crop) per ha | 43 (60 kg bag) | soybean price (60 kg bag) | $7.84 |
| (ii) Av. maize production cost (irrigated crop) per ha | $521.24 | expected maize yield (irrigated crop) per ha | 120 (60 kg bag) | maize price (60 kg bag) | $5.59 |
| (iii) Av. maize production cost (rainy season crop) per ha | $329.80 | expected maize yield (rainy season crop) per ha | 80 (60 kg bag) | | |
| (iv) Av. dry bean production cost (irrigated crop) per ha | $886.11 | expected irrigated dry bean yield per ha | 43 (60 kg bag) | dry bean price (60 kg bag) | $24.16 |
| (v) Cost of 1 ha of pasture | $20.00 | | | price of 1 ha of pasture | $60.00 |

## Relevance of this case study for rural project appraisal

The results presented in this case study show that new agricultural development schemes should not be designed without *ex post* evaluation of the projects already established in the cerrados region. The farm planning and management concept described above should be considered in any project appraisal involving the cerrados rural and farm development.

The major objective of this study was to establish a data base for agricultural and rural planning for the cerrados region. The details of this data base including technical and financial coefficients and other information involving farm activities are available upon request from the first author (Veloso et al. 1995). These data, together with other case studies, will form a data base for future studies, thus helping to evaluate long-term project appraisal and feasibility of other cooperatives to be formed under similar conditions.

## References

Mueller C, Torres H, Martine G (1992) Settlement and agriculture in Brasil's forest margins and savannah agrosystems. Documento de trabalho 10, Instituto SPN, Sociedade, População e Natureza. Brasília DF, Brazil. 79 p.

USD – United States Department of Agriculture (1994) Agricultural situation: Annual report. American Embassy, AGR No. BR4634, Brasília DF, Brazil. 24 p.

Veit K P (1978) Systems dynamics in corporate long-range strategic planning. Pages 533–544 in Roberts E B (Ed.) Managerial applications of system dynamics. Productivity Press, Cambridge, MA, USA.

Veloso R F, Carvalho E R O, Goulart A M (1995) Desempenho técnico, econômico e financeiro da fazenda São Francisco no Projeto Piratinga. Série Documentos, EMBRAPA/CPAC, Planaltina DF, Brazil. 80 p.

# Factors influencing animal traction adoption in Kakamega district, Kenya

J.N. MBATA

*Department of Agricultural Economics and Business Management, Egerton University, Njoro, Kenya*
*Present address: Dept of Agricultural Economics and Rural Sociology, The National University of Lesotho, P.O. Roma 180, Lesotho, Southern Africa*

**Key words:** animal traction, probit analysis, maximum likelihood estimation

**Abstract**

One of the main issues in food crop production systems in Kenya is the reduction or complete elimination of the drudgery in farm chores and operations. One important technology that can offer solution to the above problem and intensify agricultural production is animal traction technology. This study identifies the socioeconomic factors responsible for animal traction in Kakamega District, Kenya. For primary data collection, a multi-stage random sampling technique was used to select both adopters and non-adopters in the area. The analysis was performed using the probit model via the maximum likelihood estimation technique. Also, the effect of different policy options was quantified by making use of changes in stimulus levels of adoption as a result of predetermined changes in selected socioeconomic variables. The results from the study show that animal traction adoption is influenced by mainly economic factors rather than sociological and institutional factors. In particular, the number of oxen available to farmers, availability of credit and the price of maize are the major determinants of animal traction in the study area. It is, therefore, recommended that efforts be made to increase the economic base of the farmers for widespread animal traction adoption in the area.

## Introduction

Agriculture is the mainstay of the Kenyan economy accounting for about 30% of the Gross Domestic Product (GDP). The sector employs about 75% of the total population. Agriculture and agri-based products contribute about 70% of the total export earnings. However, the contribution of the agricultural sector to the nation's GDP has been on the decline. From Table 1, it can be seen that, although the output value of agriculture has been increasing marginally in nominal terms, the percentage contribution of the sector in real terms to the country's GDP has been decreasing. In addition, it has been stated that an annual growth rate of 4.5% is required in the agricultural sector for the overall economy to achieve the targeted growth rate of 5.4% in the current plan period (Kenya 1989). Unfortunately, the best recent record in the growth rate of the agricultural sector was between 1979 and 1983 when an annual average growth rate of 3.4% was observed; between 1980 and 1984, it was only 2.6% (Godfrey 1986).

In terms of aggregate production, most of the crops are produced by small-scale farmers. Lugogo (1983) defines small-scale farmers as those who dwell in the rural areas of Kenya, own not more than five hectares of land and derive their income



*P.S. Teng* et al. *(eds.): Applications of Systems Approaches at the Farm and Regional Levels, 191–202.*

*Table 1.* Percentage contribution of agriculture to the gross domestic product (GDP), 1984–90 (*Source:* Kenya 1991).

| Year | Contribution to GDP % |
|---|---|
| 1984 | 32.1 |
| 1985 | 30.7 |
| 1986 | 31.2 |
| 1987 | 29.6 |
| 1988 | 29.4 |
| 1989 | 28.1 |
| 1990 | 25.9 |

basically from agricultural production. Apart from wheat and sisal, which are grown mainly on large farms, most crops are produced by the small-scale farmers. Ninety percent of Kenya's maize is grown by small-scale farmers. Other food crops such as sorghum, millet, rice, potatoes, beans and cassava are grown mainly on small holdings. Small-scale farmers also produce over 48% of the tea, 83% of the sugar-cane and more than 60% of horticultural crops (Lugogo 1983).

The inability of the agricultural sector to achieve the projected growth rate in recent times has been attributed to a number of factors. Such external factors as unfavourable weather conditions, insect pest and disease outbreaks and poor infrastructure militate against efficient production by small-scale farmers and limit their production possibility curves to suboptimal levels. The endogenous factors affecting farmers' production efficiency include inadequate knowledge of modern agronomic practices and insufficient use of both farm and non-farm inputs. Other factors are untimely farm operations due to a dearth of farm labour, especially during peak periods of labour demand, and low capital formation, particularly with the small-scale farmers.

## Problem statement and justification for study

One of the main issues in small-scale agricultural production in Kenya is the reduction or complete elimination of the drudgery in farm chores and operations. This stems from the inability of the small-scale farmers to meet with the seasonal demand for human labour and the non-use of farm machinery due to the small size of farm holdings. In fact, there is widespread recognition that seasonal labour bottlenecks, lack of energy and inappropriate equipment are significant constraints to higher farm productivity and efficiency among small-scale farmers (Daramola 1990).

The Food and Agriculture Organization (FAO 1975) has noted that in making use of animal traction, the farmer reduces his heavy physical work and increases the productivity of his labour and farm enterprise as a whole. Realizing the importance of animal traction technology in small-scale agriculture, the Government of Kenya, through the Ministry of Agriculture, has taken steps to encourage farmers to adopt animal traction technology through extension services, training courses on animal traction at all levels, the establishment of an ox-unit at Bukura Institute of Agriculture and by financing local research on animal traction in the country.

The use of animal traction can free labour previously used in the farm sector to other sectors of the economy. In addition, more farm labourers will have time to obtain training and adult literacy education, thus improving the quality of the farm labour force. Increased use of animal traction will also result in an increased need for animal-drawn implements that are less capital-intensive and entail less demand on the limited cash resource available to the small-scale farmers. Animal-drawn equipment can be fabricated efficiently by rural-based industries. This would lead to the promotion of small-scale industries and encourage entrepreneurship in the rural areas, as well as saving some foreign exchange used to import farm machinery.

Hitherto, the tendency has been for the government of the less developed countries to concentrate on modern technology leading to the substitution of animal traction and other simple farm implements with mechanization in agriculture (Olayide 1980). Certainly there is the need to take into consideration the changing ecosystems and to develop techniques that can effectively grapple with the dynamics of the production landscape (Daramola 1990). However, some factors such as the small size of farm holdings in most areas of Kenya, the flatness of some of these farms and the high cost of fuel and tractor spare parts due to the current Structural Adjustment Programme (SAP) in the country, make mechanization among small-scale farmers uneconomical.

Kakamega district was chosen for this study for several reasons. The district lies in the Western Province which is considered to be an important food crop producing area in the country, second only to the Great Rift Valley Area of Kenya in terms of aggregate production. A major section of Kakamega District shows an undulating topography with moderately deep soils. Kakamega District experiences a high annual rainfall of about 1100 mm per year. The soil moisture content of the area is, therefore, favourable for crop production and suited for draught animal use. Due to the high population density in Kakamega District, farm holdings average between 1.33 and 4.00 ha (MOA 1981).

## Objectives and organization of the study

Although there have been some studies on animal traction in the area, no studies as yet have tried to delineate the factors responsible for animal traction adoption among small-scale farmers. The general objective of the study therefore is to isolate the socioeconomic factors affecting animal traction adoption among small-scale farmers in Kakamega District, Western Kenya. Subsumed in this overall objective are the following specific objectives:

- to identify the socioeconomic characteristics of farmers adopting animal traction in Kakamega District of Western Kenya;
- to quantify the effect of selected exogenous socioeconomic variables on the decision of farmers to adopt animal traction technology;
- to predict how changes in these factors can alter the rate of adoption of animal traction in the study area; and

- to make recommendations based on the findings of the study for possible policy interventions that will encourage animal traction adoption to a reasonable extent in the study area.

## Conceptual framework of the study

To investigate the decision to adopt animal traction technology, the probit model was used to estimate a relationship in which the likelihood to adopt animal traction is a function of a set of predetermined variables. This is because the ordinary and generalized least-square regressions are limited when economic relationships are formulated with dependent variables that are dichotomous (Amemiya 1981). Since the dependent variable is binary (discrete), the analysis depends upon the use of qualitative choice models. In that case, it is possible to view the dependent variable as a probability that a farmer consumed a particular durable good or service or used a specific innovation.

Three statistical models available to analyse binary choice models such as to adopt or not to adopt a technology, or whether or not an individual consumed a particular durable good, are linear probability, logit and probit models (Pindyck and Rubinfeld 1981). In other words, these models are alternative specifications of qualitative choice models. However, the logit and probit models are preferred to the linear probability model when qualitative choice models are to be estimated because the latter suffers from a number of deficiencies: the variance of the disturbance term of the model is heteroscedastic and the standard errors of the least-squares parameter estimates are biased: also, the disturbance term is not normally distributed; the classical statistical tests are, therefore, inapplicable; finally, within a linear probability model, the predicted probability of adoption can lie outside the 0 to 1 boundary imposed on probabilities – even if values between 0 and 1 are viewed as probabilities, negative values or values greater than 1 cannot be interpreted (Capps and Kramer 1986; Judge et al. 1988). Empirical evidence suggests that neither logit nor probit has an advantage over the other in the case of binary choice models (Nayga and Capps 1992). Because the choice is a matter of convenience (Hanushek and Jackson 1977), the probit model is chosen in the study.

## The probit model

Given data on adoption of animal traction, $Y_t$, and exogenous variables $X_{it}$, $i=1, 2, \ldots N$ for a number of farmers $t=1, 2, \ldots M$, we adopt a probit model for the expected probability of adoption $P_t$ by a farmer with exogenous variable values $X_{it}$, $i=1, 2, \ldots N$. That is, we assume:

$$P_t=f(\beta_0+\beta_1 X_{1t}+\beta_2 X_{2t}+\ldots+\beta_N X_{Nt})$$

where $f$ is the standard normal cumulative distribution function. We estimate the coefficients $\beta_i$ by maximum likelihood from the observed data $Y_t$ and $X_{it}$, assuming a binomial error distribution (Aldrich and Nelson 1984).

*Table 2.* Descriptive statistics of the socioeconomic variables (US 1$=57 Kshs).

| Variable | | Mean | Standard deviation |
|---|---|---|---|
| $X_1$ | Age | 51.74 | 7.52 |
| $X_2$ | Farm labour size | 7.36 | 3.68 |
| $X_3$ | Level of education | 4.91 | 2.77 |
| $X_4$ | Farm size | 2.79 | 2.01 |
| $X_5$ | Farming experience | 21.80 | 8.41 |
| $X_6$ | Number of oxen | 2.41 | 1.14 |
| $X_7$ | Extension visits | 2.45 | 2.23 |
| $X_8$ | Cooperative membership | 0.59 | 0.69 |
| $X_9$ | Mechanical input | 4.71 | 3.42 |
| $X_{10}$ | Credit availability | 0.44 | 0.64 |
| $X_{11}$ | Modern input use | 0.57 | 0.84 |
| $X_{12}$ | Irrigation availability | 0.08 | 0.22 |
| $X_{13}$ | Farm income (Kshs) | 22568 | 2445 |
| $X_{14}$ | Price of maize per ton (Kshs) | 10872 | 1698 |

### *Evaluation of the effect of policy options using the probit model*

The effect of different policy alternatives on the number of animal traction adopters could also be analysed using the probit model. The procedure for achieving this assumes that substitution of a new set of exogenous variables, $X_{it}$, into the estimated probit model produces unbiased estimates of $P_t$, the probability of adoption by a farmer having the new $X$ values.

This procedure is then used for a particular situation (i.e. changes in any or combination of the explanatory variables) to obtain a new estimate of the number of farmers adopting animal traction technology as a result of change in policy.

## Data source and description

The farm-level data used for this study were obtained by structured questionnaires through personal interviews with the farmers using a multi-stage sampling technique. First, Kakamega District was subdivided into the existing nine divisions and by means of simple random sampling, five divisions were selected. The selected divisions were then subdivided into their various locations. From a total of 22 locations in the district, ten locations were selected randomly. Finally, the selected locations were further subdivided into sublocations. Two sublocations were selected from each of the ten locations. A list of all the farmers in the selected villages was obtained from the extension officers. A total of 145 farmers were selected from the list to represent the two classes of farmers in the proportion of 79 to 66 adopters and non-adopters, respectively.

Information was obtained from the selected farmers on whether they used animal traction or not and on their socioeconomic characteristics which, based on prior expectation, could influence their decision to adopt animal traction.

The prototype probit model specified for this study included one independent variable and 14 explanatory variables. The variables were specified as follows with the direction of the expected effect of the explanatory variables in the model noted in parentheses. Dependent variable $Y=1$ if the farmer used animal traction and 0 otherwise. The explanatory variables are:

$X_1$ Farmer's age (+): this was specified in years.

$X_2$ Farm labour force (–): the number of person-equivalents who work regularly in the farm. The regular labour force is usually made up of the farmer, his wife (wives) and children. Using weighting developed by Norman (1969), the labour contributed by women and children in the family was reduced to male equivalents.

$X_3$ Level of education (–): the number of years of formal education attended by the farmer.

$X_4$ Farm size (–): this was measured in hectares.

$X_5$ Farming experience: the number of years of active farming.

$X_6$ Number of oxen (+): the number of oxen and bullocks owned by the farmer.

$X_7$ Extension contact (+): the number of times the farmer visited or was visited by an extension worker in the year.

$X_8$ Membership in a cooperative or farmer's association: $X_8=1$ if the farmer is a member and 0 otherwise.

$X_9$ Access to mechanical inputs (–): the total number of hours that a farmer employed the services of a tractor or mechanically operated equivalent in his farm during the year of 1993.

$X_{10}$ Credit availability (+): $X_{10}=1$ if the farmer obtained credit from any source during the year and 0 otherwise.

$X_{11}$ Use of modern inputs: the modern inputs of interest here are fertilizer, improved seed varieties and agricultural chemicals. $X_{11}=1$ if the farmer used any or all of the modern inputs and 0 otherwise.

$X_{12}$ Irrigation availability (–): $X_{12}=1$ if any part of the farm was under irrigation, 0 otherwise.

$X_{13}$ Farm income (+): this is the amount of money (in Kenyan Shillings) the farmer received from farming activities in the previous year (1992).

$X_{14}$ Price of maize (+): this variable was included because of the partial control of the maize market during the period of study which introduced different unofficial prices as against the government-controlled producer price. Maize is the most important staple food crop in the study area.

Table 2 represents the descriptive statistics of the socioeconomic variables of the respondents. The sample looks like a good representation of the farmers in the study area. The average number of adults and average income closely resemble what have been previously reported (Kenya 1988). Also, the standard deviation of the variables from their mean values indicates the sample cuts across the different groups of small-scale farmers in the area.

*Table 3.* Full model maximum likelihood estimates of the probit model for animal traction adoption in Kakamega District, 1993.

| Dependent variable whether farmer is adopter | | | Yes = 79 | No = 66 |
|---|---|---|---|---|
| Explanatory variables | | Coefficient | Standard error | Asymptotic *t*-value |
| $X_1$ | Age | 0.0376 | 0.0167 | 2.2515 |
| $X_2$ | Farm labour | −0.1174 | 0.0942 | −1.2463 |
| $X_3$ | Level of education | −1.0420 | 0.3468 | −3.0046 |
| $X_4$ | Farm size | −0.2074 | 0.1350 | −1.5363 |
| $X_5$ | Farming experience* | −0.4618 | 0.9643 | −0.4789 |
| $X_6$ | Number of oxen | 0.2678 | 0.0742 | 3.6092 |
| $X_7$ | Extension visits | 0.4299 | 0.5553 | 0.7742 |
| $X_8$ | Cooperative membership | 0.3573 | 0.3417 | 1.0351 |
| $X_9$ | Access to mechanical input | 0.0792 | 0.0864 | 0.9167 |
| $X_{10}$ | Credit availability | 0.0912 | 0.0871 | 1.0471 |
| $X_{11}$ | Use of modern inputs | 0.6358 | 0.8023 | 0.7925 |
| $X_{12}$ | Irrigation availability* | 0.0219 | 0.0325 | 0.6738 |
| $X_{13}$ | Farm income* | 0.0726 | 0.0999 | 0.7267 |
| $X_{14}$ | Price of maize | 0.9241 | 0.4633 | 1.9946 |
| | Intercept | −1.6932 | | |
| | −2 Log likelihood Normal | 172.8800 | | |
| | Pseudo $R^2$ | 0.3540 | | |

*Excluded from the restricted model.

## Empirical results

### *Results of the probit model*

The maximum likelihood estimates for the probit analysis are exhibited in Table 3. The full model was refined and this was achieved by excluding some of the variables included in the prototype model on the basis of *a priori* and statistical considerations. In other words, the variables whose coefficients bore the wrong signs and were statistically insignificant were dropped from the restricted equation. These variables include farmers' farming experience and income that may be correlated with age and credit availability, respectively. On the other hand, the exclusion of variable $X_{10}$ could be justified on the grounds that very few farmers (less than 10%) had access to irrigation facilities. To ascertain the consistency of the model, which implies checking that the coefficients of the excluded variables are zero, the log likelihood ratio test was conducted (Aldrich and Nelson 1984).

The results of the test showed the absence of mis-specification error with the exclusion of the three variables ($X_5$, $X_{12}$ and $X_{13}$). In other words, the coefficients of these variables were indeed zero.

In the restricted model, the probit equation was estimated with 11 explanatory variables. This model performed as well as the full model as indicated by the results of the likelihood ratio test. Table 4 presents the results of the final or restricted probit model of animal traction adoption by the farmers in the study area. Based upon statistical significance and sign of the coefficients in Table 4, the results indicate that older farmers are more likely to use animal traction technology than the younger farmers. This is in conformity with the *a priori* expectations as aged farmers are

*Table 4.* Maximum likelihood estimates of the restricted model for the determinants of animal traction adoption in Kakamega District, 1993.

| Dependent variable whether farmer is adopter | | Yes = 79 | No = 66 |
|---|---|---|---|
| Explanatory variables | Coefficient | Standard error | Asymptotic *t*-value |
| Age | 0.0084** | 0.0049 | 1.7143 |
| Farm labour | −0.2985** | 0.1725 | −1.7304 |
| Level of education | −1.0149 | 1.2143 | 0.8358 |
| Farm size | −0.3942*** | 0.1492 | −2.6421 |
| Number of oxen | 0.4648 | 0.1771 | 2.6245 |
| Extension visits | 0.2917** | 0.1476 | 1.9763 |
| Cooperative membership | 0.3431 | 0.3219 | 1.0654 |
| Access to mechanical input | 0.0878 | 0.0996 | 0.8815 |
| Credit availability | 0.0596*** | 0.0244 | 2.4426 |
| Use of modern inputs | 0.1488 | 0.1527 | 0.9745 |
| Price of maize | 0.0044** | 0.0021 | 2.0952 |
| Intercept | −1.4252 | | |
| −2 Log Likelihood | 148.49*** | | |
| Pseudo $R^2$ | 0.3022 | | |
| Number of iterations | 6 | | |

***Significant at 1%; ** Significant at 5%.

generally less receptive and find it more difficult to adopt modern technologies. The results also indicate that availability of regular labour force will significantly discourage the adoption of animal traction. This finding is in line with *a priori* expectations as farmers with regular supply of labour are less likely to be affected with labour scarcity and can, therefore, use the available family labour for their farming operations. Although not statistically significant, farmers with formal education are less likely to adopt animal traction technology. Also, as the farm size increases, the likelihood of using animal traction decreases. This finding is in conformity with *a priori* expectation since, on the average, it becomes unprofitable and time-consuming to use animal traction on very large farms; the smaller the farms, the easier and more practical it is for the farmers to use animal traction. As the number of oxen owned by the farmers increases, the more likely it is for farmers to adopt animal traction. Contact with extension agents was also found to be a significant variable. It seems that efforts by the government through the use of extension agents to get farmers to adopt this practice for land preparation and cultivation are achieving the desired objectives.

The results of the restricted probit model further indicate that the farmers with access to mechanical input as expected are less likely to adopt animal traction technology though this variable was not significant. Other significant variables affecting animal traction adoption are credit availability and the maize price. The availability of credit from either formal or informal sources enables the farmer to purchase the bullocks and the necessary animal-drawn implements for his farming operations. Due to the present Structural Adjustment Programme (SAP), there has been a general rise in price of consumption, as well as production of goods, and these naturally have resulted in an increase in the price of consumption, as well as production goods,

*Table 5.* The stimulus level for selected factors affecting animal traction adoption.

| Variable | Change in stimulus level |
|---|---|
| Total labour[1] | −0.299 |
| Farm size[1] | −0.211 |
| Number of oxen[2] | 0.494 |
| Frequent extension visit[2] | 0.301 |
| Promotion of cooperative membership[2] | 0.061 |
| Credit availability[2] | 0.341 |
| Price of maize[2] | 0.362 |

[1]Variable is increased by 10%.
[2]Variable is increased by 50%.

and these naturally have resulted in an increase in the prices of livestock and the animal-drawn implements as well. The situation has tended to put the prices of oxen and the implements beyond what the average farmer can afford, unless he is assisted with credit.

The price of maize as a variable in animal traction adoption is significant at the 5% level. Because maize is the most important staple food in the country, efforts by the government to control the price of this commodity before the year 1994 (when the marketing of maize was finally liberalized) led to different unofficial prices. Farmers were ready to invest in animal traction technology to enable them to produce more of this crop which they could easily sell at prices much higher than that set by the government.

The log likelihood test showed that the model had 1% level of significance. Another goodness of fit measure is the pseudo $R^2$ which had values of 0.3540 and 0.3022 for the full and restricted models, respectively. These values lie within the expected 0.2 and 0.4 range. Also, these pseudo-$R_2$ values are reasonable considering that the data for the study were obtained from a cross-sectional survey of selected individuals from the study area.

*Effects of changes in selected variables*

To show the effect of the explanatory variables on animal traction adoption, the predicted changes on the stimulus level I have been computed for predetermined changes of explanatory variables. The idea was to find out how the number of adopters would change as a result of a predetermined change in the level of exogenous variable(s). Using the procedure outlined above, the changes in stimulus levels and associated new probabilities were calculated. It should be noted that the change in the probability as a result of a specified change in the stimulus level is not always the same. This is because the transformation from stimulus to probability is sigmoidal rather than linear. Thus, the effect of a change in the stimulus depends on the initial location on the curve (Theil 1971). Table 5 shows the predicted change on the stimulus level as a result of changes in the explanatory variables.

Judging from the relative magnitude and signs of the coefficient and the predicted change in stimulus level, it is apparent that the most important variables responsible

*Table 6.* The effects of different policy alternatives on animal traction adoption in Kakamega District, 1993.

| | Policy instrument | Sum of probability | Estimated no. of adopters | Farmers adopting (%) | Change from the original situation (%) |
|---|---|---|---|---|---|
| 1 | No change (restricted model) | 78.79 | 79 | 54.5 | – |
| 2 | Increase in extension visits | 89.64 | 90 | 62.1 | 13.9 |
| 3 | Increase in number of oxen | 97.15 | 97 | 66.9 | 22.8 |
| 4 | Liberal credit facilities | 103.78 | 104 | 71.7 | 31.6 |
| 5 | Increase in maize price | 106.94 | 107 | 73.8 | 35.4 |
| 6 | Policies 3 and 4 combined | 110.04 | 110 | 75.9 | 39.2 |
| 7 | Policies 3, 4 and 5 combined | 115.66 | 116 | 80.0 | 46.8 |

for animal traction adoption in the study area are the number of oxen owned by the farmer, frequency of extension visits, availability of credit and the price of maize.

Using these important variables, the effect of possible policy options on animal traction adoption were quantified. For example, assuming that the government policy is to encourage the extension of credit to as many small-scale farmers as are qualified, the question a policy analyst would be asking is: 'what would be the effect of liberal credit facilities on the number of adopters?' This simply involves obtaining the difference between the sum of predicted probabilities and the sum after specifying that an additional 50% of original sample that did not benefit from credit now have access to credit while leaving all the other variables as before.

Table 6 presents the sum of the predicted probabilities and the effect of different policy alternatives on the number of adopters. From the results in Table 6, it is evident that the stimulus index of animal traction adoption is more sensitive to farmers' access to credit facilities, the price of maize and the number of oxen available to farmers than it is with membership of farmers' cooperative societies, literacy level, family size and age of the farmer. In other words, economic factors are more significant in influencing farmers' decision to adopt animal traction than sociological factors. These results are the reverse of those of Falusi (1976), Daramola (1987) and Mbata (1994) for fertilizer use. The findings from this study tend to lend support to the fact that high costs of bullocks and animal-drawn implements may be responsible for the low rate of animal traction adoption in the study area.

## Policy implications and conclusion

The purpose of the study was to identify the factors responsible for animal traction adoption by farmers in Kakamega District. The results clearly show that such government policies that would encourage liberal credit to farmers, good pricing for the major crops and increase in number of bullocks owned by farmers would elicit a reasonable level of animal traction adoption among small-scale farmers. The usefulness of the probit model, however, does not only lie in identifying the factors responsible for widespread adoption of an innovation, but also in evaluating the cost of different policy alternatives. In other words, assuming that both liberal credit facilities and good pricing policies for farmers' produce would achieve the same level of adoption among farmers, the government could then evaluate the cost of implementing each policy option and opt for that alternative that would consume less of the nation's resources.

## Acknowledgements

The author gratefully acknowledges the International Foundation for Science (IFS), Sweden, for partly funding this research and also for providing the travel grant to participate in the Symposium.

## References

Aldrich J H, Nelson F D (1984) Linear probability, logit and probit models. Sage University Publications Inc, Beverly Hills, CA, USA.
Amemiya T (1981) Quantitative response models: A survey. J. of Econ. Literature 19(4):1483–1536.
Capps O, Kramer R (1986) Analysis of food stamp participation using quantitative choice models. Am. J. Agric. Econ. 68:519–527.
Daramola A G (1987) A quantitative analysis of the adoption of improved food production technology in Oyo State, Nigeria. PhD thesis, University of Ibadan, Nigeria. 144 p.
Daramola A G (1990) Prospects of animal traction in Southern Nigeria: An empirical study. Paper presented at the Fourth Workshop of the West African Animal Traction Network, 9–13 July 1990, Kano, Nigeria.
Falusi A O (1976) Application of multivariate probit to fertilizer use decision: Sample survey of farmers in three states of Nigeria. J. of Rural and Economic Development 9(1):49–66.
FAO – Food and Agriculture Organization (1975) Assistance in agricultural mechanization in Tanzania. FAO Mission Report, September–October 1974. Food and Agriculture Organization, Rome, Italy.
Godfrey M (1986) The agricultural sector in Kenya to 1990. Pages 53–62 in Special Report Economist Intelligence Unit No. 1052.
Hanushek E, Jackson J (1977) Statistical methods for social scientists. Academic Press, New York, USA.
Judge G, Griffiths W, Hill R, Lutkepohl H, Lee T (1988) The theory and practice of econometrics. 2nd Edition. John Wiley and Sons, New York, USA. 1024 p.
Kenya (1988) Economic survey. Central Bureau of Statistics, Government Printers, Nairobi, Kenya.
Kenya (1989) Sixth national development plan 1989–1993. Central Bureau of Statistics, Government Printers, Nairobi, Kenya. 262 p.
Kenya (1991) Statistical abstract. Central Bureau of Statistics, Government Printers, Nairobi, Kenya.
Lugogo J A (1983) A polyperiod risk programming analysis of small-holder farm management in Kenya. PhD dissertation, Department of Agricultural Economics, Virginia Polytechnic and State University, Virginia, USA. 155 p.
Mbata J N (1994) Fertilizer adoption by small-scale farmers in Nakuru District, Kenya. Fert. Res. 38: 141–150.
Ministry of Agriculture (1981) Annual Report Kakamega District. Nairobi, Kenya.
Nayga R M, Capps O (1992) Determinants of food away from home consumption: An update. Agribusiness 8(6):549–559.
Norman D W (1969) Labour inputs of farmers: A case study of the Zaria Province of North-Central State of Nigeria. Nigerian Journal of Economics and Social Studies 11(1):3–14.
Olayide S O (1980) Agricultural technology and Nigeria's small farmers. Pages 52–65 in Olayide S O,

Eweka J A, Bello-Osage V E (Eds.) Nigerian small farmers: Problems and prospects in integrated rural development. CARD Ibadan University Press, Nigeria.
Pindyck R S, Rubinfeld D L (1981) Econometric models and economic forecasts. McGraw-Hill Book Company, New York, USA. 630 p.
Theil H (1971) Principles of econometrics. John Wiley and Sons, New York, USA. 736 p.

# Flow modelling with ECOPATH: providing insights on the ecological state of agroecosystems

J.P.T. DALSGAARD and V. CHRISTENSEN
*International Center for Living Aquatic Resources Management (ICLARM), MCPO Box 2631, 0718 Makati, Metro Manila, Philippines*

**Key words:** agroecology, systems modelling, sustainability indicators

**Abstract**
ECOPATH is an ecological modelling software developed for the analysis of aquatic ecosystems. Incorporating elements of ecosystem theory and more recent developments within network analysis of trophic interactions, it computes a series of system attributes. Such quantified properties can also be useful for evaluation of agroecosystems and for providing insights into their ecological performance, productivity and efficiency. This has been demonstrated for integrated agriculture-aquaculture systems at both the farm subsystem and whole-farm level. The paper summarizes the major findings in modelling and analysing agroecosystems using ECOPATH. It is anticipated that this type of approach and comparison of ecological properties across different natural resource management scenarios, and of various levels throughout the agroecological hierarchy, will help identify quantifiable indicators towards ecological sustainability in agricultural development.

## Introduction

Concepts and approaches introduced to promote ecologically sound and productive farming towards sustainable agriculture have been reported in the literature and include low-external-output agriculture, organic farming, agroecological engineering, integrated production, permaculture, integrated agriculture-aquaculture, agroforestry and integrated natural resources management (Young 1989; Mollinson 1991; Reintjes et al. 1992; Lightfoot et al. 1993a; Altieri 1995; Pretty 1995).

To assess the ecological performance of farming calls for quantitative measures. Given the complexity of natural resources and their management, systems-oriented approaches are in the future likely to play a prominent role within this area of research and development.

Agricultural systems can be viewed as assemblages of plants and animals linked in multiple ways. This perspective is useful as stocks and flows which can be described, modelled and analysed, as well as assessed and compared, on a system level. Agroecosystems further have certain properties because of their parts and intra-relationships. We assume that modelling such systems and investigating their aggregate properties can provide useful agroecological insights.

### The agroecological hierarchy

A farm activity and its interaction with the environment constitutes an agroecosystem. In the wide sense, agroecosystem refers to both biophysical and socioeconomic



*P.S. Teng* et al. *(eds.): Applications of Systems Approaches at the Farm and Regional Levels, 203–212.*

aspects of farming (Conway 1985). However, in the more narrow sense adopted here, the agroecological system emphasizes the ecological and biophysical aspects of farming (Altieri 1995).

The agroecological system can be viewed at many levels such as: an individual rice plant forming part of a soil-water-plant continuum; and a population of rice plants (a rice field) host not only to rice, but also to a variety of organisms such as insects, molluscs, fish, reptiles, birds, zooplankton, phytoplankton, benthic organisms, soil microbes, macrophytes, grasses and weeds. Such field communities form part of a larger system wherein crops, trees, poultry, livestock, fish, etc. combine to make up a farm; groups of farms constitute villages located within watersheds; and so forth (Figure 1).

In theory, the agroecological system can be defined at any level throughout this nested hierarchy. Each level within the hierarchical framework shares a number of common characteristics such as stocks (or components), inputs and outputs and flows of resources between components. Thus, the agroecological system at any level identified by the observer can be viewed as an assemblage of resource components and flows. This generic feature indicates the potential for applying general analytical tools at multiple and possibly at complementary levels.

## Modelling agroecological systems

Agroecological systems are dynamic systems. It is a goal of ecological modelling to construct dynamic and structural dynamic models (Jørgensen 1994) that adequately emulate organic systems. Their complexity and dynamism even at the most basic level can be overwhelming, and where limitations to our state of knowledge prevent the application of such methods, we must look for alternatives.

An option is to initially ignore the temporal component and take a more static view. Despite a generally lower level of mathematical sophistication (linear as opposed to differential equations), this approach should not necessarily be seen as only a second-best option. A mass-balanced model can provide an accurate and useful picture of an ecological system in an average state, and hence facilitate comparisons between different states of ecosystems. The insights provided through this type of modelling can fulfil many of our requirements and modelling expectations.

ECOPATH models produce average pictures or 'snapshots' of systems. Repeated over time these 'snapshots' can be combined to provide a time series view of a system's performance – one way of recapturing some of the dynamics that are otherwise ignored through the static approach. Another useful aspect of the average state concept is the insights it may provide into aggregate system properties. Diversity, for instance, is not measured by looking at a single component or species, but is a community attribute as are cycling, stability, and sustainability. These attributes, along with other key properties, such as productivity and efficiency, may equally be addressed and assessed at the system level.

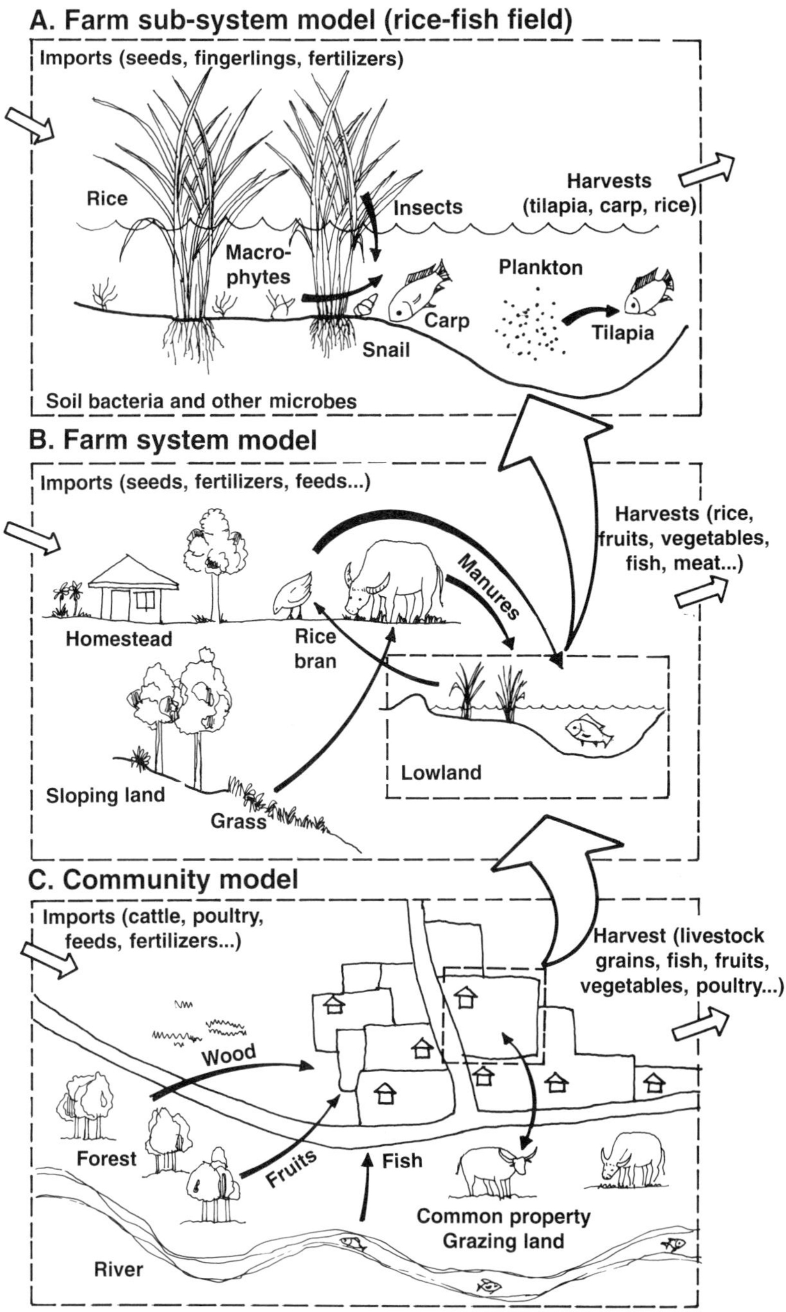


*Figure 1*. Example of a nested agroecological hierarchy.

## Features of ECOPATH

The ECOPATH approach was developed with the purpose of constructing mass-balance models of aquatic ecosystems (Christensen and Pauly 1992a,b). The utility of the non-commercial software implementing the ECOPATH approach as a structuring, analytical and synthesizing tool is documented in Christensen and Pauly (1995). Here, ECOPATH applications were focused across large- and small-scale, tropical and temperate aquatic systems including oceans, lakes and rivers, as well as fishponds and rice field ecological systems.

Aquatic and terrestrial ecosystems have many commonalities, and the mass-balance concept and basic ecological principles upon which ECOPATH is founded apply equally well to both types of environments. Through network analysis of trophic interactions, ECOPATH provides a practical means of incorporating and synthesizing some of the dominant thinking in ecological systems theory over the last decades into the modelling and quantitative analysis of ecosystems. Several attributes of development and maturity in natural ecosystems suggested by Odum (1969) are incorporated into the analytical routines. These and other systems attributes, notably ascendancy (Ulanowicz 1986), are computed as ratios, indices and summary statistics which provide a basis for evaluating the ecological state of a given system.

### *Model parameterization*

To construct an ECOPATH model, data on resource stocks and flows are required. The basic input parameters needed for each defined group of organisms are: average standing biomass (B); production per unit of biomass (P/B ratio); consumption per unit of biomass (Q/B ratio); harvests; other exports (e.g. losses); and the diet matrix giving the composition of food (i.e. energy or nutrient sources) for each group of plants and animals.

For instance, a rice plant gets all its nutrients from the detritus (soil) box, whereas a group of livestock typically will derive their food from a variety of sources such as grasses/weeds, straw and feed imports into the system. These different flows must be quantified. Finally, the model is balanced by the software so that input equals output for each system component.

### *Model outputs*

On the basis of the mass-balanced model, ECOPATH quantifies a range of ecosystem concepts and properties. These include cycling indices, pathways, mixed trophic impacts, food niche overlap, mortality coefficients, energy/nutrient transfer efficiencies through the food web, and other descriptive statistics. Model statistics, ratios and indices included are primarily those commonly applied within fisheries biology and aquatic ecology. Although not all of these are easily interpreted within or necessarily relevant to the agroecological analysis, many of the other output functions have already been explored through agroecological modelling. Furthermore, it is the intention to modify future versions of the software so as to facilitate agroecological applications.

## ECOPATH applications

A number of applications of ECOPATH to agroecosystems have already been made (Lightfoot et al. 1993b; Cagauan et al. 1993; Ruddle and Christensen 1993; Van Dam et al. 1993; Dalsgaard et al. 1995).

A cooperative study between the International Rice Research Institute, Central Luzon State University and the International Center for Living Aquatic Resources Management was made to use ECOPATH to compare rice field scenarios with and without fish (Figure 2). Preliminary results from on-station research showed that stocking rice fields with fingerlings not only produced marketable fish, but also led to a greater nitrogen efficiency in rice production (Lightfoot et al. 1993b). They also suggested that the results of using ECOPATH can assist in the identification of long-term experiments and important parameters for further study to better understand how these ecosystems work. A follow-up study by Cagauan et al. (1993) confirmed the greater ability of the integrated rice-fish system to capture and use nitrogen. A negative impact of fish on weeds, snails and insects as found by the model, showed the usefulness of fish (carp) in biological pest control.

Ruddle and Christensen (1993) found ECOPATH to represent a robust modelling approach in investigating the energy flow of a Chinese mulberry dike-carp pond farming system (Figure 3). The analysis showed the system to have very high throughput and production. Developed over hundreds of years, this is indeed a well-designed integrated agriculture–aquaculture system.

Van Dam et al. (1993) evaluated the trophic interactions in a napier grass-fed aquaculture pond in Malawi. Results shows that constructing a mass-balanced model demonstrated the gaps in our knowledge on food web structure, and on the productivity of various groups of organisms in grass-fed ponds.

Moving up one step in the agroecological hierarchy, Dalsgaard et al. (1995) used ECOPATH II as the main tool in the analysis, quantification and comparison of system attributes across four rice-based farms (Figure 4). It was found that our intuitive or qualitative sense of ecologically sound farming seems to stand up to a more quantitative scrutiny based on key indicators including diversity, nutrient cycling, stability and productive capacity.

The above-mentioned results show how the application of ECOPATH has gradually expanded and broadened to include farm enterprises, as well as the aquatic rice and fish units. Typical examples are being drawn no longer from on-station research, but increasingly from farms. Models have also been developed at different hierarchical levels and, where initial analyses focused on detailed trophic interactions at the farm subsystem level, other models are now being generated at the whole-farm system level.

An on-form study is presently underway in which four Philippine smallholder rice farm scenarios are being closely monitored for subsequent modelling, analysis

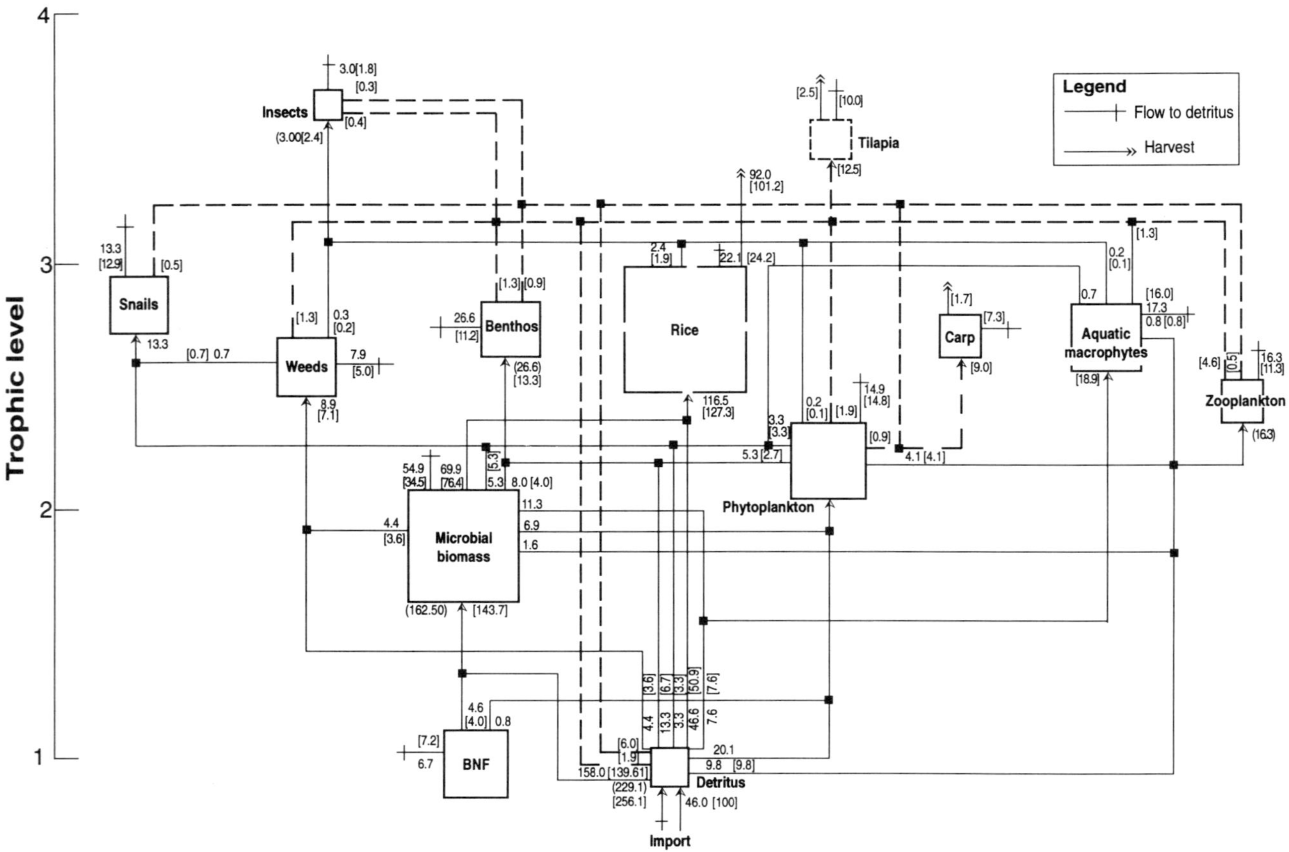


*Figure 2.* Comparison of steady-state nitrogen models of a wetland rice field ecosystem with (rice-fish model) and without fish (rice model). Rates with square brackets apply to the rice-fish model only, while other rates apply to the rice model where no rates in square brackets are given. Dotted lines show flows that are present in the rice-fish model only. All rates are in kg N ha$^{-1}$ crop$^{-1}$ (*Source:* Lightfoot et al. 1993).

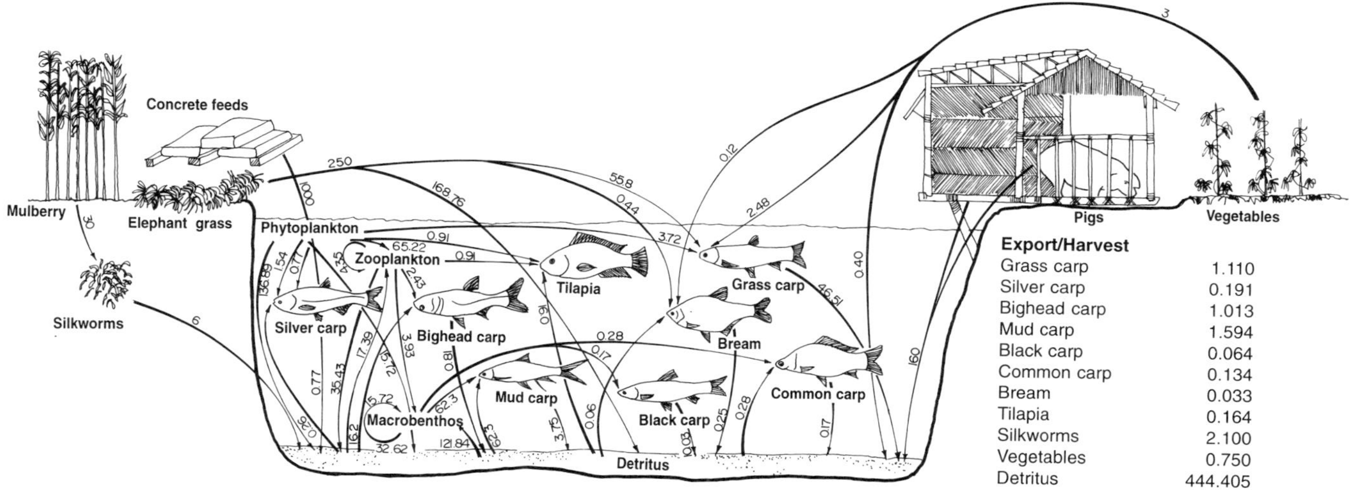


*Figure 3.* A quantitative representation of the flows in the integrated dike-pond polyculture system of the Zhujiang Delta, South China. Flows are expressed in ton $ha^{-1}$ $yr^{-1}$ (*Source:* Ruddle and Christensen 1993).

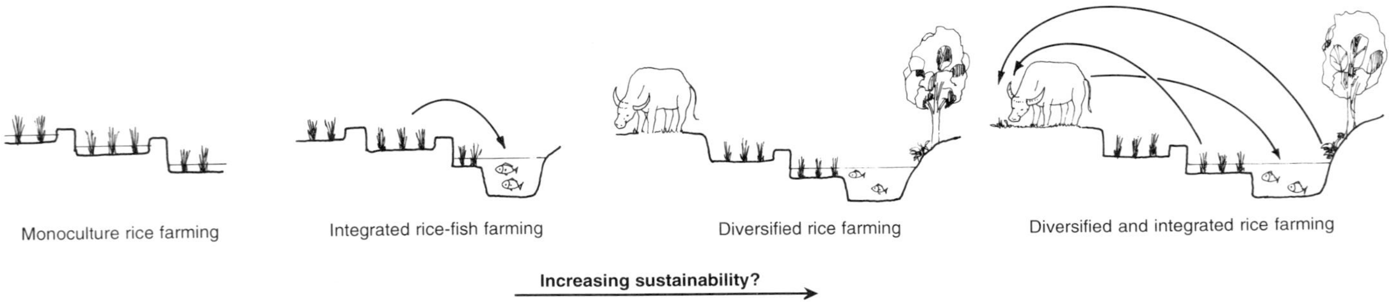


*Figure 4.* Ecological sustainability in rice farming.

and comparisons of productivity and nutrient (nitrogen) efficiency, as well as nutrient cycling, diversity and other agroecological attributes (Dalsgaard 1996; Dalsgaard and Oficial 1996a,b,c). The farms range from a low-external-input to a high-external-input system with and without integrated aquaculture, livestock, poultry, vegetable and tree components. It is expected that these types of investigations of aggregate system properties at the farm level will assist in the identification of quantifiable indicators of ecological sustainability, and in deriving general guidelines for sustainable, i.e. productive and ecologically sound, integrated farming systems.

## Conclusions

The general broadening of the application of ECOPATH from aquatic towards terrestrial-based systems suggests that the approach could become, in the future, an example of a general tool unifying fields of fisheries, science, limnology, aquaculture, farming systems and agroecological research (Christensen and Pauly 1995).

At present, the mode of application of ECOPATH in agroecological analysis has been descriptive. Added facilities in new versions of ECOPATH (Appendix 1) could eventually open up possibilities for using the software in a more predictive mode. Combined with other modelling facilities and tools which simulate system behaviour and responses to different forcing functions, ECOPATH could become a very useful quantitative instrument in the exploration of feasible paths towards an ecologically sustainable management of natural resources.

## Acknowledgements

Wholehearted thanks are due to Dr. Clive Lightfoot, Information Centre for Low External Input and Sustainable Agriculture (ILEIA), for fruitful and inspiring cooperation. The work of the authors is funded by the Danish International Development Assistance (Danida). This is ICLARM Contribution No. 1164.

## References

Altieri M A (1995) Agroecology, the scientific basis of alternative agriculture. Westview Press, Boulder, CO, USA. 428 p.

Cagauan A C, Dela Cruz C R, Lightfoot C (1993) Nitrogen models of lowland irrigated ecosystems with and without fish using ECOPATH. Paper presented at the Regional Workshop on Integrated Rice-Fish Research and Development, 6–11 June 1993, Sukamandi Research Institute for Food Crops, Sukamandi, West Java, Indonesia. 15 p.

Christensen V, Pauly D (1992a) ECOPATH II – a software for balancing steady-state ecosystem models and calculating network characteristics. Ecol. Modelling 61(3/4):169–185.

Christensen V, Pauly D (1992b) A guide to the ECOPATH II software system (version 2.1). ICLARM Software 6. International Center for Living Aquatic Resources Management, Manila, Philippines. 72 p.

Christensen V, Pauly D (1995) Fish production, catches, and the carrying capacity of the world oceans. NAGA ICLARM Q. 18(3):34–40.

Conway G R (1985) Agroecosystem analysis. Agric. Administration 20:31–55.

Dalsgaard J P T (1996) Agroecological sustainability and ecosystem maturity. Ecol. Modelling (submitted).

Dalsgaard J P T, Lightfoot C, Christensen V (1995) Towards quantification of ecological sustainability in farming systems analysis. Ecol. Eng. 4:181–189.

Dalsgaard J P T, Oficial R T (1966a) Agroecological systems analysis of four smallholder rice farm scenarios: Productivity and efficiency comparisons. Agric. Syst. (submitted).

Dalsgaard J P T, Oficial R T (1996b) Agroecological systems analysis of four smallholder rice farm scenarios: Modelling, quantifying and comparing ecological attributes. Agric. Syst. (submitted).

Dalsgaard J P T, Oficial R T (1996c) Quantitative analysis of agroecological systems with ECOPATH. ICLARM Technical Report No. 53. International Center for Living Aquatic Resources Management, Manila, Philippines (in press).

Jørgensen S E (1994) Fundamentals of ecological modelling. Developments in environmental modelling. 2nd Edition. Elsevier, Amsterdam, The Netherlands. 628 p.

Lightfoot C, Bimbao M A P, Dalsgaard J P T, Pullin R S V (1993a) Aquaculture and sustainability through integrated resources management. Outlook Agric. 22(3):143–150.

Lightfoot C, Roger P A, Cagauan A G (1993b) Preliminary steady-state nitrogen models of a wetland rice-field ecosystem with and without fish. Pages 56–64 in Christensen V, Pauly D (Eds.) Trophic models of aquatic ecosystems. ICLARM Conf. Proc. 26. International Center for Living Aquatic Resources Management, Manila, Philippines.

Mollinson B (1991) Introduction to permaculture. Tagari Publications, Tyalgum, Australia. 198 p.

Odum E P (1969) The strategy of ecosystem development. Science 104:262–270.

Pretty J N (1995) Regenerating agriculture: Policies and practices for sustainability and self-reliance. Earthscan Publications Ltd., London, UK. 320 p.

Reintjes C, Haverkort B, Walters-Bayer A (1992) Farming for the future: An introduction to low-external-input and sustainable agriculture. Information Centre for Low External Input and Sustainable Agriculture (ILEIA), Macmillan, London, UK. 250 p.

Ruddle K, Christensen V (1993) An energy flow model of the mulberry dike-carp pond farming system of the Zhujiang Delta, Guangdong Province, China. Pages 48–55 in Christensen V, Pauly D (Eds.) Trophic models of aquatic ecosystems. ICLARM Conf. Proc. 26. International Center for Living Aquatic Resources Management, Manila, Philippines.

Ulanowicz R E (1986) Growth and development: Ecosystem phenomenology. Springer Verlag, New York, USA. 203 p.

Van Dam A A, Chikafumbwa F J K T, Jamu D M, Costa-Pierce B A (1993) Trophic interactions in a napier grass (*Pennisetum purpureum*)-fed aquaculture pond in Malawi. Pages 65–68 in Christensen V, Pauly D. (Eds.) Trophic models of aquatic ecosystems. ICLARM Conf. Proc. 26. International Center for Living Aquatic Resources Management, Manila, Philippines.

Young A (1989) Agroforestry for soil conservation. CAB International, Wallingford, UK. 276 p.

## Appendix 1

ECOPATH version 3.0 (a Windows-based version released in 1996) incorporates an 'EcoRanger' module featuring a least-squares approach for construction of trophic models. This module introduces uncertainty into ECOPATH analysis by opening up the possibility for entering a range and mean/mode values for all the basic parameters, for the biomasses, consumption rates and production rates, as well as ecotrophic efficiencies and diet compositions.

Random input variables are then drawn using user-selected frequency distributions for each parameter type and the resulting model is evaluated (based on user-defined criteria, and physiological and mass balance constraints). The process is repeated in a Monte-Carlo way and of the model runs that pass the selection criteria, the best-fitting one is chosen using the least-squares criterion. The EcoRanger module thus introduces a statistical approach to the model fitting and, importantly, it helps the user in selecting the model that best fits a given set of constraints (Christensen and Pauly 1995).

# On the articulation of simulation and heuristic models of agricultural production systems

P.G. COX[1], K.A. PARTON[2], A.D. SHULMAN[3] and P.E. RIDGE[1]
[1]*Agricultural Production Systems Research Unit (APSRU), P.O. Box 102, Toowoomba, QLD 4350, Australia*
*Present address: ICRISAT, Patancheru, Andhra Pradesh 502 324, India*
[2]*Department of Health Services Management and Public Health, School of Health, University of New England, Armidale, NSW 2351, Australia*
[3]*Graduate School of Management, University of Queensland, Brisbane, QLD 4072, Australia*

**Key words:** model categories, simulation, heuristics, research design, re-engineering

**Abstract**
Good models are perceived as accurate, simple and having explanatory power. In practice, these are all bundled together. But the balance between these three attributes, and the strength of their expression, change depending on the use to which the model is put: whether for a technical, a practical or an emancipatory purpose. The relative emphasis on the accuracy of technical models may be misplaced if they are to be used to facilitate performance in a practical situation or to promote emancipation by providing a framework that allows individuals to work through their own problems. Farming practice issues may be better handled by building upon farmers' own heuristic models, with relatively greater emphasis on simplicity. Similarly, emancipatory issues can be more appropriately handled through educational models that stimulate and facilitate insight into system behaviour, with relative emphasis on explanatory power. The issue of the articulation of different model categories becomes important as models are used to design and engineer changes in the models used by others for different purposes, and as work is done simultaneously with different model categories to achieve a particular purpose.

## Introduction

Simulation models of crop and soil processes are being developed by many research groups (e.g. Thornton and Hoogenboom 1994; Thornton et al. 1995; McCown et al. 1996). These are primarily mechanistic models which estimate crop yield deterministically from soil and climate data. They can be used either to re-create experimental results ('looking back') or to predict outcomes from an anticipated change in management practice ('looking forward') on the basis of assumptions about the probabilistic behaviour of future weather. There are various justifications for modelling: it captures what we know about the biophysical processes that underpin crop production; it serves to integrate existing scientific knowledge in a way that can be interrogated to draw out additional understandings; the knowledge thus gained is then available for critique by others; and it can be used to support decision making by those responsible for the management of agricultural production. It is the last of these, the development of decision-support systems for use by farmers, that is frequently used in Australia to justify a substantial investment in model development (e.g. Woodruff 1992; McCown et al. 1996).



*P.S. Teng* et al. *(eds.): Applications of Systems Approaches at the Farm and Regional Levels, 213–228.*

Several authors have expressed concern recently about the use (abuse) of information technology (e.g. Talbott 1995; Cox 1996). The arguments of these authors are similar: that computers are about the past, and that explicit deterministic modelling is an inefficient (and, indeed, ineffective) way to bring about future change. Cox (1996) uses the term 'pathological' to describe this situation and suggests some of the reasons for the 'unwarranted escalation of organizational commitment' to it. Yet, in reality, what has gone wrong? And what can we do about it?

Agricultural production systems are managed effectively and efficiently using heuristics (Murray-Prior 1994; Cox et al. 1996b). The systems may very well be complex at a biophysical level, but it is our ability to conceptualize, identify and manipulate critical processes to make a difference at a management level that counts. Farmers' rule sets are customized to technological components, locations and farming systems. Occasionally, engagement within scientists' models may provide insights not already perceived by farmers. But, for the most part, farmers' rule sets have provided stable and adaptive models with which to manage production systems characterized more by multiple paths (equifinality) and uncertainty in outcome (multi-possibilities) than by complexity in the opportunities available for management and in the understanding needed to make a difference to outcomes. They provide an elegant solution to the problem of managing open systems, which are relatively forgiving of mistakes (exhibit homeostasis and multi-possibilities) because the carryover effect of a mistake in one season can often create an opportunity in the following season (e.g. N applied to a failed crop).

There are, of course, exceptions (e.g. land degradation such as the progressive development of dryland salinity) because the consequences of particular actions were unforeseen, or even if foreseen were heavily discounted because of the rate at which these biophysical processes take place compared with day-to-day economic transactions. Differences in the planning horizon may require different kinds of models. These situations provide arguments for policy interventions of various sorts (e.g. legislation governing the quality of water leaving irrigation enterprises; tax subsidies for land conservation measures), to which farmers' models will respond. These situations do not constitute a general criticism of the adequacy of a farmer's model, but are the result of working within other people's frameworks associated with the need of various actors to negotiate different perceptions of the system boundaries. The model we need to operate with, in order to achieve a desired outcome most effectively, is not always the most immediate one within this nested system of models.

Thus, different people construe different categories of model, for different purposes, from different perspectives within a landscape of possible models. Scientists and farmers use different models of agricultural production systems: simulation and heuristics are examples. This paper suggests criteria for gauging their appropriateness, for different purposes, within this landscape. Second, the implications of a model mismatch are discussed. Within this mismatch, it becomes clearer that the meanings and uses people attribute to a model are unique to the historical and perceived future position the person stands in when working with it. The meanings arise within this relationship. In this sense, models do not carry meanings – people read meanings into their engagements with the model. It is this 'reading into' that can be constrained or facilitated by the parameters of the model (what gets included gets attended to). While at one extreme anything can be read into a model, in practice the readings are constrained and opened by the physical parameters of the model. Third, the question of how the different position-based categories of models within the landscape are related is explored. Fourth, we suggest some opportunities for different model categories to influence the development and performance of other model categories used by scientists and farmers.

## Characteristics of a good model

Casti (1992) is quite clear about what constitutes a good model in *science*: accuracy, simplicity and explanatory power. *Accuracy* refers to the agreement between predicted and observed effects[1]. *Simplicity* is about being economical in the number of variables included in the model, and in our specification of the relationships between them. *Explanatory power* is about insights gained through the use of the model. We have no quarrel with these general categories, although it is often difficult to say that a model is accurate, simple and explains things well. These statements refer to comparisons between different formulations or competing model structures for a particular purpose and from the point of view of a particular model-user.

Casti's typology maps directly onto three fundamental anthropological interests – the *technical*, the *practical* and the *emancipatory* (Flood and Jackson 1991). Roughly, we often see claims for accuracy equated with technical (i.e. scientific) models, with the best models being the simplest consistent with the accuracy required; simplicity with practical (i.e. farmers') models; and explanatory power with emancipatory (i.e. social change) models. This shifts the focus from model structure to model function. For any one purpose, it does make sense to ask whether a particular model structure is more or less appropriate for achieving that purpose than another for a particular person. Models are valued because of their contribution to all three functions.

The evolution of the entire system of models will depend on the way these various categories are articulated, often unconsciously, by a particular user. This has become problematic because of what has been called the 'category mistake' by Ryle (1963). Ryle illustrated what he meant by a category mistake by telling a story about someone who was taken on a tour of a university including the lecture theatres, laboratories and libraries. After the tour, the visitor turned to the guide, somewhat confused and said: 'You have shown me lots of things, but where is the university among these?' The visitor was making a category mistake, assuming the university to be another part, rather than the whole. Similarly, those who expect meaning to be a property of a model mistakenly believe meaning to be a part that can be separately identified, rather than the outcome of an interaction between the model and the human agent. The problem of a category mistake in agricultural modelling is that

[1]Casti uses the term 'accuracy' to describe one of the characteristics of a good model. A distinction is sometimes made between accuracy and precision, see Wise (1995). If this distinction is maintained, some of our comments refer to precision rather than to accuracy.

meaning is taken as belonging to the category of the physical world, when it actually belongs in the human world.

## Implications of a model mismatch

Adequate attention must be given to issues of balance between different intended functions (whether practical, technical or emancipatory), and the articulation of model categories (each with different strengths) by different users to address a particular purpose. A mismatch can arise because we develop an inappropriate model in relation to the way we intend it to be used, or we apply a particular model where an alternative structure would have been more effective. In either case, the cost can be substantial in terms of lost opportunities to describe, to learn and to effect change in the way agricultural production systems are managed.

This was part of the argument for Farming Systems Research (FSR): to expand the notion of an agricultural system to the production and consumption issues faced by the farm family in its social context, not just the biophysical transformations that underpin crop or animal production. FSR was a response to the inappropriateness of much early agricultural research in poor countries and the development by professional researchers of inappropriate technologies for use by farmers. Our critique of crop modelling shows similar lines, and for the same reasons – modellers, just like their 'production technology' predecessors, are using inappropriate criteria for model development, use and evaluation if their intention is to use these tools in a management setting. That is, they are not engaging with potential end-users in a manner which increases the probability that the scientific model will affect the ways farmers see and act. This failure to engage represents a waste of resources unless it can be re-directed into more effective activities, and more efficient ways of bringing about desired change. The opportunity costs of a model mismatch can be high.

## Horses for courses

With this in mind, we posit a general design space based on three dimensions. This represents an expansion of Casti's properties of a good model, for each purpose. These dimensions are:

- Complexity and scope (accuracy).
- Stability and adaptability (simplicity).
- Transparency and re-usability (explanatory power).

It is the balance between these three dimensions that we are trying to get right. Because of their different positions within the model landscape, farmers' models and scientists' models handle these trade-offs differently.

### *Complexity and scope*

To some extent, we can physically increase complexity and scope simultaneously, but there are computation restrictions on model size (which lead to instability).

Enlarging the scope often requires increasing complexity; but it is possible to add complexity without any intention to enlarge the scope of a model. These changes and permutations also lead to difficulties in interpretation through lack of transparency. The same physical model is perceived differently by different audiences. Generally, a model is perceived as either broad (wide scope, many variables considered at a given level of organization) or deep (few variables considered at one level, but explained by hierarchical dependence on levels of explanation below that). Typically, farmers' models are broad in scope but of low hierarchical complexity; scientific models are usually narrow in scope but much richer in hierarchical complexity. Simulation models fail because of their attempt to capture hierarchical complexity and broad (well, broader!) scope simultaneously.

*Stability and adaptability*

Models can be stable (reliable, constant, unchanging or slowly changing over long periods of time) or unstable, and/or adaptable to changing circumstances, evolving in structure and formulation as our perception of the world to which they refer constantly changes. Stability implies that the models are not sensitive to short-term perturbations in inputs and/or errors, and that the response to changed circumstances is damped (not erratic) in order to avoid exaggerated responses to random fluctuations in the environment. It also implies stability in the relevant knowledge that the user brings to interaction with the model. Adaptability implies that models can be changed to fit changing opportunities and circumstances, as we become more certain that substantial shifts in the environment are permanent (not just noise), and additional opportunities become available.

In general, scientific principles are quite stable, particularly across large numbers of observations. But scientific (crop and soil simulation) models are often relatively unstable: they can fall over because of missing data, or use of input data outside the bounds for which the model is calibrated, or because of configuration requirements of which the user is unaware, e.g. the ordering of specific activities in the model; or they can generate results that are not understood – partly because of their complexity, partly because of the intrusion of site-specific noise. Simulation models can react in unexpected ways and it is often hard to sort out the signal (significant meaning) from noise (empty of meaning). Nor do they adapt at all readily – the rules for gravity or water use efficiency do not change very quickly; indeed, it is important that they do not within the context of their use.

Farming practice can sometimes change rapidly and purposefully, e.g. adoption of zero tillage in response to severe soil erosion occurring during a drought, or the deregulation of financial and grain markets. Farmers' models must be able to cope with this sort of situation; they must evolve quite rapidly as opportunities for action unfold. On the other hand, perhaps associated with their openness and simplicity, they are relatively stable and not thrown severely off course by a perverse environment. They do respond to the recent past, e.g. growing cotton on wide rows, or with different row configurations, because the seasons seem to be less reliable and this has proven to be right through four consecutive years of drought. Thus, even though the scientific principles on which they are based are stable, simulation models are typically unstable and non-adaptive to changing management issues; farmers' models are stable and adaptive in relation to the same issues.

### *Transparency and re-usability*

With a transparent model, the important input variables (i.e. those important in relation to a specific function or purpose), and their relationship to output, are accessible to the model user, available for inspection and critique, and modifiable in response to personal knowledge, new scientific knowledge (i.e. externally acquired), and changing conditions. Indeed, transparency is often cited as a criterion of a good model (Sneiderman 1987; Shulman and Penman 1992). The obverse of this is that such models are readily internalized and replaced by other (often tacit) constructs; the models, as explicit structures operating independently of the user, are eventually discarded or ignored because they are no longer needed.

This has been the fate of some agricultural decision-support systems (DSS). It is sometimes argued that DSS have been successful to the extent that they were used, but discarded once the information contained in them has been assimilated into an operational heuristic. One reason for this discardability suggested by Cox (1996) is that the model output (particularly the relationships between inputs and outputs) is compressible. This means that the insights can be captured more simply, e.g. by a rule-based system. The assumptions on which simulation models are based are not transparent, but these models often have low re-usability as a guide to action because the output can be reconstructed in other ways or merely confirms current practice.

In general, farmers' models ('rules of thumb') are transparent: the assumptions on which they are based are open, discussible and discussed; they are available for embellishment, critique and upgrading, so the model structures are infinitely re-usable. And, unlike many scientist-generated models, they are often stated in the negative, e.g. 'don't plant before Christmas'. As such, farmers' heuristics emancipate, encouraging active experimentation that is not dismissed: knowledge about the things that do not work is also valued. Practical knowledge needed for managing a property does not require many of the formal model structures of the researcher. Skilled researchers would claim to be able (using their models as a basis and filling in the interpretative gaps themselves) to produce insights useful for influencing the farmers' model. But it does not follow that the farmer can use, or will want to use, or needs to use the same model structure as the researcher. This is particularly the case when the practitioner only needs to apply that bit of the model that is locally relevant. This is not to say that some re-structuring of farmers' models may not yield immediate and long-term benefits.

If this typology of models, and the dimensions we have used to categorize them, are accepted, the distinctive nature of simulation and heuristic models is readily apparent. The transfer of simulation models to a farming context (with an implied shift from technical to practical, from description to understanding and emancipation) means that they do not address the issues faced by practical land managers on their terms. This has serious implications for the economic rate of return on investment in agricultural research, development and extension (R,D&E) to use one model structure in an attempt to change another. A mismatch can be a source of challenge. But it can also lead to confusion and dependence, particularly when the model user does not understand the model formalism used by others: scientists and farmers often do not appreciate the purpose of each other's models, but may accept their appropriateness on trust.

We recognize the existence of different purposes (or courses). Scientists and farmers face different courses: they have different jobs to do, and go about them in different ways. We also recognize that they may have different tools (or horses) including different models. Much previous agricultural R,D&E attempted to influence the course faced by one group (farmers) by introducing new tools which would shape new purpose. The assertion of simulation models as a basis for DSS for use by farmers seeks to change the models farmers already use to manage existing technology. This is part of a wider project of substituting knowledge for physical inputs, such as fertilizer or water, in order to obtain a better match between resource use and potential benefits. But the issue of which horse for which course, and how we can exploit one horse-course combination (whether science or farming) to influence the performance of another horse-course combination remains open. We take this up in the next section.

## Horse trading

We usually get a polite acknowledgement of 'horses for courses' from the professional research community (and indeed from farmers), but in practice different configurations of horses and courses are possible, and different problems do require different solutions. The argument runs: farmers' models are very effective, efficient, elegant and enjoyable for what they want to do (and, therefore, are relevant too). The researchers' models are grounded in a different culture, have a different history, are chasing different things. If they keep to their own worlds, no problems. Where they start to overlap and try to engage each other, without recognizing these differences, the probability of fruitful engagement is limited.

There are several ways to attempt to overcome this difficulty. We characterize six ways of developing horses for courses (see Table 1). These differ in the way they manage the relationship between different horses (models) and different courses (purposes). They are grouped in pairs. In the first pair, the farmers' purpose is not questioned; in the second pair, the researchers' purpose is not questioned; in the last pair, the purpose of both groups of actors is questioned. Within each pair, the first one works with existing tools; in the second, the focus is on the development of new and better tools.

*Table 1.* Six strategies for matching horses with courses (FP=farmer's purpose; SP=scientist's purpose; unQ=unquestioned; Q=questioned).

| Strategy | FP | SP | Tool | Relationship between horses and courses |
|---|---|---|---|---|
| 1 Work with farmers' models of decisions | unQ | – | old | *One horse, one course at a time* (course is well defined, dealing with routine decisions; little formal scientific input) |
| 2 Use professional models to generate heuristics | unQ | – | new | *Change farmers' course by changing researchers' horse* (use of alternative formulations of researchers' models to provide a more evocative formalism for addressing farmers' issues) |
| 3 Deconstruct simulation models | – | unQ | old | *One horse, many courses* (recognizes researchers' model as a standard, but re-interpreted by researchers for wider use) |
| 4 Develop decision-support systems based on simulation models | – | unQ | new | *The chameleon solution* (researchers' models can be made to look like farmers' models) |
| 5 Participatory technology development | Q | Q | old | *Change researchers' horse by changing course it runs on* (participatory approaches to technology design encourage researchers to change horses) |
| 6 Work with multiple models simultaneously | Q | Q | new | *Many horses, many courses* (because we recognize the value of working with multiple representations and are determined to become proficient in using them) |

### *(1) Work with farmers' models of decisions*

The problem is not to get farmers to behave like scientists (they have different objectives), but to get scientists to engage with farmers in the development of models if they are to have any impact on the behaviour of farmers. One way of doing this is to build jointly upon the heuristic models of farmers – to work with farmers' models. Previous work reported by Cox (1996) suggests that this can be entirely appropriate, e.g. defining spray thresholds for *Heliothis* on cotton on the basis of simple linear time-dependent segments captured much of the benefit claimed for more complex model formulations (a quadratic function of day-degrees from sowing; or as a function of square and boll numbers). Nothing was gained by moving to more complex model formulations, so why bother? In this case, the real issue was about the feasibility of defining the size of pest populations rather than the precise definition of pre-determined thresholds for comparison. We could have started and ended here, especially since the thresholds used for comparison were determined outside the context of the simulation model. Similar situations occur in agronomy, e.g. determination of fertilizer requirements by calculating the amount of nitrogen needed to achieve a target yield. A move to a more complex model formulation may or may not be justified when the horse runs up against a hurdle that it cannot jump over. But in many situations, the farmers' horse jumps as well as does the scientists'.

*(2) Use of professional models to generate heuristics*

The characteristics of this approach is that we accept the justification for using different professional (researchers') models if we are to match the issues (objectives, constraints, opportunities) faced by farmers: new tools are needed. Simulation models may be good at describing the behaviour of biophysical components of farm systems (plants, soils), but overlook the importance of constraints in the way it is all put together, and the dynamics of adaptation, e.g. comparison of cotton and sorghum done in isolation overlooks the opportunity to convert an early failure with cotton to a late-planted sorghum crop. For that, researchers will need to use a different model structure, one geared to the generation and evaluation of novel possibilities.

Mathematical programming (MP) models have several characteristics that recommend them for this role. Their activity analysis structure enables them to be constructed in a manner that more closely resembles farm decision making. As a consequence, the heuristics generated within an MP model have a more natural correspondence to farmer-generated heuristics.

One example of the correspondence between MP models and the farmers' models is the direct manner in which the whole-farm aspects are incorporated (Hazell and Norton 1986). In the MP approach, an individual farm activity is represented on a unit basis as a vector of resource requirements and outputs. For example, one hectare of winter wheat would require certain amounts of land, labour, machinery, etc. at various times during the production process, and produce outputs including grain and crop residues. By bringing together many such individual activities, the whole-farm production (and consumption) process is mimicked closely. This enables cross-activity resource impacts to be observed more easily than with simulation models.

Another advantage of the MP approach is its ability to analyse agricultural land use within its socioeconomic context. Because it is an optimizing procedure, an objective function representing a variety of farmer goals can be analysed. This endogenizing of farmer decision making is another aspect of MP that enables it to match farmers' heuristics more closely. Moreover, financial constraints, such as debt repayment or taxation obligations imposed on the business, can be analysed. Similarly, the MP model can be used to assess changes to the policy environment facing agriculture.

The modelling process becomes more difficult when risk is introduced. Non-embedded risk, where it is assumed 'that it is realistic to model the system as if all decisions are made initially and then the uncertainty unfolds subsequently in terms of risky consequences of the choice taken' (Hardaker et al. 1991, p. 10), has received most attention. Embedded risk, where farmers make tactical adjustments to plans as information becomes available during a production period, has been largely overlooked. Nevertheless, further modelling work in the area of embedded risk seems appropriate (Dorward and Parton 1996). MP approaches to embedded risk, such as discrete stochastic programming (Rae 1971a, b), are possible but suffer the curse of dimensionality. Simulation has also been used to examine embedded risk problems (McCown et al. 1991). A promising line of research is that of Dorward (1991, 1994) which suggests that there is scope for combining optimizing, simulation and heuristic models. But even if we do that, the marginal value of more complex model formulations may still be small (Pannell et al. 1996).

### *(3) Deconstruct simulation models*

Another strategy to effect the articulation of simulation and heuristic models is to deconstruct the output of professional models. This can be done in two ways: through the examination of the results of multiple simulation runs in the hope of picking out simple relationships which can be framed as rules; or by highlighting the underlying scientific relationships on which the model is built. Both approaches appear inefficient: the first because the search space has not been specified very tightly, and few clues about the appropriateness of potential patterns are provided (in terms of objectives and constraints); the second because it is unclear, if this approach is adopted, why the cause-effect relationships were buried in the model in the first place. However, the lack of critical engagement of the end-user within the development of the model reduces the probability that either of these two approaches will work. But the act of translation, and the negotiation of meaning that this entails, can itself be salutary.

Thus, the deconstruction of the SIRATAC pest management model, as part of the development of the *entomoLOGIC* package, led to the specification of recommendations for insecticide use based on approximately 60 rules (Cox 1996). This figure exaggerates the complexity of the rule base because only subsets of these were used at any one time depending on the stage of the resistance management strategy. This level of complexity hardly justified the development of an expert system to handle it.

We can catalogue model output (data base simulation model runs), but the question remains of how to identify response surfaces on the basis of these standard runs. Heuristics can be used to restrict the search space. But in general this makes little sense: because the data base results are not geared to the analysis of particular issues; they are often based on prior assumptions about the world, e.g. that its behaviour is determined by some combination of crop/water/N (factors handled by the models); it fails to distinguish embedded and non-embedded risk (Dorward and Parton 1996); and it fails to exploit what understanding we already have about thresholds in requisite precision, for both inputs and outputs. The model is not matched to the realization of a specific objective, subject to specific constraints. Rather, the objectives and constraints are chosen on the basis of what the model can do.

### *(4) Develop decision-support systems based on simulation models: the chameleon solution*

A model like APSIM (McCown et al. 1996) could be configured (through software) to impersonate another category of model. Certainly, the APSIM Manager module can incorporate management rules (heuristics), and some of the implications of different rule sets can be compared within the model framework. The interface could be written to emphasize these rules. But the difficulties associated with the interpretation of model output remain. There are problems with the different scope of the two model categories: the wrong comparisons may be attempted. For example, a suite of crop models can be used to set up a yield comparison between sorghum and cotton for specified starting conditions. But a single season comparison of this sort is meaningless unless it takes into account the way in which one choice changes future options. Thus, the choice of cotton may preclude a subsequent crop, and the economics of sorghum plus an opportunity crop of chickpeas versus cotton looks very different to sorghum versus cotton. Although, in principle, this sort of comparison could be attempted using an APSIM-like model over multiple time periods, it still says nothing about the management skills required to manipulate farming activities with this level of precision. Even though a comparison of the effectiveness of alternative rule sets may be attempted, this comparison may well be overridden by practical considerations (e.g. labour availability, or the herbicide strategy in place), rather than based on detailed comparisons of 'objective' distributions using stochastic dominance analysis. If such constraints are included in the problem definition, the appropriate model structure will be different.

APSRU has attempted this kind of activity as part of an on-farm research project in Queensland. In general, the models have confirmed the appropriateness of the rules farmers already use to do with planting dates for different crops. A positive aspect is that the model served as a tool for initiating dialogue between scientists and farmers. But the possibility of making researchers' models look like farmers' models is cosmetic. To the extent that it does succeed, it seems to reinforce a degree of conservatism in farm practice that the authors find disturbing. Experiments are done with a model which we suspect underestimates the probability of extreme events, but it does this with greater precision than is commonly utilized in farmers' models. The appearance of precision, at the farm level, is illusory (at least until there is agreement about the materials, instruments, methods and values that reach out into the larger culture, see Wise 1995); and it restricts attention to opportunities that are located inside the boundaries of the model. This is the dilemma Talbott (1995) refers to when he talks about the polarity of accuracy and meaning. The chameleon solution is a retrograde step.

### *(5) Participatory technology development*

In some situations, the course is poorly defined – we may not know what the issue is, or where the potential users are on the model landscape. We need first to find out, then to identify, what can be done. This negotiation of the problem will shape the course (we will be doing something new), but professional scientific input is brought into immediate contact with the issue at the farm level, working with farmers directly to define and address the issue. We can attempt to change the farmers' models by introducing additional (scientific) concepts, e.g. trends in water use efficiency as a measure of the health of the production system, or local benchmarking. We are dealing with farmers' models, but deliberately engineering the introduction of novel (particularly scientific) points of view. These have to prove their work within the context of the farmers' own models, and may eventually be incorporated in them. But also, the farmers' purpose dictates changes in the researchers' models. There is a single (joint) course and a single end-point.

This can sometimes be achieved by changing scale. Again, our example is from pest management: the Insecticide Resistance Management (IRM) strategy is an attempt to deliberately manage the progressive development of resistance to specific (and valued) insecticides in an insect population by voluntary restrictions on their use at certain times of the year in summer crops in eastern Australia (Cox 1996). This is a negotiated (social) technology that has proved highly effective. We are chasing the development of novel technological components through negotiated arrangements between different actors, leading to evolutionary change in the heuristics of land managers. The scientific horse is compelled to run on a different course, but scientific input is accepted into this process – in the IRM case, insights from professional entomology (simple models, interpreted and communicated) were essential inputs to the design of the technology.

*(6) Work with multiple models simultaneously*

The issue here is not about the choice of the best model formulation to capture an objective natural world, but simultaneous use of several model formulations to address an issue from the various perspectives of different actors. One way of doing this is the tool kit approach proposed by Walker et al. (1995) for improved natural resource management in the Herbert River catchment. Simulation, programming and rule-based models (and others, e.g. Monte Carlo simulation, decision trees, GIS) are all available for use to address a particular issue with a client; they are all tools in the tool kit of the professional researcher to be taken out and applied as appropriate. This sees R,D&E as synonymous with professional consultancy (or perhaps as a consultant to the consultants). It introduces a sense of balance in R,D&E, in an attempt to avoid unwarranted escalation of organizational commitment to a particular model category/approach. Simulation models may supply default estimates of a coefficient included in one cell of a spreadsheet model. But, the achievement of pluralism in modelling may require the threat of a more radical change in research methodology. It is here that the dance between researchers and farmers becomes most intense. The process recognizes the emergent articulation of multiple compatible purposes – there is no single end-point. It can be sequential (episodic) and cyclical; and, unlike participatory technology development, does not require continuous joint development of a common tool.

## Discussion

Models have many purposes: to help think through an issue; to identify the most important factors that are likely to influence the outcome (the boundary of an issue); to engage people in dialogue, and negotiations of meanings that (i) there is an issue, and (ii) that there are opportunities for a solution; and to draw out the practical implications for action. What constitutes an issue depends on who you talk with. The choice of model also depends on what you are talking about.

Crop simulation models do much of the above for scientists engaging with other scientists, but the factors incorporated and the boundaries drawn may be entirely inappropriate for the practical manager. Even for scientists, they should be focused on an issue that needs to be resolved; usually they are not. Does integrating many bits of specialist knowledge into crop simulation models necessarily lead to new insights? Scientists argue that there is no other way of integrating all (*sic*) the relevant environmental factors. This calls for an act of faith that we are finding increasingly difficult to accept. One of the problems with simulation models is not that they are integrated, but that they are partial; not that they do not take into account the right factors to 'explain' a situation (i.e. describe it), but that they ignore the distinction between controllable and uncontrollable factors and hence fail to provide guidelines for action.

This argument has several ramifications. First, we value multiple models for use by different people for different purposes; different models will be used anyway, and it seems perverse to insist on a unitary model structure ('one size fits all') irrespective of the context of model use. Second, recognition of different model structures for different purposes emphasizes that practitioners already use powerful representations that they have adapted to address the issues they face. Assertion of a different category of model has to be evaluated in comparison with these existing processes, not in any absolute sense. Third, models should be valued as process tools at least as much as products (whether descriptive or analytical), i.e. negotiation of a model structure by different actors can facilitate communication and generate novel options. The model can then be discarded. The existing simulation technology seems intent on describing an objective world of biophysical processes that is almost inaccessible to practitioners and is, in any case, mostly outside their control. The practical world is occupied largely with continual re-adjustment to ever-changing circumstances. The articulation of the models of researchers and farmers is, at present, accidental.

The evaluation of alternative model formulations is a critical issue. The bald assertion that a model captures the best available scientific knowledge just will not do if we are intent, as we believe we should be, in bringing about change in the way agricultural production systems are managed. We have to be concerned about the marginal contribution of alternative model formulations for different uses. There are rapidly diminishing marginal returns to increasing complexity. For many purposes, we may have already gone past the optimal level of complexity (because of a computational problem to do with a combinatorial explosion of input specifications; because we are not restricting the search space to isolate likely solutions; and because the input requirements are too onerous to collect, and hence are poorly specified). We have gone too far chasing marginal improvements in precision (required to satisfy a scientific audience) at the expense of explanatory power, creating dependence when we should be empowering. We endorse the judgement of Pannell et al. (1996): '...in comparing simple and complex models for farmer use, there is no trade-off between costs and benefits; simple models are better on both counts.' Rather than spending inordinate amounts of time and effort in the decision-making process, it would seem at least as useful to spend time implementing decisions and dealing with their ramifications (Pfeffer 1992). This argues for greater practical engagement by researchers.

The criteria for success in modelling activities are thus changing: towards greater emphasis on achieving outcomes (changes in the behaviour of others), not outputs (residual artefacts of our own behaviour); and towards acceptance of modelling as a process for achieving outcomes, not a product as an end in itself. The more immediate question we face is how two very different categories of model (those of scientists and farmers) can be articulated in order to bring about purposeful change in the behaviour of both communities. We have proposed various ways of overcoming this problem as researchers: working with farmers' models; using alternative research model formulations (such as mathematical programming); using translation models (to move between model categories); using sequential processes (construct/deconstruct); or all of these as part of a tool kit approach. In some situations, it seems pointless to question the purpose of either farmers or scientists; in other situations, this will be essential in order to bring about the kind of changes we want. In some situations, existing tools will suffice; in others, new tools are urgently needed. No one solution will be everywhere the best. In all cases, what we do will depend on the issues we face.

We see particular value in two approaches: participatory technology development, which emphasizes the joint design and evaluation of novel technological components by both researchers and farmers, and the joint development of models to do this. And the development of heuristic generators within an alternative model framework (not simulation, although simulation may provide first estimates of some of the coefficients in such a model structure; possibly MP).

But, more generally, we argue for the simultaneous use of, and fluency in, multiple representations by professional researchers intent on effecting change in the way in which agricultural production systems are managed. The first thing is to recognize the value of farmers' models, and the limitations of researchers' own models, for action. We need to demonstrate how professional models can lead to practical innovation rather than just providing analytical ('objectively derived') justification for knowledge already incorporated into practical models. Part of this cultural change will be brought about by participation in model development, starting over from scratch; part by taking responsibility for deconstructing model output into practical guidelines; part by learning to construct entirely new categories of model. We argue the need to re-engineer R,D&E processes if we are to realize the potential of researchers' models to change production practice (Cox et al. 1966a). Models become tools for managing within the communication process as we begin the delicate task of re-articulating two cultures that have grown apart.

## Acknowledgements

This research project was financed in part by a research grant from the Land and Water Resources Research and Development Corporation, the Grains Research and

Development Corporation, and the Rural Industries Research and Development Corporation of Australia.

## References

Casti J L (1992) Reality rules. Picturing the world in mathematics. II. The frontier. John Wiley and Sons, New York, USA. 424 p.

Cox P G (1996) Some issues in the design of agricultural decision support systems. Agric. Syst. (in press).

Cox P G, MacLeod N D, Ridge P E, Shulman A D (1996a) Reengineering agricultural R,D&E to support management decision-making: Problems and prospects. Pages 168–171 in Proceedings of the 8th Australian Agronomy Conference. 29 January–2 February 1996, University of Southern Queensland, Toowoomba, Queensland. The Australian Society of Agronomy Inc., Toowoomba, Queensland, Australia.

Cox P G, Shulman A D, Ridge P E, Foale M A, Garside A L (1996b) An interrogative approach to system diagnosis: An invitation to the dance. J. Farming Syst. Res. Ext. (in press).

Dorward A (1991) Integrated decisions rules as farm management tools in smallholder agriculture in Malawi. J. Agric. Econ. 42(2):146–159.

Dorward A (1994) Farm planning with resource uncertainty: A semi-sequential approach. Eur. Rev. Agric. Econ. 21:309–324.

Dorward A, Parton K A (1996) Quantitative whole farm models and embedded risk in complex, diverse and risk prone agriculture. J. Agric. Econ. (in press).

Flood R L, Jackson M C (1991) Creative problem solving: Total systems intervention. John Wiley and Sons, Chichester, UK.

Hardaker J B, Pandey S, Patten L H (1991) Farm planning under uncertainty: A review of alternative programming models. Rev. Marketing Agric. Econ. 59(1):9–22.

Hazell P B R, Norton R D (1986) Mathematical programming for economic analysis in agriculture. Macmillan, New York, USA. 400 p.

McCown R L, Wafula B M, Mohammad L, Ryan J G, Hargreaves J N G (1991) Assessing the value of a seasonal rainfall predictor to agronomic decisions: The case of response farming in Kenya. Pages 383–410 in Muchow R C, Bellamy J A (Eds.) Climatic risk in crop production: Models and management for the semiarid tropics and subtropics. CAB International, Wallingford, UK.

McCown R L, Hammer G L, Hargreaves J N G, Holzworth D P, Freebairn D M (1996) APSIM: A novel software system for model development, model testing, and simulation in agricultural systems research. Agric. Syst. 50:255–271.

Murray-Prior R (1994) Farmers as scientists: A personal construct theory interpretation of hierarchical decision models. Paper presented at 38th Annual Conference of the Australian Agricultural Economics Society, 7–11 February 1994, Victoria University, Wellington, New Zealand.

Pannell D J, Malcolm B, Kingwell S (1996) Are we risking too much? Perspectives on risk in farm modelling and farm management. Am. J. Agric. Econ. (in press).

Pfeffer J (1992) Managing with power. Politics and influences in organisations. Harvard Business School Press, Boston, MA, USA. 391 p.

Rae A N (1971a) Stochastic programming, utility, and sequential decision problems in farm management. Am. J. Agric. Econ. 53(3):448–460.

Rae A N (1971b) An empirical application and evaluation of discrete stochastic programming in farm management. Am. J. Agric. Econ. 53(4):625–638.

Ryle G (1963) The concept of mind. Penguin, Harmondsworth, Middlesex, UK. 316 p.

Shulman A D, Penman R (1992) Developing information infrastructures that work. Pages 1–7 in Service delivery communications in the 1990s. Department of Transport and Communication, Canberra, Australia.

Sneiderman B (1987) Designing the user interface: Strategies for effective human-computer interaction. Addison-Wesley, Reading, MA, USA. 448 p.

Talbott S L (1995) The future does not compute. Transcending the machines in our midst. O'Reilly & Associates, Sebastopol. 481 p.

Thornton P K, Hoogenboom G (1994) A computer program to analyze single-season crop model outputs. Agron. J. 86:860–868.

Thornton P K, Hoogenboom G, Wilkens P W, Bowen W T (1995) A computer program to analyze multiple-season crop model outputs. Agron. J. 87:131–136.

Walker D H, Johnson A K L, Bellamy J A (1995) Development of a decision support system for improved natural resource management in the Herbert River catchment. Paper presented at the National Conference on Downstream Effects of Land Use, 26–28 April 1995, Central Queensland University, Rockhampton, Australia.

Wise M N (Ed.) (1995) The values of precision. Princeton University Press, Princeton, NJ, USA. 372 p.

Woodruff D R (1992) WHEATMAN: A decision support system for wheat management in subtropical Australia. Aust. J. Agric. Res. 43:1483–1499.

# Towards data standards for enterprise and farm-level analysis

J.W. HANSEN[1], P.K. THORNTON[2], J.W. JONES[1] and B.M. JACOBSON[1]

[1]*University of Florida, Agricultural and Biological Engineering Department, P.O. Box 110570, Gainesville, FL 32611-0570, USA*

[2]*International Fertilizer Development Center, P.O. Box 2040, Muscle Shoals, AL 35662, USA*

**Key words:** data standard, simulation, hierarchy, enterprise, farm, farm model, crop model

**Abstract**

In this paper we discuss data standards for enterprise and farm-level analyses, and illustrate the use of such data by an enterprise profit calculator and by a farm simulation model. We take a 'bottom-up' approach that builds incrementally on approaches and data standards that have proven useful for field-level analyses. A schedule of field operations links field-level production process to enterprise-level resource accounting. Prices convert resource use to a monetary basis. Each combination of single- versus multiple-season and deterministic versus stochastic analysis suggests a different approach to handling price data. The social and cultural factors that impinge on farm decision processes are not fully understood. We can, however, obtain useful information from analyses that are based on simplifying assumptions and modest sets of data. At its simplest, farm-level analysis may consist of aggregating enterprise production or returns across the farm. At the next level, farm analysis accounts for constraints of various resources and for off-farm income and expenditure, requiring an initial inventory of resources. Dynamic production, consumption and investment decisions add complexity at the third and most general level. Since each level of analysis builds on the models and data requirements of simpler levels, agronomic data can be shared by enterprise studies, and enterprise data used for farm-level analyses. The scope of validity of some types of data is broad enough to allow data bases to be shared among different locations. The ability to share data among applications or locations adds value to data.

## Introduction

The development of data standards for field crops was a milestone in the advancement of system simulation as a tool for addressing a range of agricultural and environmental problems. However, several categories of problems must be addressed at higher system levels (Jones et al. 1996). Extending analysis from the field level to the enterprise or farm levels places increasing demands on conceptual and mathematical models which, in turn, impose a greater demand for data. The additional effort required can be justified only if the marginal value of the information produced exceeds the marginal cost of the data and analysis. Sharing data sets in standard form among applications and across locations reduces the cost and increases the value of the information required for enterprise and farm analyses.

The purpose of this paper is to encourage a 'bottom-up' approach to developing data standards for systems analysis at the enterprise and farm levels. We first discuss the concept of data standards and examine the development of field crop data



*P.S. Teng* et al. *(eds.): Applications of Systems Approaches at the Farm and Regional Levels, 229–245.*

standards for lessons that we can apply to enterprises and farming systems. We then discuss specific information required for analyses at the enterprise and farm levels. Software tools illustrate the use of data at these system levels. Finally, we discuss opportunities for sharing standardized data sets across applications and locations.

## Data standards

Our discussion of data standards begins with some preliminary considerations: prerequisites, benefits and guidelines. There are at least two prerequisites to a data standard. First, a data standard cannot be drawn up in a vacuum; it must relate to some purpose or tool, in the same way that data collection in the absence of a hypothesis is logically absurd. Data are not collected for their own sake. They are valuable only in the context of a model, a hypothesis, an application. Second, there can be no unilateral declaration of a data standard. A set of variables, methods for their collection and formats for their storage, only becomes a useful standard if other people see value in it and collect their data in conformity to it. These considerations are practical because they force the would-be standardizer to answer the fundamental question: what is this data standard actually for?

Data standards offer two significant benefits: they can enhance data usefulness and they can enhance data quality. Usefulness includes completeness, accessibility, and exchangeability. For field trials grown in a variety of locations by many different people, a data standard can help ensure that information collected is consistent in scope and format and adequate for its intended purpose. Data sets then become accessible to people other than the individual experimenter, and can be easily exchanged among different applications. It is also sometimes possible to take previously collected field data, and by making the data conform to the standard, the original data can be used for purposes for which the experiment was never designed. The value thus added to historical field trials may be substantial, and may even presumably exceed the original worth of the trial.

In addition to usefulness, a standard can help ensure data quality. Data in a common format can easily be checked for consistency, and standard error checking procedures can be developed and applied to data sets conforming to the standard. A well-constructed data standard also provides a means of backtracking from final results to original measurements (D.W.G. Van Kraalingen, AB-DLO, Wageningen and M.C. Van Den Berg, IRRI, Los Baños, pers. commun.), so that analyses and data manipulation can be verified. These advantages are important in view of recent moves towards multi-partner, multi-location agricultural research consortia whose mandates encompass ecoregional scales. While on-line data bases and information systems can facilitate the distribution of information, appropriate standards can help preserve data integrity and data transparency in a situation that could easily degenerate into a 'free-for-all' situation in their absence.

Several guidelines for developing data standards (D.W.G. Van Kraalingen and M.C. Van Den Berg, pers. commun.) may be enumerated. First, the standard should be completely independent of the platform on which the data are to be stored, and

*Table 1.* A field-level crop model data standard (*Source:* Hunt et al. 1994; Rabbinge 1993).

| | Potential | Attainable | | Actual |
|---|---|---|---|---|
| | *Level 1* (defined by weather and $CO_2$) | *Level 2a* (limited by water) | *Level 2b* (limited by nutrients) | *Level 3* (reduced by pests and various stresses) |
| **Weather** | Maximum and minimum temperature; solar radiation | *Level 1* and rainfall | *Level 2a* | *Level 2a* |
| **Soil** | | Critical water contents (lower, drained upper, saturated); bulk density; initial water content; root weighting factors | *Level 2a*; pH; organic C; initial $NH_4^+$ and $NO_3^-$; relative mineralization rate; total, organic, and available P | *Level 2b*; may include soil toxicities and soil-borne pathogen levels |
| **Crop** | Genetic coefficients (development, potential grain characteristics) | *Level 1* | *Level 1* | *Level 1*; may include pest populations, crop susceptibility, and physiological coupling points for pests and stresses |
| **Management** | Date of planting; planting geometry | *Level 1*; irrigation dates, amounts and method | *Level 2a*; N and P fertilizer dates, type, amount, methods; residue application dates, amounts, C:N ratio and P amounts | *Level 2b*; may include pesticide applications, residue management, and rotation effects |

the data should be accessible without recourse to commercial software. Second, a data standard should be concise; it should be complete for its envisaged purposes, but not redundant. Third, it should also be consistent; the syntax should be unambiguous. Finally, extending the standard should also be possible as tools and analytical capabilities change through time.

### *Field crop data standards*

The International Benchmark Sites Network for Agrotechnology Transfer (IBSNAT) developed a data standard that has proven particularly useful. The purpose of IBSNAT was to evaluate and use process-level crop simulation as a tool for transferring agricultural technology across locations and time more efficiently than can be done through traditional field trials (Harrison et al. 1990). The organizers recognized the importance of collecting a consistent set of data using standard methods and storing the data in a common format across the international network of collaborators.

IBSNAT started standards to provide agronomic data for developing, testing, and applying the suite of field crop simulation models contained in the software package, Decision Support System for Agrotechnology Transfer (DSSAT) version 3.0 (Tsuji et al. 1994). This standard, the so-called Minimum Data Set (MDS), consists of a set of data that may be collected from field trials, entered into a particular format, and used to run crop models (among other uses). The data set consists of weather, soils, management and genetic data. IBSNAT published standards for experimental design and data recording procedures (IBSNAT 1984), field and laboratory methods for measurements (IBSNAT 1990), and input and output file formats (IBSNAT 1986). The MDS standard has been extensively modified over the last five years, primarily with respect to its format (Hunt et al. 1994). It is now undergoing additional revision, with the aim of providing a more widely applicable standard format for collecting, reporting, storing, and analysing field trial information (ICASA 1995).

The development of the IBSNAT MDS was closely linked to the idea of levels of analysis. Nix (1984) proposed a sequence of levels of analysis, each with its own data requirements, based on the detail and temporal resolution of the crop models. An alternative scheme (De Wit and Penning de Vries 1982; Rabbinge 1993) was based on factors that limit production. The practical development of the IBSNAT crop models followed a similar general scheme. Historically, crop models capable of simulating potential production in an environment were constructed first. Such models have a modest input data requirement; water and nutrients are assumed to be non-limiting so soil data are not required (*level 1*, Table 1). These potential production models were then modified by the addition of a water balance, which placed heavier demands on input data (*level 2a*). Nitrogen and phosphorus processes have now been added to these models (*level 2b*), which place additional demands on input data. Pest and disease routines (*level 3*) might be the next sensitivities to be added. For management-oriented crop models, there are clearly defined sensitivities that the models must have if they are to provide useful information to users. On the other hand, as Table 1 makes clear, increasing model sensitivity is bound to increase data input requirements. Balancing the competing forces of data requirements and model usefulness at any level of analysis is a challenging task.

### *The enterprise and farm levels*

The MDS is a data standard for analyses of agronomic systems at the field plot level. Higher levels of aggregation require different types of standard for different types of analysis. Agricultural systems span parallel ecological, economic, and social system hierarchies (Figure 1). Field-level analysis focuses on agricultural ecosystems, where the ecological processes of crop or livestock production occur. The enterprise level adds an economic component in the form of a set of managed resources. A social component – a farmer or farm household – comes into play at the farm level. We can, thus, define a farming system in terms of its interacting ecological, economic, and social components; farm-level analyses must consider all three.

Traditional disciplines tend to isolate the ecological, economic, and social subsystems of agriculture. Although advocates of systems approaches have long held that a system cannot be understood by studying its components in isolation, system-oriented research has also neglected to integrate these subsystems fully, focusing more on the field and enterprise levels than the farm level. This is not surprising.

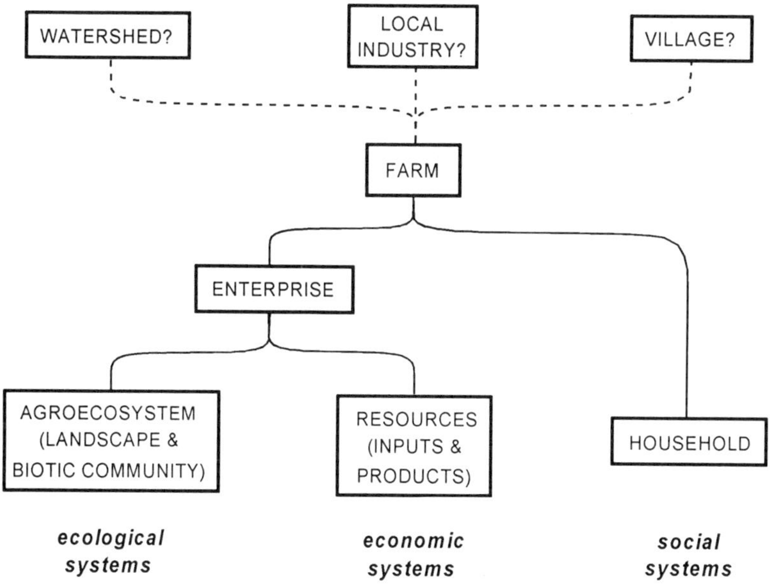


*Figure 1.* Parallel ecological, economic and social hierarchies of agricultural systems.

The ecology of smallholder farming systems in the tropics can be very complex. In addition, it is at the farm level that the social hierarchy has its most profound influence. Agricultural research methodology is currently not well equipped to deal with such ecological and social complexity. On the one hand, the bulk of agronomic research relates to monocropping systems in temperate regions. On the other, we simply do not understand very well why and how farmers make decisions. If, however, we are to make progress in understanding and predicting the impacts of external change on farming systems, then it is critical that future models start to account for the complex ecological, social, and economic factors that impinge on farming system behaviour.

### *A bottom-up approach*

Little progress seems to have been made on data standards for economic and social components of agricultural systems, despite some treatment in the literature (Oram 1987; Ritchie and Dent 1994) and attempts by the Farming Systems Research community in the late 1980s and an International Agricultural Research Center initiative in the early 1990s. Much more effort seems to have been expended on tools than on data (Anderson et al. 1985; Bouma and Beek 1994). Efforts to specify data standards and design data bases for farm-level system analyses can follow one of two approaches (Ritchie and Dent 1994). A top-down approach starts with higher-level systems, such as the region or watershed, and attempts to identify the range of variables that could influence enterprise- and farm-level systems. A major problem with this approach is knowing when to stop collecting data. The top-down approach (Occam's Razor in reverse) is also not particularly amenable to modelling; while it may be true that everything is, ultimately, connected to everything else, this is not a particularly fruitful insight for the modeller who is attempting to explain the maximum amount of variation in system response with the minimum number of variables.

The second method is the incremental bottom-up approach used by the developers of data standards for field crops. This is the approach of Table 1; the investigator starts with a model considered appropriate to a level of analysis, and then attempts to reduce the set of variables to the minimum required to describe the system reasonably accurately at that particular level of analysis (Nix 1984). The bottom-up approach is more amenable to modelling, but a major danger is that there may be overarching social factors governing behaviour in the system that remain unperceived until the observer is higher up the system hierarchy (Ritchie and Dent 1994). Accounting for such social factors in a modelling framework may not be easy.

In our view, the bottom-up approach presents the lesser of two evils. In the following sections of the paper, we start with the field level and add to the framework the minimum set of processes and data needed to represent enterprises, and then the minimum set needed to represent farming systems, given a few well-defined assumptions. The goal is a set of building-blocks: data bases and models that will prove useful across a range of applications and environments. However, given our current knowledge, we doubt that it is feasible to identify a set of economic and social variables that could meet the full range of purposes across all locations.

## The enterprise level

Resource accounting is the most important process that the enterprise level adds to field level systems. The result of enterprise resource accounting is usually expressed as net monetary returns per unit of land area. The implicit assumption is that land is the only resource that is fixed. Alternatively, results can be expressed in terms of some other limiting resource, for example return on labour.

Replicating an enterprise analysis with stochastic inputs of weather and price data yields a distribution of returns. Stochastic dominance methods can then be used to compare the potential acceptability of alternative crops or production methods (Thornton and Hoogenboom 1994). Extending an enterprise analysis in time allows one to examine trends in returns that result from the interplay between production trends due to changes in soil or climate and price trends due to changes in markets. By combining long-term and stochastic enterprise analysis, one can examine changes in the distribution of returns, or systemic changes in the probability of a negative return (Thornton et al. 1995).

### *Agroecological production processes*

Data sets for enterprise analysis build on minimum data sets for agroecological production processes. A data set for an agroecological production process must characterize the physical environment, biotic community, and management operations. A data set used for simulating long-term management must represent sequences of

*Table 2.* Format of the operations output file.

| Variable | Name | Format* |
|---|---|---|
| Date (year, day of year) | DATE | 1I5 |
| Days after planting | CDAY | 1I5 |
| Operation type: | OPTY | 2C4 |
| TILL = tillage | | |
| RESD = residue application | | |
| PLNT = planting | | |
| CHEM = chemical application | | |
| FERT = fertilizer application | | |
| IRRI = irrigation | | |
| HARV = harvest | | |
| Operation option: | OPOP | 5C1 |
| A = automatic | | |
| R = on reported dates | | |
| D = specified days after planting | | |
| M = at maturity | | |
| Operation method code | OPME | 1C5 |
| Amount of material resource produced | RAMT | 0R91 |
| Unit of material resource | RUNT | 1C5 |
| Name of material resource | RNAM | 1C5 |
| Crop code | CC | 4C5 |
| Cultivar code | CLTV | 1C6 |

*Format descriptions consist of: number of leading spaces, variable type (C = character, I = integer, R = real), field width, and (if real) number of decimal places.

```
*OPERATIONS

!Timing....  Operation.......  Material resource use...........
@DATE  CDAY  OPTY  OPOP  OPME     RAMT  RUNT   RNAM     CC CLTV
82056    -1  RESD     R AP002  -1000.0 kg/ha RE001     MZ IB0035
82057     0  PLNT     R PM001    -27.2 kg/ha PM001     MZ IB0035
82063     6  IRRI     R IR009    -13.0    mm WATER     MZ IB0035
82074    17  FERT     R AP009    -56.0 kg/ha FE001     MZ IB0035
  .
  .
  .
82178   121  HARV     M HA001  13647.9 kg/ha HA001     MZ IB0035
```

*Figure 2.* A portion of the operations output file from a simulated maize irrigation × N fertility experiment, Gainesville, Florida, 1982.

production operations, and of crop or livestock populations that inhabit a field. The DSSAT MDS provides a suitable basis for simulating a field crop enterprise. Improving the representation of sequences will enhance the usefulness of the MDS for simulating enterprise management. Comparable data standards do not yet exist for livestock enterprises.

*Operations*

Although growth of a crop or animal is a continuous process, the production process affects farm resources at discrete times during operations such as tillage, planting, applications of fertilizers and pesticides, irrigation, harvest, post-harvest processing, and marketing. All of the economic information derived from dynamic models of crop or livestock production processes can be represented by a schedule of operations that includes the material balance associated with each operation. The timing of operations is important because prices of inputs and products may change during a production cycle. It is particularly important for farm decisions where cash flow may be an important constraint and where resource use conflicts among enterprises must be resolved through scheduling.

In the MDS, field operations can be represented either as timed events whose date is specified in advance, or as state events that occur in response to some specified set of field conditions. State events include planting when soil moisture is within an acceptable range and harvesting in response to the attainment of maturity. While timed events can be derived from model input, the timing of state events can only be determined from model output. Because the current DSSAT output files do not provide a complete schedule of operations, we developed an operations output file (Table 2; Figure 2) as a link between production process models and economic accounting or budgeting models.

*Matching operations to resources*

The operations output file gives the balance of material resources, such as planting material, chemicals, irrigation water and harvest product, used or produced in the production process. Requirements for timed resources, such as equipment and labour whose availability is expressed in units of time, are determined after the production process has been simulated. A data base links each combination of operation type, method, and crop or livestock species to requirements for a particular set of timed resources. There may be several combinations of operator and power unit that could accomplish a particular operation. The ability to match operations to timed resources after production processes are simulated is important in a farm analysis because several enterprises may compete for the same set of timed resources and several combinations of operator, power unit, and implement could accomplish the same operation. The actual choice will then depend on availability.

*Prices*

Prices are the basis for converting material, labour, and equipment use into cash flow. In more complex systems, we can generalize prices to represent the exchange or conversion rate between any pair of resources. For example, the 'price' of labour may include wages, a meal, or a reciprocal labour agreement.

Four combinations of single or multiple seasons and deterministic or stochastic enterprise analyses are possible. Each suggests a different scheme for representing and sampling prices. First, a deterministic analysis of a single season calls for a single price. Second, a deterministic, multiple season analysis would use a historical series or, possibly, a single realization of a generated series of prices. The series may or may not be deflated. For each replicate of a stochastic, single season analysis, a price is sampled from a probability distribution. The stochastic, multiple season case is the most complex, but the most general. Our approach has been to fit deflated, detrended, deseasonalized, historical price series to a multivariate autoregressive moving average (ARMA) model (Box and Jenkins 1970), then to combine the deterministic and stochastic components to generate a different realization of prices for each replicate.

Good historical records are the basis for estimating prices whatever the type of analysis. Monthly records provide a good basis for accounting for seasonal price patterns. A historical measure of inflation such as a consumer price index is valuable for deflating a price series.

*Illustration: an enterprise profit calculator*

Jacobson et al. (1995) developed a decision support system (DSS) to aid farmers in selecting the best management practices for their fields. Rather than recommending a particular practice, the DSS reports the agronomic, economic, and environmental impacts of alternative management practices. The DSS incorporates a profit calculator to evaluate economic impacts of management practices. This profit calculator is a specific implementation of the general scheme for using data for enterprise-level analysis, adapted to highly mechanized soybean farmers in the USA.

The profit or net return associated with a management practice consists of the income from selling the products minus the cost of operations, resources used, land rental (if any), and base costs excluding equipment costs (e.g. buildings, property taxes, etc.). Net return, $N$ (\$ ha$^{-1}$), is given by:

$$N=\sum_i (Y(i)\times P_Y(i))-\sum_j (C_{Op}(j)+Q_M(j)\times P_M)-C_{Rnt}-C_B$$

where $i$ is an index of a crop product (e.g. grain or residue), $P_Y$ is product price (\$ kg$^{-1}$), $Y$ is product yield (kg ha$^{-1}$), $j$ is an index of an operation, $C_{Op}$ is operation cost (\$ ha$^{-1}$), $Q_M$ is the quantity of material input (e.g. water, seed, or pesticide) used (kg ha$^{-1}$), $P_M$ is the price of the input (\$ kg$^{-1}$), $C_{Rnt}$ is any land rental cost (\$ ha$^{-1}$), and $C_B$ is any additional base cost associated with the production scenario (\$ ha$^{-1}$). The costs associated with a particular operation include labour cost, and the fuel, repair, and depreciation costs of using equipment and any separate power unit that it may require:

$$C_{Op}=T_{Use}(k)[C_{Use}(k)+C_{Use}(l)+(Q_F(k)+Q_F(l))\times P_F]+T_L\times P_L$$

where $k$ is the index of the equipment required, $l$ is the index of any associated power unit, $C_{Use}$ is cost of using equipment (\$ h$^{-1}$, excluding labour and fuel), $T_{Use}$ is time to complete operation (h ha$^{-1}$), $P_F$ is price of fuel (\$ L$^{-1}$), $Q_F$ is quantity of fuel used (L h$^{-1}$), $P_L$ is price of labour (\$ h$^{-1}$), and $T_L$ is labour required to complete the operation (man-h ha$^{-1}$). The cost of using equipment (excluding fuel and labour) is:

$$C_{Use}=\frac{(1+R)P_{Eq}}{T_{Eq}}+\frac{P_{Eq}\times I}{2T_{Au}}$$

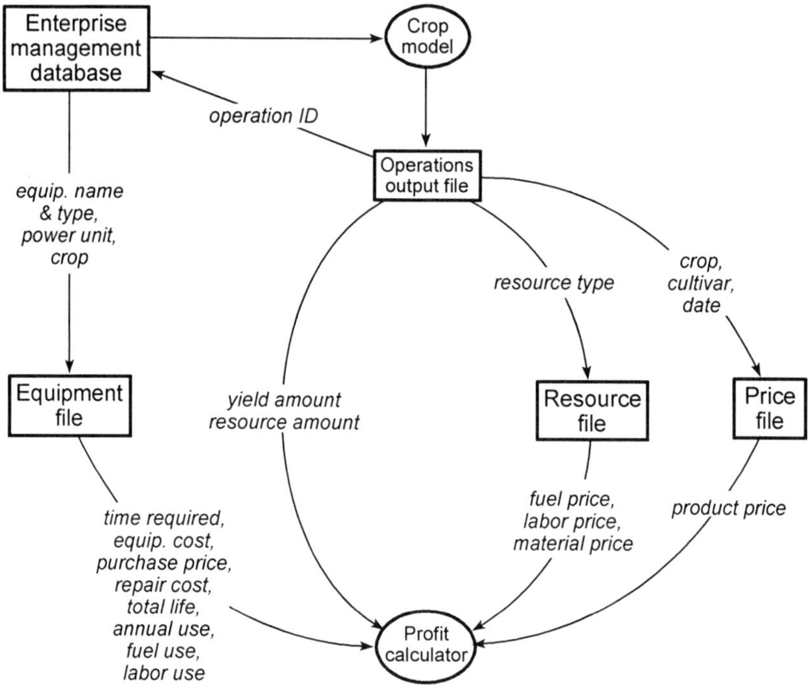


*Figure 3.* Flow of information in an enterprise profit calculator (*Source:* Jacobson et al. 1955).

where $P_{Eq}$ is the purchase price of the equipment ($), $R$ is repair cost as a fraction of purchase price, $T_{Eq}$ is expected life of equipment (h), $I$ is interest rate on an average investment (yr$^{-1}$), and $T_{Au}$ is expected annual use (h yr$^{-1}$).

We constructed these equations so that users could get all of the input data from readily available sources: crop model outputs, economic bulletins produced by state extension economists, and common farmer knowledge. The DSS uses this information to produce the standard management input data file used by the crop models. Figure 3 shows the flow of information from the management input file to the final profit calculator. The DSS provides the resulting projection of net income, along with the agronomic performance and environmental impacts to aid the farmer in selecting best management practices for their particular circumstances.

## The farm level

The emergence of a social subsystem (a farm household) brings an increase in complexity in moving from an enterprise to a farming system. Interactions between farm resources can add another layer of complexity. However, a range of assumptions and simplifications about the social and economic subsystems of a farm is possible. The balance between the cost of an analysis and the value of the information obtained

*Table 3.* Proposed levels of farm analysis.

| Level | Assumptions | Data required |
|---|---|---|
| 1 Unconstrained | Fixed management; all accounting on a monetary basis; production constrained only by the amount of land | Data sets for all enterprises, and field sizes |
| 2 Resource-constrained | Fixed enterprises but flexible inputs; limited by resources in addition to land; includes household resource consumption | *Level 1*; resource inventory, sources, sinks and linkages; household consumption rules |
| 3 Adaptive | Flexible enterprises and inputs; limited by resources in addition to land; includes household resource consumption | *Level 2* and management alternatives; additional data vary with decision model |

should guide the assumptions and simplifications. The cost of obtaining data is an important part of the cost of an analysis.

We describe three levels of farm analysis, unconstrained, resource-constrained, and adaptive, analogous to the levels of crop simulation (Table 3). They serve as a framework for discussing minimum data sets. Jones et al. (1996) discuss these levels in more detail.

### *Level 1: Unconstrained Farm Analysis*

The simplest level of farm analysis treats a farm as simply a collection of enterprises. Management follows a fixed set of strategies, such as a particular set of crop rotations. Results may be expressed as biological outputs or as cash flow. At this level, farm production is constrained only by the availability and productivity of land. These assumptions approximate the situation when a farm is fully integrated into a well-developed market from which all inputs other than land can be purchased or hired, and when feasibility or policy restricts enterprise or land-use options.

Data required for *level 1* farm analysis consist only of the required data sets for all farm enterprises and field sizes. Since resource accounting is expressed as cash flow, enterprise results must be expressed on an absolute basis rather than per unit of land area.

### *Level 2: Resource-constrained Farm Analysis*

Whereas the first level of farm analysis attempts to deal rigorously with only the ecological production processes, the second level – *resource-constrained* – also attempts to deal with the economic processes of resource accounting and allocation in a rigorous manner. At this level, resources other than land constrain production as enterprises compete for a fixed endowment of labour, equipment or crop residues, for example. Amounts of inputs and timing of operations may be constrained by availability of material, labour, or equipment inputs, or by the flow of cash needed to purchase or hire these inputs. For a fixed set of enterprises, resource-constrained farm production or returns will generally be less than unconstrained (*level 1*) as a result of the resource constraints.

Besides *level 1* data requirements, analysis at this level requires an initial inventory of all resources used for production or consumed by the household, specification of possible linkages between resources, rules, or priorities for resolving resource conflicts among enterprises, and rules for consumptive use of resources.

Although analysis at *level 2* avoids characterizing the decision-making process of particular farmers, rules for allocating scarce resources imply a decision model. We can relax the assumption that the enterprise mix is fixed and assume that decisions about both the mixture of enterprises and allocation of resources among competing enterprises are made with the objective of maximizing profit. The analysis could then benefit from tools such as linear programming. This would add to data requirements as the analyst would need to characterize the range of potential management alternatives.

### *Level 3: Adaptive Farm Analysis*

The third level of analysis expands its focus to include the social component of a farming system as it attempts to deal with the farmer's decision process. This is the most general case and the most complex to represent in a dynamic model. A wide range of approaches to modelling farmers' production decisions exists, each with its own set of assumptions and requirement for information. Ellis (1993) provides an excellent summary of models of farm decision-making that are based on utility maximization. Other options include lexicographic utility and heuristic models (Day 1983; Dent and Thornton 1990; Edwards-Jones and MacGregor 1994). Although data requirements vary widely among the different approaches to modelling farm production and consumption decisions, all require a set of alternative management strategies. Most require information about the goals or preferences of individual decision makers, for example the degree of risk aversion that determines the trade-off between profit and security. Some approaches consider a range of personal and social factors that may influence decisions.

The human component of a farming system is complex and the decision process is not fully understood. However, relatively simple models are available that assume farmers optimize only a few important criteria (e.g. profit and security) subject to constraints. The trade-off between predictive power and simplicity offered by these models may be acceptable for many purposes. The curse of dimensionality may limit the application of continuous optimization to modelling farm decisions in a comprehensive, dynamic farm simulation model. However, selecting the optimum combination of management strategies from a small list of discrete options is quite feasible.

### *Illustration: Farming System Simulator*

The Farming System Simulator (FSS) is an object-oriented, dynamic, discrete-event, whole-farm simulation model (Hansen 1996). We developed it to characterize sustainability, expressed as long-term viability (Hansen and Jones 1996), of smallholder

*Table 4.* Sections in the farm scenario file.

| Section | Purpose |
|---|---|
| SCENARIO | Specifies title, units, other files, and simulation control |
| OUTPUTS | Controls farm-level file output |
| ANALYSES | Controls graphic display and file output of resource box plots, resource distribution plots, and sustainability time plots |
| RESOURCES | Specifies all farm resources. Includes resource name, type, and units. Other information varies with resource type |
| LINKAGES | Defines relationships between resources, including the price of exchanging one for the other. Linkage types are either fixed costs, variable costs, or components |
| OPERATION REQUIREMENTS | Specifies priorities, a time window, and a set of timed resources required for each combination of operation type, crop, and method |
| SCHEDULED OPERATIONS | Schedules operations which are not included in the experiment file or returned by crop models. Includes operation type, crop, method, date, and type and amount of material used. Date can be absolute, incremented each simulation year, or relative to another operation |
| LANDSCAPE | Identifies homogeneous fields and their hillslope positions and characteristics. Points to information in the experiment file |
| STRATEGIES | Specifies sequences of crop management activities and identifies a schedule of additional operations |
| ENTERPRISES | Links management (i.e. strategies) with landscape (i.e. fields) |
| PRODUCTION DECISIONS | When developed, will specify criteria for selecting strategies for each enterprise each simulation year |
| CONSUMPTION DECISIONS | Specifies household subsistence and discretionary consumption of a set of resources |

hillside farms in southwestern Colombia. The current version of FSS can simulate potential farm production and some of the processes associated with the resource-constrained level of farm analysis. Although it simulates household resource consumption, resource balances and interactions among resources, it does not adjust management, either based on maximizing a farmer's objective or in response to within-season resource constraints. We present FSS to illustrate a scheme for organizing and using information for analysis at the farm level.

*Input files.* Six types of data files – a scenario file, a price file, an experiment file, a soil file, weather or climate files, and one or more genotype files – provide FSS with all the information needed to simulate a set of stochastic replicates of a farm scenario. The experiment, soil, weather or climate, and genotype files make up the DSSAT MDS, and are documented in Tsuji et al. (1994). The price file represents each price as either constant values or as a set of parameters of a time-series model used to generate sequences of future values.

The farm scenario file is divided into sections (Table 4). The RESOURCES section specifies the structure of a farm by defining its state variables. The object-oriented hierarchy of resource classes – consumable, credit, timed, capital, machine, seasonal and aggregate – provides a robust mechanism for representing different types of resources. LINKAGES determine the possible interactions among farm resources. Resource linkages determine what resources a particular resource can access to meet a deficit, dispose of a surplus, or charge fixed costs. Every resource linkage is assigned a price from the price file. The OPERATION REQUIREMENTS section provides a link between the operations output file and resource accounting by identifying alternative sets of timed resources that each type of operation requires. Through the SCHEDULED OPERATIONS section, FSS provides a flexible means of handling operations that the crop models do not consider, such as post-harvest processing, marketing, and management and production of crops, for which no dynamic model is available.

*Flow of information.* During each simulation year, FSS first forecasts monthly prices and stores them for use in interpolating daily values as needed. Second, for each field, FSS calls external crop models to simulate the crop and fallow activities for the current year. It passes the appropriate MDS to the crop model, then it reads field operations from the operations output file and stores them in an event queue. Third, any additional, scheduled operations are retrieved from each enterprise and inserted into the event queue. Fourth, the event queue executes its accumulated events. The event queue checks conditions for farm failure with each event that it executes. If event execution fails, then the current replicate is suspended. Finally, any user-requested output is written to files.

Household consumption and all resource transactions occur in response to execution of events. The flow information from a field operation event to its effect on farm resources (Figure 4) is similar to the scheme used by the enterprise profit calculator (Figure 3). FSS uses a data base of operation requirements to match an operation to a set of requirements for timed resources (e.g. labour and equipment). Each resource class possesses methods for checking its availability, updating its fixed costs, charging variable costs, and 'buying' from or 'selling' to other resources. Fixed and variable costs associated with a particular resource are represented as links to other resources, with price inputs determining exchange rates between resources.

The result of resource accounting is the time at which farm resources are no longer adequate to cover fixed costs and a minimum level of subsistence consumption. Taken across stochastic replicates, the probability distribution of this time to failure provides an indicator of sustainability of the farm under a given set of assumptions (Hansen and Jones 1996).

## Opportunities for data sharing

We can increase the marginal benefits of collecting and storing data by sharing them among several applications. Each level of analysis, from field to enterprise to unconstrained, resource-constrained, and adaptive farm analysis, builds on the component models and data requirements of the previous level (Table 5). Information is shared in the vertical direction when enterprise analyses build upon previous efforts to compile minimum data sets for field crops, and farm analyses build on enterprise data sets. Horizontal data sharing can occur when data are relevant to a geographical region that is wider than the specific field or farm under study. Some data, such as field layout and resource endowment, must be collected for each farm that is to be

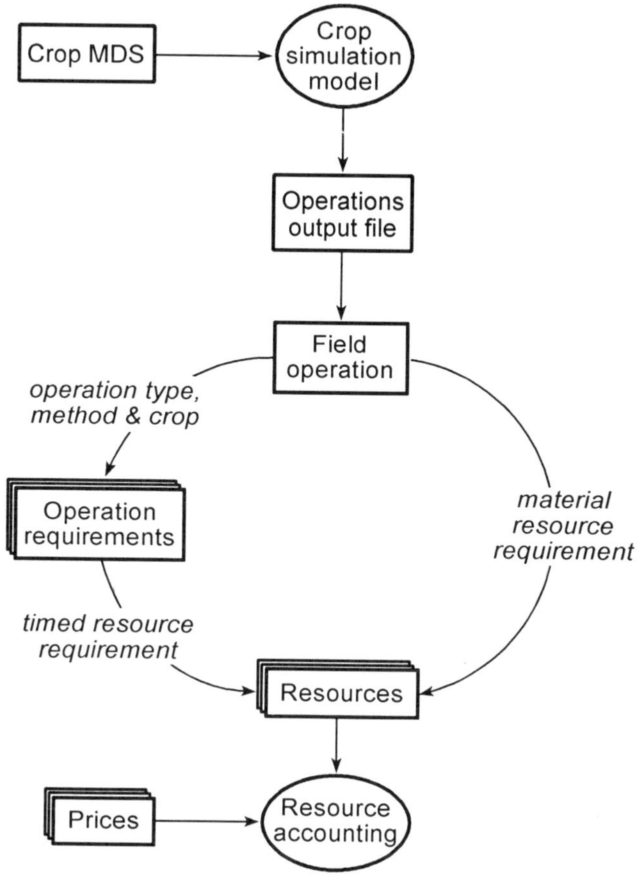


*Figure 4.* Flow of information in a farm model (FSS) from a field operation to resource accounting (*Source:* Hansen 1996).

studied. However, other data, including prices, resource requirements for operations, and policy and sometimes social constraints, may be valid across a region and can be shared among different locations and applications.

## References

Anderson J R, Dillon J L, Hardaker J B (1985) Socioeconomic modelling of farming systems. Pages 77–88 in Remenyi J V (Ed.) Agricultural systems research for developing countries. ACIAR Proceedings 11. Australian Centre for International Agricultural Research, Canberra, Australia.

Bouma J, Beek K J (1994) New techniques and tools. Pages 23–30 in Fesco L O, Stroosnijder L, Bouma J, Van Keulen H (Eds.) The future of the land: Mobilising and integrating knowledge for land use options. John Wiley and Sons, Chichester, UK.

Box G E P, Jenkins G M (1970) Time series analysis: Forecasting and control. Holden-Day, San Francisco, CA, USA. 592 p.

Day R H (1983) Farm decisions, adaptive economics, and complex behavior in agriculture. Pages 18–49

*Table 5.* Summary of data requirements that emerge at enterprise and farm levels of system analysis.

| Type | Data required |
|---|---|
| | **Enterprise level:** |
| Prices | Historical monthly prices and consumer price indices |
| Operation requirements | Labour, power units and implements required ($h\,ha^{-1}$ land, $h\,Mg^{-1}$ material processed, or h $operation^{-1}$) for each combination of operation type, method, and crop |
| | **Farm level 1 (Unconstrained):** |
| Field layout | Location and area of each field |
| | **Farm level 2 (Resource-constrained):** |
| Resource endowment | Inventory of all monetary resources, debt, equipment, permanent sources of labour, and material inputs and outputs |
| Resource linkages | Potential sources and sinks for each resource, and rates of exchange (i.e. prices) between resources |
| Non-agricultural income and expenditures | Household expenditures, tax rules, and any off-farm employment |
| Resource allocation rules | Priorities or decision rules for dealing with within-season resource contraints |
| Alternative strategies | Some representation of the set of potential production strategy alternatives |
| | **Farm level 3 (Adaptive):** |
| Decision criteria | May include risk aversion, drudgery aversion, goal priorities, or goal weighting |
| Social, policy, and other constraints | |

in Baum K H, Schertz L P (Eds.) Modeling farm decisions for policy analysis. Westview Press, Boulder, CO, USA.

Dent J B, Thornton P K (1990) Whole-farm systems simulation using biological crop models. Pages 47–58 in Proceedings of the International Symposium: Decision Support System for Agrotechnology Transfer, Las Vegas, NV. 16–19 October 1989. Part I: Symposium proceedings. International Benchmark Sites Network for Agrotechnology Transfer (IBSNAT) Project, Department of Agronomy and Soil Science, College of Tropical Agriculture and Human Resources, University of Hawaii, Honolulu, HI, USA.

De Wit C T, Penning de Vries F W T (1982) L'analyse des systèmes de production primaire. Pages 20–23 in Penning de Vries F W T, Djitèye M A (Eds.) La productivité des pâturages Sahéliens. Agr. Res. Rep. 918, Pudoc, Wageningen, The Netherlands.

Edwards-Jones G, MacGregor M J (1994) The necessity, theory and reality of developing models of farm households. Pages 338–352 in Dent J B, MacGregor M J (Eds.) Rural and farming systems analysis: European perspectives. CAB International, Wallingford, UK.

Ellis F (1993) Farm households and agrarian development. 2nd Edition. Cambridge University Press, Cambridge, UK. 309 p.

Hansen J W (1996) A systems approach to characterizing farm sustainability. PhD dissertation, University of Florida, Gainesville, FL, USA. 265 p.

Hansen J W, Jones J W (1996) A systems framework for characterizing farm sustainability. Agric. Syst. 51:185–201.

Harrison S R, Thornton P K, Dent J B (1990) The IBSNAT project and agricultural experimentation in developing countries. Exp. Agric. 26:369–380.

Hunt L A, Jones J W, Hoogenboom G, Godwin D C, Singh U, Pickering N, Thornton P K, Boote K J, Ritchie J T (1994) General input and output file structures for crop simulation models. Pages 35–73 in Uhlir P F, Cater G C (Eds.) Crop modeling and related environmental data, a focus on applications for arid and semarid regions in developing countries. CODATA, Paris, France.

IBSNAT – International Benchmark Sites Network for Agrotechnology Transfer (1984) Technical Report 1: Experimental design and data collection procedures for IBSNAT. 1st Edition. Department of Agronomy and Soil Science, College of Tropical Agriculture and Human Resources, University of Hawaii, Honolulu, HI, USA.

IBSNAT (1986) Technical Report 5: Documentation for IBSNAT crop model input and output files for the decision support system for agrotechnology transfer (DSSAT V.2.1), Version 1.0. Department of Agronomy and Soil Science, College of Tropical Agriculture and Human Resources, University of Hawaii, Honolulu, HI, USA.

IBSNAT (1990) Technical Report 2: Field and laboratory methods for the collection of the IBSNAT minimum data set. 1st Edition. Department of Agronomy and Soil Science, College of Tropical Agriculture and Human Resources, University of Hawaii, Honolulu, HI, USA.

ICASA – International Consortium for Agricultural Systems Applications (1995) Exploring concepts and methods for ICASA systems tools and applications, Report of Workshop Progress, 26 February – 5 March 1995 Gainesville, Florida. Agricultural and Biological Engineering Department, University of Florida, Gainesville, FL, USA.

Jacobson B M, Jones J W, Welch S M (1995) Decision support system to assess agronomic, economic, and environmental impacts of soybean and corn management. ASAE Paper No. 952696. American Society of Agricultural Engineers (ASAE), St. Joseph, MI, USA.

Jones J W, Thornton P K, Hansen J W (1996) Opportunities for systems approaches at the farm scale. Pages 1 – 18 in Teng P S, Kropff M J, Ten Berge H F M, Dent J B, Lansigan F P, Van Laar H H (Eds.) Applications of systems approaches at the farm and regional level. Kluwer Academic Publishers, Dordrecht, The Netherlands (this volume).

Nix H A (1984) Minimum data sets for agrotechnology transfer. Pages 181 – 188 in Proceedings of the International Symposium on Minimum Data Sets for Agrotechnology Transfer, 21 – 26 March 1983 ICRISAT Center, India. Patancheru, A. P. 502 324, India. International Crops Research Institute for the SemiArid Tropics, Patancheru, India.

Oram P (1987) Combining socioeconomic data with biophysical environmental data. Pages 261 – 270 in Bunting A H (Ed.) Agricultural environments. CAB International, Wallingford, UK.

Rabbinge R (1993) The ecological background of food production. Pages 2 – 29 in Crop protection and sustainable agriculture. Ciba Foundation Symposium 177. John Wiley and Sons, Chichester, UK.

Ritchie J T, Dent J B (1994) Data requirements for agricultural systems research and applications. Pages 153 – 166 in Goldsworthy P R, Penning de Vries F W T (Eds.) Opportunities, use, and transfer of systems research methods in agriculture to developing countries. Kluwer Academic Publishers, Dordrecht, The Netherlands.

Thornton P K, Hoogenboom G (1994) A computer program to analyze single-season crop model outputs. Agron. J. 86:860 – 868.

Thornton P K, Hoogenboom G, Wilkens P W, Bowen W T (1995) A computer program to analyze multiple-season crop model outputs. Agron. J. 87:131 – 136.

Tsuji G Y, Uehara G, Balas S (Eds.) (1994) DSSAT Version 3, Vol. 2. International Benchmark Sites Network for Agrotechnology Transfer, University of Hawaii, Honolulu, HI, USA.

PART II

# Regional level

# Integrating policy and technical issues for international research on agriculture and the environment using systems approaches

R. RABBINGE

*Department of Theoretical Production Ecology, Wageningen Agricultural University, P.O. Box 430, 6700 AK Wageningen, The Netherlands*

*Netherlands Scientific Council for Government Policy (WRR), P.O. Box 20004, 2500 EA The Hague, The Netherlands*

**Key words:** integration, multidisciplinary, sustainability

**Abstract**

Systems approaches were introduced in the 1970s and have been increasingly popular in agricultural research. Their role has enabled an intensification of integration, and bridged gaps between basic and applied sciences and their use in practical agriculture. The broadening of objectives further requires better integrative tools. Systems approaches are used in rural development, natural resource management and food production studies.

## Introduction

Systems approaches used in agricultural research have evolved during the last few decades. They developed from a speciality for a few into a widely accepted approach in the agricultural and environmental sciences. That development was necessary in response to a changing demand but was made possible by the development of new integrating techniques and methodologies. Thus, it changed from a technique that was only used to gain insight and understanding into an approach that helps to explore options to derive various technologies and may be used as a powerful tool in strategic policy making.

In this paper, some historical developments are described. Various applications of systems approaches are discussed and the changes in international agriculture research during past decades are analysed. Finally, examples of systems approaches as integrative tools are described. The example concerns the widely accepted notion of sustainable agriculture, discussed and operationalized in various ways.

## A short history of systems approaches in agricultural research

Some 30 years ago, the first models of crop growth were developed and used to gain insight into crop growth and development. The models were integrative instruments and bridged the gap between the disciplinary specialists and the applied generalists in agricultural research.



*P.S. Teng* et al. *(eds.): Applications of Systems Approaches at the Farm and Regional Levels, 249–262.*

That function of simulation and systems analysis was very clear in the early explanatory models. Various hierarchical levels of understanding may be bridged with these models. The physical, chemical and physiological processes that determine crop growth are integrated in the models, which leads to explanations of the performance and production of crops. Such analyses are useful to determine the potential growth and production when the prevailing climatic and basic physiological, phenological, optical and geometrical characteristics of crops are introduced. These analyses may also be used to determine attainable yields, i.e. the growth when growth-limiting factors such as water and nutrient limitation are taken into account. A detailed analysis with such models helps to identify why there are differences between actual, attainable and potential yield, and how such limitations may be overcome. It is also very functional for the identification of the quantitative meaning of various processes such as maintenance respiration or partitioning. Since the first models were developed, more and other approaches have been developed.

Explanatory approaches are, in many cases, combined with descriptive approaches. For example, most models of crop growth are explanatory for the basic processes such as photosynthesis, stomatal behaviour, respiration and transpiration, but do not consider in much detail the morphogenesis of crops. Relations between form and function are still poorly understood and that explains why development is, in many cases, introduced as a forcing function. Explanatory models are still poor when they are used for prediction of yields; they are functional in the explanation but only useful in prediction when the conditions of the growing crop are well determined and are introduced through descriptive forcing functions.

More recently, the objectives of systems approaches were considerably broadened. Predictive approaches were developed and explorative studies initiated. These studies required other models, descriptive, analytic models for predictions and multiple goal linear programming models, for example, for the development of explorative studies.

## Aims of systems approaches

The objectives of systems approaches should be very clearly distinguished: conceptualization, explanation, organization, prediction, exploration and planning.

Conceptualization may be used in many situations to structure the thinking on the endogenous and exogenous relations of a system. Conceptual models are used in many situations, also when quantification of basic processes is not yet possible. Explanatory models require the distinction between various hierarchical levels of understanding. Most modelling exercises aim at the development of such explanatory models. Organization of knowledge, of concepts and of thinking is done as well by explanatory models as by descriptive models. Descriptive models are widely used and do not need to distinguish levels of understanding. They help to organize relations at the systems level and to gain insight in the way various external factors correlate or describe phenomena that occur at the systems level.

Predictive purposes are often formulated with models that are descriptive. These models are, in most cases, better for that purpose than the explanatory models. The major reason for that difference in accuracy is due to the absence of the appropriate input data on the forcing functions in the explanatory models.

Exploration of possibilities for crop growth and feasibility studies on the consequences of changes, for example in water availability, carbon dioxide concentration or temperature, may be executed with explanatory models or summaries of those models. The summary models are very appropriate for such purposes and normally have the required accuracy, representability and reliability. Models that are used to explore options for development may be based on such crop growth models and these models may be used in land-use studies. These models are of a completely different nature. They do not explain, but confront various objections and constraints. On the basis of exploratory studies, planning exercises may be started. Such planning exercises require more and other inputs, but may use the outcomes of exploratory model studies on what can be grown, where and what inputs are needed at which moment.

All these studies are based on the perceived need to integrate knowledge from different disciplines to solve well-defined problems. That is the mission of production ecology.

**Production ecology (development of a mission statement)**

The expected increase in world population from 5.4 billion to 11 billion by the middle of the next century and continued economic growth will together require a substantial increase in agricultural production. This, in turn, will increase demands on the natural resources of soil, air and water, which are already being depleted and degraded in many parts of the world. Agriculture needs to meet a rising demand for marketable output, while satisfying ever tighter constraints with respect to the toxicological safety of their products and the impact of production techniques on man, nature, environment and landscape.

These issues require a comprehensive and integrated scientific analysis of socially relevant options for agricultural production activities at crop, farm and regional levels. This analysis is to be focused on developing sustainable agricultural production systems. The term 'sustainable' is used in the sense that use of natural resources – soil, water and air – does not lead to irreversible deterioration.

The concept of 'production ecology' integrates knowledge of the basic processes in living production systems in such a way that sustainable land use and natural resource management respond to well-defined socioeconomic, ecological and agricultural objectives and constraints.

The term 'production' refers to the interaction of energy and matter at different trophic levels. Part of this energy accumulation consists of products useful to man or animal. Animal production and the effects of animals on crops and agroecosystems are integral parts of this research.

This research requires the synthesis of knowledge from various disciplines on primary production into a coherent framework which is then used to develop, implement and evaluate location-specific options for profitable agricultural production systems associated with sustainable land use and natural resource management.

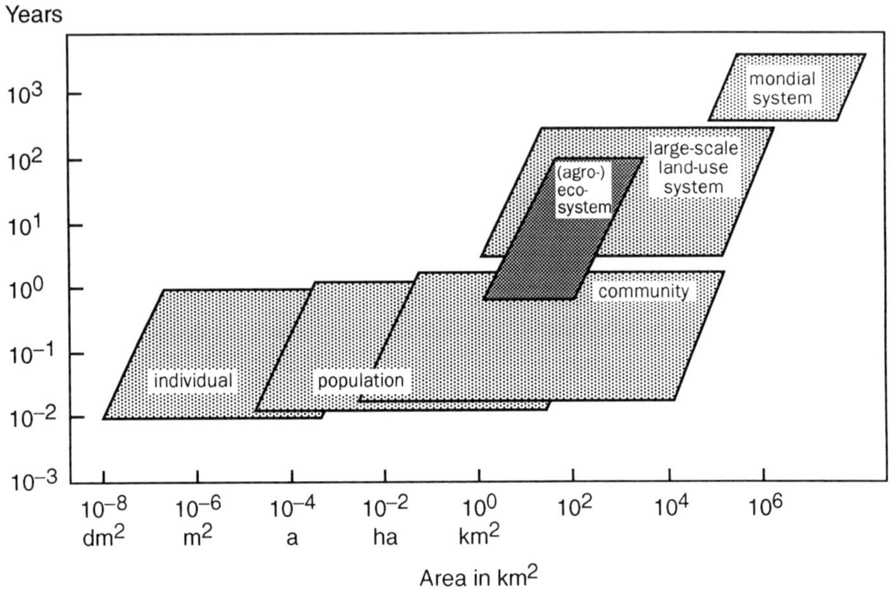


*Figure 1.* Spatial and temporal scales in agroecosystems development.

There is need to substantially increase our ability to control the biological interactions in our farm systems. Production ecology provides understanding of and ability to manage such factors as:

- Pest-predator balance, for lowering losses and reducing the need for pesticides.
- The shift of genetic make-up of weeds which makes them resistant to common control methods.
- The use of crop diversity and residue management to enhance soil quality and efficiency of nutrient flow, and to achieve better synchrony of seasonal soil nutrient flux with crop demand; this leads to greater efficiency of resource use, less environmental loading, and enhanced production potential.

Various scales of research in time and space can be distinguished, varying from the individual plant to global production systems. Figure 1 shows the spatio-temporal scales of primary production which can be distinguished in production ecology. Elements at a certain level are used to answer questions on a higher integration level. Studies of higher integration levels reveal missing elements at lower levels. Processes below the individual plant level are studied when information is needed to study higher integration levels. At the ecosystem level and higher integration levels, interactions between ecosystems and biophysical and socioeconomic variables are taken into consideration. Agricultural production systems are found at the ecosystem level or higher.

Production ecology studies quantitative relations between environmental and genetic factors and growth, and the development of agroecosystems under various

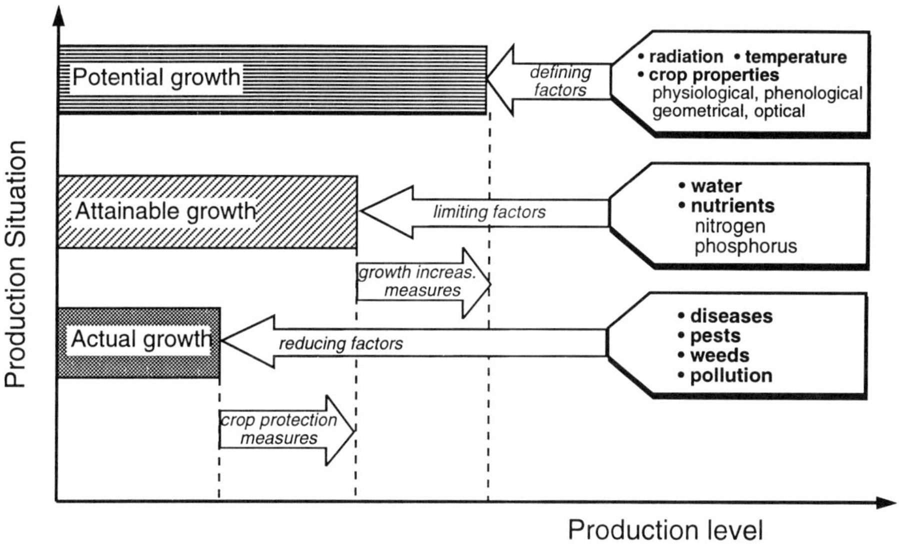


*Figure 2.* Factors determining crop growth (*Source:* Rabbinge et al. 1993).

ecological and socioeconomic conditions. In agricultural production growth-determining, growth-limiting and growth-reducing factors can be distinguished (see Figure 2).

*Growth-determining factors*

These determine the growth potential realized when crops grow with ample supply of water and nutrients. Growth-determining factors include site-specific environmental variables determined by meteorological conditions which depend on location and season, and on species-specific characteristics of physiology, phenology and geometry of leaves and roots. Situations where potential growth rates are reached are rare; less than 1% of the world's agriculture takes place under such conditions.

*Growth-limiting factors*

These comprise abiotic resources such as water and nutrients, which limit the growth rate of the crop to a value below the maximum when their supply is suboptimal. The associated growth level is called 'attainable'. Management of growth-limiting factors focuses on optimal fertilization practices which are observed in close association with natural or man-made water regimes in soils that can be influenced by management practices.

*Growth-reducing factors*

These reduce attainable growth to actual growth. Reducing factors are both biotic (plant pests and diseases, weeds) as well as abiotic (pollution). Research focuses on biotic stress factors. Management of growth-reducing factors aims at dealing with pests, diseases and weeds, emphasizing integrated pest management practices based on biological control mechanisms.

Production ecology and systems approaches are increasingly accepted in international agricultural research, as is illustrated below.

## Evolution of international agricultural research and the increasing need for systems approaches

During its short history, the Consultative Group on International Agricultural Research (CGIAR) has played an important role in international research. Systems approaches were more and more adopted during the various phases which are distinguished below. The description of the various phases is necessarily simplified and based on a schematized representation of the work of the CGIAR. The phases start in the late 1950s and early 1960s.

### *Crop improvement through better use of genetic material*

In this phase, the need for better varieties was vital. IRRI (Philippines) and CIMMYT (Mexico) were created with a clear mandate: to develop improved varieties for the most important grain crops, wheat, rice and maize. That task was very successful and more and more national agricultural research systems became involved and took part of the responsibility for fulfilling this task. Systems approaches were limited; the first crop growth models were developed. These models played a role in explanation and integration of knowledge. Their application in the development of breeding aims or the design of ideotypes was still very limited. It was not before the late 1970s and early 1980s that such applications of crop growth models became possible and accepted.

### *Development of technologies and agronomic innovations*

Improvement of varieties was not sufficient. Appropriate agricultural measures were needed. Seed bed preparation, soil fertility and soil management, irrigation and crop protection were seen as important pillars in crop improvement and increasing productivity. Those developments resulted in better agricultural methods enabling higher productivity and food production. It was the adequate response to the need for new and more productive methodologies, and it was very successful.

### *Farming systems research, and research tailored to farmers' needs*

This phase recognized that changes in agricultural systems required more knowledge of the way farming systems function. Socioeconomic conditions and possibilities and better technologies dictate agricultural development. Technologies are instrumental in that process. The need for tailor-made technologies was felt clearly and required readjustment of research activities and a better division of responsibilities. Many farm household models were developed in this phase. They were all of a descriptive nature and very rarely combined with explanatory crop models.

*Awareness of environmental side-effects and the need to contain them*

Awareness of negative side-effects of agriculture, especially in fragile environments, grew in the 1970s, calling for adoption and changes in agricultural practices. Resource use management, soil conservation and erosion control research became necessary.

*Integrating strategic, basic, applied or participatory research and tailoring it to ecoregions*

More recently, the need to tailor approaches and technologies to specific circumstances was recognized. Ecoregional approaches were developed to meet that demand. Such studies are in general at a high aggregation level. They explore the possibilities and constraints for development and change in biophysical and socioeconomic terms. To achieve this change in research direction, institutions had to be adjusted, and sustainable development and food security included in the mission of the CGIAR. To achieve that mission, linkages with Advanced Research Organizations (AROs) and National Agricultural Research Systems (NARS) had to be strengthened.

Thus, in its twenty-five-year history, the CGIAR saw its agenda change from supply-oriented research to more demand-oriented research, from general to specific research, from technology push to technology pull, and from concrete research products (varieties) to approaches, problem articulation and regional solutions. Its role as focal point in the tripartite collaboration with NARS and AROs became increasingly important.

That role as moderator, coordinator and integrator required extensive use of systems approaches. The early crop modelling exercises were tolerated by non-modellers, in later phases when the integrating and priority setting power of the models became clear, modelling was accepted, but in the present phase where natural resource management and ecoregional studies have been initiated, systems approaches have become indispensable.

One response of the CGIAR centres to the change in demand was to rely more heavily on farming systems research. As the name implies, this sort of research involves looking at farming as a system rather than focusing on a single crop. In particular, farming systems research recognizes that farmers make decisions based on social and economic factors and takes these factors into account. Farming systems teams included a significant proportion of social scientists. This contributed to increasing their numbers and strengthening their role in the centres. Descriptive models were developed and applied.

In the second half of the 1980s, 'farming systems perspective' permeated the

CGIAR. In practice, looking at farming as a system required the centres to make greater efforts to understand soil, water, weeds and climate. By all accounts, a majority of farming systems programs in the centres were reborn as resource management programs using a farming systems approach.

The dialectic between productivity and natural resources research, of course, did not proceed in a tidy and linear fashion, as this account might suggest. The dynamics varied from centre to centre and the partisans did not divide on absolutist lines. No one, for example, took the position that preserving the resource base was unimportant. Nevertheless, the shift in priorities in the 1980s has gone in the direction of resource research.

From the perspective of a long-time participant in the CGIAR, a change in outlook towards sustainability in recent years was most clearly apparent when the CGIAR decided to bring in centres that specialized in agroforestry, forestry and irrigation, all of which have a strong natural resource management emphasis. The need for systems approaches increased during the various phases. The types of models that had to be developed broadened, and the unifying function through concepts exemplified in these models increased.

## Sustainable agriculture: an illustration of systems approaches in a problem-oriented study

The concept that agricultural production must be sustainable, has always been inherent in good agricultural practice, where farmers produce in an economic and sociopolitical environment conducive to long-term conservation and use, and do not want their source of survival and generation of income to disappear. The question of sustainability arises when the resources used for production are placed under stress. This is now happening widely in developing countries where increasing population pressure continues to strain production resources, and also in developed countries due to excessive irrigation, and fertilizer and pesticide application.

An agricultural production system is not sustainable if it leads to declining productivity; degrades the resource base, particularly the soil and water resources; and is not economically viable nor socially acceptable. Research on 'natural resource management' is a key to sustainability, such research being directed at understanding and preserving the resource base on which agriculture depends. Concern about the need for such research may be traced back to the 1960s, and from the mid-1980s onwards became a central theme.

A benchmark in the political etymology of the word 'sustainability' was established with its conspicuous use in 'Our Common Future', the report of the World Commission on Environment and Development (1987). By then, the CGIAR had anchored sustainability as a formal goal in its mission statement. Sustainable agriculture was defined as the 'successful management of resources for agriculture to satisfy changing human needs while maintaining or enhancing the quality of the environment and conserving natural resources'. While the concept of sustainable agricultural production has gained currency and broad acceptance, three aspects that are not always recognized in the public debate, should be emphasized.

*Sustainable agriculture has to be interpreted according to socioeconomic and environmental condition*

The agenda for sustainable agriculture development will vary from one ecoregion to another and will be influenced by factors such as the level of socioeconomic development, share of population dependent on agriculture, and rates of change in productivity. For this purpose, production systems can be grouped into three general categories:

- Intensive cereal and root crop systems. These are located in the best soil and water resource areas, and produce the bulk of the world's food. Early CGIAR research focused squarely on these areas, maintaining and increasing their long-term yield potential, developing methods to contain nutrients at high flow rates, and reducing the need for high pesticide loading, an area still central to current CGIAR mandates.
- Highly productive, vertically integrated livestock, speciality crop and estate crop systems. These fall mostly within private sector influence and have had little CGIAR involvement. Their agenda for sustainability is somewhat different from that of intensive cereals.
- Highly diverse systems on fragile soils or in areas subject to moderate to severe growth-determining or growth-limiting factors. Such areas are coming under widespread and increasing population pressure, with issues of poverty, productivity, resource maintenance and social stability being dominant. The sustainability agenda in these areas is far more complex.

In terms of total degraded land area, depending on the classification used, as much as 95% of unsustainable agriculture today takes place on land falling into the last category, while only 5% occurs on land that falls in the first two categories. This indicates the magnitude of the problem which – as explained in the next section – is characterized by a spiral of unsustainability.

*Research on sustainable agriculture should continue to focus on productivity and efficiency, within a framework of both short- and long-term environmental concerns that take into account inter- and intra-generational considerations*

Research into improved productivity and efficiency must be based on insight and knowledge of basic processes (physical, chemical and physiological) that occur in living systems. More sharply focused up-front research is required to understand how agricultural systems can be manipulated for ecological, agricultural and economic aims, and how efficiency gains may differ according to the aim pursued.

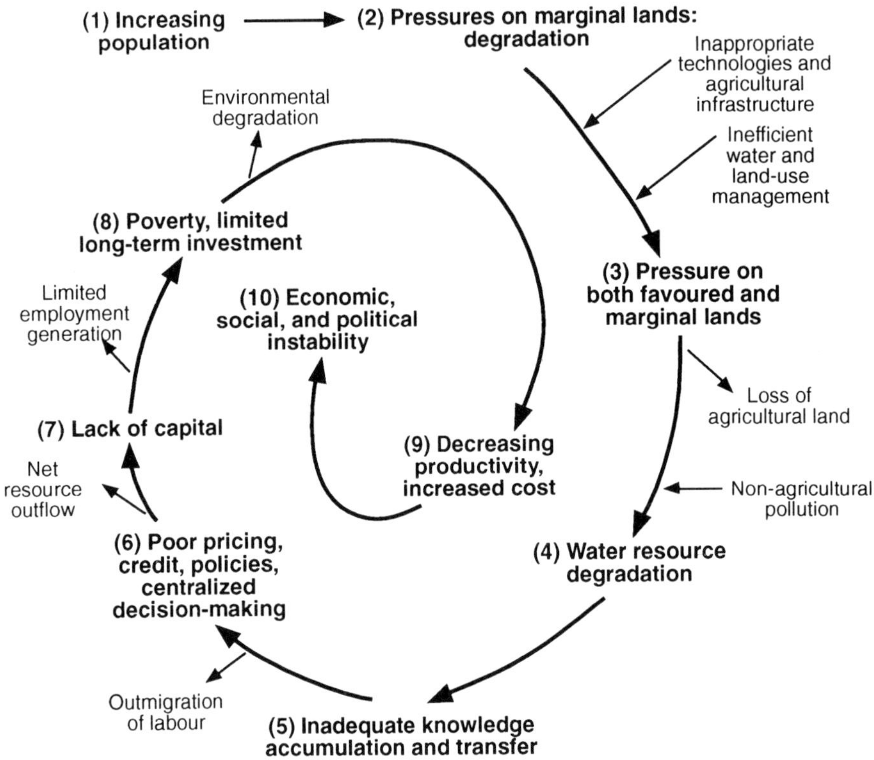


*Figure 3.* The spiral of unsustainability (*Source:* adapted from Rhoades and Harwood 1992).

*Policies for future agricultural growth and food production must take into consideration poverty and equity concerns in all their ramifications*

Careful analysis for policy formulation must consider:

- social, economic and environmental trade-offs;
- effect on soil degradation and declining availability of water resources;
- long-term availability of essential chemical and other inputs;
- the dynamics of future changes in sources and cost of world energy supplies; and
- the possible impact of climate change on the agroecological potential of different regions.

The identification of appropriate policies for land tenure, market structure and infrastructure investment is most urgent in resource-poor areas where population pressure and poverty are increasing.

## The 'spiral of unsustainability'

With increasing world population, pressure on natural resources is mounting. Land degradation, including erosion, is proceeding rapidly. Surface water resources are increasingly pre-empted by urban use while supplies available to agriculture are diminishing, leading to loss in production, more poverty and even heavier demands on remaining resources.

Visualizing this sequence as a downward spiral of diminishing resource availability, deteriorating environmental quality and disintegrating social and economic structures which, in turn lead to even greater instability, emphasizes the cumulative impact of destructive factors (Rhoades and Harwood 1992). Figure 3 depicts this spiral, flags the factors driving it, and their interaction. This 'spiral of unsustainability' can effectively be broken through forces driving the spiral and the momentum redirected towards sustainability under conditions of more effective resource use, greater productivity and improved quality of life. To arrest these forces, certain principles must be followed when designing and programming research to reverse the unsustainability spiral. These principles are:

- *Principle One:* Contributions of research organizations are valuable to the degree that they attack key problems in the spiral of unsustainability. Research can only succeed if other closely linked problems/processes are simultaneously addressed. For example, erosion control engineering must be linked to studies of cropping systems and economic incentives.
- *Principle Two:* The systemic nature of the unsustainability spiral demands that scientists conducting the research be part of broad multidisciplinary teams. A prerequisite for correcting symptoms of unsustainability (such as declining fertility of marginal lands, increased salinity of irrigated lands or decreasing biodiversity of forest ecosystems) must be a deeper, more sophisticated, analysis involving all relevant sciences.
- *Principle Three:* An appropriate research scale is at the watershed level. Although research at the plot, field, cropping system or regional level is legitimate and will provide necessary inputs into the research agenda, such inputs must be related to all factors driving the spiral of unsustainability and operating on a larger scale.
- *Principle Four:* Some processes in the unsustainability spiral are beyond the direct influence of agricultural research scientists (e.g. urban sprawl, global commodity trade, national land policies). Nevertheless, there is a continual obligation to raise awareness among donors, policy makers and government officials that technical problems cannot be solved in isolation. Policy research has a special role to play in sensitizing national policy leaders to the options and likely benefits of policy reforms, closely associated with socioeconomic and biophysical research on production systems.
- *Principle Five:* Politically acceptable mechanisms must be devised for tapping into local traditional knowledge resident with both genders, and for imparting to local communities scientific knowledge that may involve the introduction of technologies, policy adjustments or institutional reforms developed elsewhere.
- *Principle Six:* Research on sustainable agriculture must be conducted as efficiently as possible. Hence researchers should not work in isolation, but need

*Table 1.* Comparison of approaches to traditional cropping systems, farming systems and natural resource management research (*Source:* IBSRAM 1994).

| Parameter | Traditional cropping systems research | Farming systems research | Natural resource management research |
|---|---|---|---|
| Temporal | Annual cycle | 1–3 year cycle | 5–25 year cycle |
| Spatial | Field-plot | Field-village | Watershed-region |
| Beneficiary | Farmer | Family | Multiple |
| Technology | Component | From production system | Natural resource system |
| Target | Self-sufficiency | Profits | Monetary/non-monetary intergenerational equity |
| Role of farmer | Recipient of technology | Provider of information | Participatory |
| Policy | Input/prices | Marketing | Multiple |
| Environment concern | Minimal, on-site | Marginal, on-site | Maximum, on- and off-site |

to benefit from common work plans which reach across well-selected global sites where long-term experiments are being conducted. This principle implies development of a global network of researchers, organizations and projects in which all understand how their efforts fit into a global plan to reverse the spiral of unsustainability.

- *Principle Seven:* Research on sustainability issues necessarily involves a longer-term perspective than conventional agricultural research. Hence, it must focus on specific points on the unsustainability spiral, with built-in provision for objective external review of progress and endorsement of future carefully targeted research plans.
- *Principle Eight:* Any proposals for changes leading to sustainability, and reversing the spiral of unsustainability, must have ownership at the local, national, regional and international level. International input into the research phase of the process must be structured so as to enhance and support such local and national identification.

## Research for sustainable agriculture

Any land-use system is unsustainable if it leads to irreversible biophysical changes in the ability of the land to produce equally well in a future cycle of similar land use, or if the costs of preventing and reversing the degradation are not economically affordable and socially acceptable.

Reversing the unsustainability spiral requires an integrated research strategy involving new methods and institutional arrangements significantly different from earlier production approaches. Natural resources are no longer perceived by the scientific or policy communities as merely the medium to produce more food through high yields of plants and animals, but rather in terms of local and global ecosystem functioning. In addition to production output, ecosystem maintenance, biodiversity, water recharge, clean air and bequeath value are important topics in the research effort.

Compared to crop management, natural resource management is more complicated for both farmers and scientists. Natural resource science involves relationships which are highly interactive. One land-user can affect the health and production of many others. This complexity of natural resource systems raises the issue of linkages within sociopolitical hierarchies and scales of intervention. Sustainability raises new issues such as time and spatial dimensions, social hierarchies, and societal versus individual benefits, and therefore requires new approaches to solving problems. Although lessons can be learnt from farming systems research, its goals and earlier approaches differ from those of natural resource management research (see Table 1).

*A production-ecological approach*

While research programs do consider some aspects of sustainability, the need for new approaches to sustainability research may not have been given full attention. Creating sustainable agricultural and natural resource management systems from unsustainable ones can only be achieved by *finding solutions to whole-system deficiencies, as well as to component problems.* It is futile to attack technical problems – as has been common in the past – without, at the same time, addressing the overall pattern of degradation caused by socioeconomic pressures. Human behaviour driven by poverty, population dynamics and inappropriate economic and land policies is the underlying cause of land and water degradation. The CGIAR agenda for research on sustainable agriculture must include social, economic and policy issues, as well as studies of biophysical processes.

Sustainability research must aim at all levels, including farmer's fields, cropping systems, farming systems, and regional and supra-regional levels. At higher levels, choices of land use become important as they imply trade-offs among socio-economic, environmental and biotechnical benefits. Integration of natural resource management in land use studies requires systems approaches.

Such systems approaches, or better and more appropriate, a 'production-ecological approach', from the outset will recognize and build into a research program environmental factors that bear on production. These factors can be grouped as described above into three categories: growth and yield defining factors such as radiation, climate and crop properties; growth and yield limiting factors such as nutrient and water availability; and growth and yield reducing factors such as pests and diseases. Assessing interactions among these factors may lead to better insights and understanding of how agroecosystems function and how their continuity in time and space can be assured. In this approach, productivity increase and sustainability are not conflicting, but complementary goals, with one being a condition for the other to materialize.

In this context, a recent study by the Netherlands Scientific Council for Government Policy (1994) entitled 'Sustained Risks: A Lasting Phenomenon' should be mentioned. That study also flags the need for more studies on how to guarantee food production and food security for a growing world population. The study shows convincingly how the right combination of land-use options and production technologies can stimulate productivity while minimizing deleterious environmental side-effects.

## Future of systems approaches

The examples above demonstrate how the evolution and increased use of systems approaches have been internalized by many scientists. It is no longer a speciality for few but a commonly accepted working method by many. Disciplinary scientists, production ecologists and the new generation of agronomists are familiar with systems approaches and use it. That will increase in the future as it is virtually impossible to address all the various wide-ranging research objectives without such systems approaches.

Food production, rural management, natural resource management and many other objectives require systems approaches. The tools are there, models of the fourth generation are now available and many different instruments are now mature enough for application. Why not use them?

## References

IBSRAM – International Board for Soil Research and Management (1994) Soil, water, and nutrient management research – a new agenda. CGIAR document MT/94/07. IBSRAM, Thailand.

Netherlands Scientific Council for Government Policy (1994) Sustained risks; a lasting phenomenon. Report No. 44. Sdu, The Hague, The Netherlands. 205 p.

Rabbinge R, Van Latesteijn H C, Goudriaan J (1993) Assessing the greenhouse effect in agriculture. Pages 62–79 in Environmental change and human health. Ciba Foundation Symposium 1975. John Wiley and Sons, Chichester, UK.

Rhoades R E, Harwood R R (1992) A framework for sustainable agricultural development: synthesis of a workshop discussion. In Sustainable agriculture in Asia. Asian Development Bank/Winrock International, Philippines/Arkansas, USA.

World Commission on the Environment and Development (1987) Our common future. Report of the Brundtland Commission, Oxford University Press, UK. 400 p.

# Rice supply and demand in Asia: a socioeconomic and biophysical analysis

M. HOSSAIN
*International Rice Research Institute, P.O. Box 933, 1099 Manila, Philippines*

**Key words:** income elasticity, population growth, yield frontier model, attainable yield, yield loss

**Abstract**
High- and middle-income Asian countries may not need further increase in rice production to meet domestic demand, as their population growth is slow and per capita rice consumption has been declining. Demand pressure will, however, remain strong in low-income countries that contain the bulk of Asian populations, because of continuous high population growth rates and positive income growth. Sustaining the historical growth in rice supply will be difficult if research and extension efforts target only the irrigated ecosystem, as farmers have almost achieved the maximum attainable yield for that ecosystem. The yield gap is large in the dominant rainfed ecosystems because of abiotic stresses and suboptimal use of inputs due to risks in rice cultivation.There is a need to strengthen systems-oriented rice research to develop crop management practices that avoid abiotic stresses, and high-yield varieties that can withstand temporary submergence and prolonged droughts.

## Introduction

Asia accounts for 90% of the world's production and consumption of rice because of its favourable hot and humid climate. Rice is grown in numerous tiny farms primarily to meet family needs. Hence, the marketable surplus is small and prices fluctuate widely with droughts, floods and typhoons. Maintaining self-sufficiency in production and stability in prices are important political objectives in most Asian countries. Less than 5% of the rice production is traded in the international market. China, India, and Indonesia account for three-fourths of the global rice consumption. The world rice market is highly sensitive to shortfalls in domestic production in any of these countries.

With growing economic prosperity and urbanization, per capita rice consumption has started declining in the middle- and the high-income Asian countries. But nearly a fourth of the Asian population is still poor and has considerable unmet demand for rice. Also, the Asian population is growing at 1.8% per year, and may not stabilize before the middle of the next century. The demand for rice is projected to increase by 70% over the next 30 years, primarily due to rapid population growth. It will be difficult to increase rice production to meet this demand as land, labour and water are becoming scarce with increasing competition from the fast-growing non-farm sectors. As technical progress is getting out of stream, the low-income countries will find it difficult to sustain self-sufficiency in rice production. The middle- and high-income countries will have problems sustaining farmers' interest in rice cultivation.



*P.S. Teng* et al. *(eds.): Applications of Systems Approaches at the Farm and Regional Levels, 263–279.*

The affluent Asians in these countries can access rice from the world market, but the prospect of generating a surplus for export outside Asia is limited. These considerations will compel high-income countries to protect the domestic rice industry and to raise prices needed to maintain the balance in incomes of rice farmers and urban workers.

This paper focuses on the forces that govern rice production and consumption to highlight the challenges that Asian rice-consuming countries will face in sustaining food security over the next few decades.

The information for this paper is drawn mostly from the Rice Supply and Demand Project (RSDP) implemented jointly by the International Rice Research Institute (IRRI) and the International Food Policy Research Institute (IFPRI) in collaboration with scientists from national institutions (Ahmed and Siddiqui 1995; Huang et al. 1995a,b; Isvilanonda and Poapongsakorn 1995; Kako et al. 1995a,b; Kumar and Rosegrant 1995; Shahabuddin and Zohir 1995; Sudaryanto and Simatupang 1995).

## Determinants of demand

The growth in demand for a staple grain depends on (i) the level of per capita income, (ii) the rate of growth of population and (iii) the change in prices relative to substitute crops.

At low levels of income, when meeting energy needs is a serious concern, rice is considered a luxury commodity. With increases in incomes, people tend to substitute low-cost sources of energy such as coarse grains, cassava and sweet potato, for rice. But at high levels of income, rice becomes an inferior good (Ito et al. 1989). As incomes rise further, consumers go for a diversified diet and substitute rice for high-cost quality food with more protein and vitamins, such as vegetables, bread, fish and meat. Growing urbanization which accompanies economic growth leads to changes in food habits and the practice of eating outside home further reduces per capita rice consumption. Japan, China, and Taiwan have already passed through these phases and have experienced a decline in per capita rice consumption after reaching a high level several decades earlier. Malaysia and Thailand are now experiencing the same phenomenon. But these high- and middle-income countries – where per capita rice consumption has been declining – account for less than 10% of total Asian rice consumption. The income threshold at which consumers start substituting rice for higher-quality and more varied foods has not yet been reached in large countries such as India, Bangladesh, the Philippines and Vietnam. These countries account for more than 40% of total rice consumption and dominate the growth in consumption of rice. The per capita grain consumption in many of these countries is still lower than the peak level reached in Republic of Korea and Japan during their early phase of development (Table 1). With increased incomes and alleviation of poverty, the per capita rice consumption may increase further in these countries.

In order to project the effect of the growing economic prosperity in Asia and the change in relative prices on future trends in rice consumption, we estimated parameters of the demand function for major rice growing countries of Asia. The country

*Table 1.* Per capita consumption of cereal grains and rice (kg $person^{-1} yr^{-1}$) in major rice growing countries, 1990–92 (*Source:* FAO 1994).

| | Cereal grains | Rice |
|---|---|---|
| China | 200 | 91 |
| Myanmar | 199 | 194 |
| Indonesia | 169 | 134 |
| Korea, Rep. | 169 | 105 |
| Bangladesh | 163 | 158 |
| Philippines | 137 | 88 |
| Thailand | 135 | 126 |
| India | 133 | 73 |

*Table 2.* The demand response to incomes and prices for rice, estimates for selected Asian countries (*Source:* International Rice Research Institute (IRRI) and International Food Policy Research Institute (IFPRI 1995) Rice Supply and Demand Project).

| Country | Percent increase in demand from 1% increase in incomes | Percent increase in demand from 1% increase in prices |
|---|---|---|
| Bangladesh | 0.41 | −0.20 |
| Indonesia | 0.11 | n.a. |
| China | 0.09 | −0.26 |
| Thailand | 0.08 | −0.61 |
| Philippines | 0.08 | −0.93 |
| India | 0.06 | −0.23 |
| Korea, Rep. | −0.11 | n.a. |
| Japan | −0.25 | −0.17 |

studies estimated an Almost Identical Demand System (AIDS) model which uses a two-stage budgeting framework to model household behaviour. In the first stage, the household decides how much of the available income is to be allocated among food and other household needs. In the second stage the household decides the allocation of the predetermined expenditure on food among various items depending on prices, the level of incomes and the relevant demographic and locational variables that may influence tastes and preferences. The estimated parameters showing the income and price response of demand for rice are reported in Table 2.

The estimates show that per capita rice consumption will continue to decline with economic growth in the high- and middle-income countries of Asia such as Japan and Republic of Korea. For other countries, the demand response to income is still positive, although very small – except in countries where poverty is pervasive such as in Bangladesh. If the level of income doubles, the per capita rice consumption is expected to increase by 6–11% in India, Thailand, the Philippines, China and Indonesia.

The demand response to its own price is also small, which is typical of staple grains. A 10% increase in the price of rice will reduce its consumption in the range of 1.7% to 2.6%. The exceptions are Thailand and the Philippines where the price response to demand is fairly large, indicating that with an increase in rice prices, large-scale substitution of rice for other food items will take place. The price response of demand also suggests that a 10% shortage of supply would lead to about 50% increase in rice prices.

The major boost in demand for rice over the next few decades will, however,

*Table 3.* Projections of population in major rice producing and consuming countries in Asia, 1995 to 2025 (*Source:* Bos et al. 1994).

| Country | Population in 1995 (million) | Annual growth rate (% per year) | | Projected population in 2025 (million) | Percent increase 1995–2025 |
|---|---|---|---|---|---|
| | | 1995–2000 | 2020–2025 | | |
| China | 1199 | 0.9 | 0.5 | 1471 | 23 |
| India | 934 | 1.7 | 1.0 | 1370 | 47 |
| Indonesia | 192 | 1.4 | 0.8 | 265 | 38 |
| Pakistan | 130 | 2.7 | 1.6 | 243 | 87 |
| Japan | 125 | 0.3 | −0.3 | 124 | −1 |
| Bangladesh | 121 | 1.8 | 1.1 | 182 | 50 |
| Vietnam | 74.1 | 2.0 | 1.2 | 117 | 58 |
| Philippines | 69.2 | 2.2 | 1.2 | 115 | 66 |
| Thailand | 60.5 | 1.3 | 0.7 | 80.8 | 34 |
| Myanmar | 46.8 | 2.1 | 1.1 | 72.9 | 56 |
| Korea, Rep. | 44.8 | 0.8 | 0.3 | 52.9 | 18 |
| Asia (excluding China) | 2244 | 1.8 | 1.1 | 3389 | 51 |

come from population growth. Table 3 shows the most recent projection of Asian population made by the World Bank (Bos et al. 1994). The population is still growing at 1.8% per year, at which it is expected to double within a period of 40 years. Although the growth rate has been decelerating, it is not going to stabilize even by 2025, when the growth rate will decline to 1.1% per year. China is likely to dominate the population growth in Asia. Although demographers project a relatively slow growth of population in China in view of its past progress in population control, with economic liberalization China might find it difficult to sustain its policy of one-child family and hence the growth may be higher than the projected figure. Excluding China, Asian population is expected to increase by 51% over the next three decades. The lower the level of income of a country, the higher the projected growth of population. The projection also shows that most of the additional population will be located in urban areas. The marketed surplus of rice has to increase substantially to meet the demand from the inflating urban population.

Recent projections made by the International Food Policy Research Institute (IFPRI) on the basis of the income and own and cross-price elasticities of demand and the projected changes in population, incomes and prices, show that the demand for rice will increase by about 1.8% per year over the 1990–2020 period (Rosegrant et al. 1995). This means that over the next 30 years, rice consumption will increase by nearly 70%, and Asian rice production must increase to about 840 million tons by year 2025, from the present level of about 490 million tons, if rice prices are to be maintained at current levels.

## Attainable rice yield

Prior to the 1960s, the growth of rice production was slow and the growing food needs were made mainly through expansion of area. Only for *japonica* rice grown in temperate zones of East Asia were farmers able to increase rice yield with new varieties developed by scientists. With the introduction of IR8, a semi-dwarf fertilizer-responsive photoperiod-insensitive variety, rice scientists were able to shift the yield potential of *indica* varieties from 3.5 to about 10 ton $ha^{-1}$ in the tropics and subtropics (Khush 1995). Over the next two decades, scientists incorporated resistance to major insects and diseases in the modern varieties, reduced the duration of the crop maturity period from over 150 days to around 100 days, and improved the quality of the grains. As farmers gradually replaced traditional varieties by these modern cultivars, rice yields and production started increasing. Since the mid-1960s, rice production has increased by 2.6% per year, enough to meet the population and income growth-induced demand for food. About 80% of the increase in production came from the increase in rice yield (Hossain 1994).

However, there are concerns that the past trends in yield may not continue in future (Cassman and Pingali 1995). Experimental station data from IRRI farms show little or no improvement in yield over the 1966–1987 period. The lack of progress at shifting the yield frontier has led to the concern of yield stagnation and thus the capacity of Asia to meet future food needs.

*Yield frontier*

What are the limits to increase in rice yield given the existing technologies and technologies in the pipeline? To answer this question, Dey and Hossain (1995) applied a stochastic frontier production-function model on the data obtained from field trials conducted under the IRRI International Network for Genetic Evaluation of Rice (INGER) and estimated the yield potential under different agroecological and environmental situations. The function was estimated separately for different rice ecosystems. The sample size consisted of 7734 INGER trials for irrigated early rice varieties, 7758 trials for irrigated medium-duration late rice varieties, 1122 trials for rainfed lowland rice, and 2164 trials for rainfed upland rice. The model was estimated in the following form:

$$\begin{aligned} \ln Y = {} & \ln a_0 + a_1 \ln x_1 + a_2 \ln x_2 + a_3 \ln x_3 + a_4 \ln x_4 \\ & + 1/2\, a_5 (\ln x_1)^2 + 1/2\, a_6 (\ln x_2)^2 + 1/2\, a_7 (\ln x_3)^2 \\ & + 1/2\, a_8 (\ln x_4)^2 + a_9 \ln x_1 \ln x_2 + a_{10} \ln x_1 \ln x_3 + a_{11} \ln x_1 \ln x_4 + v - u \qquad (1) \end{aligned}$$

where:

$Y$ = rice yield (ton $ha^{-1}$);
$x_1$ = nitrogen use (kg $ha^{-1}$);
$x_2$ = rainfall during the growing season (mm);
$x_3$ = average minimum temperature during the growing season (°C);
$x_4$ = sunny hours during the growing season (days/week);
ln = natural logarithm of the variables.

The two error terms $v$ and $u$ are assumed to be independent of each other, where $v$

*Table 4.* Maximum likelihood estimates of stochastic production frontier for high-yielding rice varieties.

| Variable | Rice varietal type | | | |
|---|---|---|---|---|
| | Irrigated early variety | Irrigated late variety | Rainfed lowland | Rainfed upland |
| Intercept | −0.370** | 5.26*** | −3.792*** | −2.368** |
| | (0.183) | (0.521) | (1.129) | (1.209) |
| ln (nitrogen) | 0.306*** | −1.925*** | 3.128*** | 0.992* |
| | (0.097) | (0.277) | (0.556) | (0.564) |
| ln (rainfall) | 0.005 | −0.463*** | 0.173 | 0.391*** |
| | (0.031) | (0.059) | (0.1317) | (0.120) |
| ln (sunny hrs) | −0.567*** | −1.008*** | −1.342*** | −0.280 |
| | (0.036) | (0.058) | (0.146) | (0.230) |
| ln (average min. temp.) | 0.985*** | 1.627*** | −0.831*** | 1.072*** |
| | (0.765) | (0.1157) | (0.216) | (0.197) |
| 1/2 ln (nitro) ln (nitro) | 0.079*** | 0.500*** | −0.820 | −0.051 |
| | (0.027) | (0.071) | (0.150) | (0.146) |
| 1/2 ln (rainfall) | 0.021*** | −0.197*** | 0.018 | −0.015 |
| ln (rainfall) | (0.002) | (0.003) | (0.013) | (0.012) |
| 1/2 (sunny hrs) | 0.187*** | 0.355*** | 0.080** | 0.147* |
| ln (sunny hrs) | (0.010) | (0.013) | (0.035) | (0.090) |
| 1/2 ln (average min. temp.) | −0.115*** | −0.143*** | 0.002 | −0.420*** |
| ln (average min. temp.) | (0.025) | (0.008) | (0.105) | (0.110) |
| ln (nitrogen) ln (rainfall) | 0.015** | 0.115*** | −0.071*** | −0.084** |
| | (0.007) | (0.013) | (0.034) | (0.040) |
| ln (nitrogen) ln (sunny hrs) | 0.050*** | 0.078*** | 0.256*** | 0.009 |
| | (0.008) | (0.013) | (0.027) | (0.040) |
| ln (nitrogen) ln | −0.182*** | −0.293*** | 0.198*** | −0.095*** |
| (average min. temp.) | (0.014) | (0.024) | (0.036) | (0.032) |
| $\lambda$ | 2.927*** | 3.252*** | 6.855*** | 5.05*** |
| | (0.092) | (0.103) | (0.858) | (0.423) |

***Significant at the 0.01 level; ** at the 0.05 level; * at the 0.10 level. The estimated standard errors are in parentheses.

is the two-sided, normally distributed random error, and $u$ is the one-sided efficiency component with a half normal distribution. If $u_i>0$, the yield is lower than its maximum possible value given by the frontier and the variety/entry used is not the highest-yielding one. Thus, if $u_i=0$, the yield is highest given the level of input and environmental factors within which the experiment was conducted. The model suggested in equation (1) is a restricted translog type (Antle and Capalbo 1988). The model considers average N use, average minimum temperature, rainfall and hours of sunshine, and their square terms and some of their interaction terms. We did not include the interactions between rainfall and average minimum temperature, between average minimum temperature and sunny hours and between rainfall and sunny hours as these interactions are not biologically meaningful for rice production.

Table 4 reports the maximum likelihood estimates of the parameters of the stochastic frontier for varieties grown in different ecosystems. The variance ratio parameters (s.d. of $u$/s.d. of $v$) proposed by Battese and Corra (1977) are statistically greater than zero, implying that variation in actual yield from maximum yield between experimental sites mainly arose from differences in genetic characteristics, rather than random variability. Positive and significant coefficients of the square terms of

*Table 5.* Simulated yield of irrigated rice under alternative climatic conditions and N use.

| N use (kg ha$^{-1}$) | Rainfall (mm season$^{-1}$) | Sunny hrs (h week$^{-1}$) | Average temperature (°C) | Yield (ton ha$^{-1}$) | |
|---|---|---|---|---|---|
| | | | | Early | Late |
| 60 | 350 | 45 | 15 | 5.96 | 6.41 |
| 80 | 350 | 45 | 15 | 6.74 | 7.13 |
| 100 | 350 | 45 | 15 | 7.45 | 7.96 |
| 120 | 350 | 45 | 15 | 8.10 | 8.88 |
| 120 | 300 | 45 | 15 | 8.16 | 9.41 |
| 120 | 400 | 45 | 15 | 8.06 | 8.39 |
| 120 | 450 | 45 | 15 | 8.01 | 7.96 |
| 120 | 350 | 25 | 15 | 7.04 | 6.84 |
| 120 | 350 | 35 | 15 | 7.40 | 7.50 |
| 120 | 350 | 40 | 15 | 7.75 | 8.18 |
| 120 | 350 | 45 | 10 | 8.7 | 9.37 |
| 120 | 350 | 45 | 20 | 7.6 | 8.42 |
| 120 | 350 | 45 | 25 | 7.2 | 8.02 |

*Table 6.* Simulated yield of rainfed lowland and upland rice under alternative climatic conditions and N use.

| N use (kg ha$^{-1}$) | Rainfall (mm season$^{-1}$) | Sunny hrs (h week$^{-1}$) | Ave. minimum temperature (°C) | Yield (ton ha$^{-1}$) | |
|---|---|---|---|---|---|
| | | | | Rainfed lowland | Rainfed upland |
| 60 | 800 | 30 | 20 | 5.10 | 4.14 |
| 80 | 800 | 30 | 20 | 6.14 | 4.09 |
| 100 | 800 | 30 | 20 | 6.75 | 4.04 |
| 60 | 500 | 30 | 20 | 4.79 | 4.24 |
| 60 | 600 | 30 | 20 | 4.90 | 4.20 |
| 60 | 700 | 30 | 20 | 5.01 | 4.17 |
| 60 | 800 | 25 | 20 | 5.13 | 3.96 |
| 60 | 800 | 35 | 20 | 5.10 | 4.32 |
| 60 | 800 | 40 | 20 | 5.10 | 4.49 |
| 60 | 800 | 30 | 10 | 5.17 | 5.58 |
| 60 | 800 | 30 | 15 | 5.13 | 4.80 |
| 60 | 800 | 30 | 25 | 5.08 | 3.60 |

N use indicate that nitrogen use in irrigated rice (80 kg ha$^{-1}$), has not yet reached the stage of diminishing returns for irrigated rice. But it has already reached that stage in rainfed lowland and upland ecosystems with an average use of 60 kg ha$^{-1}$.

Tables 5 and 6 report the results of a simulation exercise to see the impact of nitrogen use and other climatic factors (rainfall, average minimum temperature and sunny hours) on rice yield frontiers in different rice ecosystems. Results show that the yield frontier varies substantially due to the change in the N use level for all ecosystems except upland rice. Within the modal climatic condition, increase in N use from 60 to 120 kg ha$^{-1}$ would increase rice yield by about 2.0 tons for early-maturing varieties, and 2.5 tons for medium-maturing varieties. The advanced breeding lines for the rainfed lowland ecosystem that appeared in the INGER trials are highly fertilizer responsive, the yield increases by 3.2 tons with increase in N use from 40 to 100 kg. But farmers are unlikely to use higher doses of fertilizer in rainfed environments because of its risky nature. Too much or too little rainfall reduces nitrogen use efficiency. The upper limit of N use in rainfed ecosystems is about 60 kg $ha^{-1}$. For the upland ecosystem, rice yield is not responsive to fertilizer use.

Average minimum temperature has a significant impact on rice yield. The yield frontier would decrease with the increase in average minimum temperature. Seshu (1988) reports that within the critical limits of temperature, lower temperatures favour higher yield because of higher net photosynthesis. Low temperature reduces photorespiration and increases the growth duration, including that of the ripening period. Because of the longer ripening period resulting from the low temperature, the accumulated radiation during that important period increases resulting in higher rice yield. Seshu and Cady (1984) observed a significant negative correlation between the mean temperature during the 30-day period after flowering and grain yield in the INGER irrigated trials. We should mention here that the temperatures of all the sites of INGER trials used in the analysis fall with the critical limits of temperature.

The increase in the duration of sunshine from 25 to 40 hours per week increases rice yield except in rainfed lowland, from 1.3 tons for medium-duration irrigated rice to 0.5 tons for the upland rice. This suggests that the rice yield is considerably lower in humid compared to subhumid areas because of longer duration of cloud cover during the monsoon season. It also suggests that the yield is higher for rice grown in the dry season than in the wet season.

Rainfall did not show significant impact on rice yield in any of these ecosystems. Rainfall is not a limiting factor in irrigated environments. Even in rainfed lowland and upland conditions, the rainfall pattern is often very unstable – sometimes too high causing submergence and sometimes too low causing drought. In experimental plots, supplementary irrigation is often provided whenever there is a moisture stress even for rainfed crops. As a result, it is not surprising that there is insignificant impact of rainfall on rice yield for rainfed lowland and upland ecosystems.

*Simulation of maximum attainable yield*

As different national rice breeding programmes often use INGER as a source of genetic material, the yield frontier established from the INGER trials may be used to estimate maximum attainable yield of rice varieties in the pipeline. Table 5 shows the yield potential of different types of rice in tropical and subtropical environments, estimated from the parameters of stochastic production frontiers at optimum levels of environmental factors for the specific ecosystem, and at highest levels of N use (marginal productivity of N = 0).

It is estimated that with an average use of 120 kg N $ha^{-1}$, the maximum attainable yield of irrigated varieties would be about 9.4 ton $ha^{-1}$ in temperate countries (China, Japan and Republic of Korea), approximately 8.4 ton $ha^{-1}$ in subtropical countries, and

*Table 7.* Maximum attainable rice yield (ton ha$^{-1}$) for rice ecosystems and agroecological zones (*Source:* estimated from parameters in Table 4).

| Rice ecosystem | N use (kg ha$^{-1}$) | Agroecological zones | | |
|---|---|---|---|---|
| | | Temperate | Subtropics | Tropics |
| Irrigated | 120 | 9.4 | 8.4 | 7.2 |
| Rainfed lowland | 60 | – | 4.8 | 5.1 |
| Rainfed upland | 60 | – | 4.2 | 3.6 |

*Table 8.* Maximum attainable yield and yield gaps in selected Asian countries.

| Country | Percentage of rice area under ecosystem | | | | Maximum attainable yield (ton ha$^{-1}$) | Current (1991–93) (ton ha$^{-1}$) | Required yield to sustain food security (ton ha$^{-1}$) |
|---|---|---|---|---|---|---|---|
| | Irrigated | Rainfed lowland | Flood-prone | Upland | | | |
| Bangladesh | 35 | 34 | 23 | 8 | 5.4 | 2.70 | 5.8 |
| China | 93 | 5 | 0 | 2 | 7.6 | 5.96 | 8.9 |
| India | 45 | 33 | 7 | 15 | 5.9 | 2.69 | 5.4 |
| Indonesia | 72 | 7 | 10 | 11 | 6.4 | 4.38 | 7.4 |
| Philippines | 61 | 35 | 2 | 2 | 6.3 | 2.85 | 5.4 |
| Thailand | 7 | 86 | 6 | 1 | 5.3 | 2.14 | 2.3 |
| Vietnam | 53 | 28 | 11 | 8 | 6.1 | 3.45 | 5.6 |
| Myanmar | 18 | 52 | 24 | 6 | 5.1 | 3.15 | 3.7 |

about 7 ton ha$^{-1}$ in tropical countries (Table 7). In estimating the yield potential for irrigated rice, we have assumed that greater proportion of irrigated area would be under short-duration early rice varieties. As the costs of irrigation and other inputs are increasing and as farmers are increasingly diversifying their farm by growing non-rice crops before or after rice, this assumption is a reasonable one. Maximum attainable yields of rainfed lowland varieties in tropical and subtropical countries are estimated to be around 5.1 ton ha$^{-1}$ and 4.8 ton ha$^{-1}$ respectively. The maximum attainable yield of rainfed lowland rice, which is usually grown during the monsoon period, is generally lower in subtropical countries because of higher moisture stress. The maximum attainable yield estimated for the rainfed lowland ecosystem is for favourable conditions; it does not consider the effect of temporary submergence and moisture stress that inflict rice cultivation in unfavourable rainfed ecosystems. Maximum attainable yields for upland varieties in tropical and subtropical countries are about 3.6 and 4.2 ton ha$^{-1}$, respectively.

*Yield gaps.* Table 8 estimates the maximum attainable yields of the major rice growing countries in Asia on the basis of the existing distribution of rice growing areas under different ecosystems and compares them with the highest yield achieved at the national level. The results show that there is still large potential to increase rice yield in Asia in the medium term by developing appropriate rice varieties suitable for different local conditions and by disseminating those varieties through better extension services and better infrastructure.

Is the yield potential large enough to meet the growing food needs over the next quarter-century? To answer this question, we estimated the required rice yield in year 2020 using the following assumptions: (i) to sustain food security from domestic production the country must produce 300 kg of staple grains per capita, an assumption used by FAO; and (ii) there is no further possibility of increase in rice growing area, which will remain at current levels. The estimates are also shown in Table 8. Among the countries listed in Table 8, only Thailand and Myanmar will be in a comfortable position in meeting the rice needs of their population. Technically they can generate enough surplus to meet some of the shortage of other countries. Vietnam and India will be in a tight situation, even if they can exploit the full potential of the technologies in the pipelines. The countries that are likely to face severe shortages are China, Indonesia, the Philippines and Bangladesh, unless there is further investment in transforming the rice area from unfavourable to favourable ecosystems and/or technological breakthroughs in shifting the yield potential for each of the rice ecosystems.

## Constraints to achieving full yield potential

*Biophysical constraints*

Several biotic and abiotic factors limit rice yields in the farmers' fields. They include adverse climate and soils, insects and diseases and other pests such as weeds, rodents, etc. Information was collected on the perceptions of knowledgeable farmers and extension workers regarding the probability of occurrence of different yield-reducing factors, the extent of area affected when they occur and the extent of yield loss in the affected area for sample locations representative of different agroecological zones. A number of national studies supported by the Rockefeller Foundation have estimated yield losses caused by different biotic and abiotic factors at the individual country level. These yield loss estimates were weighted by rice area under different ecosystems to estimate overall yield loss for the Asian region. We classified the rice growing areas into seven homogeneous agroecological zones and four rice ecosystems and estimated the proportion of area covered by each group ($W_i$). We then estimated the production loss per hectare for each of the groups $i$ for a particular problem $j$, ($P_{ij}$). The estimates of loss for Asia for a specific problem, $P_j$, was then obtained as the weighted average for the agroecological zones or ecosystems; thus:

$$P_j = \sum_{i=1}^{n} W_i P_{ij}$$

Results of the extrapolation exercise are given in Table 9. For the irrigated ecosystem, yield loss due to all technical constraints is 962 kg ha$^{-1}$, about 20% of the average yield in that ecosystem. For rainfed lowland and flood-prone ecosystems, yield loss due to technical constraints is about 33% of the average yield, and in the upland ecosystem is more than 40% of the average yield. For Asia as a whole, all technical constraints causes a yield loss of about 23% of the average rice yield – 833 kg ha$^{-1}$

*Table 9.* Yield loss due to various technical constraints in different rice ecologies of Asia (*Source:* Evenson et al. 1996).

| Constraints | Irrigated ($kg\,ha^{-1}$) | Rainfed lowland ($kg\,ha^{-1}$) | Flood-prone ($kg\,ha^{-1}$) | Upland ($kg\,ha^{-1}$) | Average for Asia | |
|---|---|---|---|---|---|---|
| | | | | | Loss ($kg\,ha^{-1}$) | Loss (%) |
| Biotic factors | | | | | | |
| Diseases | 69 | 146 | 18 | 70 | 83 | 3.1 |
| Insects | 108 | 166 | 16 | 65 | 110 | 2.3 |
| Other pests | 29 | 88 | 21 | 120 | 52 | 1.4 |
| Abiotic factors | | | | | | |
| Water | 400 | 288 | 429 | 227 | 358 | 9.9 |
| Soil | 356 | 75 | 13 | 80 | 229 | 6.4 |
| Total | 962 | 763 | 496 | 563 | 833 | 23.1 |
| Loss as % of yield | 19.6 | 33.2 | 33.1 | 40.2 | 23.1 | – |

in absolute terms. Abiotic constraints were found more important than biotic constraints for all ecosystems. Yield loss due to climatic constraints like submergence, drought and cold, ranges from 227 $kg\,ha^{-1}$ (20% of average yield) for upland to 429 $kg\,ha^{-1}$ (28% of average yield) for the flood-prone ecosystem. Obviously, drought is the most important abiotic constraint in the upland ecosystem, and submergence in the flood-prone ecosystem.

Soil-related problems were found important for the irrigated ecosystem but insignificant for others. Intensive cultivation in an irrigated ecosystem causes nutrient deficiency problems. Yield loss due to insects and disease is more important for the rainfed ecosystem, as the rainfed environment favours the growth of plant pathogens and insects. Yield loss due to weed infestation is highest for uplands.

Scientists have been able to reduce yield losses from biotic constraints by developing multiple resistance against important pests in rice varieties. Continued research is needed to maintain such resistances and to develop resistance against new pathogens and insects. Although the problems are being addressed using biotechnology, there has been little progress in reducing yield losses from abiotic stresses. Because of these technical constraints, farm yields will remain substantially lower than the yields reported in Table 7.

*Socioeconomic constraints.* The maximum attainable yield reported in Table 7 assumes that farm-level inputs such as labour, fertilizer and water (for the irrigated ecosystem) are non-limiting. The farmer will use the inputs up to the level where their further increase will no longer increase yields; in economics jargon, the marginal productivity of an input at that level is equal to zero. The farmer, however, has to pay for the inputs. In order to maximize profits, he will use the inputs only up to the level where the marginal productivity of the input is equal to its cost. If he uses more, he will incur losses as the incremental return would be lower than the cost of the input. Thus, as long as the cost of the inputs is positive, the economically optimum yield will always be lower than the technically optimum yield. Also, if the

*Table 10.* Input supply and yield response to rice prices: estimates from dynamic supply response model (*Source:* IRRI/IFPRI, Rice Supply and Demand Project).

| Country | Percent increase in fertilizer from 1% increase in rice prices | Percent increase in rice yield from 1% increase in rice prices |
|---|---|---|
| Bangladesh | 0.20 | 0.06 |
| Thailand | 0.22 | 0.18 |
| Pakistan | 0.30 | 0.04 |
| Vietnam: | | |
| Northern region | 0.42 | 0.02 |
| Southern region | 0.17 | 0.09 |
| Indonesia: | | |
| Java | 0.43 | 0.05 |
| Off-Java | 0.27 | 0.08 |

input cost had increased in relation to output prices, the farmer would use the input in lower amounts reducing the profit-maximizing yield. If the input prices were lower and output price higher, the economically optimum yield would move closer to the technically optimum yield. Thus, the crop yield at a point of time depends on the prevailing input-output prices, which are functions of demand-supply imbalances of output and inputs and government policies.

If the supply of rice lags behind the increase in demand, the rice prices will increase. The resultant increase in the marginal value products may encourage farmers to use inputs in larger amounts that will raise the yield and contribute to a reduction in the demand-supply imbalances.

To study the magnitude of price response for rice, a number of country studies under RSDP used a dynamic supply-response model developed by McGuirk and Mundlak (1991). This is a sequential decision-making model that uses a choice-of-technique framework (Rosegrant and Kasryno 1992). Based on formation of multi-period household decision making, the model estimates three sets of supply equations using a recursive simultaneous system. The first block of equations determines the allocation of resources to quasi-fixed factors (that cannot be changed in the short run) such as land and irrigation, depending on expected net returns in alternative economic activities. The second block of equations determines the allocation of predetermined land and water resources to crops that complete for these inputs, depending on their relative prices. The third block of equations determines the input supply and yield response of particular crops to the input-output prices, given the predetermined levels of quasi-fixed inputs and allocation of area under different crops. The system yields better estimates of supply parameters than the single equation estimates of area and yield response to price and non-price variables for specific crops. The response of rice yield and supply of modern inputs to rice prices obtained from the model can be seen from Table 10.

The results suggest that the supply response to prices for rice is typically small. A 10% increase in price would increase rice yield in the range of 0.4% to 1.8%. The response comes mainly from fertilizer use, which would increase in the range of 2%

*Table 11.* Estimates of fertilizer use in irrigated and rainfed rice cultivation in selected Asian countries (NPK $ha^{-1}$) (*Source:* IRRI 1995).

| Countries | Irrigated modern | Rainfed modern | Rainfed traditional |
|---|---|---|---|
| China | 368 | – | – |
| Bangladesh | 173 | 109 | 41 |
| Vietnam | 173 | – | 15 |
| India | 172[a] | – | 32[b] |
| Philippines | 114 | 62 | 24 |
| Cambodia | 83 | – | 18 |

[a]Average for Punjab and Tamil Nadu.
[b]Average for Bihar and Madhya Pradesh.

to 4% in response to a 10% increase in rice prices. But the output elasticity of fertilizer is small as land, water and labour are still dominant inputs in rice production activity.

What is the potential for reducing the yield gap through further increase in the use of chemical fertilizers? A number of micro-studies carried out in Asia shows that when irrigation is reliable farmers use fertilizers in optimal amounts (Hossain and Singh 1995). For the irrigated ecosystem, the variation in fertilizer use among Asian countries is relatively small (Table 11). In most countries in Asia, chemical fertilizers were popularized among farmers with large amounts of subsidies. In recent years, however, Asian governments have started withdrawing subsidies from this input, causing an upward pressure on farm-level prices. This trend in prices may reduce the level of fertilizer use and yield of rice in the irrigated ecosystem, which can only be offset by an increase in technical efficiency in the use of nutrients.

In the rainfed ecosystem, however, farmers use fertilizer in suboptimal amounts, and the rice yield could be substantially increased if farmers could be induced to move towards optimal levels of application. The main reason behind the low use in this ecosystem is that farmers are *not* assured of the returns from investment in fertilizer because of the risks of crop failures from floods, temporary submergence and prolonged droughts. Since most farmers live around subsistence levels, they are risk averse and make input allocation decisions on the basis of minimum assured returns. Unless rice science helps stabilize the yield, it will be difficult to encourage farmers to increase fertilizer use in the rainfed ecosystems.

Another important factor that might constrain increasing rice yield in the irrigated ecosystem, particularly in the middle- and high-income countries, is the growing scarcity of labour in rural areas (Hossain 1995b). The rapid rural–urban migration, the increase in labour productivity and the escalating wage rate in the fast-growing non-agricultural sector puts an upward pressure on the rural wage rate which affects the costs of production and the profitability of rice cultivation. The growth in agricultural wage rates from 1961 to 1991 for selected Asian countries is shown in Table 12. In the early 1960s, the difference in wage rates across countries was only marginal. In the low-income countries such as Bangladesh, India and the Philippines, the agricultural wage rates increased only marginally, but there was an escalation in wage rates in Japan and Republic of Korea. Thus, with economic progress, the farmer

*Table 12.* Long-term trend in wage rates (US$ $d^{-1}$) in selected Asian countries (*Source:* IRRI 1995).

| Country | 1961 | 1971 | 1981 | 1991 |
|---|---|---|---|---|
| Bangladesh | 0.46 | 0.44 | 0.86 | 1.39 |
| Philippines | 1.39 | 0.59 | 1.51 | 2.16 |
| Korea, Rep. | 0.82 | 1.86 | 10.84 | 32.59 |
| Japan | 1.22 | 8.19 | 24.16 | 51.93 |

looks for technologies that not only increase yields, but also save labour and increase labour productivity. The experience of Japan, Republic of Korea and Taiwan, China shows that substitution of labour through mechanization has *not* helped in sustaining farmers' incentives in rice cultivation in the face of escalating wage rates; the governments had to protect domestic markets and increase in rice prices. If these countries open their rice markets to comply with GATT agreements, their rice prices will decline. This will reduce farm profits and divert land and labour from rice cultivation to other economic activities. Consolidation of small holdings into big farms, such as in the USA and Australia, will be needed to avert this trend and to maintain the balance in incomes between rice farmers and labour households in non-farm activities.

A major source of increase in rice yield in the past was public and private sector investment on irrigation, flood control and drainage that converted rainfed into irrigated ecosystems to facilitate the adoption of modern rice varieties and improved farming practices. Theoretically, the potential for further increase in rice yield through this source is still large, as only 55% of Asian rice land is irrigated (IRRI 1995). The scope for further conversion of rainfed into irrigated ecosystems is, however, becoming limited. Water is becoming a scarce resource with increasing demand for human consumption, industrial use and generation of power. The cost of irrigation has increased substantially, as easy options for irrigation development have already been exploited. Also, environmental concerns regarding adverse effects of irrigation and flood control projects on waterlogging, salinity, fish production and the quality of groundwater have been growing. Already there has been drastic decline in investment for the development and maintenance of large-scale irrigation projects in many Asian countries (Rosegrant and Svendsen 1993).

Economic prosperity and industrial progress is leading to rapid urbanization and concentration of people in a few large cities. By 2025, some 53% of the people in Asia will live in urban areas compared to 30% in 1990. An important implication of growing urbanization is that some of the fertile agricultural land has to be diverted to non-agricultural uses to meet the demand for housing, factories and roads. Also, with urbanization, growing incomes and associated changes in food habits, the markets for vegetables, fruits and livestock products have been growing stronger. The growing market for livestock products will increase the demand for grains for animal feed (corn) and fodder. There will be economic pressure to reduce the area under rice cultivation to accommodate these relatively high-value crops. The area under rice cultivation has already started declining in China, Java, Indonesia, the Philippines, Thailand and parts of India (IRRI 1995).

A deceleration in growth in rice supply has already set in due to the plateauing of yield in the irrigated ecosystem, and the slowing down of investment for developing and maintenance of irrigation infrastructure. The annual growth in Asian rice production was 3.1% per year during 1965–85; it decelerated to 1.7% per year during 1985–94. The recent growth has failed to outpace the population growth in many Asian countries, and the earlier gains in the per capita rice intake are being gradually eroded. This is reflected in the upward trend in rice prices over the last two years.

## Conclusions

To meet the growing demand for food and to sustain food security for low-income people in low-income countries, rice production has to be raised by another 70% over the next three decades. This means raising the rice yield from the current level of 3.7 to 6.3 tons $ha^{-1}$ by 2020 if these countries can maintain the rice growing area at current levels (an optimistic assumption).

With existing technologies and technologies in the pipeline, there is considerable unexploited potential that, if tapped, could increase rice yield substantially. A stochastic yield frontier model fitted with the data from INGER trials estimated maximum attainable yield for the irrigated ecosystem at 7.2 ton $ha^{-1}$ for the tropics and 8.4 ton $ha^{-1}$ for the subtropics, under non-limiting input use conditions. For the rainfed lowlands, the maximum attainable yield was estimated at 5.1 ton $ha^{-1}$ in the tropics and 4.8 ton $h^{-1}$ in the subtropics. With the present allocation of land under different ecosystems, China, Indonesia and Bangladesh will develop substantial deficits in rice production by 2020; while India, Vietnam and the Philippines will have a tight supply–demand system, even if they could fully exploit the technical potential. Only Thailand and Myanmar will be in a comfortable position.

For the irrigated ecosystem, the rice yield will be difficult to raise from the current levels of 5–6 ton $ha^{-1}$. About 20% of the rice yield is lost annually due to biotic and abiotic constraints, and the fertilizer use is at the optimum level. Reduction of subsidies on fertilizer and water, growing scarcity of labour and escalating wage rates will reduce the use of these inputs, further constraining increases in yield. In this ecosystem, the rice farmer will look for technologies that increase labour productivity and farm profits rather than those that increase only yields. If the middle- and high-income countries open their market, the price in the domestic market will decline, and even a 50% shift in the yield potential would not be able to sustain farmers' interest in rice farming. In this scenario a large amount of land will go out of rice farming.

The potential for raising yield in the rainfed ecosystem is still vast, as the current yield is only about 2.0 ton $ha^{-1}$ (compared to 5.0 ton attainable yield), and in nearly 45% of the land rice is grown under rainfed conditions. Many traditional varieties have developed traits through centuries of evolution that enable them to withstand temporary submergence and prolonged droughts that cause large year-to-year fluctuations in yield for this ecosystem. Rice scientists have so far had limited success in identifying these traits and incorporating them in high-yielding modern cultivars (Zeigler and Puckridge 1995). Where the rainfall is unreliable and the drainage is poor, farmers still grow traditional varieties, and use fertilizers in sub-optimal amounts in modern varieties, due to the uncertainty of returns from investments. These are the major factors behind the low-yield and large-yield gap in the rainfed ecosystems. If rice research succeeds in incorporating traits that help withstand the abiotic stresses, or improved systems management helps escape these stresses, the modern varieties will be adopted more extensively in the rainfed system. This will help increase the stability of yield and reduce risks in rice cultivation; and thereby provide incentives to farmers to apply inputs in optimum amounts that will contribute to further increase in rice yields.

Research problems in unfavourable rainfed ecosystems are more challenging and the probability of research success more uncertain than those for the relatively homogeneous irrigated ecosystem. To make headway on them will require mobilizing the best of science and the best of scientists in national agricultural research systems, international centres and advanced institutions through partnership research. Research relevance and effectiveness will depend on close collaboration among research partners, selection of key sites, the mechanism of partnership research and enabling research environments that encourage scientists to work on problems with uncertain outcome. Recent advances in molecular biology, systems models and Geographic Information Systems will hopefully encourage scientists to put more effort in addressing the more difficult and complex problems of rainfed rice cultivation than they have done in the past.

## References

Ahmed M, Siddiqui R (1995) Projections and policy implications of medium and long-term supply and demand for rice in Pakistan. Paper presented at the Final Conference on Projections and Policy Implications of Medium- and Long-term Rice Supply and Demand, 24–26 April 1995, Beijing, China.

Antle J M, Capalbo S M (1988) An introduction to recent developments in production theory and productivity measurement. In Antle J M, Capalbo S M (Eds.) Agricultural productivity, measurement and explanation. Resources for the Future, Washington DC, USA.

Battese G E, Corra G S (1977) Estimates of a production frontier model, with application to the pastoral zone of Eastern Australia. Aust. J. Agric. Econ. 21:69–79.

Bos E and associates (1994) World population projections, 1994–95 edition. The Johns Hopkins University Press, Baltimore, London.

Cassman K, Pingali P L (1995) Intensification of irrigated rice systems: Learning from the past to meet future challenges. Geojournal 35(34):299–306.

Dey M, Hossain M (1995) Yield potentials of modern rice varieties: An assessment of technological constraints to increase rice production. Paper presented at the Final Workshop of the Projections and Policy Implications of Medium- and Long-term Rice Supply and Demand, 24–26 April 1995, Beijing, China.

Evenson R, Herdt R W, Hossain M (Eds.) (1996) Rice research in Asia: Progress and priorities. CAB International, Wallingford, UK (in press).

FAO – Food and Agriculture Organization (1994) Agrostat database. FAO, Rome, Italy.

Hossain M (1994) Recent development in Asian rice economy. Challenges for rice research. Paper presented at the Workshop on Rice Research Prioritization in Asia, 21–22 February 1994, International Rice Research Institute, Los Baños, Philippines.

Hossain M (1995a) Sustaining food security for fragile environments in Asia: Achievements, challenges and implications for rice research. Paper presented at the International Rice Research Conference, 13–17 February 1995, International Rice Research Institute, Los Baños, Philippines.

Hossain M (1995b) Economic prosperity in Asia: Implications for the demand for rice research. Paper presented at the International Genetics Symposium, Manila, Philippines.

Hossain M, Singh V P (1995) Fertilizer use in Asian agriculture: Implications for sustaining food security and environment. Paper presented at the Workshop on Plant Nutrition Management, Food Security, Sustainable Agriculture, and Poverty Alleviation in Developing Countries, 16–17 May 1995, Viterbo, Italy.

Huang J, Rozelle S, Rosegrant M W (1995a) Supply, demand, and China's future grain deficit. Paper presented at the Final Conference on Projections and Policy Implications of Medium- and Long-term Rice Supply and Demand, 24–26 April 1995, Beijing, China.

Huang J, Rosegrant M W, Rozelle S (1995b) Public investment, technological change, and agricultural growth in China. Paper presented at the Final Conference on Projections and Policy Implications of Medium- and Long-term Rice Supply and Demand, 24–26 April 1995, Beijing, China.

IRRI – International Rice Research Institute (1995) World rice statistics, 1993–94 edition. IRRI, P.O. Box 933, Manila, Philippines.

Isvilanonda S, Poapongsakorn N (1995) Rice supply and demand in Thailand: The future outlook. Thailand Development Research Institute, Bangkok, Thailand.

Ito S, Peterson E W F, Grant W R (1989) Rice in Asia: Is it becoming an inferior good? Am. J. Agric. Econ. 71(1):32–42.

Kako T, Gemma M, Ito S (1995a) Projections and policy implications of supply and demand for rice in South Korea. Paper presented in the Final Conference on projections and Policy Implications of Medium- and Long-term Rice Supply and Demand, 24–26 April 1995, Beijing, China.

Kako T, Gemma M, Ito S (1995b) Projections and policy implications of supply and demand for rice in Japan and Taiwan. Paper presented in the Final Conference on Projections and Policy Implications of Medium- and Long-term Rice Supply and Demand, 24–26 April 1995, Beijing, China.

Khush G S (1995) Breaking the yield frontier of rice. Geojournal 35(3):325–328.

Kumar P, Rosegrant M W (1995) Projections and policy implications of medium- and long-term rice supply and demand in India. Paper presented in the Final Conference on Projections and Policy Implications of Medium- and Long-term Rice Supply and Demand, 24–26 April 1995, Beijing, China.

McGuirk A, Mundlak Y (1991) Incentives and constraints in the transformation of Punjab Agriculture. Research Report No. 87. International Food Policy Research Institute, Washington DC, USA.

Rosegrant M W, Kasryno F (1992) Supply response with endogenous technology: Food crops in Indonesia. Paper presented at the Planning Workshop on Projections Implications of Medium- and Long-term Rice Supply and Demand, 25–27 March 1992, International Rice Research Institute, Los Baños, Laguna, Philippines.

Rosegrant M W, Svendsen M (1993) Asian food production in the 1990s: Irrigation investment and management in Asia. Food Pol. 13–32.

Rosegrant M W, Agcaoili M, Perez N (1995) Rice and the global food economy: Projections and policy implications of future food balances. Paper presented at the Final workshop Projections and Policy Implications of Medium- and Long-term Rice Supply and Demand, 24–26 April 1995, Beijing, China.

Seshu D V (1988) Impact of major weather factors on rice production. Paper presented at the workshop on Agrometeorological Information for Planning and Operation in Agriculture, 22–26 August 1988, Calcutta, India.

Seshu D V, Cady F B (1984) Response of rice to solar radiation and temperature estimated from international yield trials. Crop Sci. 24:649–654.

Shahabuddin Q, Zohir S (1995) Projections and policy implications of medium and long term rice supply and demand: Country report for Bangladesh. Bangladesh Institute of Development Studies, Dhaka, Bangladesh.

Sudaryanto T, Simatupang P (1995) Projections and policy implications of medium and long term rice supply and demand in Indonesia. Paper presented at the Final Conference on Projections and Policy Implications of Medium- and Long-term Rice Supply and Demand, 24–26 April 1995, Beijing, China.

Zeigler RS, Puckridge D W (1995) Improving sustainable productivity in rice-based rainfed lowland systems of South and Southeast Asia. Geojournal 35(3):307–324.

# The limits of consumption and production of food in China in 2030

J.C. LUYTEN[1], SHI QUINGHUA[2] and F.W.T. PENNING DE VRIES[3]

[1]*Agricultural and Biological Engineering Department, University of Florida, P.O. Box 110570, Gainesville, FL 32611-0570, USA*

[2]*Jiangxi Agricultural University, Nanchang 330045, People's Republic of China*

[3]*DLO-Research Institute for Agrobiology and Soil Fertility, and C.T. De Wit Graduate School for Production Ecology, P.O. Box 14, 6700 AA Wageningen, The Netherlands*

**Key words:** food production, food demand, food security, diet, natural resources, food technology, SIMFOOD

**Abstract**

Maximum global food demand is estimated for the 15 major regions of the world by multiplying population size and average human diet. Low, medium and high population growth scenarios are considered, as well as a vegetarian, moderate and affluent diet. The food demand is compared with the maximum food production in each region, which is computed from (i) the area with soils suitable for cropping and grazing, (ii) the climatological conditions, (iii) the amount of irrigation water available and (iv) the farming system used. Two alternative farming systems are considered: (i) production-oriented agriculture with high hectare yields due to ample, but efficient, use of fertilizer and biocides, and (ii) environment-oriented agriculture where legume crops provide all nitrogen and no biocides are used, but with lower hectare yields. We focus on comparisons of food demand and supply in East Asia, comprising China, Japan and Korea. Since China accounts for more than 85% of the population of this region and for 94% of its territory, it is a fair first approximation to apply our data for East Asia to China. While the global total demand for food can be easily met by the global total production potential, there are extraordinarily large differences between regions and continents. In Asia as a whole, and in China separately, food production cannot meet food demand in all scenarios. A moderate or affluent diet remains out of reach of a large part of the Asian population, even with maximum use of all natural resources, unless a much more efficient food technology emerges or massive food imports take place. With production-oriented agriculture on all its suitable land, China can offer its people a moderate diet. Although our basic data are not accurate and our conclusions have a fair degree of uncertainty, we draw attention to the limited scope for raising food production with current technology, because the supply of suitable land is virtually exhausted and the supply of irrigation water is limited, intensive agriculture is already practised on a large scale, and the yield ceiling for current crops could be reached within decades. Furthermore, a direct competition might develop for agricultural land between production of food and feed, and production of energy for rural development. Major research efforts are required to exploit fully current food production technology and to improve it significantly.

## Introduction

The United Nations projects that the global human population will stabilize around 2040 at a level of 9.5 billion: almost double that of 1990 (UN 1992). Their low, medium and high population estimates for China are 1.3, 1.5 and 1.8 billion, respectively. The amount of plant biomass required for the daily food supply for an individual



*P.S. Teng* et al. *(eds.): Applications of Systems Approaches at the Farm and Regional Levels, 281–293.*

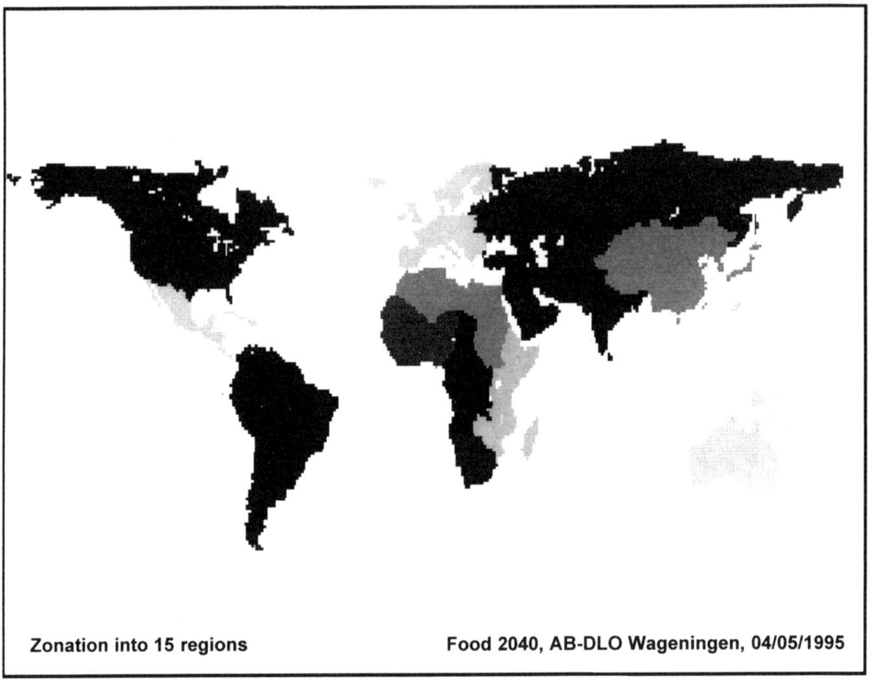


*Figure 1.* World map of the 15 major regions.

is twice or thrice the amount needed today if a richer diet is adopted. Can the earth provide enough food by socially acceptable ways of farming, and without damaging the natural resource base?

This issue has previously been discussed, for example the total carrying capacity of the globe has been estimated between 5 billion (Brown and Kane 1994; WRI 1994) and 30 billion persons (Buringh et al. 1975). Better knowledge of soils, fresh water resources and crop performance, plus the addition of rangeland production, allow us to upgrade these estimates. Our objective is to examine whether natural resources allow food security, two generations from now, in the 15 major regions of the world (Figure 1), and particularly in China, using agroecologically sustainable farming systems.

A full technical report on methods and data of this study was published by Luyten (1995); a global overview was presented earlier (Penning de Vries et al. 1995); results for China are included in a broad exploration of alternative scenarios for food, energy and natural resources for industry (WRR 1994).

## Outline of the approach

Six features have been combined. Potential food demand is (i) the product of population size, and (ii) per capita consumption. Potential food production is approximated,

*Table 1.* Population data for the 15 regions of the world; Low, Medium, High esimates for 2040, and current population sizes (*Source:* UN 1992). China's population accounts for approximately 85% of the 'East Asia' region, while Japan and Korea account for about 10% and 5%, respectively.

| | Region | Population (million people) | | | |
|---|---|---|---|---|---|
| | | 1990 | 2040 | | |
| | | | Low growth | Medium growth | High growth |
| 1 | South America | 296.7 | 481.1 | 557.8 | 662.9 |
| 2 | Central America | 151.4 | 249.7 | 295.7 | 346.8 |
| 3 | North America | 275.9 | 274.1 | 327.9 | 397.6 |
| 4 | Northern Africa | 140.6 | 276.6 | 342.9 | 418.5 |
| 5 | Western Africa | 193.7 | 466.4 | 634.9 | 797.6 |
| 6 | Central Africa | 70.1 | 190.1 | 240.4 | 285.6 |
| 7 | Eastern Africa | 196.9 | 537.0 | 678.7 | 841.6 |
| 8 | Southern Africa | 40.9 | 88.7 | 100.3 | 122.6 |
| 9 | Oceania | 26.5 | 32.2 | 37.2 | 44.5 |
| 10 | Southeast Asia | 444.8 | 657.7 | 819.6 | 1005.2 |
| 11 | East Asia | 1335.6 | 1502.8 | 1770.4 | 2098.3 |
| 12 | South Asia | 1200.6 | 1964.5 | 2408.1 | 2888.5 |
| 13 | West Asia | 131.8 | 248.9 | 323.6 | 398.8 |
| 14 | (former) USSR | 288.6 | 323.2 | 369.0 | 418.8 |
| 15 | Europe | 498.4 | 436.6 | 497.6 | 563.4 |
| | World | 5292.5 | 7729.6 | 9404.1 | 11290.7 |

*Table 2.* Food composition in three diets. Note: the data have a different meaning of 'grain use per capita' used by many authors (e.g. Crosson and Anderson 1992; Brown and Kane 1994). Those figures refer to cereals *sensu strictu,* whereas our definition includes the opportunity cost (in grain equivalents (GE)) of all other food crops as well, such as tuber, oil seed and leguminous, fruits and vegetables.

| Diet | GE (kg $cap^{-1}$ $yr^{-1}$) | Energy (kJ $d^{-1}$) | Animal protein (g $d^{-1}$) | Plant protein (g $d^{-1}$) |
|---|---|---|---|---|
| Vegetarian | 475 | 10.0 | 8.6 | 66.7 |
| Moderate | 875 | 10.0 | 31.2 | 50.0 |
| Affluent | 1530 | 11.5 | 63.2 | 28.9 |

taking into account four natural resources: crops, land, water and climate. Crop production per unit area is quantified for (iii) production-oriented, and for (iv) environment-oriented production systems. The area of potential arable land and grassland (v) is less than the total land area because a large part is unsuitable for modern, mechanized agriculture. Water supply for irrigation is limited in some regions (vi).

For the future size of the regional populations (Table 1), we followed a recent UN report (UN 1992) that provides low, medium and high projections. The amount of plant biomass needed to produce food depends on the composition of the diet. We performed our calculations for three different diets: a healthy vegetarian diet, a moderate diet (with some meat) and an affluent diet (with a large proportion of meat and luxury products). These diets correspond with biomass production of 475 kg, 875 kg and 1530 kg grain equivalents (GE) $yr^{-1}$, respectively (Table 2). Current per

*Table 3.* Typical yields of cereal crops and permanent grassland (GE, ton ha$^{-1}$ yr$^{-1}$) in temperate and tropical zones for production-oriented (PO) and environment-oriented (EO) production systems. In the tropics, up to three arable crops per year are grown.

| Grain | PO | EO |
|---|---|---|
| *zone* | | |
| Tropics, irrigated | 16–20 | 4–6 |
| Tropics, rainfed | 4–12 | 2–4 |
| Temperate, irrigated | 7–11 | 2–4 |
| Temperate, rainfed | 4–9 | 2–3 |
| Grassland: | PO | EO |
| *zone* | | |
| Tropics, irrigated | 21–26 | 7–11 |
| Tropics, rainfed | 5–16 | 4–7 |
| Temperate, irrigated | 9–14 | 4–7 |
| Temperate, rainfed | 5–12 | 4–5 |

capita food consumption in China is about 550–600 kg GE yr$^{-1}$ (He 1990). The three population projections and three diet levels combine into nine food demand scenarios.

We address two types of farming, based on contrasting views with respect to the 'best' use of land (WRR 1994). In the first view, one aims at maximum land productivity: agriculture is characterized by a high degree of mechanization, high use of fossil fuel, and ample N-fertilizer and biocide use. It is assumed that proper farming techniques permit an ecologically sustainable agriculture, while environmental damage remains at acceptable levels by recirculation of nutrients. Losses of inputs to the environment are minimal per unit of food produced. We call this the 'production-oriented' (PO) production system. In the second 'environment-oriented' (EO) view, the environment is put first: the production system is designed to minimize loss per unit area and impact on the environment. This is realized by replacing all nitrogen (N) fertilizer by biological N-fixation, elimination of biocides, intensive recirculation of nutrients and 'local' consumption of products. In this system, crop yields are lower, but less fossil energy is used and farming can be more integrated with nature conservation. In both farming systems, availability of potassium (K) and phosphorus (P) is assumed to be optimal. Farmers are assumed to be always skilled, well-informed and use the best technology available. Though mixed forms of land use are possible, and are in fact commonplace in China, we consider only the PO and EO extremes in order to quantify the possibilities for different production techniques.

A simple crop growth model is used to simulate crop production, irrigation water requirements and N-fixation per unit area. We base our potential production calculations on two crop types (cereal and grass) for reasons of simplicity; a temperate variety of the cereal ('wheat') and a tropical one ('rice') are distinguished to account for climatic adaptation. Typical yields for both farming systems are shown in Table 3. In the north of China, only one crop per year of wheat can be grown, while in the south up to three rice crops per year can be produced by transplanting the seedlings and by keeping the turn-around period between crops to a minimum. The use of

*Table 4.* Total land area according to the model (in Mha), and the average fraction of the land suitable for modern agriculture, either for arable farming or for grassland, by region. Note: land suitable for arable farming is also suitable for grassland. China's territory accounts for approximately 94% of the 'East Asia' region, while Japan and Korea account for about 4% and 2%, respectively.

| | Region | Total area (Mha) | Suitability for grassland | Suitability for arable farming |
|---|---|---|---|---|
| 1–3 | S, C and N America | 3504 | 0.693 | 0.396 |
| 4–8 | N, W, C, E and S Africa | 2858 | 0.767 | 0.353 |
| 9 | Oceania | 787 | 0.768 | 0.360 |
| 10 | Southeast Asia | 347 | 0.580 | 0.223 |
| 11 | East Asia | 1104 | 0.523 | 0.214 |
| 12 | South Asia | 646 | 0.600 | 0.305 |
| 13 | West Asia | 413 | 0.660 | 0.145 |
| 14–15 | Europe plus (former) USSR | 2544 | 0.440 | 0.220 |
| | World | 12203 | 0.637 | 0.312 |

rangeland for food production is neglected in many studies (e.g. Buringh et al. 1975). However, in China there is more than 1.5 times as much land only suitable for grazing than there is for cropping, so that rangelands can add significantly to food production (Table 4). To achieve maximum production, grasslands are fertilized in PO-agriculture or with legumes in EO-agriculture, but always rainfed.

We define suitability of land for modern farming as the fraction of the area where crops or grass can be grown without soil-related restrictions. Arable cropping is much more demanding in terms of suitability of soils than grass production. The basis for our assessment of the suitability of soil and terrain for modern farming is the NASA data base (Zobler 1986). Land suitability is very different among regions (Table 4). From the global land area of 12200 Mha, only a fraction 0.31 is suitable for arable cropping, while an additional fraction of 0.33 is suitable for grassland. It is striking that the fraction of land suitable for arable farming is below the global average in the four Asian regions: 0.15–0.3; an additional fraction of 0.25–0.5 is also suitable for grassland. Yet, vast areas in China are totally unsuitable for any form of modern agriculture.

We used data on the water discharge of 95 major river basins ($> 3\times10^4$ km$^2$). Their total watershed area occupies about 60% of the land surface and carries currently 90% of the total population. Total fresh water in rivers was adjusted to account for unavoidable flood runoff (about half of the total), corrected for projected household and industrial water use (assuming first priority for these uses) and for the current extraction for irrigation schemes (Delft Hydraulics 1992). Our assumption of maximum efficiency of use of limiting resources, such as water, is translated into a fraction of 0.5 actually transpired (and contributing to production), 0.25 transport losses and 0.25 returned to the river and preventing salinization. We assumed that water cannot be transported beyond a river basin, although this could sometimes be feasible (e.g. water from the Yellow River for the North China plain; Wang 1993). Our assumptions imply the presence of modern irrigation schemes wherever there is surface water and suitable soil and climate.

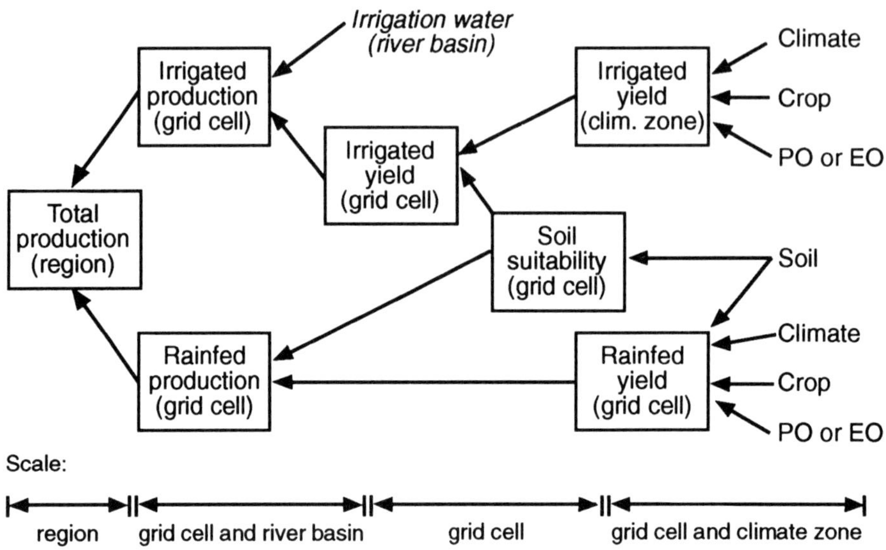


*Figure 2.* Steps and aggregation levels to compute maximum food production by region.

*Table 5.* Maximum food production (GE, $10^9$ kg yr$^{-1}$) in production-oriented (PO) and environment-oriented (EO) production systems. Total potential food production (first column) is made up of yield from irrigated crops (second column), rainfed crops (third column) and rangeland (column 1 − (2 + 3)).

| | Region | PO | | | | EO | | | |
|---|---|---|---|---|---|---|---|---|---|
| | | Total prod. | Irrigated grain | Rainfed grain | Rainfed grass | Total prod. | Irrigated grain | Rainfed grain | Rainfed grass |
| 1–3 | S, C and N America | 28644 | 15182 | 2853 | 10611 | 10940 | 5607 | 0 | 5333 |
| 4–8 | N, W, C, E and S Africa | 19747 | 10611 | 2346 | 6791 | 7749 | 3928 | 342 | 3478 |
| 9 | Oceania | 4137 | 1020 | 1131 | 1986 | 2238 | 821 | 275 | 1142 |
| 10 | Southeast Asia | 3670 | 1394 | 0 | 2276 | 1185 | 368 | 0 | 817 |
| 11 | East Asia | 4056 | 1949 | 0 | 2108 | 2261 | 740 | 0 | 1521 |
| 12 | South Asia | 3442 | 1594 | 581 | 1268 | 1836 | 931 | 53 | 851 |
| 13 | West Asia | 1245 | 772 | 33 | 440 | 658 | 305 | 1 | 352 |
| 14–15 | Europe plus (former) USSR | 7316 | 2656 | 957 | 3704 | 3807 | 1488 | 0 | 2319 |
| | World | 72256 | 35175 | 7899 | 29182 | 30673 | 14188 | 671 | 15814 |

Maximum food production is approximated in a series of steps (Figure 2), based on the aggregation of yields, soil suitability and irrigation water availability from small units (grid cells of 1° latitude by 1° longitude), applying the concept that it is better to use basic data first, and to aggregate subsequently, than to start with average values, because of many non-linear responses (De Wit and Van Keulen 1987).

To compare expected food consumption and potential production, we express both in grain equivalent (GE). GE is a theoretical weight unit. In the production process, it refers to the quantity of dry grain that would be produced if only one type of crop were grown (a cereal), plus the amount of grain that need not be produced because of feed (grass) harvested from land unsuitable for arable farming; feed requires conversion via animals for human consumption. In the consumption process, GE refers to the amount of cereals needed as raw material for the food consumed, plus the 'opportunity cost' to grow food that cannot be produced via 'grain' (e.g. fruit).

## The regional and global food situation

### *Maximum production*

In absolute terms, South America has by far the highest food production potential of all regions (Table 5), due to its very large area of suitable land (much of which is now under rain forest) and its high capacity for irrigation. A second substantial potential area for irrigated cropping is Central Africa, while East Africa comes third. Clearly, there are enormous potentials in areas with a low current population pressure.

There are large regional differences in the fraction of land that can be irrigated. With maximum use of all fresh water resources, 0.8 of the area suitable for cropping of four Asian regions under production-oriented farming can be irrigated, and in East and Southeast Asia all arable land could be fully irrigated. The non-irrigated rangelands are vast and receive sufficient rain to produce as much food (GE) as the irrigated arable fields. The large attention in China given to animal production in the future is, therefore, not only warranted from the demand side, but also from the production side (Simpson et al. 1994).

The maximum annual food production in the four Asian regions together is 12.4 Gt of GE under PO agriculture, of which 45% from arable crops and the rest from fertilized rangelands (Table 5); agriculture without use of N-fertilizer yields at most 5.9 Gt GE (40% from arable crops). Natural resources are then used to a maximum degree. Large differences exist among the Asian regions. Maximum productivity in the EO production system is lower than in PO systems (Table 3). In South Asia this is partly compensated by higher production from a larger irrigated area (lower productivity areas need less irrigation water). However, in East Asia there is ample water to irrigate all land suitable for cropping in production systems. It is a challenge for China, therefore, to develop its irrigated cropping to a maximum degree of productivity.

### *Supply versus demand*

The ratio of potential food supply and demand is a rough measure of regional food security (Table 6). Ratios vary from values below 1.0 to over 100. Their absolute values are not easy to interpret in terms of food security, because our calculations are coarse, and ignore annual fluctuations in production, economic inequity and large-scale food transport between regions. Moreover, these ratios are really maximum

*Table 6.* Ratios of potential food supply and demand in 2040, for the production-oriented (PO) and environment-oriented (EO) systems, by region. For the food demand, the medium and two extreme combinations of population size and human diet are used.

| | Region | PO | | | EO | | |
|---|---|---|---|---|---|---|---|
| | | Min. demand | Med. demand | Max. demand | Min. demand | Med. demand | Max. demand |
| 1 | South America | 89.2 | 41.7 | 20.0 | 30.1 | 14.1 | 6.8 |
| 2 | Central America | 15.6 | 7.2 | 3.5 | 6.8 | 3.1 | 1.5 |
| 3 | Northern America | 49.3 | 22.3 | 10.5 | 25.0 | 11.3 | 5.3 |
| 4 | Northern Africa | 13.7 | 6.0 | 2.8 | 8.1 | 3.5 | 1.7 |
| 5 | Western Africa | 16.0 | 6.4 | 2.9 | 6.8 | 2.7 | 1.2 |
| 6 | Central Africa | 83.2 | 35.6 | 17.1 | 29.6 | 12.7 | 6.1 |
| 7 | Eastern Africa | 22.0 | 9.4 | 4.3 | 7.4 | 3.2 | 1.5 |
| 8 | Southern Africa | 31.0 | 14.8 | 6.9 | 14.6 | 7.0 | 3.3 |
| 9 | Oceania | 270.7 | 126.9 | 60.6 | 146.5 | 68.7 | 32.8 |
| 10 | Southeast Asia | 11.8 | 5.1 | 2.4 | 3.8 | 1.7 | 0.8 |
| 11 | East Asia | 5.7 | 2.6 | 1.3 | 3.2 | 1.5 | 0.7 |
| 12 | South Asia | 3.7 | 1.6 | 0.8 | 2.0 | 0.9 | 0.4 |
| 13 | West Asia | 10.5 | 4.4 | 2.0 | 5.6 | 2.3 | 1.1 |
| 14 | (former) USSR | 29.5 | 14.0 | 7.0 | 16.0 | 7.6 | 3.8 |
| 15 | Europe | 13.5 | 6.4 | 3.2 | 6.5 | 3.1 | 1.6 |
| | World | 19.7 | 8.8 | 4.2 | 8.4 | 3.7 | 1.8 |

*Table 7.* Ratios of potential food supply and demand in China in 2030, with production-oriented agriculture, for each combination of population size and diet.

| | Population growth | | |
|---|---|---|---|
| | Low (1.3 billion) | Medium (1.5 billion) | High (1.8 billion) |
| Vegetarian diet | 4.9 | 4.1 | 3.5 |
| Moderate diet | 2.6 | 2.2 | 1.9 |
| Affluent diet | 1.5 | 1.3 | 1.1 |

numbers, as it is implied that the 'composition' of food demand (fraction of plant and animal products) matches the 'composition' of the maximum supply (fraction of food and feed production); if not, the ratios will fall.

Yet, the values of the ratios of Table 6 carry an important message. If a potential food supply/demand ratio below 2.0 signals that food security may be difficult to realize at the household level, it appears that the Asian regions are always in or near the danger zone. This is in contrast to most of the other regions in the world. The situation is even more unfavourable than the ratio suggests, since the 'option' of a fully vegetarian diet would in fact exclude the contribution to food of rangelands (Table 5).

### *Food security in China*

The region of East Asia consists of China (including Hong Kong and Taiwan), Japan, the Democratic People's Republic of Korea and the Republic of Korea and

Mongolia. Production-oriented farming has been practised in China for many years because of its high population pressure. China's population is about 86% of the East Asian region; the maximum food production (PO) is only 74% of the region's potential. China's potential food security situation is, therefore, tighter than the average for the region (Table 6). For a population growth scenario halfway between the UN 2030 medium and high estimates (He 1990), China could offer its population a moderate, but not an affluent diet, even when all suitable land and water are used in production-oriented agriculture (Table 7). Large-scale environment-oriented agriculture is insufficiently productive, and appears to be impossible. This creates an enormous challenge to maximize land productivity in an ecologically sustainable manner.

## Discussion

In China, the area of arable land, currently around 96 Mha, decreases by about 0.5 Mha $yr^{-1}$ by soil degradation and expansion of cities, so that less than 78 Mha may be left in 2030 (He 1990; Xu 1994). We did not take this decline into account in Tables 4, 5 and 6. However, if this loss of land is translated into a proportional loss of food production capacity, the ratios in Table 6 for East Asia would still be 25% lower!

The global average increase in the demand for food and feed in the next 50 years is about six-fold. Although the full range does not have to be realized in all regions (some regions grow slowly and have already a rich diet; other regions may prefer a vegetarian or moderate diet), an extra 150% to 250% of present food and feed production will probably be required for the global population. At least half of this increase is due to an expected transition to a more 'expensive' diet when people become more affluent. It is highly desirable, therefore, to develop technology to produce 'expensive' food (in terms of GE) considerably more efficiently, particularly for animal products. Farming of fish and shrimps are activities currently under-exploited (ICLARM 1992). China is already increasing rapidly its consumption of meat, particularly pork, for which consumption reached 21.3 kg per capita in 1991 (Simpson et al. 1994). The fact that production of pork, and poultry, is energetically much cheaper than that of beef provides a little more leeway for food security than Table 6 suggests.

A large increase in energy consumption is expected in developing countries in the coming decades (WRI 1994). There is a distinct possibility that biofuels (rapeseed, pine plantations) will have to contribute in meeting this demand as other sources may run out and/or produce too much $CO_2$ (fossil fuels), or may be unacceptable (nuclear energy). While biofuel can be produced very well (trees, oil crops, straw), part of its production competes with agricultural production. For China, particularly where there is little land to spare, biofuel production might pose another threat to food production. We had no data to consider whether the large areas in China usuitable for either crops or grass could eventually be used for energy crops. An explorative study on this issue is urgently needed.

Simulated grain yields per unit area under production-oriented agriculture correspond to that on best experimental fields (Buringh et al. 1975); these data have been confirmed many times. In tropical Asia, seasonal rice yields of more than 10 ton $ha^{-1}$ can be obtained (Kropff et al. 1992, 1993), and with close to 5 crops per two years, annual rice production may reach 25 ton $ha^{-1}$. Experiments in China showed seasonal rice yields of 11 ton $ha^{-1}$ (Liu and Zhang 1990). Although plant breeding continuously improves the available genetic stock, top yields at experimental farms have not really increased for decades (Yoshida et al. 1972; Brown and Kane 1994). The stagnation of top yields is an important issue that is being addressed (IRRI 1989, 1993). Fortunately, recent indications are that inability to reproduce previous record yields was only a temporary set-back (Kropff et al. 1993), and one might be optimistic that higher yields are possible if innovative breeding efforts are successful. However, in China the situation requires more than just maintaining top yield levels (Penning de Vries 1991). IRRI intends to breed a rice variety with a yield potential of about 15 ton $ha^{-1}$ (compared to the current level of about 11 ton $ha^{-1}$), which uses 50% less fertilizer and is also suitable for saline conditions (IRRI 1993).

Current total irrigated area in Asia is about 1.5 $Mkm^2$ (FAO 1993), while an additional 0.7 $Mkm^2$ is potentially irrigable (Crosson and Anderson 1992). These are below our estimates for Asia (4.6–5.5 $Mkm^2$), mainly because we assume use of all available water at maximum efficiency. We do not consider in our study the economics of new irrigation systems or the rate with which these can be implemented. It is clear, however, that our estimates of potential food production cannot be realized without special and major efforts to increase available water for irrigation and water use efficiency. Options for this have been discussed elsewhere (Delft Hydraulics 1992).

Simulated yields for environment-oriented farming systems are more difficult to compare with observed values, as experimental data for this type of agriculture are scarce. Indirect evidence confirms the values that we computed: average wheat yields in The Netherlands around 1900, when little fertilizer was used but cereals were grown in rotation with legumes, were about 2 ton $ha^{-1}$ (Spiertz et al. 1992). In Thailand, with a low rate of fertilizer application (FAO 1993), average rice yields in rice-legume cropping systems are about 1.8 ton $ha^{-1}$, which for 2–3 crops per year leads to production levels shown in Table 3. Our yield estimates are probably conservative since in the cases quoted, the soil P level was probably sub-optimal for N-fixation. Predicted crop yields depend heavily on the assumed N-fixing capacity of leguminous crops, which is subject to discussion (Caporali and Onnis 1992). It has been suggested that biotechnology might increase this significantly (e.g. by incorporating N-fixation capacity in cereals). However, we do not anticipate a future increase in N-use efficiency as selection pressure for this property has already been high for a long time.

The results of the soil suitability classification, and hence the method applied, have a major influence on the results of this study, while water-holding capacity and soil depth are important under rainfed conditions. Our knowledge of soils across the world shows important weaknesses with respect to the extent of coverage, extrapolation of point observations to grid cells, definition of soil characteristics compatible with crop models and handling of preferential flow in soil profiles. Efforts in this direction are currently underway (Batjes 1990). The distribution of suitability among and within regions could show considerable deviation, changing the results. For example, when we reduced suitability by 0.1 in all grid cells during a sensitivity analysis, rainfed production in PO-systems declined by 15%, while irrigated production was almost unchanged; total production decreased by 9%. It may be argued that areas with much sloped land (as in China) that is unsuitable for PO farming could still be used for EO farming, boosting total production. When we eliminated the negative effects of slope on agricultural practices, the irrigated area (and hence production) increased by 6% and rainfed production by 1%, leading to an overall global increase in potential food production of only 3%. The large relative increase is in the Asian regions, underlining the importance of productive and sustainable hillside agriculture.

Soils can respond slowly to management. Its water-holding capacity may increase (more organic matter) or decrease (erosion). The relative importance of water-holding capacity was examined in a sensitivity analysis by raising it 67%. More rainwater would then be retained for crops, and with a fixed amount of available water, a larger area could be irrigated. In the production-oriented farming system, total rainfed production would increase by 14% and the total global production would go up by the same percentage. In the environment-oriented farming system, water is often not a limiting factor, and a higher water-holding capacity would increase total production only by 5%. Effects of land degradation and erosion, which reduce the area suitable for agriculture, were not addressed.

We assumed that all farmers use the known 'best technical means' of production and resource use. Although this may sound 'conservative' as better techniques will emerge, it already implies that an enormous amount of adaptive research is needed to adapt known technologies to local conditions.

The environment-oriented agriculture demands significantly more P in the soil than production-oriented systems, as N-fixation is relatively P-demanding, and it covers a much larger area due to the lower productivity. It should be explored whether the natural sources of P reserves are sufficient to permit this type of agricultue at a global or regional scale.

## Conclusions

This exploratory analysis indicates that natural resources are available to increase food production at the global level and can meet global potential food demand whatever the growth and consumption scenarios. However, strong spatial variability exists in regional potential demands and in their potential production capacities. The Asian regions, which will carry almost half of the global population in 2030, will have to use all land (East Asia, China) and/or water (South Asia, India) as efficiently as possible for agriculture.

China's potential food security situation is tighter than the average for the Asian region. The nation could offer its population a moderate, but not an affluent diet, even when all suitable land and water are used in production-oriented agriculture. Large-scale environment-oriented agriculture is insufficiently productive, and appears to be impossible. It is, therefore, a major challenge to maximize land productivity in a sustainable manner under a level of damage to the environment that is socially accepted.

Three times more biomass is required for an affluent, meat-rich diet than for a vegetarian diet. If it were possible to develop a new food technology that largely replaces animal products by plant products with an equal degree of social acceptance, regions insufficiently endowed with natural resources could still afford an affluent diet, and much land would be spared. The large areas of potential rangelands in China should be investigated with respect to their potential contribution to food production. Future competition between food and energy crops for land and water needs to be investigated.

## References


Batjes N H, (Ed.) (1990) Macro-scale land evaluation using the 1 : 1 M world soils and terrain digital database. SOTER report 90/5. ISSS/ISRIC, Wageningen, The Netherlands. 45 p.

Brown L R, Kane H (1994) Full house. The world watch environmental alert series. Norton and Co., New York, USA. 261 p.

Buringh P, Van Heemst H D J, Staring G J (1975) Computation of the absolute maximum food production of the world. Department of Tropical Soil Science, Wageningen Agricultural University, Wageningen, The Netherlands. 59 p.

Caporali F, Onnis A (1992) Validity of rotation as an effective agroecological principle for sustainable agriculture. Agric. Ecosyst. Environ. 41:101–113.

Crosson F, Anderson J (1992) Global food, resources, and prospects for the major cereals. World Bank, Washington DC, USA. 250 p.

Delft Hydraulics (1992) Global water availability for future irrigated agriculture. Internal report. Delft Hydraulics, Delft, The Netherlands. 142 p.

De Wit C T, Van Keulen H (1987) Modelling production of field crops and its requirements. Geoderma 40:253–265.

FAO – Food and Agriculture Organization (1993) Yearbook production 1992. FAO, Rome, Italy.

He K (1990) The basic position of food and its developing strategy. Pages 1–3 in The policy of developing strategy of food in China. Agricultural Ministry of China, Beijing, China.

ICLARM – International Centre for Living Aquatic Resources Management (1992) ICLARM's strategy on living aquatic resources management. ICLARM, Manila, Philippines. 124 p.

IRRI International Rice Research Institute (1989) IRRI towards 2000 and beyond. IRRI, P.O. Box, 933 Manila, Philippines. 95 p.

IRRI (1993) Rice research in a time of change. IRRI's medium-term plan for 1994–1998. IRRI, P.O. Box 933, Manila, Philippines. 79 p.

Kropff M J, Cassman K G, Van Laar H H (1992) Quantitative understanding of the irrigated rice ecosystem and yield potential. Pages 97–113 in Virmani S S (Ed.) Hybrid rice technology: New developments and future prospects. Selected papers from the International Rice Research Conference. International Rice Research Institute, P.O. Box 933, Manila, Philippines.

Kropff M J, Cassman K G, Penning de Vries F W T, Van Laar H H (1993) Increasing the yield plateau in rice and the role of global climate change. J. Agric. Meteor. 48(5):795–798.

Liu J R, Zhang D Y (1990) The effect of mixed application of organic and inorganic fertilizers in paddy soil. Acta Agric. Univ. Jiangxiensis 2(1):37–42.

Luyten J C (1995) Sustainable world food production and environment. Internal report No. 37, AB-DLO, P.O. Box 14, 6700 AA Wageningen, The Netherlands. 252 p.

Penning de Vries F W T (1991) Improving yields: Designing and testing of very high yielding varieties. Pages 13–19 in Penning de Vries F W T, Kropff M J, Teng P S, Kirk G J D (Eds.) System simulation at IRRI. IRRI Research Paper Ser. 151. International Rice Research Institute, Los Baños, Philippines.

Penning de Vries F W T, Van Keulen H, Rabbinge R (1995). Natural resources and limits of food production in 2040. Pages 65–87 in Bouma J, Kuyvenhoven A, Bouman B A M, Luyten J C, Zandstra H G (Eds.) Eco-regional approaches for sustainable land use and food production. Kluwer Academic Publishers, Dordrecht, The Netherlands.

Simpson J R, Xu C, Miyazaki A (1994) China's livestock and related agriculture, projections to 2025. CAB International, Wallingford, UK. 474 p.

Spiertz J H J,Van Heemst H D J, Van Keulen H (1992) Field crop systems in Northwestern Europe. Pages 357–371 in Pearson C J (Ed.) Ecosystems of the world, No. 18. Elsevier, Amsterdam, The Netherlands.

UN – United Nations (1992) Long-range world population projections: Two centuries of population growth (1950–2150). United Nations Departments of International Economic and Social Affairs, New York, USA. 35 p.

Wang T D (1993) A systems approach to the assessment and improvement of water use efficiency in the North China plain. Pages 193–205 in Penning de Vries F W T, Teng P S, Metselaar K (Eds.) Systems approaches for agricultural development. Kluwer Academic Publishers, Dordrecht, The Netherlands.

WRI – World Resources Institute (1994) World Resources 1994–1995. WRI, Oxford University Press, New York, USA. 401 p.

WRR – Netherlands Scientific Council for Government Policy (1994) Sustained risks: A lasting phenomenon. Report to the Government No. 44. WRR, Sdu Publishers, The Hague, The Netherlands. 205 p.

Xu C (1994) Needs and priorities for the management of natural resources: A large country's perspective. Pages 199–211 in Goldsworthy P R, Penning de Vries F W T (Eds.) Opportunities, use, and transfer of systems research methods in agriculture to developing countries. Kluwer Academic Publishers, Dordrecht, The Netherlands.

Yoshida S, Parao F T, Beachell H M, (1972) A maximum annual rice production trial in the tropics. Int. Rice Comm. Newsl. 21:7–32.

Zobler L (1986) A world soil file for global climate modelling. NASA Technical Memorandum 87802. NASA, Scientific and Technical Information Branch, USA. 32 p.

# Adjustment of rice supply to the evolution of the consumption pattern in Indonesia (1972–1993): mid- and long-term perspective and implications

M. BENOIT-CATTIN[1] and B. IRAWAN[2]

[1]*Centre de coopération Internationale en Recherche Agronomique pour le Développement, Unité de Recherche 'Prospective et Politique Agricole', CIRAD-URPA, BP 5035, Montpellier, Cedex, France*
[2]*Center for Agro-Socio Economic Research, CASER, Bogor, Indonesia*

**Key words:** rice supply, Indonesia, consumption pattern

**Abstract**
Due to the Green Revolution, rice production in Indonesia has rapidly increased and, in 1984, the country reached self-sufficiency in rice. The revolution, on the other hand, increased farmers' incomes which, in turn, has generated a change in consumption pattern. The food consumption pattern in Indonesia has become more diversified, moving to high-income elasticity products, while rice consumption has shifted to a tasty rice. In response to the consumption pattern evolution and to the changes in relative prices, farmers tend to cultivate more non-rice crops and the rice crop itself is including more tasty varieties of rice. This trend could threaten rice self-sufficiency: while the area available for rice cropping would be reduced, the high-yielding varieties of rice generally disseminated are of low eating quality. The anxiety is enhanced by the fact that fertilizer is relatively intensively used, so that there is no room left for intensification of rice production. One of the solutions is the improvement of the potential yield of rice varieties in various environments, as well as its quality.

## Introduction

The Indonesian government has been committed to self-sufficiency in rice since the early 1970s. Considering the economic significance of rice, a farm data survey confirmed that producing rice domestically is economically and socially more favourable than importing rice. However, due to pests and epidemics, drought, agroclimate and world market impacts, rice production in Indonesia is unstable. To increase rice production, as well as its production stability, the government intervenes in the economics of rice.

The effort to increase rice production was undertaken by encouraging farmers to intensify rice farming. Since 1965, the government has launched many rice intensification programs, which are principally made up of five components: improvement of water management, improvement of planting methods, improvement of land preparation, use of a modern rice variety, and use of fertilizers and pesticides. To eliminate the constraints of the farmers' budgets in adopting the new production technology, the government provides subsidized agricultural credit. Moreover, farmers are also stimulated by being provided subsidized input prices, and by price control of rice, to maintain their incomes.



*P.S. Teng* et al. *(eds.): Applications of Systems Approaches at the Farm and Regional Levels, 295–307.*

*Table 1.* Growth of production, area and yield of selected food crops in Indonesia, 1968–92.

| Growth (% yr$^{-1}$) | Rice | Corn | Cassava | Soybean |
|---|---|---|---|---|
| Production | | | | |
| 1968–84 | 5.06 | 4.78 | 1.72 | 2.42 |
| 1985–92 | 2.76 | 5.85 | 3.33 | 9.21 |
| 1968–92 | 4.73 | 4.84 | 1.88 | 5.38 |
| Area | | | | |
| 1968–84 | 1.01 | 0.59 | −0.71 | 0.73 |
| 1985–92 | 1.19 | 2.33 | 1.54 | 5.31 |
| 1968–92 | 1.07 | 0.74 | −0.61 | 3.19 |
| Yield | | | | |
| 1968–84 | 3.79 | 4.19 | 2.49 | 1.69 |
| 1985–92 | 1.79 | 3.52 | 1.78 | 3.91 |
| 1968–92 | 3.41 | 4.09 | 2.43 | 2.19 |

Due to comprehensive government intervention which, in turn, induced economic and technological incentives for rice farming, farmers increased their rice production either by intensifying or expanding rice farming. Consequently, during the last two decades, rice production has increased by an average 4.7% per year, mainly from yield increase and area increase, at annual growth rates of 3.4% and 1.3%, respectively. This progressive increase in rice production, in turn, enabled the country to reduce gradually the importation of rice. In 1970, Indonesia was the main rice importer in the world, but by 1984 rice self-sufficiency had been achieved. Moreover, since rice farming contributes significantly to labour absorption and to income, the increased rice production enabled an increase in per capita income, either through increased farm revenue or via the extra income generated by non-agricultural investments.

Whether or not Indonesia will be able to maintain rice self-sufficiency in the future has been a critical policy issue recently. The issue appeared because, in fact, the country has resumed importing rice at 55 000 tons in 1987, rising to 268 000 tons in 1989 and 610000 tons in 1992, after around zero importation in the period leading up to 1986. For 1995, due to drought, the amount of imported rice was estimated to be around 600 000 tons. In order to slow down this increase in rice importation, an increase in domestic rice production is needed. In line with this, policies will need to focus on an effort to increase rice yield, considering its significant contribution to past production increase. This paper attempts to describe the possibility to maintain rice self-sufficiency, particularly with respect to the rice intensification effort.

## Trend in the production of rice and other major food crops

The most impressive growth rate of rice yield in Indonesia was reached between 1979 and 1983, when yields rose by 6.5% per year, with a 1.1% growth of area (Table 1). The exceptionally strong growth of yield during that period was primarily the result of the Special Intensification (INSUS) programme launched in 1979. Under this intensification programme, farmers in irrigated areas were organized on a 50 ha production basis, and decision making pertaining to cultural practices was done within the group. All culture practices were done simultaneously by farmer groups. With this institutional innovation, economies of scale in production activities were captured, by increase in farming efficiency, particularly for the management of water distribution and insect control.

Since 1985, after rice self-sufficiency had been achieved, the rate of the growth of rice production gradually decreased from 5.1% per year for 1968–84, to 2.8% per year for 1985–92. Most of this decrease in production growth was due to the decrease in the rate of yield growth, while area growth was relatively stagnant at about 1% per year. The situation is not favourable for any future rice production increase, because increasing the per hectare yield will then be the main contributor to the increase in production. Total rice production needs to rise by about 3% per year in order to match the growth of consumption (Timmer 1990). This is a major challenge for policy makers.

While the growth rate of rice production has declined, the rate of production growth of other food crops, particularly soybean, corn and cassava, has increased. With the exception of soybean, most of the increased production growth for other food crops resulted from area expansion. In the last decade, the area expansion of these crops was higher than for rice. This shows that farming of other food crops is more competitive than rice, particularly for soybean, where area growth was 5.3% per year.

Most of the rice production in Indonesia is from wetland rice crops, which contribute more than 80% of the total production. During the last two decades, the average yield growth of wetland rice crops declined from 3.56% for 1968–84, to 1.34% for 1985–92. This decrease in the rate of yield growth occurred both in Java and other regions. In Java, the major island to produce rice, particularly wetland rice, the growth of yield declined from 3.9% to 1.6%. Since the annual growth of area also declined from 1.55% to 0.01%, the rate of production growth dropped from 5.5% to 1.6%. A similar situation also occurred in the outer islands: the rate of yield growth declined from 3.14% to 1.36%. However, since the rate of area growth declined only slightly, the decrease in production growth was relatively small, from 5.6% between 1968 and 1984, to 3.6% between 1985 and 1992.

## Need for an improvement of rice quality

The demand for food is determined by income level. As incomes increase, the food demand also increases with respect to its quantity, quality and diversity. During the last two decades, the average income per capita in Indonesia has risen by around 5–6% per year. SUSENAS (Survey of Social and Economic National) data indicate that this income increase, in turn, has induced change in the food expenditure pattern of the population. In 1981, about 39% of the consumer's food expenditure was concentrated on cereals and tuber products, particularly rice, but in 1987 and 1993 this proportion of food expenditure declined to 32% and 29%, respectively. There was a corresponding increase in the contribution of protein products, particularly meat and eggs, fruit and prepared food, which in total increased from 13% in 1981 to 23% in

*Table 2.* Rice consumption per capita in Indonesia, in 1981, 1987 and 1993 (in parentheses: average price for each category of rice) (*Source:* Central Bureau of Statistics, 1981, 1987, 1993).

| Consumption | 1981 | 1987 | 1993 |
|---|---|---|---|
| Rural area | | | |
| Consumption per capita (kg $yr^{-1}$) | 118.2 | 118.6 | 121.1 |
| Domestic rice (%) | 44 (230.1) | 46 (387.9) | 52 (598.1) |
| Special rice (%) | 50 (210.9) | 52 (366.1) | 46 (568.1) |
| Imported rice (%) | 5 (244.1) | 1 (440.6) | 1 (697.0) |
| Glutinous rice (%) | 1 (307.7) | 1 (428.6) | 1 (692.3) |
| Urban area | | | |
| Consumption per capita (kg $yr^{-1}$) | 112.9 | 109.8 | 106.5 |
| Domestic rice (%) | 44 (255.5) | 56 (410.7) | 60 (652.1) |
| Special rice (%) | 36 (229.1) | 42 (384.2) | 38 (621.0) |
| Imported rice (%) | 19 (262.5) | 2 (435.9) | 2 (689.7) |
| Glutinous rice (%) | 1 (285.7) | – | – |

1993 for the rural areas, and for the urban areas from 25% to 40%. The food consumption pattern has shifted from low-income elasticity products to higher-income elasticity products.

The changes in the consumption pattern apply also to rice consumption. In rural areas, the income increase has paralleled an increase in quantity eaten, but in urban areas, rice consumption has tended to decline. In rural areas, the annual average per capita rice consumption has continuously increased from 118.2 kg in 1981, to 121.1 kg in 1993, while it has declined from 112.9 kg to 106.5 kg in urban regions. However, as indicated in Table 2, both in rural and urban regions, rice consumption has tended to shift to a more expensive rice, with a higher tendency in urban regions to do so. Figure 1 confirms that this phenomenon resulted from the increase of consumers' income.

The situation above indicates the need for an improvement of rice quality in addition to its production increase in order to match the evolution of rice demand. In the globalization era, the domestic market is expected to have a high degree of integration into the world market. Within a competitive market, the need for quality improvement for domestically produced rice is getting higher in order to maintain rice self-sufficiency. If domestic rice producers are not able to fulfil consumers' demand for better-quality rice, this demand may then have to be met by rice importation.

## Improvement of rice varieties and their dissemination

The significant contribution of improvement of rice varieties to increasing rice production in Indonesia has been widely recognized. During the last two decades, the government released 91 improved varieties of rice which were developed either by IRRI (International Rice Research Institute, 22 varieties) or by the Agency for Agricultural Research and Development (AARD, 69 varieties). Considering its contribution to total production, most of the varieties released (71 varieties) were for the wetland rice crop.

In terms of potential yield and growth duration, the varieties released during the

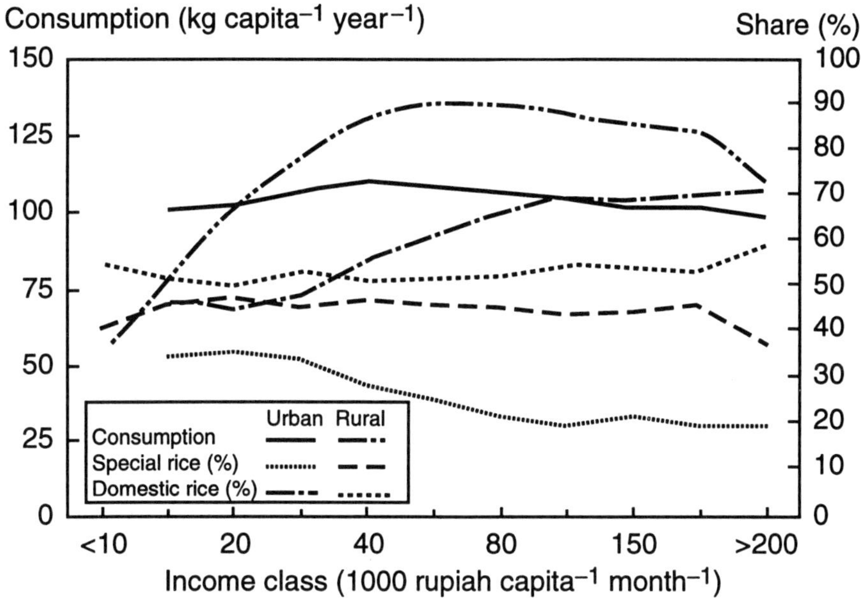


*Figure 1.* Rice consumption in relation to income class in urban and rural areas, 1993 (*Source:* CBS 1993).

*Table 3.* Characteristics of improved rice varieties for wetland areas released in Indonesia between 1972 and 1993 (*Source:* original data from Agency for Agricultural Research and Development 1993).

| Characteristics | 1972–80 | 1981–85 | 1986–89 | 1990–93 |
|---|---|---|---|---|
| Total variety | 22 | 27 | 12 | 7 |
| Good taste variety | 9 | 11 | 5 | 2 |
| Average potential yield (ton $ha^{-1}$) | 4.56 (4.0–5.0) | 4.74 (4.0–5.5) | 5.21 (4.5–5.5) | 5.28 (5.0–5.5) |
| Average growth duration (days) | 133.7 (120–145) | 122.1 (115–132) | 119.9 (110–140) | 118.5 (115–127) |

last two decades showed an improved performance from year to year. Compared to 1972–80, the potential yield of rice varieties released in 1990–93 was higher by 16%, and its growth duration was shorter by 13% (Table 3). However, since 1986 this capacity increase has been relatively stagnant. The potential yield and the growth duration of varieties released in 1986–89 and in 1990–93, were not significantly different. In addition, most of the varieties released during the last period have no taste. It seems that simultaneous improvement of yield potential, growth duration and quality is increasingly difficult. To maintain rice self-sufficiency in the future, this situation is not favourable, because varietal improvement is a fundamental need for rice production increase in Indonesia.

During the last two decades, the adoption level of improved rice varieties has progressively increased from 46% of rice area in 1972 to 88% in 1990. As indicated in

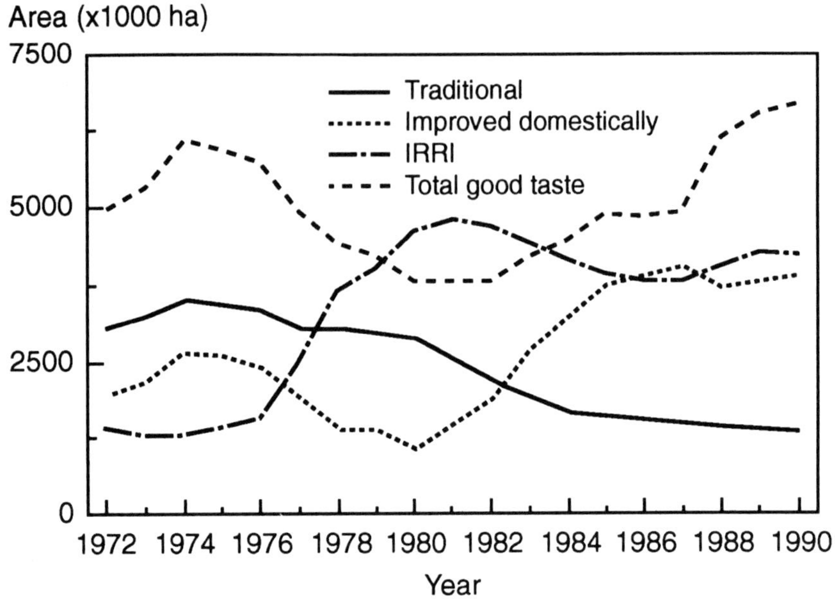


*Figure 2.* Adoption rate of rice varieties in Indonesia, 1972–90 (*Source:* data from Directorate of Food Crop Production).

Figure 2, the adoption rate of 'good taste' varieties continuously increased from 42% of rice area in 1982, to 70% in 1990. Before 1982, the contribution of local 'good taste' varieties to rice area gradually declined, while the opposite situation occurred for varieties developed by IRRI. Due mainly to its superiority of potential yield and resistance to pests and diseases compared to the domestic varieties, the contribution of IRRI varieties to the rice area increased from 19% in 1973 to 54% in 1981. However, since 1982 the adoption rate of IRRI varieties gradually decreased and its contribution has been replaced by varieties domestically improved. This substitution of varieties occurred because on one side the potential yield of varieties domestically improved tends to increase, and on the other side the varieties developed by IRRI generally have no taste. Among 22 IRRI varieties adopted in Indonesia, there were only 4 'good taste' varieties, particularly for varieties released since 1985, and among them only the variety IR64 has recently been widely adopted by farmers.

The evolution above indicates that there is a change in the farmers' preferences for improved rice varieties. Up to 1980, the potential yield of a variety was a greater concern to the farmers than its taste in determining their decision to adopt the introduced rice varieties, but the opposite situation has occurred over the last decade. This implies that taste should be emphasized in future programmes to improve rice varieties in order to match farmers' preferences. This effort is also essential to match an increasing demand for rice quality on the consumer market because taste represents an important variable determining rice quality.

The high adoption rate of improved varieties indicates that only little scope remains for increasing rice production by disseminating improved varieties. This is especially so in Java, where 95% of rice land is under improved varieties. One of the options is to replace earlier modern varieties with the latest ones (varieties released after 1985), which have a 10% higher potential yield. This option is relatively large for both Java and off Java regions because the adoption rate of varieties released after 1985 is quite low. In Java, these covered 43% of the rice area in 1990, while in the outer islands only 11%. Since most of the latest varieties have no taste, however, this option may turn out to be of little value.

**The use of fertilizer and rice farming productivity**

To enhance the dissemination of improved rice varieties in Indonesia, fertilizer subsidy has been an important policy instrument since 1970. By implementing this, it was expected that farmers would be motivated to apply the recommended rate of fertilizer introduced through extension programmes. Due to this government intervention, the use of fertilizers in rice farming has increased by 8.8% per year during the last two decades. The increase of fertilizer use and the improvement in other culture practices, in turn, increased rice yield at an annual growth rate of 3.3%.

The improvement in farming has reduced the gap between the potential yield and the yield achieved by farmers. Assuming variety is the major contributor to potential yield, the average of potential value at the aggregate level, roughly, might be estimated from the proportion of each variety's area multiplied by potential value of each variety. Under this assumption, Figure 3 shows that the gap between the potential yield and yield achieved by farmers for the wetland rice crop has decreased during the past two decades. In 1972, the average farmer yield was at about 60% of potential value, and by 1990 it was around 90%.

The progressive increase in the use of fertilizers has been associated with decreasing efficiency, particularly for the wetland rice crop in Java. The farm level studies done by CASER (1984), Erwidodo (1984), Irawan and Rachman (1988) and Rachman (1988) have confirmed this. Those studies concluded that farmers in Java have generally been using fertilizers at 10–20% above the recommended rates and the marginal productivity of fertilizers used is close to zero.

Motivated by a need to improve efficiency and under severe budgetary constraints, the government launched its commitment to phase out the fertilizer subsidy starting in 1987. Between 1987 and 1991, the government annually raised the price of fertilizers by 20–25%. These increases in prices, however, have had little effect on reducing the farmers' fertilizer use level. A monitored farm-level study (Rahma and Quizon 1995) showed that in west and east Java, there was no significant decrease in the use of fertilizers by farmers over six seasons monitored between 1989 and 1991. Elsewhere fertilizers used on wetland rice crops continually increased at a decreasing rate, while in the Java region they tended to decline at a slow rate.

At a national level, the total fertilizers (Urea, TSP, KCl) used on the wetland rice crop in 1988–91 ranged from 330 to 340 $kg\,ha^{-1}$, and this represents more than 95%

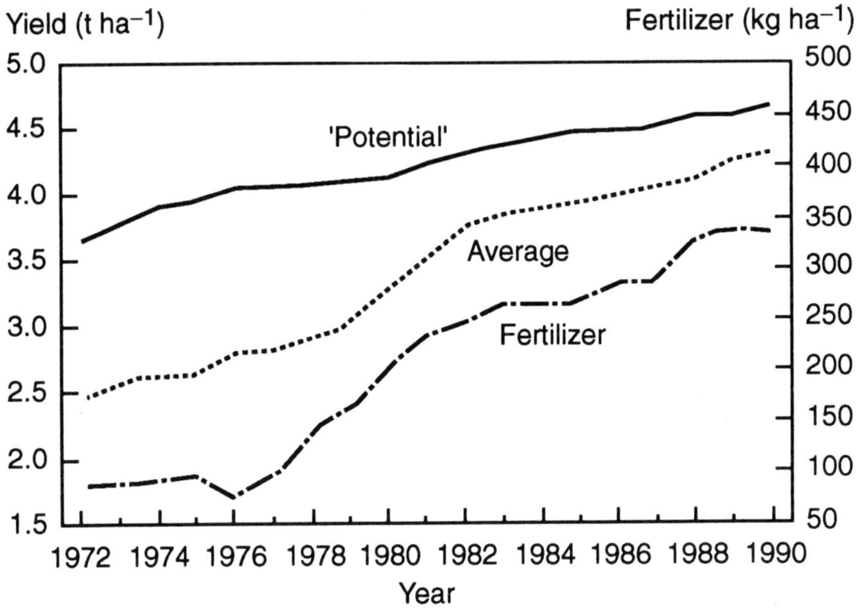


*Figure 3.* Fertilizer use, average achieved and 'potential' yields (see text) in wetland rice crops in Indonesia, 1972–90 (*Source:* data from Central Bureau of Statistics).

of the recommended rates. These figures indicate that it is unlikely that yields can be increased much beyond the current level via further intensification. In any case, reducing fertilizer input might be the best solution to improve fertilizer efficiency. For outer islands, there is the possibility to boost the yield growth by increasing the use of fertilizers because rice farmers in those regions have been using fertilizers up to 10–30% below the recommended rates.

## Irrigated area development

Investment in irrigation has been a major contributor to the rapid growth in rice production over the last two decades. The importance of irrigation development in Indonesia is reflected by the increasing share of irrigation investment in the total government expenditure from 43.1% in 1979–83, to 49.1% in 1984–88 (Pasandaran and Taryoto 1992). This investment policy is reasonable because irrigated land is a major contributor to rice production in Indonesia. About 85% of the rice harvest is from irrigated land and its yield is more than double that of the non-irrigated land.

Irrigated areas in Indonesia are classified as technical, semi-technical and simple irrigation systems. Because of their high yield as well as their cropping intensity, technical and semi-technical irrigated areas are the major contributors to rice production. For the 1988 situation, Pasandaran and Rosegrant (1995) estimated that 39.4% and 17.9% of the rice production resulted from technical and semi-technical irrigated areas, which occupy 23.7% and 12.2% of the total areas, respectively.

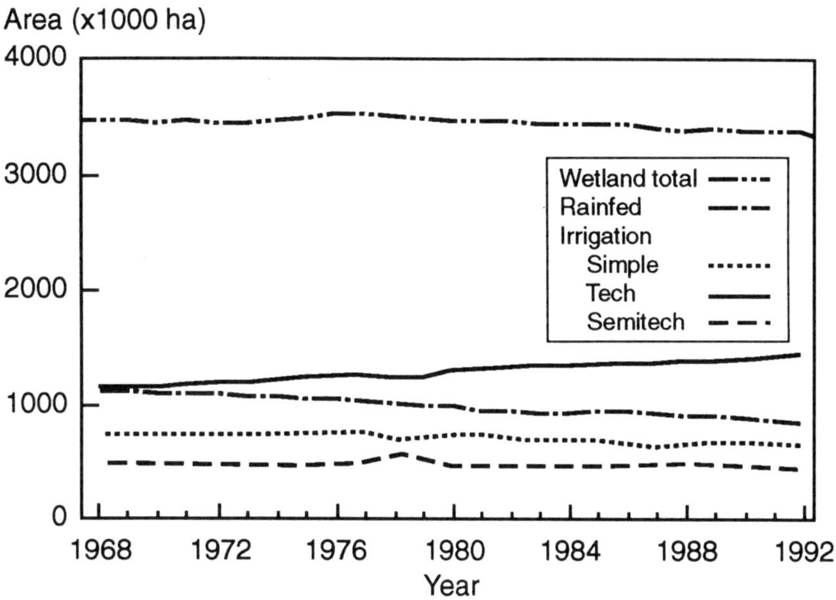


*Figure 4.* Wetland and irrigated area in Java, 1968–92 (*Source:* Central Bureau of Statistics).

More than 75% of the technical irrigated areas and about 55% of the semi-technical irrigated areas are located on Java, while this island contributes to 40–45% of the simple irrigated areas. This high degree of concentration of irrigation in Java implies that agricultural areas on the island are more favourable to rice farming than those on the outer islands.

Between 1968 and 1992, the irrigated areas in Indonesia increased at an annual rate of 2.5%. However, since rice self-sufficiency was achieved in the mid-1980s, the irrigated areas have grown at a decreasing rate. During 1974–83, the annual growth rates of irrigated areas in Java and outer islands, on average, were respectively 0.31% and 3.20%, but between 1984 and 1992 declined to 0.06% and 2.01%. Pasandaran and Rosegrant (1995) pointed out that large losses in government revenues due to declining oil prices since the early 1980s are a major contributor to the decline of irrigated area growth. The success of rice production, coupled with declining world prices of rice and the rising cost of new irrigation investments, have led the government to rearrange its investment priorities. Since the rice price decrease induced a decrease in social and economic gain from irrigation investment (Mann 1990), instead of the construction of a new irrigation system, the government's priority was on the maintenance and improvement of the existing ones in order to keep the efficiency of the irrigation investment.

Another contributor to the decreasing rate of the growth of irrigated area, particularly in Java, is the conversion of some areas to non-agricultural uses. Although there have been substantial new investments in the irrigation system, the conversion of the existing areas to non-agricultural uses has resulted in little annual net increase of the total irrigated areas. In 1981–86, for example, the conversion of wetland areas in Java and Bali amounted to 118% of the total additional wetlands that have been built during the period (Sudaryanto et al. 1992). Consequently, wetland areas in Java continuously decrease (Figure 4) at a growing rate from −0.11% per year in 1974–83, to −0.33% per year in 1984–92.

The decrease of wetland areas in Java implies that a bigger contribution from this island to maintain rice self-sufficiency in the future cannot be expected. More agricultural resources, as a consequence, need to be developed in order to boost the growth of rice production and, in this case, the priority of irrigation investment for outer islands might be a prospective alternative. The development of irrigation on these islands is necessary not only to increase rice yield and the rice crop intensity by supplying more water, but also for accelerating the adoption of improved technologies by farmers. The latter is determined by the impact of water availability on the risks of production.

## Rice price policy

Rice price policy has a long history in the Indonesian food system. The major objectives of the policy are to maintain a floor price attractive enough in order to create an economic incentive for increasing rice production and to stabilize the domestic price of rice. The stability of the prices for rice has been a primary concern of the food pricing policy in Indonesia because of the economic significance of rice. Some argue that maintaining stability in the price of rice is frequently a greater concern for policy makers than the price level because it involves both the political and the economic stability of the nation. Since rice is the major staple food for most of the population and contributes significantly to the nutritional need (it contributes to around 53% and 45% of the total calorie and protein intakes), the stability of the price of rice is closely tied to food security in the country.

In pursuit of food security and domestic price stability, BULOG (National Logistic Board) is the government institution charged with maintaining an adequate domestic supply of rice at a stable price through the buffer stock mechanism. Each year, the government sets a ceiling price in the retail market and a farm-level floor price, in such a way that these prices are compatible with the consumers' income and attractive enough for the farmers to increase rice production. To support the operational costs of BULOG, the government subsidizes the interest rate of credit borrowed by this institution. In its first 22 years of operation until 1989, this institution received a highly subsidized interest rate of 6% per year.

In operating a buffer stock for interseasonal price stabilization, BULOG purchases rice in the surplus period (peak harvest period), when the price drops below the announced floor price, and sells its stock to consumers in the deficit period, when rice prices climb up. This interseasonal transfer of rice is not only done by BULOG, but also by farmers and private traders. For the situation in 1990, Ellis (1993) estimated that 21% of rice surplus in the peak harvest season were transferred by

BULOG and most of the surplus (59%) were transferred by traders. This interseasonal transfer by traders is not closely related to the reason for seasonal price margins, which is the usual approach by economists to stockholding. Traders store rice and unhusked rice mainly for operational purposes such as the continuity of unhusked rice supply into mills, and the regular supply of rice to consumers.

Due to the BULOG operation on the rice market, the rice price in the domestic market is more stable than in the world market. Between 1972 and 1991, the coefficient of variation of the annual price of rice in the domestic market was only 15.9%, much smaller than the 38.9% of the world market. This shows that the government has been successful in insulating domestic prices from long-term fluctuation in the world market price changes. The government has also been successful in maintaining price stability in the short term because the average coefficient of variation for the monthly price, for the same period, was only 5.08% for the retail market and 7.21% for the producer market. In addition, the ratio of annual farm-level price to the announced floor price was 1.09 on average, which means that farmers have received a price above the floor price.

Undoubtedly, the implemented price policy has brought an economic incentive for farmers to increase rice production. The implementation of this policy, however, has also increased the government's expenditures, and this budgetary burden will increase as rice production increases. The situation has become worse, particularly since rice self-sufficiency was achieved, because the self-sufficiency condition enables the accumulation of a rice surplus which should be handled by BULOG. Due to the increasing surplus and rice importation restrictions, it was reported that BULOG's operation to maintain price stability has faced an increasing deficit from 26.7 billion rupiahs in 1984 to 123.4 billion rupiahs in 1986. This deficit might have been avoided by more access to the world market because world market prices declined during the same period which, in turn, led to a higher domestic price.

In response to this situation, in 1989 the government changed the concept of self-sufficiency in rice, from absolute self-sufficiency to sufficiency on trend. With this new concept, the trade regulation on rice is more flexible for BULOG because this institution is now allowed to import or export rice as long as it is done to stabilize the price of rice. This policy, however, gives no possibility to BULOG to improve its financial situation because since 1987 the rice price in the world market has tended to decrease gradually. In addition, the government has also eliminated the subsidy on BULOG's operation by raising the interest rate of credit from 6% in 1989 to 16% in 1990 and to 19% in 1991. This situation, in turn, has reduced BULOG's profits from rice operation. Pearson (1993) estimated that as the interest rate of credit increased, BULOG's profits from rice operations have declined from 46 billion rupiahs in 1989 to 7 billion rupiahs in 1991, and for the following years, it was estimated that the implementation of the policy of price stabilization would lead to a deficit, unless there were an additional government subsidy.

Those financial difficulties, in turn, tend to reduce the ability of BULOG to maintain the announced floor price, as well as a stable price at the farm level. This situation is reflected by the increase of the coefficient of variation of the monthly price at the farm level from 4.26% between 1980 and 1985 to 10.26% between 1986 and 1991. In addition, the ratio of the monthly price during the main harvest period (April to July) to the announced floor price has also declined. Between 1980 and 1985, these ratios ranged from 1.07 to 1.21, but between 1986 and 1991 it ranged only from 0.81 to 0.95. It shows that instead of receiving more than the floor price announced, as it was during the first period, the farmers received a lower price during the second period.

The decrease of price stability at the farm level reflects a decrease in the economic incentive to rice farming. This situation is not favourable for the future increase of rice farming, because when they grow rice, farmers respond to the price they expect, and the price expected will decline as price stability decreases. However, the situation seems to be unavoidable because a policy of price stabilization for rice involves some substantial costs to the country. Moreover, these costs may have some alternative opportunities in other investments or programs that might offer higher pay-offs.

## Conclusions and policy implications

The current situation of the rice economy in Indonesia shows that this country is facing some problems to maintain the rice self-sufficiency which was achieved in 1984. Total production growth is limited by the decrease in the rate of yield growth of rice. Problems are more severe for Java, the main producer of rice in the country. The lowland rice area in Java tends to decrease continuously because of the conversion of agricultural land to non-agricultural use, irrespective of new investments in the irrigation systems. In addition, farmers in this area are already using fertilizers beyond their optimal level. So it seems that any further additions using current fertilizers will not increase rice outputs.

However, rice self-sufficiency should be maintained, considering the permanent growth of the population and rice consumption, especially in rural areas, as well as the significant contribution of rice to the calories and protein intake. This implies that the recovery in the growth of rice production at a sustainable rate is needed. In line with this, in order to stimulate farmers to increase their rice production, efforts to create some incentives on technological aspects such as the improvement of varieties and crop management, need to be emphasized instead of economic incentives such as trade regulation and pricing policy. Indeed, the latter approach tends to incur a substantial social cost to the country as the production increases. In addition, this policy is not useful for the future economic situation, which will tend to evolve to a liberalized market.

From the side of production technology, the efforts to increase rice yield should be clearly determined. In line with this, the improvement of the quality of the varieties, as well as the dissemination of the latest improved varieties, should be considered. The need to improve the rice varieties implies devoting more resources to genetic and biotechnological research, in which the taste consideration should be emphasized as well as growth. This farmers' preference also suggests that the dissemination of the latest available improved varieties, those released after 1985, should be undertaken with a selection of the tasty varieties in order to facilitate their adoption by farmers.

The performance of Java in the growth of production is worse than that of the outer islands. This implies that it will be hazardous for the country to put excessive expectations on Java's contribution in the future to maintain rice self-sufficiency. Considering the level of technology applied by the farmers, which is relatively advanced, efforts to increase rice yield in Java seem to be limited. The opposite situation occurs in the outer islands and, in these regions, the wetland areas have tended to increase continuously. This suggests that there is some space in the outer islands to boost rice production, either by intensifying rice farming or by increasing the use of production inputs. Although some consider that expanding the production outside Java is hazardous because of the lack of infrastructure and inadequate support facilities, such claims still have to be proven. On the contrary, these regions may collectively be a 'sleeping giant'.

## References


AARD – Agency for Agricultural Research and Development (1993) Agricultural research statistics. AARD, Jakarta, Indonesia. 211 p.

CASER – Center for Agro-Socio Economic Research (1984) Analysis of fertilizers marketing in four provinces in Indonesia (in Indonesian). CASER, Bogor, Indonesia. 157 p.

CBS Central Bureau of Statistics (1981) National socio-economic survey (SUSENAS): Expenditure for consumption of Indonesia. CBS, Jakarta, Indonesia. 217 p.

CBS (1987) National socio-economic survey (SUSENAS): Expenditure for consumption of Indonesia. CBS, Jakarta, Indonesia. 233 p.

CBS (1993) National socio-economic survey (SUSENAS): Expenditure for consumption of Indonesia. CBS, Jakarta, Indonesia. 259 p.

Ellis F (1993) Indonesia rice marketing study: Aggregate rice market and role of BULOG. Indones. Food J. IV(7):23–27.

Erwidodo (1984) Determinant factors of rice yield in irrigated area in Jatiluhur, West Java (in Indonesian, English summary). J. Agro Ekon. 3(1):44–57.

Irawan B, Rachman B (1988) The efficiency of rice farming in West Java (in Indonesia). Pages 97–109 in Kasryno F, Suryana A, Djauhari A (Eds.) The evolution of rural economic to balanced economic. Center for Agro-Socio Economic Research, Bogor, Indonesia.

Mann A R (1990) Agricultural diversification. Policies and issues from East Asian experience. The World Bank, Washington DC, USA. 48 p.

Pasandaran E, Taryoto A H (1992) Irrigation policy in the context of crop diversification in Indonesia. Paper presented at the Seminar on Southeast Asia Agriculture in Transition: Implication for Food Policy. Department of Agriculture, Republic of Indonesia and International Food Policy Research Institute, 9–11 September 1991, Jakarta, Indonesia. 14 p.

Pasandaran E, Rosegrant M W (1995) Irrigation investment in Indonesia: Trends and determinants. J. Agro Ekon. 14(1):1–26.

Pearson S R (1993) Financing rice price stabilization in Indonesia. Indones. Food J. IV(7):83–96.

Rachman H P (1988) The estimation of profit function and analysis of rice farming efficiency in irrigated area in West Java (in Indonesian, English summary). J. Agro Ekon. 8(2):41–53.

Rahma H, Quizon J (1995) Determinants of fertilizer use among rice farmers in Java. Center for Policy and Implementation Studies, Jakarta, Indonesia. 43 p.

Sudaryanto T, Hermanto D, Pasandaran E, Rosegrant M W (1992) Food situation and outlook for Indonesia. International Food Policy Research Institute in collaboration with Center for Agro-Socio Economic Research, Bogor, Indonesia. 197 p.

Timmer C P (1990) Food price stabilization: The Indonesia experience rice. Harvard Institute for International Development, Cambridge, MA, USA. 174 p.

# Increasing Indonesian soybean production: testing policies in a lowland farming system

I. MARTY, F. GÉRARD, ERWIDODO, F. LANÇON and D. DEYBE

*Centre de coopération Internationale en Recherche Agronomique pour le Développement, Jl Merdeka 145, Bogor 16111, Indonesia*

**Key words:** Indonesia, Java lowlands, agricultural diversification, soybean production, soybean policies, farming system analysis, model MATA, alternative policies, policy simulations

**Abstract**

The complexity and diversity of the agricultural sector make the evaluation of the impact of economic policies on farmers a difficult task. We present a methodology based on a precise representation of farmers' behaviour using farming system analysis, and test the model on the soybean production in Java. Testing agricultural policies allows us to evaluate their possible impact not only on farmers (shift of production, income evolution, etc.), but also on the total production in the area described.

## Introduction

Since the beginning of the 1980s, the demand for soybean as food and feed in Indonesia has increased rapidly. In ten years, the human consumption of soybean-derived products has doubled from 4 kg to 8 kg per capita. During the same period, soybean use for feed increased even more rapidly from around 30 000 tons to 450 000 tons. National production has not been able to match this expansion of the soybean market. In 1985, the government launched a special program to boost soybean production. Apart from soybean self-sufficiency, the ultimate goal of this program was also to use soybean production intensification as a means to increase farmers' incomes in rainfed areas that have benefited less from the modernization of rice cultivation. Soybean cultivation was also considered an opportunity to optimize irrigated land use where water availability or pest management prevent farmers from growing a second or third rice crop. Ten years after the implementation of this policy for soybean production support, the degree of achievement has not met policy-makers' expectations: after a significant impact on the development of production in the late 1980s, self-sufficiency has not been achieved, although national soybean production has grown steadily from 500 000 tons to more than 1 500 000 tons in 1993. In 1993, soybean use for feed relied entirely on the import of seeds or meals, and imports represented 20% of soybean food processors' requirements. In June 1995, the government almost eliminated the remaining non-tariff trade barrier on soybean meal import to remove any constraints hampering the development of the booming chicken feed industry.

The soybean production development policy relies mainly on a price support mechanism based on import regulation. Within the current momentum towards a



*P.S. Teng* et al. *(eds.): Applications of Systems Approaches at the Farm and Regional Levels, 309–324.*

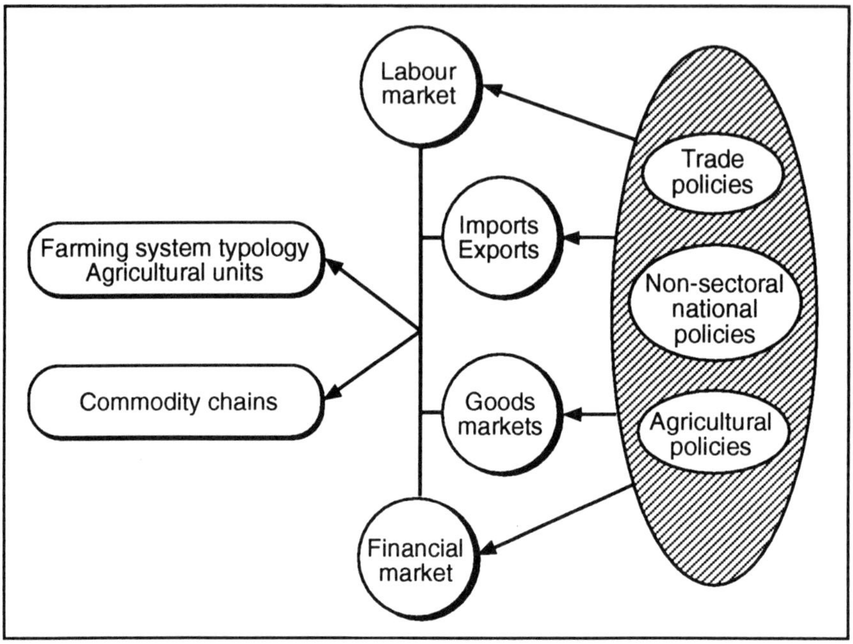


*Figure 1.* Links between macroeconomic level and farming system.

more open economy, this strategy has become more and more criticized by policy analysts within and outside Indonesia (World Bank 1992; Pangestu 1995). As there is a widening gap between agricultural and non-agricultural income, Indonesian authorities are concerned at maintaining specific support to farmers. Recent poor rice production performances show that food self-sufficiency is not sustainable and that policy adjustments are needed to increase rice production growth to ensure a stable socioeconomic environment.

## Methodology: modelling economic agents' behaviour

### *A set of dynamic and recursive models to represent the main farming systems*

Most models do not provide a satisfying description of the impact of policies on the agricultural sector at the household level because they represent either the field or plot level and neglect the links between agents, or they are formulated at a very aggregated level (see existing model as Rosegrant et al. 1987; SOW 1988, 1991; Nanseki et al. 1992; Thorbecke 1991; Altemeir and Bottema 1991).

The methodology proposed in this paper is based on a precise representation of farming systems through the opportunities and constraints they face, and on which an objective function is applied. Opportunities and constraints are determined by agroclimatic and socioeconomic conditions, and by each farmer's endowments in fixed factors and liquidity. In part, these general conditions are determined by macroeconomic variables and agricultural policy measures. To take them into account, the farming system modules are linked with a macroeconomic module determining the socioeconomic environment in which farmers' decisions take place. The modules are recursive and dynamic to allow each year to be linked with preceding years, and thus to take into account the importance in current decisions of past results, which partly determine present endowments of the farm in terms of liquidity and factor ownership. Farming system modules are linked together with markets: labour and land markets at the village level, agricultural product markets at the national level (Figure 1). In the version used here, prices are exogenous but the present paper is part of a broader project using the MATA model for policy testing (see Gérard et al. 1994).

All these model characteristics allow us to address policy effects both at the farm level and, after aggregation of all types of farms, at the national level. The model also shows immediate impact and time lag effects. These two features are important: first, because the heterogeneity in the farm situation will lead to different impacts of a given policy on farm revenue. It is useful to evaluate these differences in order to compensate the negative impacts by appropriate measures. Second, because the reactions of the agents are not instantaneous and their behaviour can delay impacts (e.g. on the environment), it is important to be able to evaluate both short- and long-term effects.

## Model description

The model reproduces farmers' behaviour to evaluate their response to policy and to estimate the impact on economic characteristics of the farm and production level. The first step is to identify the main farming systems, the second to represent the decision-making process of each representative farmer by an appropriate set of possibilities and constraints.

First, a zonation is set up by using statistical analysis on a set of variables crossing agroclimatic and socioeconomic data. Then, to achieve a good representation of the farming systems of the region studied, additional factors are defined by using farm-level data, so as to better reflect the different production factor endowments (land holding, etc.) under which the farmers make their decisions (see Marty 1994). The typology allows us to find representative farms and to select some of them according to the policy measure to be tested. Then mathematical programming models (non-linear) are built to describe each selected farm.

It is assumed that each farmer chooses from a set of activities and techniques those that maximize wealth utility under simultaneous constraints. Wealth is defined as the total value of the assets at the end of the year, including land, equipment, livestock, cash and savings. In order to consider the farmer's risk attitude, wealth utility is defined as increasing with expected wealth and decreasing with its variance (Markowitz 1992). Activities and techniques are defined according to agroclimatic conditions, the state of technology and general socioeconomic context. The constraints concern land availability, inputs, mechanization and traction requirements, labour,

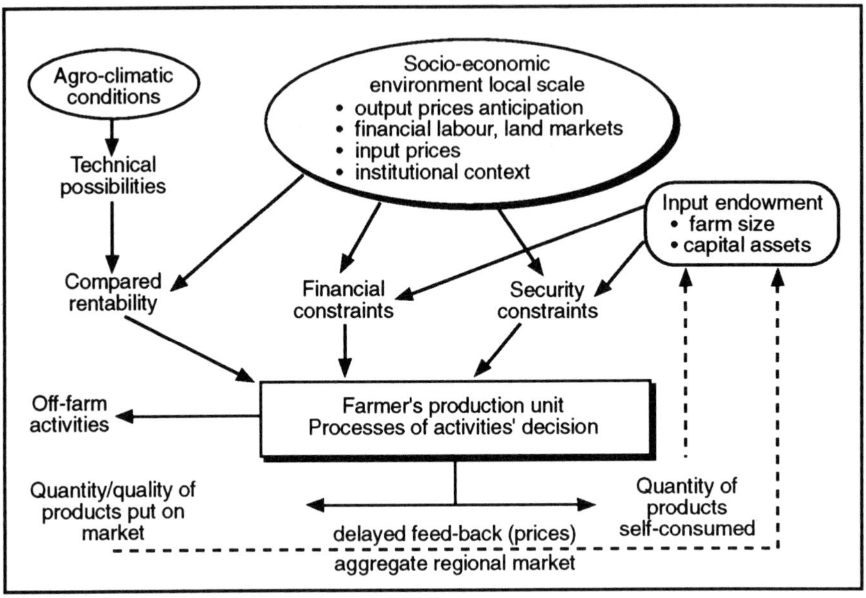


*Figure 2.* Farm behaviour model.

animal availability and feed requirements, financial availability for investment, and cash flow. The socioeconomic context is represented by degree of access to the production factors and output markets, the prices on these markets, availability of credit, relation between farms and markets (agricultural labour, land, animals and equipment markets).

The decision process leads to land allocation, livestock activity level, investment and borrowing, and labour allocation between on-farm and off-farm activities (Figure 2). Decision processes are based on expectations of prices and yields, in order to consider time lags in agricultural production, information imperfection of agricultural markets and risk. At the end of each period, real prices and yields are computed by applying a random coefficient on an average value. The production level is then calculated with 'real variables', as well as the end-of-period farm endowments. In this way, the results of each year are used as exogenous parameters for the next period. The methodology was used for a model of French agriculture by Boussard and Gérard (1992) and by Deybe et al. (1993) for a village model in Burkina Faso. Two recursive links are considered in the model: on one side, past results of the farm and current state of the financial market determine liquidity constraints, consumption and fixed input endowment; on the other, computed prices are used to correct the next period's profitability and risk expectations. As mentioned by Gérard (1991), different possibilities for formulating price expectations have been proposed in the literature.

To obtain the regional impact of a given policy measure, data at farm level are multiplied by the scale parameters defined by the typology. Regional production and employment impact are thus quantified.

## Soybean farming systems in Java lowlands

Multifactor analysis was used to make an agroecological zoning of Indonesia. The results define three agroecological zones for food crops: (i) irrigated lowlands, with more or less water; (ii) rice-based rainfed areas, non-irrigated, usually in more sloping areas; and (iii) dryland areas, which are mainly uplands, where corn and cassava often replace rice as the main food crop.

Also, three climatic periods, corresponding to three cropping seasons, can be identified, mainly based on rainfall, with the temperature being quite steady all year long: a wet season, a first dry and a second dry season.

As the farming systems are very diverisifed in Indonesia and as the method applied here requires a detailed analysis, we chose to start the applications in a limited area of Indonesia, Java lowlands. The reasons for this choice are multiple. First, Java represents 60% of the total population, but only 7% of the area. Moreover, as the density of population on Java has already reached 814 inhabitants per square kilometre, policy impact on farmers' income and, consequently, on rural migration is an important concern for policy makers. Second, Java produces about 60% of the food crops in Indonesia, especially the lowlands with 90% of the rice and 60% of the soybean production of Java. Third, in Java lowlands the macroeconomic environment for farmers is homogeneous.

### *Farming systems in Java lowlands*

Water availability is a main constraint to cropping patterns, influencing not only the number of cropping seasons possible during one year, but also the type of crops and the level of yields. Thus, we can further differentiate irrigated areas based on water availability and the level of water management as either high, intermediate or low.

In Indonesia, we have to consider mainly technically irrigated areas, in which water supply is the result of some infrastructure development such as dams, and simply irrigated areas in which water comes mainly from springs and its control is less efficient. Two main types of cropping systems exist in wetlands: rice monocropping, and combined rice-secondary crops (soybean, corn, groundnuts, etc.).

The irrigated 'sawah' (rice fields where the level of water can be controlled), with high level of water control produce the highest yields in Java. Mainly concentrated in the rich volcanic soils of lowland Java (Heytens 1991), the characteristic cropping pattern consists of two crops of rice, often followed during the second dry season by a secondary crop among which soybean is a main choice for farmers as well as corn, or vegetables if marketing is possible. A third non-rice crop seems to be far more common than a third rice crop, due to labour and pest-control constraints. Farmers in irrigated areas use 'green revolution' varieties of rice, as well as high level of fertilizer. Their yields reach more than 5.5 $ton\,ha^{-1}$.

In the area of moderate water control, water availability often allows two crops of rice, but with yields lower than in sawah especially during the first dry season. A third crop is also common.

*Table 1.* Main lowland farming system characteristics (*Source:* Kawagoe et al. 1990).

| Level of water control | High | | | Moderate to low | | | | Rainfed | |
|---|---|---|---|---|---|---|---|---|---|
| Type of farming systems | Big | Less | Medium | Med1 | Med2 | Litl1 | Litl2 | Litl1 | Litl2 |
| Area controlled* (ha) | 2.4 | 0.95 | 1.05 | 1.2 | 1.2 | 0.7 | 0.7 | 0.35 | 0.35 |
| Active persons per farm | 3.2 | 3.6 | 3.2 | 2.5 | 2.5 | 2.7 | 2.7 | 2 | 3 |
| Type of land** | irt | irt | irt | irt | irt | irt–irs | irt–irs | rai | rai–dry |
| Crop area cultivated (%) | | | | | | | | | |
| Rice | 95 | 94 | 96 | 44 | 56 | 60 | 70 | 47 | 30 |
| Soybean | 2 | 1.3 | 1.5 | 36 | 8 | 36 | 6 | 3 | 2 |
| Corn | | | | 19 | 36 | 4 | 24 | 30 | 29 |
| Other | 3 | 4.7 | 2.5 | | | | | 20 | 39 |
| Hand tractor | yes | yes | yes | yes | yes | yes | yes | no | no |
| Yearly income per cap (million rupiah) | 1.5 | 0.2 | 0.34 | 1.1 | 1.2 | 0.3 | 0.35 | 0.24 | 0.28 |
| % off-farm income | 14 | 38 | 30 | 13.5 | 13.5 | 30 | 29 | 34 | 35 |
| % animal in farm wealth | 0.01 | 6.3 | 0.3 | 3.5 | 3.5 | 8.1 | 7 | 0.2 | 0.3 |

*Land 'controlled' is land 'owned' plus land 'rented in' minus land 'rented out'.
**irt = 'technically irrigated', irs = 'simple irrigated', rai = 'rainfed land' and dry = 'dryland'.

In areas of low water control, secondary crops are more developed, the level of rice yields during the two dry seasons making the other crops more competitive. During the wet season, poor drainage makes the cultivation of non-rice crops nearly impossible. In these areas, traditional rice varieties can be found. The presence of springs in fields can also assist cultivation of rice during the other seasons.

Rainfed farmers have to rely on rain to cultivate their land and wait for the monsoon rainfall to grow rice; they face more risk than farmers in irrigated areas. As the water control is low, high-yield varieties of rice are less used. The level of fertilizer is also lower and the yields seldom reach more than 4.5 ton $ha^{-1}$. Due to a lack of water during the dry season, rice cannot be grown then, and the soybean is a common crop. During the second dry season, there is usually no cultivation due to drought.

The differences in yields between high, moderate and low level of water control are not very important, but the cropping patterns are more diversified in moderate and low water control, since rice is a riskier crop, especially during the second season. A secondary crop which needs less water, is more profitable and less risky.

The soybean yield varies between 800 and 1400 kg $ha^{-1}$ (Kawagoe et al. 1990; Irawan and Lançon 1992), depending on soil conditions, water availability at the beginning of the crop and input used by farmers. Even if during the last ten years, soybean yields have been increasing in Indonesia, they are still far below the levels in the important producing countries: more than 2 ton $ha^{-1}$ in south and north America (2.5 ton $ha^{-1}$ in USA) and 1.3 ton $ha^{-1}$ in China (FAO Production Year Book 1992). This is also the case with corn yields but the amount of inputs, fertilizers, pesticides and labour used relative to the yields, are low enough to make this crop competitive with imported corn on the national market.

Even if most of the agricultural activities are still mainly done by hand, animal traction and hand tractors are now used for land preparation for rice in most irrigated areas but not in rainfed areas. Hand tractors are often rented by rich farmers or by a local firm, sometimes owned by groups of farmers. This mechanization allows farmers to plant earlier the second crop and, thus, improve potential yields of the third season crops that, if planted also earlier, get more rain.

Animals are of little importance in most of Java wetlands and make less than 10% of farmers' wealth. Off-farm income is an important part of farmers' income, between 20% and 30% of their time being spent on other than on-farm agricultural activities. These non-agricultural activities are highly diversified among farmers, depending mainly on the proximity of urban areas, and capital availability. They include regular work and some seasonal activities. Only the difficulty of finding off-farm activities, rather than the on-farm needs of labour, seems to limit the former.

In the three types of land, we found nine main types of farming systems based on production-factor availability and production attained (Table 1). Even if farmers' objectives are the same, their endowments as well as their water, land, labour and liquidity constraints, their access to markets, and their risk aversion all differ. Farmers in rainfed areas, depending on rainfall, are more averse to risk than are farmers in irrigated areas. Therefore, their cropping patterns are more diversified. Large-scale farmers in irrigated areas with a high degree of water control are highly specialized in rice growing and not very averse to risk.

The area controlled by households is between 0.3 and 2.4ha, but some farmers rent some or all of this. Farmers in rainfed areas control less land than those in irrigated areas. As mentioned before, rice is the predominant crop, especially in areas of high water control, where rice occupies more than 90% of the total cultivated area. Off-farm income is about a third of the total income for the poorest farmers (rainfed area and landless farmers).

The nine farming systems described above are not the only existing farming systems of Java wetlands, but they give the main characteristics of the production systems. It will always be possible to make a detailed analysis and test the different policies on other farming systems. Already, the selected farmers produce 95% of wetland rice, and 100% of wetland corn and soybean.

### *Soybean production constraints*

In terms of fostering soybean production, the Agency for Agricultural Research and Development (S. Hadiwigeno, Indonesia, 1991, pers. commun.) has identified three major sources of growth, namely: (i) area expansion in both irrigated sawah (soybean following rice) and dryland areas; (ii) yield improvement (including stabilizing yields over time and across sites and narrowing yield gaps); and (iii) reducing production losses and post-harvest losses. Considering technological progress so far, area expansion is expected to provide the largest contribution to soybean production growth in the near future. But in Java wetlands, where agricultural land use is already at a maximum, an expansion of soybean area would be at the expense of other crops. In relation to this, it is critically important to establish efficient systems of production and distribution of good-quality seed.

Soybean is considered riskier than other crops. It is also considered to be relatively labour and capital intensive among the secondary crops. Moreover, soybean production in the tropics is also known to be very susceptible to pest and disease infestation. This situation makes the crop not as attractive as other crops, despite relatively high prices. Previous studies on soybean production identified five major technical constraints: pests and diseases, plant nutrition, variety of seed, soil moisture, and plant population. The relative importance and seriousness of each constraint vary from place to place and from season to season, which, therefore, makes it impossible to propose simple recommendation packages (CGPRT 1992). This suggests that any attempt to evaluate packages of technology has to take into account local conditions.

It is apparent now that efforts for increasing soybean production will mainly depend on the ability to reduce production constraints and risks. Investment in research, particularly in improving yield and reducing costs, is essential. In addition, the government should make continuing efforts, particularly through extension, for improving cultivation practices.

## Past, current and future policy for increasing soybean production

### *Soybean production and price support policy*

In the early 1980s, the soybean deficit increased very rapidly. From 1983 to 1985, soybean local seed production represented less than 70% of the total seed supply compared with 85% to 90% in the previous years. At a time when imports were constrained by a tight monetary policy, this situation was a real problem. To reduce promptly a part of the deficit, the government implemented a program based on production targets assigned to various locations.

Within a long-term perspective, a soybean production support policy, largely inspired from the rice experience, was concurrently designed. In fact, Indonesia achieved rice self-sufficiency within the same period of time, after 15 years of continuous efforts. This strengthened public institutional confidence to master rural development issues, and heavily influenced the content of subsequent agricultural policies.

The soybean policy was based on two components: development and adaptation of technical packages, along with a reinforcement of extension services dedicated to soybean, and the establishment of a favourable and attractive economic environment through a price policy.

On the production support side, the programs were basically similar with those adopted for promoting rice production: BIMAS programs, designed to increase production through the use of improved seeds, fertilizers, pesticides, water management, improved cultural practices and the development of farmer cooperatives; INMAS, designed to help farmers with capital or access to it and to use modern inputs and cultivation practices; the INSUS scheme, an improvement of BIMAS scheme, which encouraged farmers with contiguous planting areas to cooperate and make joint decisions about seeds, planting times, and crop choices in addition to rice; and Upaya Khusu (UPSUS), a special effort that set the target of area expansion, and paid a direct

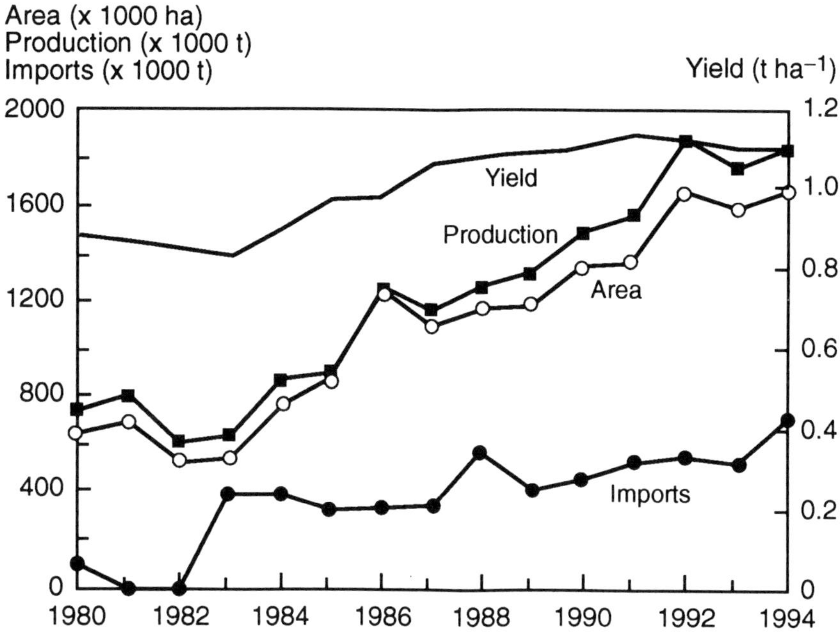


*Figure 3.* Indonesian soybean production performances and imports.

role in program implementation and supervision, particularly on improved seed distribution and credit facilitation. Between 1981 and 1991, no less than 15 soybean varieties were released by the Ministry of Agriculture, and numerous packages of technology were adapted to fit the current production environments, that is irrigated, rainfed or dry land (CGPRT 1992).

On the price support side, unlike the case of rice, BULOG (National Logistic Board) intervention in domestic soybean markets has been relatively slight and indirect. During 1980–92, two instruments were used: a floor price on soybean and an import control. The floor price was ineffective, and therefore is no longer used, since the farm-gate prices of soybean were always well above the announced floor price. Controls over imports and distribution are now the principal means by which the government influences the soybean market. BULOG is the sole importer of soybeans but, in practice, issues importing and processing contracts (for meal crushing) to the private sector. The level of authorized imports maintains the domestic price of soybean well above the import parity price. This policy is at the expense of consumers. Shifts in relative price structure between rice, soybean and, especially, maize (the major competing crop) explain to a large extent soybean's attraction for farmers: between 1980 and 1991, the average soybean price appreciated at an annual rate of 4.5% relatively to the price of rice.

These policies had a significant impact on soybean production which increased annually at a 10% growth rate from 1980 until 1991. But as shown in Figure 3, the major part of this increase was due to area expansion, whereas yield increases were only 3% of the total production increase. This increase in planted area results from various factors ranging from a shift in cropping patterns in favour of soybean during the dry season, to cultivation of fallow rice fields when water availability or pest management do not permit another rice crop, but mostly from opening new production zones outside Java.

After regular yield increases in the first years of the soybean policy implementation, soybean average yields have levelled off since the end of the 1980s. This is due to: (i) a decline in the marginal yield increase during the intensification process (i.e. once the major components of the package were adopted by farmers, the yield growth slowed down); and (ii) a reduction in farmers' capacity in further intensifying their cropping techniques, in particular in irrigated areas with high cropping intensity indexes and where rice production intensification had already mobilized a large share of production factors.

Moreover, soybean production intensification is hampered by the high risks (yield variability) attached to the adoption of new technologies under current cropping conditions. These risks are mainly related to the limited efficiency of pest control and to difficulties faced by farmers in water management (CGPRT 1992).

Considering the limited applicability at farm level of soybean high-yielding production techniques developed in agricultural research stations, the reinforcement of the current soybean policy will depend on economic and/or technical options that contribute to farmers' ability to manage such a risky crop.

In summary, the government uses a mixture of policy instruments to influence soybean production and marketing, and BULOG is the key implementer of these policy decisions. In addition, the government devotes considerable resources to the development and dissemination of improved technologies. Some policy instruments can be considered as supply shifters (e.g. fertilizer subsidies and technological innovations), while others encourage output expansion along supply functions (e.g. indirect producer price supports through import restriction).

### *Description of policies simulated*

Considering the policies implemented until now, and the government goals concerning soybean production, we ran different policy simulations as described in Table 2 for each farm type. The analysis of the results is performed first at the regional level (i.e. the global impact on Java's lowland production) and then at farm level (i.e. impact on wealth and income of the different types of farmers represented).

In order to compare the results of the different policies with a possible evolution based on actual trends, the base runs were defined with the following parameters: increasing soybean and corn price by 5% per year, increasing population at 2% per year and increasing off-farm opportunities of labour and wages. In the base-run simulations, the evolution of the farming systems is quite stable as it is in reality, allowing us to confirm our assumptions on the different farmers' behaviour (risk aversion, maximization of expected wealth utility).

*Table 2.* Policy simulation and base run description.

| *Scenario* | *Description* |
|---|---|
| 1 | Base Run: (BR) |
| | (a) Soybean and corn prices are increasing at 5% annual rate relative to other crops prices |
| | (b) Labour wages both on-farm and off-farm are increasing at 5% rate |
| | (c) Off-farm activities available are increasing at 5% |
| | (d) Annual population growth rate is 2% |
| *Differences to Base Run* | |
| 2 | Soybean Price Increase: (SPI) |
| | Soybean price increases 10% yearly for four years |
| 3 | Technological Innovation on Soybean: (TIS) |
| | Reduction in variability on soybean gross margin |
| 4 | Technological Innovation on Soybean and Soybean Price Increase: (TISSPI) |
| | Soybean price increases as scenario 2 and reduction in variability on soybean as scenario 3 |
| 5 | Technical Innovation on Rice: (TIR) |
| | New rice seeds make yield increase 50%, no changes in input uses except labour* |
| 6 | Technical Innovation on Rice and Soybean Price Increase: (TIRSPI) |
| | Rice yield increases 50% as in scenario 5, soybean price increases as in scenario 2 |
| 7 | Technological Innovation on both Rice and Soybean: (TIRTIS) |
| | Rice yield increases 50%, reduction in variability on soybean gross margin |

*Because of yield increase, the increased work for the harvest is more important (+50%).

Because of the difficulties with soybean production support in Indonesia, we first focus on price policy with a reinforcement of the existing price policies (scenario 2, SPI). Scenario 3 simulates a decrease in the risk in soybean gross margins. Scenario 5 (TIR) is defined to release the land constraint: if rice cultivation could take place on less land because of an increase in yields the land constraint would be less effective. A new variety of 'super rice' is presently being developed by IRRI, making this simulation a viable one. In scenario 7, we keep the 'super rice' of simulation 5 and add an important decrease in the expectation of rough margin deviation for soybean, i.e. decreases in yield variability and in price instability. The reason for this simulation is that if a soybean price incentive policy is unable to promote increase in production (as shown in simulations 2 and 6), the reason may be the risks associated with the activity.

### *Modelling and evolution prospects of the farming systems*

Two issues were studied concerning the evolution of these farming systems: evolution of the production side; and evolution of the farmers' wealth and agricultural profit.

*Land allocation and soybean production.* The simulation policies which showed the highest impact on soybean production in Java lowlands were those which would allow a diminution of risk on gross margin associated with this crop, as scenarios 3, 4 and 7 (Figure 4).

Such a change in the economic environment of farmers can be obtained by a reduction of variability either of yield or price. Scenario 3 (TIS) tests the impact of technology to decrease the yield variation. The increase would be around 130000

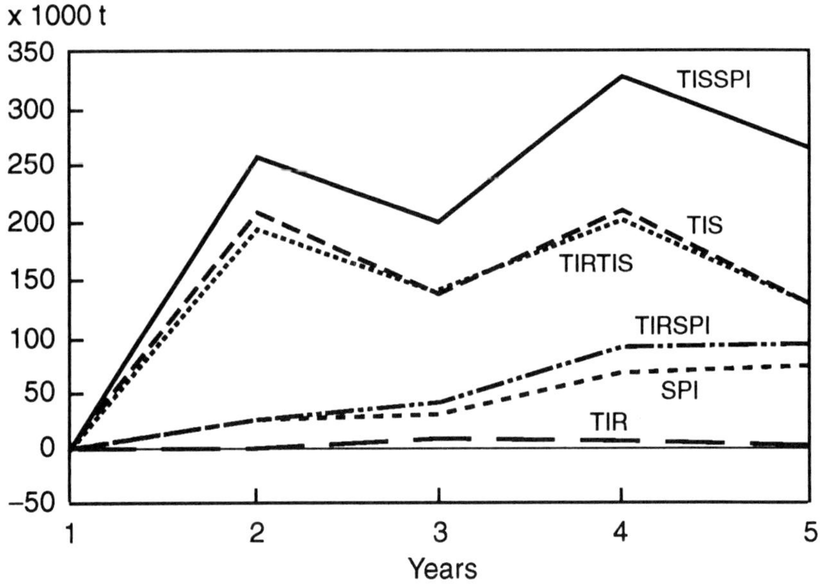


*Figure 4.* Soybean production evolution in total Java wetlands – differences to base run (see Table 2).

tons, which represents a 35% increase in regional production. The impact is even higher if a policy increasing soybean price, through a reinforcement of import restriction, is combined with the technical innovation as in scenario 4 (TISSPI). But the impact of a policy focusing only on price increases would be quite low. The difference with the base-run is only 66000 tons with a level of prices 46% higher than the actual level, which is already far above international price level (around 50%). Thus, such a policy might be costly in terms of consumer welfare. The relatively high income and price elasticities of soybean demand, combined with a very low household budget share for soybeans, suggest that there is considerable latitude in policy for the use of prices in stimulating soybean production. However, domestic prices for soybeans are far above world prices, already drawing attention of soybean exporters as the new GATT agreement has been signed. Upward movement in price as a policy direction appears limited. The impact of this price policy would be slightly higher if the land constraint were released at the same time as a technical innovation on rice as shown in scenario 6 (TIRSPI). But the three scenarios with rice improvement (TIR, TIRSPI, TIRTIS) show that at regional level of analysis, the land constraint is not the main one hindering soybean production development.

At the farming system level, the results of policies are different from one farming system to another, even if they all have the same objective.

In the higher water control, large farming system, the richest farming system represented in the study, policies which include technical improvement on rice yields are the most efficient to increase soybean production. The latter would increase by

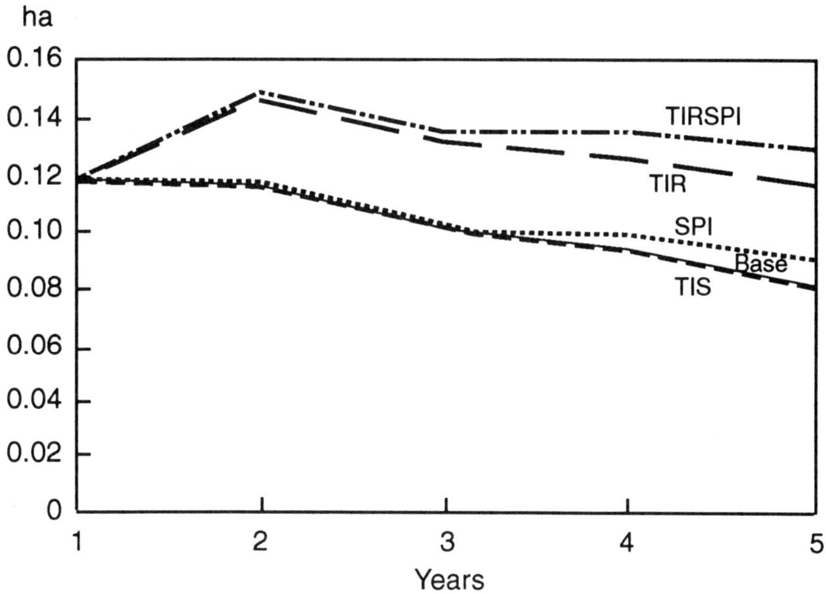


*Figure 5.* Soybean area evolution: high level of water control – large-scale farmer (see Table 2).

43% and by more than 60% if it were combined with a soybean price increase as in scenario 6 (Figure 5).

Land is the main constraint to the development of secondary crops in these large-scale farming systems. These policies would have little impact in absolute value: per farm the increase would be only 0.035 ha more land planted to soybeans. Rice is the dominant activity and the most profitable crop to be grown in these farming systems (there is little competition between rice and soybean, rice being an absolute obligation during the first season for farmers, due to both water constraints and profitability considerations). Risk is not a major constraint because the main crops in competition with soybean are vegetables, which are much more risky than soybean, but usually preferred because they are far more profitable.

The farming systems with moderate level of water control and medium size would react both to the technical innovation on soybean yield and to the soybean price increase (Figure 6). The technical innovation on soybean (scenario 3) would lead to a more than 50% increase of the area devoted to this crop. The technical innovation on rice would have impact only if combined with a risk reduction on soybean gross margin or a price increase of this crop. The increase of soybean production would be at the expense of corn production. For the price policy, the same restriction as in the regional level analysis holds: the impact is really important after four years, which means prices around 100% above those of the international market, which is not acceptable from the consumers' point of view. With a price increase of 20%, the increase in area devoted to soybean would be only 6%.

In the rainfed farming system, the poorest represented in the study, the impact of

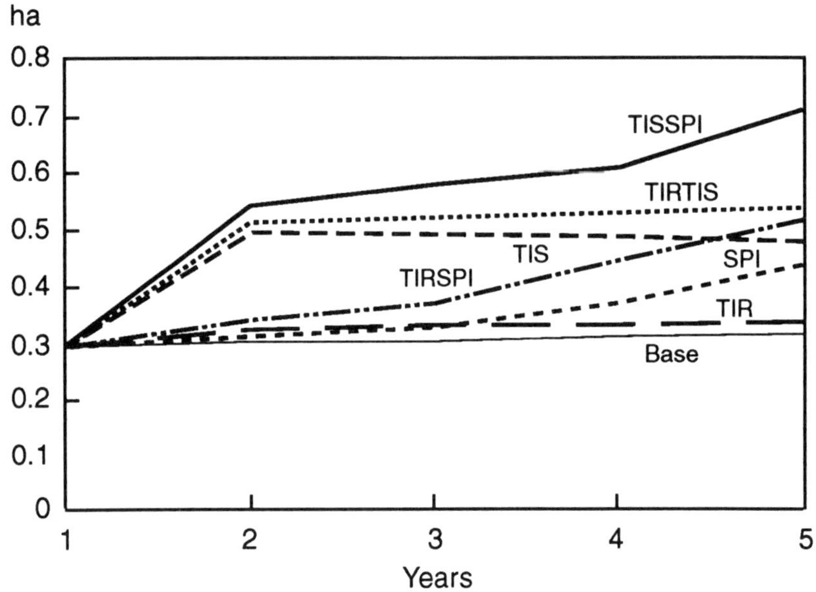


*Figure 6.* Soybean area evolution: moderate level of water control – medium-scale farmer (see Table 2).

the soybean price increase would be most important (Figure 7). There would be an increase of around 250% of the area devoted to soybean after four years of the SPI scenario which means a price increase of 46%. However, the absolute amount is still low – only 0.04 ha more per farm would be cultivated with soybean each year – because of the low proportion of soybean in the total area of the farm. The policy of diminishing risk on soybean would have low impact in absolute value: 0.01 ha more yearly would be devoted per farm to soybeans. In these farming systems, the risk on each crop is important; thus the reaction to a reduction of gross margin variability is weak. The scenario with technical improvement on rice would have no impact at all on soybean production, since soybean is cultivated in a different season than rice. In the case of a release of the land constraint, more rice would be grown.

*Wealth and agricultural profit.* The analysis of the results on wealth and agricultural profit show that there is very low impact of the policy focusing on soybean due to the small share of soybeans in total revenue. The simulations with the technical innovation on rice do show an important impact on wealth and agricultural profit for some farming systems. One interesting result is that the richest farming systems are much more able to profit from this innovation than the poorest. The high water control, large farming systems would increase their agricultural profit by 83% with this innovation, whereas the landless farmers would increase their income by only 24%. For the poorest farming systems represented in the study, in rainfed areas, the increase would be around 3%.

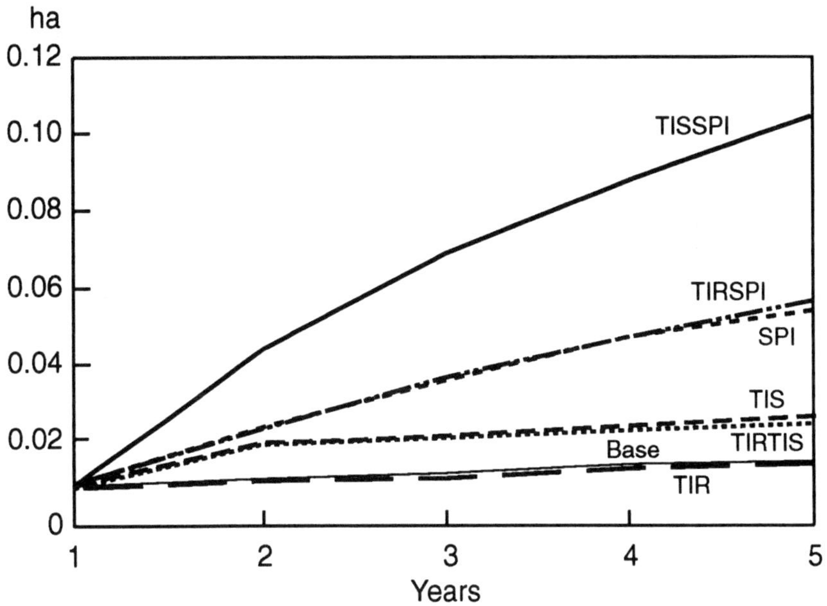


*Figure 7.* Soybean area evolution: rainfed area – small-scale farmer (see Table 2).

## Conclusion: future policy to increase soybean production in Java wetlands

Considering farmers' current situations and constraints and the soybean development program implemented by the government, a policy aiming at reducing soybean risk seems to have a much greater projected impact than price policy on Java wetland soybean production. However, at farm level, the effects of these policies are not identical:

- Large farmers located in technically irrigated plains of Java would react mainly to land constraint alleviation.
- Medium-scale farmers located in semi-technically or simply irrigated areas would respond mainly to risk reducing policy.
- Small farmers in rainfed areas would be more sensitive to price policy.

These different reactions confirm the relevance of such a micro-macro approach. Risk-reducing policies can be achieved either through yields or price stabilization. Yield variability reduction will need further research on pest management and culture practices.

Nevertheless, an analysis of Sumatra and Sulawesi farming systems, where land is not a major constraint, may show a higher response of farmers to an increase in soybean price.

Policy impacts on farmers' income also differ among farm types. A comprehensive estimation of policy effects on labour absorption will need a simulation at both farm and commodity chain level, including processing activities.

## References

Altemeier K, Bottema T (1991) Agricultural diversification in Indonesia: Price responses and linkages for the food crop sector, 1969–1988; an outlook to 2000. Working Paper No. 11. CGPRT Center, Bogor, Indonesia. 136 p.

Boussard J M, Gérard F (1992) Risk aversion and chaotic motion on agricultural markets. Working Paper URPA No. 17. Centre de Coopération International en Recherche Agronomique pour le Développement, Paris, France. 17 p.

CGPRT – Centre for Coarse Grains, Pulses, Roots and Tuber Crops (1992) Soybean yield gap analysis project (SYGAP). Final report. CGPRT Centre, Bogor, Indonesia. 120 p.

Deybe D, Ouedraogo S, Gérard F (1993) Agriculture durable: Analyse à l'aide d'un modéle dynamique et récursif. Agricultural Economic Staff Paper AE 93-10, Pullman, Washington State University, USA. 19 p.

Gérard F (1991) Instabilité des prix agricoles et influence de l'incertitude sur les comportements économiques: Essai sur les problèmes associés à la régulation de l'offre. Thèse de Doctorat, Université de Paris I Panthéon-Sorbonne, Paris, France. 567 p.

Gérard F, Boussard J M, Deybe D (1994) MATA: A multilevel analysis tool for agricultural policy. Working Paper URPA No. 23. Centre de Coopération Internationale en Recherche Agronomique pour le Développement, Paris, France. 30 p.

Heytens P (1991) Rice production systems. Pages 38–57 in Pearson S, Falcon W, Heytens P, Monke E, Naylor R (Eds.) Rice policy in Indonesia. Cornell University Press, Ithaca, NY, USA.

Irawan B, Lançon F (1992) Cultural practice differences and its impact on the efficiency of wetland soybean: The case of Japanan village, East Java. J. Agro-Ekonomi 11(2):58–78.

Kawagoe T, Fujita K, Yokoyama S, Sudana W, Zakaria A K (1991) Role of secondary crops in employment generation: A study in a rainfed lowland village in Java. CGPRT No. 23. CGPRT Centre, Bogor, Indonesia. 88 p.

Markowitz H M (1992) Portfolio selection: Efficient diversification of investments. Blackwell, New York, USA. 384 p.

Marty I (1994) Farming system typology in Japanan, East Java province, a principal component analysis approach. Working Paper. CGPRT Center, Bogor, Indonesia. 12 p.

Nanseki T, Morooka Y, Zakaria A K, Kosugi S (1992) Comparative advantage of soybean in an upland area of West-Java: Mathematical programming approach. Working Paper No. 12. CGPRT Centre, Bogor, Indonesia. 88 p.

Pangestu M (1995) Indonesia in a changing world environment: Multilateralism vs regionalism. Indonesian Q. 23(2):121–137.

Rosegrant M W, Kasryno F, Gonzales L A, Rasahan C, Saefudin Y (1987) Price and investment policies in the Indonesian food crop sector. International Food Policy Research Institute, Center for Agro-Economic Research, Bogor, Indonesia. 55 p.

SOW – Stichting Onderzoek Wereldvoedelvoorziening (1991) Long term trends of agricultural supplies in Indonesia. Research Report RR-91-01. Center for World Food Studies, Amsterdam, The Netherlands. 67 p.

SOW (1988) Agriculture in Repelita V: A review of policy issues in Indonesia through 1993. Center for World Food Studies, Amsterdam, The Netherlands. 100 p.

Thorbecke E (1991) Adjustment, growth, and income distribution in Indonesia. World Dev. 19(11): 1595–1614.

World Bank (1992) Indonesia: Agricultural transformation challenges and opportunities. Report No. 10504-IND. Washington DC, USA. 203 p.

# Simulation and systems analysis tools for crop yield forecasting

B.A.M. BOUMAN[1], C.A. VAN DIEPEN[2], P. VOSSEN[3] and T. VAN DER WAL[2]

[1]*DLO-Research Institute for Agrobiology and Soil Fertility, P.O. Box 14, 6700 AA Wageningen, The Netherlands*

[2]*DLO-Winand Staring Centre for Integrated Land, Soil and Water Research, P.O. Box 125, 6700 AC Wageningen, The Netherlands*

[3]*Joint Research Centre of the European Commission, Space Applications Institute, 21020 Ispra, Italy*

**Key words:** yield forecasting, crop growth model, GIS, remote sensing

**Abstract**

System-analytical tools such as crop growth models, GIS, and remote sensing are increasingly being explored for their use in crop monitoring and yield forecasting. The key to the optimum use of these techniques lies in their mutual combination, and in their combination with other sources of information. This paper describes the Crop Growth Monitoring System (CGMS), developed for the Joint Research Centre of the European Union for seasonal crop status monitoring and yield forecasting. The core of CGMS consists of a Geographic Information System (GIS), a set of data bases (weather, crop, soil) and a systems-analysis component: a weather module, the WOFOST crop growth simulator and a statistics module. CGMS has been operational at JRC since 1994 and the outputs are used in the preparation of its monthly crop bulletins. Options are discussed for the improvement of crop growth simulation by the use of optical, thermal-infrared and radar remote sensing. Methods are presented by which crop parameters estimated from remote sensing can be used to adjust model simulations.

## Introduction

Timely and accurate crop yield forecasts on regional to (supra-)national scales are increasingly becoming important in both developing and developed countries. Yield forecasts are defined here as the within-season estimates of what crop yield will be at the end of the growing season (as opposed to yield assessment of actually realized yields after harvest). In many developing countries, crop production forecasts have become an essential component of monitoring and warning systems for food security (Gommes 1995). In Europe, the reforms of the Common Agricultural Policy of the European Union (EU) and the conclusion of the Uruguay Round of the GATT discussions in 1994 are expected to lead to increased variation in agricultural production volumes and market prices. Therefore, both policy makers and commodity trade and transport industries need reliable and up-to-date information on expected yields and production volumes for their decision making. In 1994, the FAO and three 'agencies' of the EU, namely the Directorate-General for Agriculture, the Joint Research Centre (JRC) and the European Statistical Office (EUROSTAT), jointly organized a Seminar in Villefranche-sur-Mer, France, to address problems related to crop yield forecasting. The most widely used methods so far for operational yield forecasting are based on



*P.S. Teng* et al. *(eds.): Applications of Systems Approaches at the Farm and Regional Levels, 325–340.*

empirical, statistical or sampling techniques. The empirical approach is based on obtaining evidence from subjective sources, such as expert judgements or farmer enquiries, e.g. as in The Netherlands (Kuipers 1995). Statistical approaches use regression equations that express yield as function of various, mostly meteorological, yield determining factors, e.g. as for the whole of the EU by EUROSTAT (Palm and Dagnelie 1993). Sampling methods are based on actual measurements of yield components in farmers' fields, combined with statistical methods for processing the sampling data. Because samples are generally collected close to harvest time, they are more used for yield assessment after harvest than for actual forecasting, e.g. as in Baden Württemberg in Germany (Stadler 1995). Recently, other methods for yield forecasting are being explored that are based on new and objective techniques such as simulation and systems analysis. For instance in the EU, a ten-year project is underway at the JRC for the improvement of agricultural statistics of its member states. This project, commonly known as MARS (Monitoring Agriculture with Remote Sensing) aims at, among others, the use of deterministic crop growth models, Geographic Information Systems (GIS) and remote sensing to forecast yields of the most important crops of its 12 member states[1] (Meyer-Roux and Vossen 1994). Reasons to explore new techniques for yield forecasting are that current forecasts suffer from a lack of consistency across regions and countries, are subjective in many cases, and are not delivered on time by all member states (Heath 1990). In this paper, we discuss the Crop Growth Monitoring System (CGMS; Vossen 1992, 1995) of the MARS project that integrates crop growth modelling and GIS for yield forecasting. Next, some possibilities of improving this system by introducing information obtained from remote sensing are discussed.

## The Crop Growth Monitoring System (CGMS)

CGMS was developed for and implemented according to specifications provided by JRC of the European Commission under a contract by SC-DLO, assisted by AB-DLO and QRay-Agrimathica during 1990–94. CGMS became operational at the JRC at Ispra (Italy) in 1994 as part of the Advanced Agricultural Information System of the MARS project. The function of CGMS is to monitor the agricultural season conditions over the whole of the EU with time intervals of ten days, and to make quantitative yield forecasts at the national (NUTS-0) level. The crops addressed in CGMS are sugar beet, potato, winter wheat, spring barley, grain maize, rice, field bean, soybean, oilseed rape and sunflower.

### *CGMS components*

CGMS consists of three main components: a GIS, a set of data bases and a system-analytical part. The flow of data and operations are schematically illustrated in Figure 1.

[1]When MARS started in 1988, there were 12 countries in the EU: UK, Ireland, Denmark, Germany, Netherlands, Belgium, Luxemburg, France, Spain, Portugal, Italy and Greece. Therefore, in this paper, the EU refers to the 'old' EU-12 of before 1994.

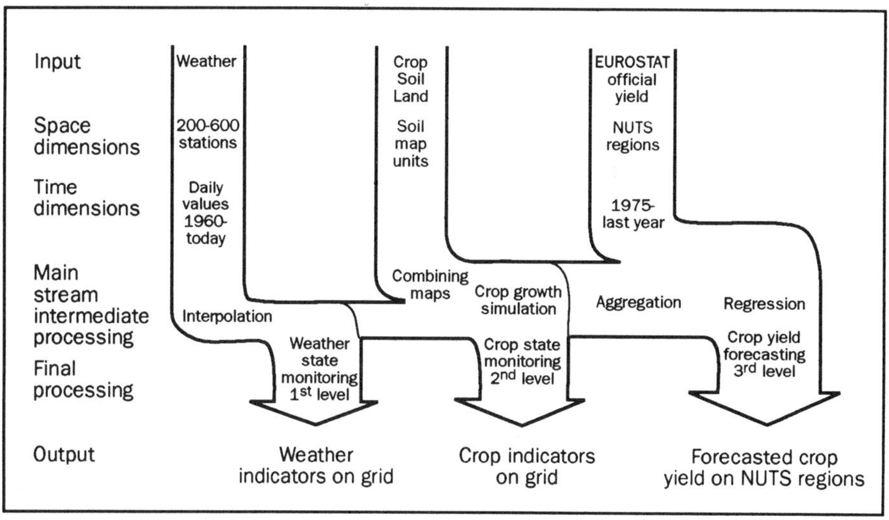


*Figure 1.* Schematic illustration of inputs, outputs, flow of data and processing steps in the Crop Growth Monitoring System (CGMS).

*GIS.* The GIS part is the ARC/INFO geographical information system, around which user-interfaces were built for the manipulation of input and output data.

*Data bases.* The data bases are built in ORACLE and contain input data needed by the system-analytical component of CGMS and generated output data. There are data bases on soil, weather, crop and yield statistics that cover the whole of the EU. The *soil data base* contains information on available water capacity and soil depth, derived from the digital soil map of the EC (FAO classification, scale 1 : 1 million). The data are needed as input by the water-balance module of the crop growth model (see below). The *weather data base* contains historic daily meteorological data over a period of 15 to 30 years for about 360 stations in the member states of the EU and adjacent countries. The weather data are directly used as indicators of progress of a current growing season, and as input for the crop growth model of the system-analytical component. The *crop data base* contains information on crop calendars and crop characteristics needed as input for the crop growth model. Crop characteristics were derived for different locations in the EU from literature and by calibration of the crop growth model on experimental data sets. The *yield statistics data base* contains historical yield statistics from 1975 onwards, drawn from EUROSTAT's REGIO data base. These yield statistics are used in the statistical module of the system-analytical part of CGMS to calculate crop yield forecasts.

*System-analytical part.* The system-analytical part of CGMS consists of three modules: a meteorological module, a crop growth module and a statistical module.

The *meteorological module* takes care of the processing of daily meteorological data that are received in real time from some 650 stations of the EU. First, it performs quality control, formatting and replacing missing values. Second, derived parameters needed as input by the crop growth model or for the assessment of 'alarm' conditions are calculated, such as solar radiation (from cloud cover or sunshine duration), vapour pressure or potential evapo-transpiration. Third, the data are interpolated to a regular grid of 50×50 km, representing spatial units of uniform weather conditions.

The *crop growth module* consists of the dynamic simulation model WOFOST (Van Diepen et al. 1989; Supit et al. 1994) in which crop growth and development are calculated from knowledge of the underlying physiological processes. In CGMS, model outputs related to crop growth are called crop indicators, e.g. biomass and Leaf Area Index (LAI). The paradigm in the use of WOFOST in CGMS is that agricultural production results from the interaction between weather, crop, soil and farming operations. Within a given region, crop and soil characteristics and farm management are relatively constant over years, and the function of WOFOST is to quantify the effect of varying weather conditions on crop growth. The working hypothesis is that real yields are influenced by weather in the same way as described by the crop model. The calculated crop indicators are related to two theoretical levels of production conditions: potential and water-limited production. Potential production represents the absolute yield ceiling for a given crop in a given year or a given site, and is determined by solar radiation, temperature and crop characteristics. This ceiling can only be reached with a high input of fertilizers, irrigation (when needed) and pest, disease and weed control. Water-limited yield represents the yield ceiling without irrigation, where crop growth may be limited by rainfall during part of the growing season. For the water-limited situation, WOFOST contains a module for the calculation of the water balance of the soil. In CGMS, WOFOST is run on a daily basis for each so-called 'simulation unit', i.e. unique combinations of weather, soil and crop (mapping) units.

In the *statistical module*, crop indicators calculated with WOFOST are related to historical yield statistics through a regression analysis in combination with a time-trend function, for at least 15 years of simulated and historical data (Vossen 1992, 1995; Vossen and Rijks 1995):

$$\text{Statistical yield} = \alpha + \beta\,(\text{trend}) + \gamma\,(\text{crop indicator}) + \text{error} \qquad (1)$$

The crop indicators used are 'total above-ground dry weight' and 'dry weight storage organs'. Alternative or additional indicators could be Leaf Area Index, crop growth development stage, water use and soil moisture. When the addition of any of the crop indicators does not improve the regression analysis, only the time trend is used and $\gamma = 0$. The regression analysis is performed at the level of NUTS-0 and NUTS-1. The resulting regression equations per crop and per region serve as the actual forecasting algorithms.

*Figure 2.* (Opposite) Example of output of CGMS: forecast winter wheat yields in the EU.

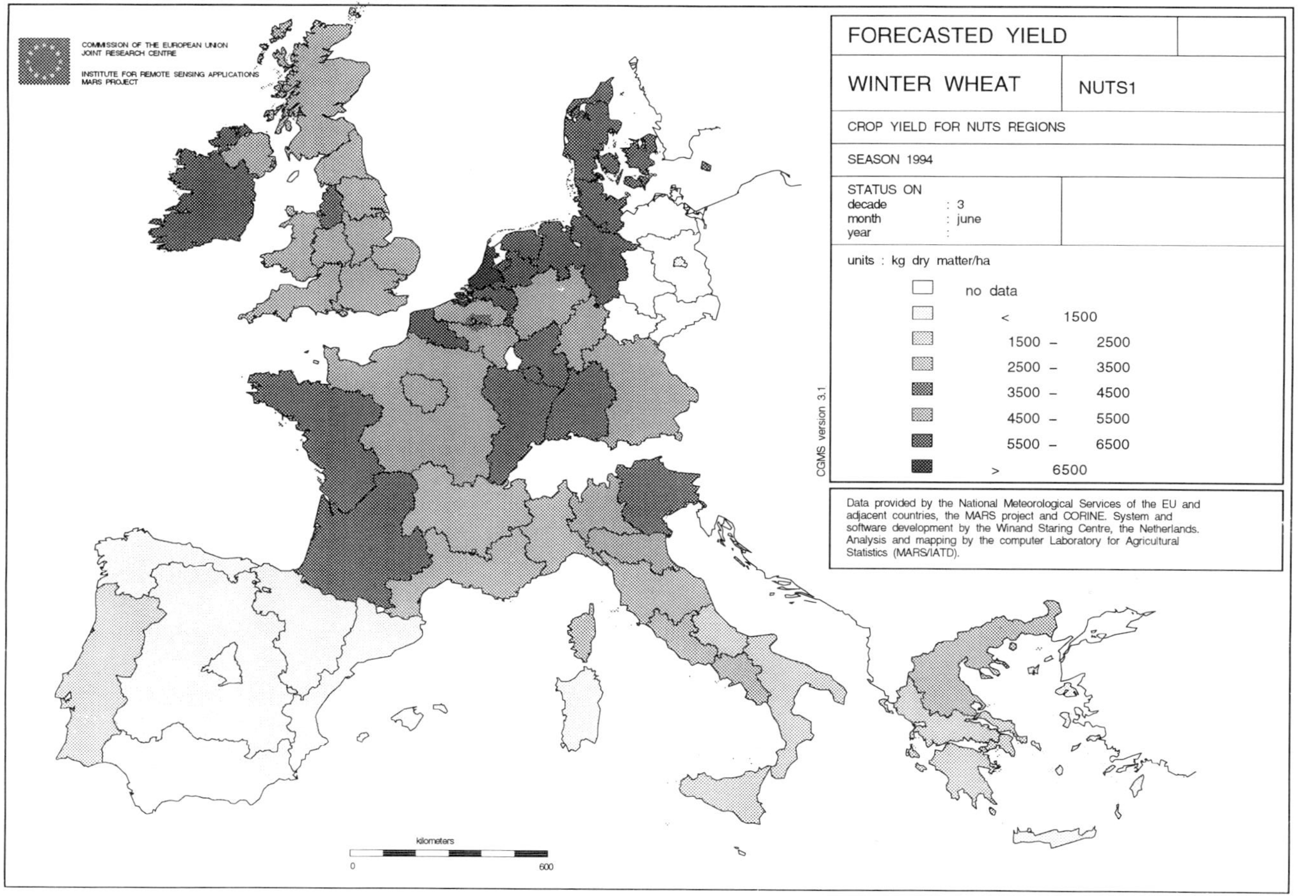

COMMISSION OF THE EUROPEAN UNION
JOINT RESEARCH CENTRE
INSTITUTE FOR REMOTE SENSING APPLICATIONS
MARS PROJECT
FORECASTED YIELD
WINTER WHEAT
NUTS1
CROP YIELD FOR NUTS REGIONS
SEASON 1994
STATUS ON
decade : 3
month : june
year :
units : kg dry matter/ha
no data
< 1500
1500 – 2500
2500 – 3500
3500 – 4500
4500 – 5500
5500 – 6500
> 6500
CGMS version 3.1
Data provided by the National Meteorological Services of the EU and adjacent countries, the MARS project and CORINE. System and software development by the Winand Staring Centre, the Netherlands. Analysis and mapping by the computer Laboratory for Agricultural Statistics (MARS/IATD).
kilometers
0
600

*CGMS output*

CGMS generates on a 10-day and monthly basis three types of output on the current cropping season:

(1) Maps of accumulated daily weather variables. These data are represented on grids of 50×50 km, and are used to follow the season's progress and detect any abnormalities, e.g. drought, frost ('alarm' indicators).
(2) Maps of agricultural season quality indicators. Simulated crop growth indicators are compared with their long-term means, and differences are evaluated in terms of more or less biomass than normal, or delayed or advanced phenological development. For this qualitative monitoring, the indicators calculated at the level of simulation units are aggregated to grids of 50×50 km.
(3) Maps and tables of yield forecasts (example for winter wheat is given in Figure 2). The yield forecasts are made for NUTS-0 (region) and NUTS-1 (country) levels using the statistical module of the systems-analysis component. These yield forecasts become available well before the official regional yield figures are established by the various national agricultural statistical offices.

Results of CGMS are used operationally in the preparation of the monthly MARS bulletins by JRC (Genovese 1994). The objectives of these bulletins are to inform readers on how the agricultural season in the EU is developing and to provide yield forecasts and acreage change estimations. The data presented in these bulletins are based on interpretations of CGMS data, satellite imagery, field surveys and expert knowledge. The main customers of the MARS bulletins are the EC's Directorate-General for Agriculture and the European Statistical Office (EUROSTAT).

*CGMS evaluated*

Crop production is a complex process which takes place on farms at field level. Crop yields vary over regions, farms, fields and years. Many different factors exert influence on the crop production process:

- Abiotic factors: e.g. soil water, soil fertility, weather.
- Farm management factors: soil tillage, planting density, sowing date, fertilizer application, crop protection measures, post-harvest losses.
- Land development factors: field size, terracing, drainage, irrigation.
- Socioeconomic factors: distance to markets, costs of inputs, prices of outputs, education level, skills, infrastructure.

Moreover, the influence of these factors may be completely overruled when the overall economic and political situation is unstable, or when crop-damaging catastrophes occur, by warfare, flooding, earthquakes, etc. The core of the CGMS system, i.e. the WOFOST crop growth model, considers mainly the abiotic factors (but not all), so that actual yields may be determined by other factors not included in CGMS. In particular, socioeconomic factors are completely ignored in the analysis of crop production conditions, while farm management factors are strongly generalized. A first evaluation of forecast yields by CGMS took place in 1993. Yields predicted by CGMS and by 'conventional' regression techniques were compared with long-term yield statistics. The first results of the statistical validation (De Koning et al. 1993; Vossen and Rijks 1995) indicated that for most regions and most crops, interannual variability of crop yield was most accounted for by the technological trend. However, model outputs significantly contributed to multiple determination coefficients in approximately more than half of the crop×country combinations, as compared to the results obtained by using the trend only. This does not necessarily mean that in all those cases the prediction errors also significantly decrease (De Koning et al. 1993). It does mean, however, that in a large majority of cases, the system contributes to better monitoring and yield forecasting of annual crops, especially on the national scale, than do time trend extrapolations alone. The actual user of CGMS, the JRC, found additional benefits of the model approach that were not included in the scientific evaluation: WOFOST predictions were timely (!), objective, quantitative and consistent over large areas. Moreover, a number of options are open to improve the quality of the simulations by WOFOST. One of these is the use of remote sensing data in the course of the growing season to derive information on the actual growth and development of crops. Such information could be used to adjust WOFOST, so that simulated yields are in better agreement with farmers' yields.

## Using remote sensing data in simulation models

Remote sensing is the measurement of electromagnetic (EM) radiation that is reflected or emitted from objects at the surface of the earth (such as field crops). Reflected EM waves may have been emitted by the sun (e.g. optical remote sensing) or by artificial sources (e.g. microwaves by radar). EM waves emitted by earth surfaces are measured in the thermal-infrared and passive microwave regions of the EM spectrum. Remote sensing measurements are mostly made with sensors on board satellites or aircraft. Ground-based instruments are used in field-level research. In this section, optical, radar and thermal-infrared remote sensing are explored for linking with crop growth models to improve growth monitoring and yield forecasting.

### *Optical remote sensing*

Optical reflectance of crops is determined by the interaction of solar radiation with the crop canopy. From the late 1960s, this interaction process has been extensively studied and canopy reflectance has been shown to be related to interesting crop growth variables such as Leaf Area Index (LAI) and fraction ground cover (e.g. Suits 1972; Bunnik 1978). For accurate estimation of these variables, so-called Vegetation Indices have been developed that are a combination of reflectance values in visible and near-infrared wavelength bands. From a comparative analysis among a number of VIs, the Weighted Difference Vegetation Index (WDVI) (Clevers 1989) was found to be a suitable VI for estimation of LAI and fraction ground cover:

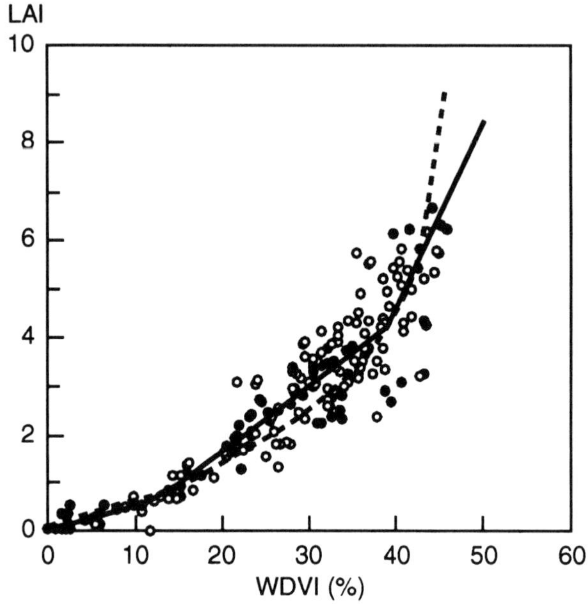


*Figure 3.* The relationship between Leaf Area Index (LAI) and Weighted Difference Vegetation Index (WDVI) for wheat. Open symbols are observations in the vegetative stage; closed symbols in the reproductive stage. The lines are regressions (exponential function (dashed) and linear segments).

$$\mathrm{WDVI} = r_{nir} - c(r_{nir}/r_{vis}), \quad \text{with } c = r_{nir,s}/r_{vis,s} \tag{2}$$

in which $r$ is the reflectance; the suffixes *nir* and *vis* indicate near infrared (about 820 nm) and a visible, either red (about 650 nm) or green (about 550 nm), wavelength band respectively; and the suffix $s$ indicates reflectance from bare soil. The constant $c$ is calculated from bare soil reflectance measurements before crop emergence. Standard relations between WDVI and fraction ground cover and LAI were derived for wheat, barley, oats, sugar beet and potato (Bouman et al. 1992). Figure 3 gives an example of the relationship between WDVI and LAI for wheat. Using these standard relations, the estimation accuracy of LAI was the same as that by using conventional techniques, i.e. 10–15% relative accuracy for LAI up to LAI values of 4.

To use the estimated values of LAI in crop growth simulation models, two methods have been developed: the 'forcing function' method and the 'simulation steering' method.

*LAI as a forcing function.* In the forcing function method, a switch can be used in the crop growth model that allows the use of external LAI values in various process calculations (e.g. assimilation, evapo-transpiration) over the use of simulated LAI values by the model itself. Such switch has been implemented in the crop growth model SUCROS (Simple and Universal CROp growth Simulator; Spitters et al. 1989; Figure 4), which is quite similar to WOFOST as used in CGMS. In a number of case

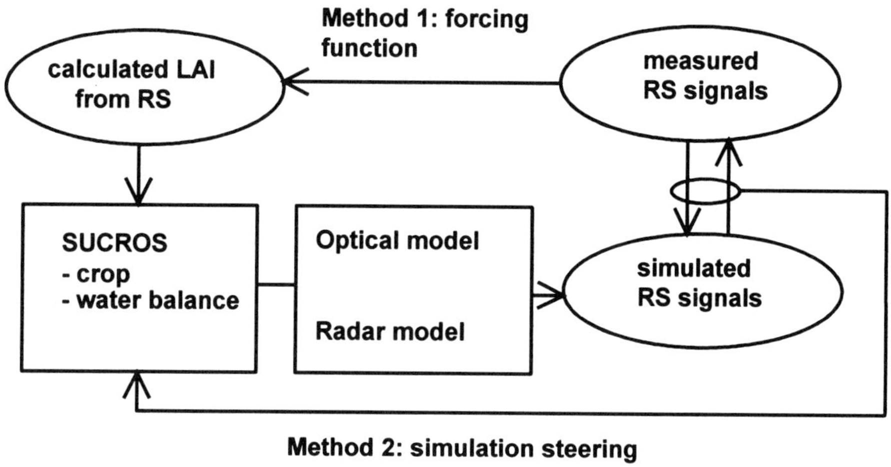


*Figure 4.* Schematic illustration of two methods of integrating crop growth simulation models with remote sensing information (see text). Boxes indicate models (SUCROS, remote sensing); circles indicate auxiliary variables; and arrows indicate flow of information.

studies, this use of LAI estimated from WDVI as 'forcing function' increased the accuracy of biomass prediction. For instance in Figure 5, results are shown for a field of sugar beet in Flevoland, The Netherlands. For winter wheat in the same area, the seasonal-average error between simulated and actual canopy biomass decreased from 1.7 ton $ha^{-1}$ using only SUCROS, to 1.3 ton $ha^{-1}$ using the 'forcing function' method for (based on seven fields in 1987 and 1988).

Two conditions must be fulfilled for the successful use of estimated LAI as a forcing function. First, a sufficiently large number of remote sensing observations needs to be available with a regular frequency (e.g. weekly observations). Daily estimated values of LAI can then be interpolated between observations without making too much error. Second, the starting-point of crop growth in time (sowing or emergence date) should be reasonably well known. Otherwise, the externally forced LAI values may not be 'in phase' with the simulated phenological development of the crop.

*Simulation steering.* When the number of remote sensing observations is limited and/or initial conditions are not well known, remote sensing observations can be used to 'steer' crop growth models by adapting the values of model parameters, so that simulated time-series of LAI match estimated time-series of LAI (Bouman 1992). SUCROS was extended with remote sensing models for optical reflectance and for radar backscatter (see next section for radar backscatter; Figure 4). Simulated remote sensing signals over time can be compared with observed time-series of remote sensing signals. The values of selected model parameters can then be 'calibrated', within realistic ranges, so that simulated remote sensing signals match the observed ones as closely as possible. Especially initial conditions that are not well known in operational context of yield forecasting, such as sowing date, may be 'calibrated'

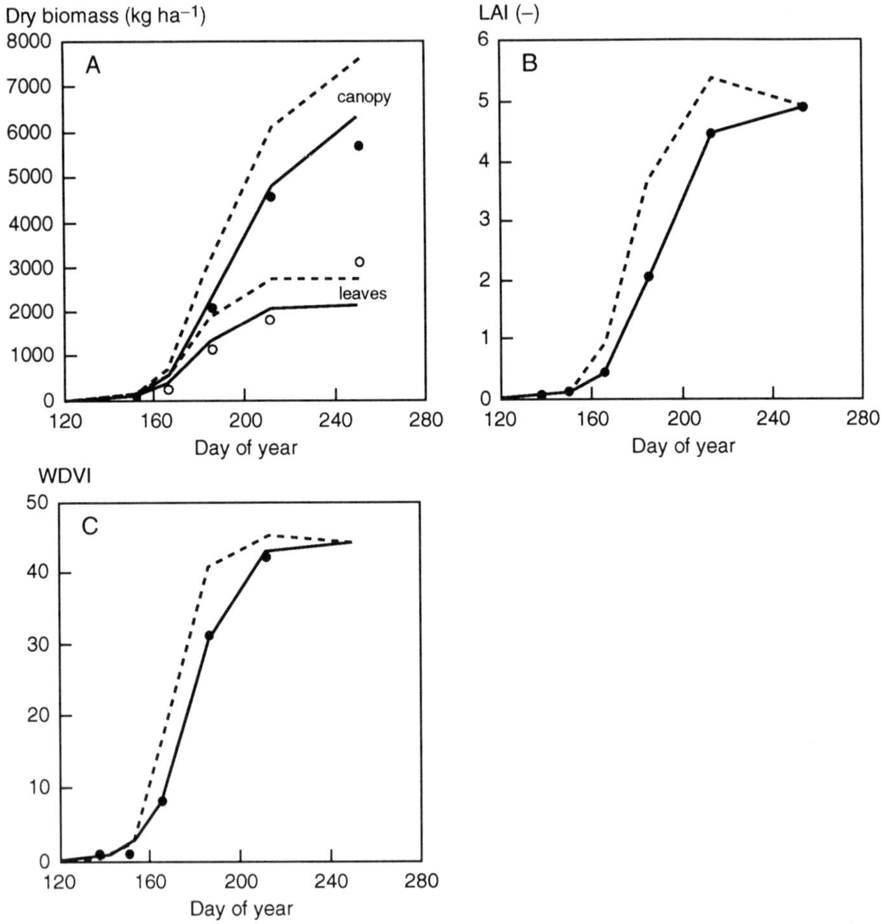


*Figure 5.* Results using estimated LAI values from optical remote sensing as a forcing function in SUCROS for sugar beet (see text): (A) simulated and observed (symbols) canopy dry biomass; (B) LAI; and (C) Weighted Difference Vegetation Index. The dotted lines are standard simulated values of SUCROS; the solid lines are simulations using estimated LAI as a forcing function.

using this technique. In CGMS, values for sowing date are taken as fixed regional-average values that are the same each year. Again using SUCROS, it was shown in a number of field-experiments that the 'simulation steering' method improved simulations of crop growth and yield predictions. For sugar beet, the seasonal-average error between simulated and actual canopy biomass decreased on the average from 0.4–0.7 ton $ha^{-1}$ using only SUCROS (but with actual sowing dates as input) to 0.2–0.5 ton $ha^{-1}$ using the 'simulation steering' method (with sowing date found by calibration). The simulations of tuber yield improved from 19% error using only SUCROS (with mean sowing date for the area as input), to 3% using the calibration procedure.

*Radar remote sensing*

Radar backscatter of crops depends on the properties of the incident microwaves, and on those of the canopy and the underlying soil. Properties of microwaves are wavelength, state of polarization and angle of incidence. Wavelengths most commonly used are the X- (≈3 cm), C- (≈5 cm) and L- (≈25 cm) bands (a relatively new band is the P-band, ≈70 cm). Microwave interaction models generally describe the scattering strength and the microwave attenuation through canopies from the dielectric constant, the volume fraction and the geometry of the various types of inclusions (e.g. Eom and Fung 1986; Ulaby et al. 1990). However, until now, such detailed models still do not describe reality sufficiently well and model simulations do not explain measurements for many crop types (Bouman and Hoekman 1993; Rijckenberg 1994). Moreover, many of the input parameters cannot be successfully related to measurable physical properties of the crop. Another class of simple physical models has been developed that treat the canopy as a collection of water droplets, e.g. the 'Cloud' model (Attema and Ulaby, 1978). The Cloud model explains radar backscatter of crops from the amount of water in the canopy and from the moisture content in the underlying top soil:

$$\gamma = C_\theta(1 - e^{-DW/\cos\theta}) + G_\theta e^{(mK-DW)/\cos\theta} \tag{3}$$

in which $\gamma$ is the radar cross section ($m^2 m^{-2}$), $W$ the crop water per unit soil surface ($kg m^{-2}$), $m$ the volumetric moisture content of top soil (%), $\theta$ the incidence angle (°), $D$ the coefficient of attenuation per unit of crop water ($m^2 kg^{-1}$), $K$ the moisture coefficient of top soil per volumetric moisture content (–), $C_\theta$ the backscattering coefficient of an optically thick vegetation cover ($m^2 m^{-2}$), and $G_\theta$ the backscattering of dry soil ($m^2 m^{-2}$).

*Biomass monitoring.* In principle, the Cloud model can be inverted to give estimates of canopy water, and derived from that, canopy biomass, which could be used as a forcing function in crop growth models. Until now, however, estimates of canopy water/biomass are generally too inaccurate to be useful for crop growth monitoring. The reason for this is that, for many crops, canopy structure has a larger effect on radar backscatter than the amount of canopy water/biomass, especially in the short wavelength bands (X- and C-band). Variation in canopy structure occurs within a season, among fields and among years because of variation in management (e.g. variety, row spacing, row direction), crop phenological development (e.g. the successive stages of tillering, shooting, flowering and ripening of cereals) and effect of weather (e.g. impact of rain, hail and wind). As a consequence of this sensitivity to canopy structure, there are often large fluctuations in temporal radar backscatter curves of individual fields, and large variation among backscatter curves of different fields that have comparable growth and development characteristics.

In an indirect approach to use radar backscatter data for crop growth monitoring, the Cloud model was added as a subroutine to the SUCROS model (Figure 4). Canopy water as input to the Cloud model is calculated from canopy biomass, and soil moisture content can either be taken from actual measurements or from simulated values by the water balance. Using the 'simulation steering' method, SUCROS can be adjusted such that simulated radar backscatter in time passes as smoothly as possible through time series of measured radar backscatter. For sugar beet, with a relatively limited variation in canopy structure over time and among fields, this approach resulted in improved accuracy of growth monitoring. However, the stability of the Cloud model parameters ($C$ and $D$ in equation (3)) in time still needs to be further studied before this method can be truly operationalized.

*Soil moisture monitoring.* The Cloud model can also be inverted to yield estimates of the moisture content in the top soil. The thickness of the top soil layer from which the moisture content can be estimated depends on the wavelength of the radar: short wavelengths (X-, C-band) penetrate the first 5–10 cm of the soil, longer wavelengths (L-, P-band) can penetrate down to about 30 or 50 cm. In the short wavelength bands, the accurate estimation of soil moisture content is as troublesome as estimating canopy water content because of the disturbing effect of the 'overlying' crop and because of the effect of soil surface structure on radar backscatter. With longer wavelengths, possibilities for soil moisture estimation appear to be better. Estimated soil moisture contents can be used as a forcing function in the soil water balance of crop growth models in the same way as external LAI values can be used in the crop growth part (see above; Figure 4). Also, the 'simulation steering' approach can be used to adjust model simulations such that predicted soil moisture contents in time (or predicted radar backscatter values) match estimated values (or measured radar backscatter, respectively) as closely as possible.

### *Thermal-infrared remote sensing*

Objects emit radiance related to their surface temperature. The maximum wavelength at which this energy is emitted depends also on the temperature (Wien's displacement law). For crops and bare soil, the frequency of emitted radiation lies in the (thermal) infrared part of the EM spectrum. Remote sensing observations in this part of the spectrum, therefore, provide measures of surface temperature. Since surface temperature is a state variable resulting from incoming and outgoing energy fluxes, it can be used to estimate surface energy fluxes. The partitioning of net available energy from incoming short- and long-wave radiation is governed largely by available soil moisture. On wet surfaces, available energy is consumed by evapo-transpiration of water, leaving little energy for heating the surface. On dry surfaces more (or all) energy is used for heating, thus resulting in relatively high surface temperatures. Surface temperature, as observed by thermal-infrared remote sensing, can, therefore, be used to estimate surface evapo-transpiration. However, this is a complex procedure since evapo-transpiration is determined by other factors as well, such as surface roughness, vegetation cover, vapour pressure deficit, etc.

*Estimation of evapo-transpiration.* A method to determine evapo-transpiration rate through the energy balance is implemented in the Surface Energy Balance Algorithm for Land surfaces or SEBAL (Bastiaanssen 1995). SEBAL solves the surface energy balance pixel-by-pixel according to micro-meteorological theories. Use is made of three derived remote sensing products: (i) surface reflectance ($r_0$) derived from optical remote sensing; (ii) surface temperature ($T_0$) derived from thermal-infrared remote sensing; and (iii) the Normalized Difference Vegetation Index (NDVI), derived from optical remote sensing. The energy balance for land surfaces reads:

$$Q^* + G_0 + H + \lambda E = 0 \quad (\mathrm{W\,m^{-2}}) \tag{4}$$

where $Q^*$ is net radiation, $G_0$ soil heat flux, $H$ sensible heat flux and $\lambda E$ latent heat flux, i.e. the amount of energy, $\lambda$ ($\mathrm{J\,kg^{-1}}$) required in the liquid-to-vapour transition of $E$ ($\mathrm{kg\,m^{-2}\,s^{-1}}$). To derive (instantaneous) evapo-transpiration rate, the surface energy balance equation can be rewritten to:

$$\lambda E = \underbrace{-(1-r_0)K\!\downarrow - \varepsilon'\sigma T_a^4 + \varepsilon_0\sigma T_0^4}_{Q^*} - \underbrace{\rho_a c_p/(r_{ah})(T_a - T_0)}_{H} - \underbrace{\rho_s c_s/(r_{sh})(T_s - T_0)}_{G} \tag{5}$$

where $r_0$ is surface reflectance, $K\!\downarrow$ ($\mathrm{W\,m^{-2}}$) global solar radiation, $\varepsilon'\sigma T_a^4$ ($\mathrm{W\,m^{-2}}$) long-wave sky emittance with $\varepsilon'$ being the apparent emissivity of the atmosphere, $\sigma$ the Stefan Boltzmann constant and $T_a$ (K) screen height air temperature, $\varepsilon_0\sigma T_0^4$ ($\mathrm{W\,m^{-2}}$) long-wave surface emittance with $T_0$ (K) the surface temperature, $\rho_a$ ($\mathrm{kg\,m^{-3}}$) air density, $c_p$ ($\mathrm{J\,kg^{-1}\,K^{-1}}$) air specific heat, $r_{ah}$ the mean aerodynamic resistance to heat transport in air, $\rho_s$ ($\mathrm{kg\,m^{-3}}$) soil density, $c_s$ ($\mathrm{J\,kg^{-1}\,K^{-1}}$) soil specific heat, $r_{sh}$ the soil resistance to heat transport.

The surface temperature is interpreted in energy balance studies as being the result of partitioning of net energy between latent heat and sensible heat. The turbulent transfer of sensible heat from a surface can be parameterized by a resistance type formulation using the difference between surface temperature $T_0$ and air temperature at screen height $T_a$:

$$H = \rho_a c_p / r_{ah}(T_0 - T_a) \quad (\mathrm{W\,m^{-2}}) \tag{6}$$

The difference between surface and air temperature is coupled linearly to surface temperature, and is obtained by inversion of the equation for sensible heat transfer and solving it for two extreme situations: one where $H \approx 0$ (wet), and one where $\lambda E \approx 0$ (dry) (Figure 6A). These extreme situations can be found in a (thermal-infrared) satellite image, and are derived from the relationship between $T_0$ and $r_0$ (Figure 6B). The fraction of energy consumed as latent heat from the total available net energy is a measure for surface evapo-transpiration. A closed canopy, with ample supply of water, can be expected to transpire optimally. Therefore, all available energy is consumed by transpiration and little is left for heating the surface (relatively low temperature). When a crop is short of water, it closes the stomata and transpiration is reduced. Now, more energy is available for sensible heat and canopy temperature increases.

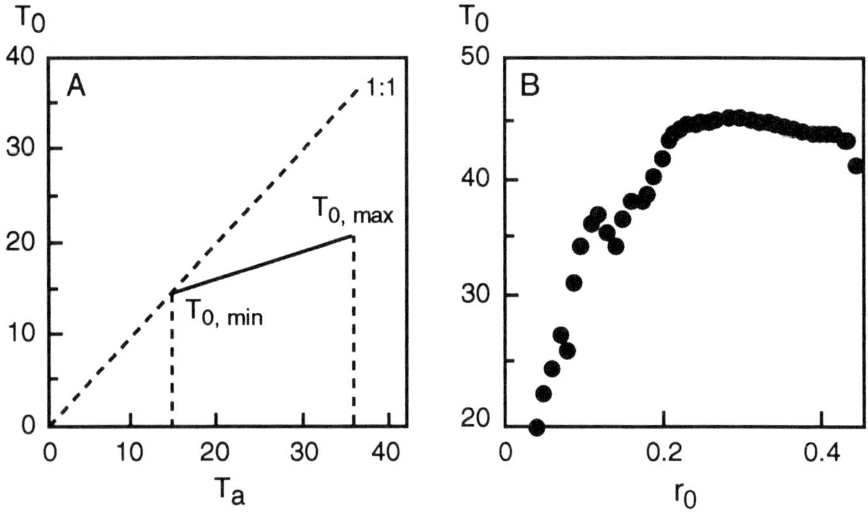


*Figure 6.* Relationship between (A) $T_0$ and $T_a$, and (B) between $T_0$ and $r_0$.

*Use of estimated evapo-transpiration in crop growth monitoring.* Although surface temperature observation through thermal-infrared remote sensing is widely used, and often very successfully, there are some problems difficult to overcome in crop water-stress monitoring. First, measured radiance is difficult to calibrate accurately, so that quantitative comparisons can well be made within images but not between images. Second, the estimation accuracy of surface energy fluxes over heterogeneous terrain still needs to be improved.

In principle, estimates of evapo-transpiration can be used as a forcing function in a crop growth model by replacing simulated values (Figure 4). However, since the estimated evapo-transpiration is a rate variable, no interpolations in time can be made and only a limited number of estimated rates can replace simulated values (with little effect on total production therefore). The magnitudes of simulated evapo-transpiration rates can be compared with estimated values as a general check on order of magnitude (absolute values). Also, regional patterns in relative evapo-transpiration rates derived from thermal-infrared images can be compared with simulated patterns (Thunnissen and Nieuwenhuis 1989).

## Conclusions and discussion

Crop growth models, GIS, and remote sensing are increasingly used in crop monitoring and yield forecasting systems. They are part of a whole set of information and analysis tools that are used to eventually predict and forecast crop yield: actual sampling and crop measurements, farmer enquiries, field-visits, expert-knowledge, weather reports and newspaper articles. The added value of these new techniques is that they are objective, quantitative and consistent over large areas. The key to the optimum use of these techniques lies in their mutual combination, and in their combination with other sources of information. Optical remote sensing data can provide estimates of light interception that can be integrated with crop growth models to adjust model simulations. A drawback for practical application of optical remote sensing from satellites is frequent cloud cover in many parts of the world. In irrigated, semi-arid areas, optical satellite remote sensing is feasible. In other areas, the use of airborne remote sensing over selected, representative test-sites could be a solution, as in the former Soviet Union (Kleschenko 1994, pers. commun.). Thermal-infrared remote sensing can provide estimates of daily evapo-transpiration rates that can be used as a check on model simulations, especially with regard to (relative) spatial patterns and regional differences. For practical applications, the same restriction applies as for optical remote sensing with regard to cloud cover. Radar remote sensing is unhindered by clouds and regular observations from the earth's surface are currently obtained from satellites (ERS, JERS). However, estimates of crop or soil properties that are useful for integrating with crop growth models have so far been successful in a limited number of cases only (e.g. for sugar beet). Future advances in microwave modelling or new techniques of radar observation, such as polarimetry and interferometry, may offer possibilities that still need to be explored.

## Acknowledgements

The Space Applications Institute of the Joint Research Centre at Ispra largely financed the development of CGMS. A large number of institutes in Europe contributed to the development of CGMS, too many to mention here. The results on remote sensing were obtained in several research projects, of which many were co-financed in the framework of the Dutch National Remote Sensing Program.

## References

Attema E P W, Ulaby F T (1978) Vegetation modelled as a water cloud. Radio Sci. 13:357–364.

Bastiaanssen W G M (1995) Regionalization of surface flux densities and moisture indicators in composite terrain: a remote sensing approach under clear skies in Mediterranean climates. PhD thesis, Wageningen Agricultural University, The Netherlands. 273 p.

Bouman B A M (1992) Linking physical remote sensing models with crop growth simulation models, applied for sugar beet. Int. J. Remote Sensing. 13:2565–2581.

Bouman B A M, Hoekman D H (1993) Multi-temporal, multi-frequency radar measurements of agricultural crops during the Agriscatt-88 campaign in The Netherlands. Int. J. Remote Sensing 14(8):1595–1614.

Bouman B A M, Van Kasteren H W J, Uenk D (1992) Standard relations to estimate ground cover and LAI of agricultural crops from reflectance measurements. Eur. J. Agron. 1:249–262.

Bunnik N J J (1978) The multispectral reflectance of shortwave radiation by agricultural crops in relation with their morphological and optical properties. PhD thesis, Agricultural University Wageningen, The Netherlands.

Clevers J G P W (1989) The application of a weighted infrared-red vegetation index for estimating leaf area index by correcting for soil moisture. Remote Sensing Environ. 29:25–37.

De Koning G H J, Jansen M J W, Boons-Prins E R, Van Diepen C A, Penning de Vries F W T (1993) Crop growth simulation and statistical validation for regional yield forecasting across the European Community. Simulation Reports CABO-TT No. 31. Centre for Agrobiological Research, Wageningen, The Netherlands. 105 p.

Eom H J, Fung A K (1986) Scattering from a random layer embedded with dielectric needles. Remote Sensing Environ. 19:139–149.

Genovese G (1994) Progetto MARS: La valutazione degli scenari agricoli. Pages 34–38 in PROTECTA, Anno VIII, Volume 3, Sedifim, Rome, Italy.

Gommes R (1995) Crop yield forecasting for food security monitoring and warning. Pages 37–42 in Proceedings of the Seminar on Yield Forecasting, 24–27 October 1995, Villefranche-sur-Mer, France.

Heath D W (1990) Introductory remarks. Pages 19–23 in Tosseli F, Meyer-Roux J (Eds.) The application of remote sensing to agricultural statistics. Joint Research Centre. Commission of the European Communities, Brussels, Luxembourg.

Kuipers N (1995) Crop estimates in The Netherlands. Pages 311–312 in Proceedings of the Seminar on Yield Forecasting, 24–27 October 1995, Villefranche-sur-Mer, France.

Meyer-Roux J, Vossen P (1994) The first phase of the MARS project, 1988–1993: Overview, methods and results. Pages 33–85 in Proceedings of the Conference on the MARS project: Overview and Perspectives. Commission of the European Communities, Luxembourg.

Palm R, Dagnelie P (1993) Tendance et effets du climat dans la prevision des rendements agricoles des differents pays de la C.E. Joint Research Centre of the EC Publication EUR 15106 F of the Office for Official Publications of the EC, Ispra, Italy. 128 p.

Rijckenberg G J (1994) Physical modeling in relation to the application of ERS-1 to agricultural studies. Pages 21–52 in Nieuwenhuis G J A, Van Rooij W W L (Eds.) Application of ERS-1 SAR data in agriculture and forestry. SC-DLO Report No. 101. SC-DLO, Wageningen, The Netherlands.

Spitters C J T, Van Keulen H, Van Kraalingen D W G (1989) A simple and universal crop growth simulator: SUCROS87. Pages 147–181 in Rabbinge R, Ward S A, Van Laar H H (Eds.) Simulation and systems management in crop protection. Centre for Agricultural Publishing and Documentation (Pudoc), Wageningen, The Netherlands.

Stadler R (1995) Harvest preview and yield forecasts in Baden Württemberg. Pages 299–310 in Proceedings of the Seminar on Yield Forecasting, 24–27 October 1995, Villefranche-sur-Mer, France.

Suits G H (1972) The calculation of the directional reflectance of a vegetation canopy. Remote Sensing Environ. 2:117–125.

Supit I, Hooijer A A, Van Diepen C A (Eds.) (1994) System description of the WOFOST 6.0 crop growth simulation model implemented in CGMS. Joint Research Centre, Commission of the European Communities, Brussels, Luxembourg. 144 p.

Thunnissen H A M, Nieuwenhuis G J A (1989) An application of remote sensing and soil water balance simulation models to determine the effect of groundwater extraction on crop evapo-transpiration. Agric. Water Manage. 15:315–332.

Ulaby F T W, Sarabandi K, McDonald K, Whitt M, Dobson M C (1990) Michican microwave canopy scattering model. Int. J. Remote Sensing. 11(7):1223 – 1253.

Van Diepen C A, Wolf J, Van Keulen H, Rappoldt C (1989) WOFOST: A simulation model of crop production. Soil Use Manage. 5:16–24.

Vossen P (1992) Forecasting national crop yields of EC countries: the approach developed by the Agriculture Project. Pages 159–176 in Proceedings of the Second Conference on the Application of Remote Sensing to Agricultural Statistics; Belgirate, Italy. Publication EUR 14262 EN of the Office for Official Publications of the EC, Luxembourg.

Vossen P (1995) Early assessment of national crop yields: the approach developed by the MARS-STAT Project on behalf of the European Commission. Pages 323–347 in the Proceedings of the Seminar on Yield Forecasting, 24–27 October 1995, Villefranche-sur-Mer, France.

Vossen P, Rijks D (1995) Early crop yield assessment of the EU countries: The system implemented by the Joint Research Centre. Publication EUR 16318 of the Office for Official Publications of the EC, Luxembourg. 180 p.

# Towards a conceptual dynamic land-use model

P.K. THORNTON[1] and P.G. JONES[2]
[1]*International Fertilizer Development Center (IFDC), P.O. Box 2040, Muscle Shoals, AL 35662, USA*
[2]*International Center for Tropical Agriculture (CIAT), Cali, Colombia*

**Key words:** land use, conceptual model, Von Thünen, IDRISI, household, geographic information system, *ex ante* assessment

**Abstract**
Non-agricultural uses and demand for food are placing increasing pressures on the land resource. The practicability of land-use modelling in addressing these pressures is limited by having to deal with human agents in agroecological systems; there are many problems in attempting to model farmers' decision-making behaviour. The benefits of appropriate land-use models are obvious enough: if the major factors that affect future land-use change are known, then steps can be taken to steer it in a direction that satisfies the often competing aims of different sections of society. On the basis that even rudimentary models of land-use change could be of real value, we describe preliminary work on a conceptual model of land use in a region based on simple and, for the most part, easily verifiable rules. We indicate future work to turn the conceptual model into a useful tool for *ex ante* policy assessment.

## Introduction

Dynamic land-use models are similar to those for technology adoption, in that everyone can see that they would be enormously useful but it is not clear how they can be constructed to be realistic and testable. Such models are among the 'holy grails' of agricultural research. The main problem is having to deal with human decision making and its effect on being able to predict or model a system. In 'hard science', as a rule of thumb, if enough is known about a set of biophysical processes, then they can be modelled using mathematical equations, and the model can, with care, be used for purposes of prediction. Breakdowns in the model can be blamed on the quality of the input data or the quality of the model's specification, or it may be that the model's limitations have been exceeded somehow. But these three problems are soluble, at least in theory. If the model is deemed to be treating a problem of sufficient severity, then it may very well be worth expending research effort and resources to refine input data, to refine or respecify the model, or to extend its applicability and sensitivity. The overarching assumption in all such activity is that hard systems can be modelled.

However, when human beings are added to a biophysical system, particularly at less aggregated levels in the hierarchy, the system's predictability plummets. This may be because relatively little is known about human decision making; decision makers tend to resist being assigned preconceived pigeon holes with labels such as 'profit maximizers'. Alternatively, it may be that systems with humans in them cannot intrinsically be known. The reason, to a large extent, is beside the point. The important



*P.S. Teng* et al. *(eds.): Applications of Systems Approaches at the Farm and Regional Levels, 341–356.*

question is: what response should agricultural scientists be making? Agriculture feeds the planet but agricultural systems are, unfortunately for the modeller, full of recalcitrant human beings.

In this paper, we discuss certain aspects of land-use modelling in an attempt to illuminate areas where future effort might be expended. In the process, we briefly describe a preliminary attempt to specify a conceptual, dynamic, land-use model, based on the notion of objects, decisions, and transactions. To date, this model is unencumbered with real-world data, but we believe that it would be straightforward to verify most of the decision rules on which the model is based. We have started to test the model in the field, within the context of one of the ecoregional initiatives of the Consultative Group on International Agricultural Research (CGIAR).

## Land-use models

The impetus for the study of land-use issues today comes from a number of sources: pressures on the land, the existence of powerful information technology tools, and the emphasis on the long term and on future generations (Fresco 1994). In much of the developing world, agriculture remains a major force in development, but increasing populations put greater pressure on the land, and only limited new land can be brought into production. Increased food production will mainly have to come from agricultural intensification. Availability of water governs the amount of additional land that can be irrigated when, in many places, water is already in scarce supply.

Land-use decisions have to be made at all levels of the agricultural sector. Land-use problems tend to be highly integrative (the result of a multitude of biophysical, economic, and sociocultural factors) and are thus difficult to study. At the national level, reasonable policy objectives might include self-sufficiency in cereals or increased foreign exchange earnings from exports. Governments must know what policy instruments will bring about particular patterns of land use. What, for example, would be the likely effects of making cheap credit available for agricultural inputs? At the farm level, land-use decisions must be taken to satisfy household criteria and goals; these might include income stability and particular levels of cash cropping to meet short- or long-term consumption and investment objectives. Which enterprise mixes will best use scarce resources under uncertainty? Can basic nutritional requirements be satisfied while reducing risk, generating certain levels of cash, or maximizing returns to capital investment? At the research and development level, land-use studies may be needed to prioritize research and extension through land-suitability studies for particular crops, or to provide information and advice to policy makers, farmers, and the private sector.

The tools that may help solve some of these complex problems and inform those who need to make the decisions have proliferated remarkably in recent years. Unprecedented computer power sits on office desks; global data bases are easily accessible, and powerful models and spatial-analysis tools can be brought to bear on land-use issues. Land-use models clearly could have a vital role to play in helping inform decision makers, in unveiling the potential conflicts that may arise in studying different and varied land-use scenarios (Fresco 1994), and in making explicit the assumptions and goals of different decision makers. The determinants of agricultural-system location are not only edaphic and climatic, but also economic and sociocultural. Much effort has been expended on models of biophysical processes, many of which are starting to be used to support decisions and resolve conflict. Conversely, models that deal adequately with human behaviour and the socioeconomic factors involved are lacking and severely limited (Carter et al. 1994).

Land-use models, in the guise of agricultural-location theory, date from the 1750s. In the classical Von Thünen model of agricultural location, the crop grown on any plot of land is that yielding the highest economic rent. Distance from the market and transport costs associated with getting the produce there are the main determinants of the level of economic rent. As distance from market increases, transport costs accumulate to the point where market price equals the costs of production plus transport. Figure 1 illustrates the case for a single market with multiple crops, showing the characteristic crop rings around the central market. The linear increase in transport costs with distance from the market determines the crop frontiers. There are several assumptions in the basic model: there is an inverse relationship between location rent and transport costs; the value of land and the intensity of land use decrease with distance from the market; the location of each crop is determined according to the economic rent; variations in land rent are related exclusively to the costs of production; and farmers are rational and seek to maximize profit (Astroth 1990).

This basic model has been greatly modified and extended since it was first developed, but represents what is often termed the economic or normative approach to land-use modelling. A large literature exists on empirical evidence to test the Von Thünen and related models, but the evidence in its support is equivocal. Another school of land-use modellers sees such normative, economic models as excessively unrealistic; farmers, especially in developing economies, simply do not make land-use decisions based solely (or even largely) on the idea of maximizing profit. This school proposes instead some kind of behavioural model. The argument is that land use in a region is the direct result of many individual farmers' decisions, and that social and cultural factors drive them as much as economic and biophysical ones do. Thus if decisions can be modelled, appropriate models of land-use patterns will follow.

An example of this type of approach is Ilbery's study (1979); he sought to explain land-use patterns in northeast Oxfordshire. He first classified farm types by land use and principal components analysis. He then attempted a typology of farmer types by ranking a range of economic and sociopersonal factors and then using cluster analysis. Finally, he constructed a matrix of farm by farmer types in an effort to isolate and clearly describe the relationships involved. But, in general, the relationship between behavioural characteristics and decision making is extremely complex; finding a framework flexible enough to be useful across a wide array of socioeconomic circumstances has proved difficult. There seems to be no common, methodological framework for the behavioural approach.

*Land-use modelling and scale*

The approach used to model land use is largely dependent on the scale involved. The economic or normative approach is appropriate to global and national studies of land use. For example, in the studies of Gallagher and Green (1984), for simulating planted crop areas in the USA, and of Zuidema et al. (1994), for simulating changes in global land cover, it is difficult to see how a behavioural approach could be applied. At a regional or watershed level, however (the level of interest here), a framework is required for integrating the economic and behavioural approaches; it should be dynamic and based on sound economic principles but also take into account the local importance of social, economic, and cultural factors in determining land use. An example is the study of Astroth (1990), who used an approach that relaxed some of the more restrictive assumptions of the Von Thünen model and introduced a risk aversion strategy that involved crop mixtures at all distances from the market. For peasant farming systems in the Punjab, both economic and behavioural factors clearly affected land use; but the model required considerable data and it was difficult to apply it widely and test it.

A suitable framework has to take account of the type of information produced and the scale at which the system is being modelled. The information a dynamic land-use model provides is essentially probabilistic, in the same way that a soils map contains probabilistic information. At a particular scale, the spatial aspect of the system being studied becomes less important; at the watershed level, for instance, land use can be dealt with in terms of characteristics or classifications of the players involved, rather than having to deal with every single plot in a watershed or region. This probabilistic interpretation of land-use dynamics is really another way of describing the issue of scale. No models are scale neutral, and model data inputs have to be matched appropriately to model outputs if meaningful analyses are to be carried out.

An appropriate scale of analysis for land-use studies depends on the problem that is being addressed and practical considerations related to data availability and methodology suitability, in view of the difficulty of dealing with the human element. But the behavioural problem cannot be ducked. The general relationships between level of analysis and (i) the quantity of data required to model biophysical process at that level, and (ii) the importance of the socioeconomic-biophysical interactions, are not of the same form. For the former, data quantity decreases with increasing level of aggregation; and for the latter, we hypothesize a bell-shaped curve; this would help to explain why the farm level is so often skipped over in moving from lower to higher levels of aggregation. A suitable analogy is the behaviour of gas molecules in a sealed container. If the number of molecules is large enough, we can describe and predict the behaviour of the gas accurately though we cannot predict the motion of one particular molecule. The same may be said to apply to humans in a national or global land-use model. Biophysical variability exhibits the same effects; at the field plot level, inter-annual cereal yield coefficients of variation are often in the range of 40–50%, whereas at the global level, it is typically about 4% (Fresco 1995).

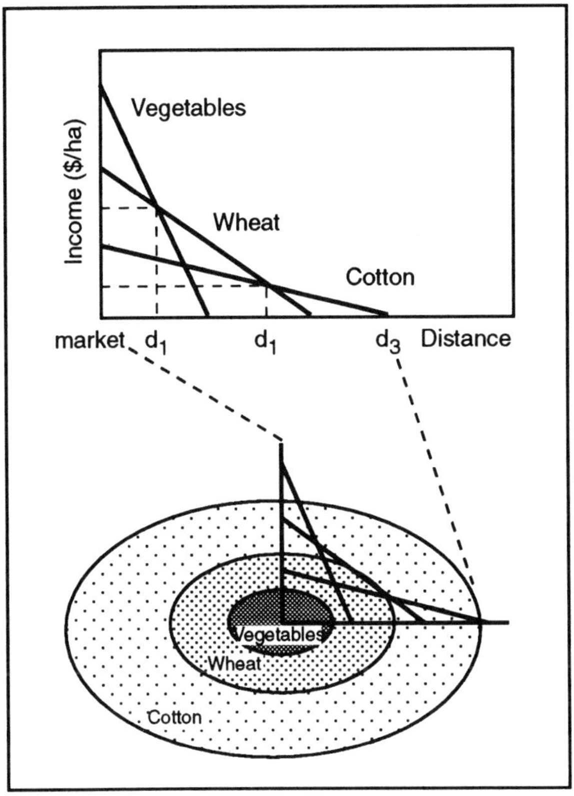


*Figure 1.* The Von Thünen three-crops-single-market model of agricultural land use (*Source:* Astroth 1990).

## A simple land-use model

We started with one of the simplest possible land-use models, the aim being first to reproduce classical agricultural location patterns such as those shown in Figure 1. This simple model is made up of four relationships:

(1) Land quality is a function of slope of the land (%) and the drainage class (1 = poor, 2 = medium, 3 = good):

$$LQ = f(\text{slope, drainage})$$

(2) Yield of crop $i$ is a function of land quality and input level (1 = low, 2 = medium, 3 = high, referring to levels of fertilizer applied, quantity of seed purchased, and amount of labour hired):

$$Y_i = f(LQ, I_i)$$

(3) Location is a function of distance of the plot of land to the nearest road

*Table 1.* Facet attributes in the simple land-use model.

| Attribute | Facet number | | | |
|---|---|---|---|---|
| | 1 | 2 | 3 | 4, etc. |
| Soil type | 1 | 1 | 1 | 1 |
| Slope (%) | 1 | 6 | 2 | 4 |
| Access distance (km) | 0.05 | 0.07 | 0.15 | 0.01 |
| Drainage class[a] | 3 | 3 | 2 | 3 |
| Distance to market (km) | 2.2 | 2.4 | 2.3 | 0.4 |
| Owner (household ID) | 0001 | 0001 | 0001 | 0002 |

[a]1 = poorly drained; 2 = moderately well drained; 3 = well drained.

*Table 2.* Crop yields (ton ha$^{-1}$) by input levels and land quality, for the simple land-use model.

| | Crop 1 | | | Crop 2 Land quality | | | Crop 3 | | |
|---|---|---|---|---|---|---|---|---|---|
| | Low | Med. | High | Low | Med. | High | Low | Med. | High |
| Low input | 0.4 | 0.7 | 1.1 | 0.6 | 0.8 | 1.3 | 0.6 | 0.8 | 1.3 |
| Med. input | 0.7 | 1.0 | 1.3 | 1.1 | 1.5 | 1.8 | 1.1 | 1.5 | 1.8 |
| High input | 1.0 | 1.4 | 1.7 | 1.3 | 1.8 | 2.2 | 1.3 | 1.8 | 2.2 |

(access or *ACC*), measured as the crow flies, and the distance along the road network to the local market (*D_M*):

$$LOC = f(ACC, D\_M)$$

(4) The gross margin per hectare of each crop is a function of yield, costs and prices ($P_i$), input level, and transport costs arising from the location of the plot:

$$GM_i = f(Y_i, P_i, I_i, LOC)$$

For this initial model, there is one decision rule: the operators of plots or facets will grow the crop that maximizes their expectations of gross margin per hectare. In the simple model, plot ownership is unimportant; in the more complicated version, each facet is assigned to a household, which may have several facets and can operate them as a managed entity. (For the simplest possible model, relationships (1) and (2) could be combined into one yield relationship that is independent of land quality, but we make more use of the idea of land quality in the model extensions described in the next section.)

We implemented this model with FORTRAN, making calls to components of IDRISI (Eastman 1993), a simple but flexible Geographic Information System (GIS) made up of a large number of independent modules that allows the user to analyse and map spatial data. The program first generates a landscape consisting of *n* plots of land; the user specifies *n*. It then constructs plots using Dirichlet tessellations around the *n* randomly selected points and randomly generates a digital terrain model (DTM). The user specifies a road network (it may be simple or complex), and the location of the local market. From this information, the program derives a number of attributes for these plots of land or facets (Table 1). The percent slope of the plot

*Table 3.* Inputs used in the simple land-use model.

| | Crop 1 | | | Crop 2 | | | Crop 3 | | |
|---|---|---|---|---|---|---|---|---|---|
| | F | S | L | F | S | L | F | S | L |
| Low input | 0 | 0 | 0 | 0 | 0 | 0 | 0 | 0 | 0 |
| Med. input | 25 | 5 | 5 | 20 | 5 | 4 | 10 | 4 | 3 |
| High input | 50 | 5 | 10 | 40 | 5 | 8 | 20 | 4 | 6 |

Note: Input used: F = fertilizer $\mathrm{kg\,ha^{-1}}$; S = seed bought $\mathrm{kg\,ha^{-1}}$; L = labour hired $\mathrm{h\,ha^{-1}}$.

is calculated from the DTM. The access distance (distance to the nearest road as the crow flies) and distance to market along the road network are calculated using the appropriate coverage and calls to various IDRISI routines.

The land use of each facet is determined by calculating the expected gross margin for each crop grown in response to each of the three levels of input (Tables 2 and 3), using costs and prices stored in a file. Transport costs are split into two components; for each facet, there is a cost involved in getting produce to the nearest road (the access cost); and then there is a road-based cost involved in transporting the produce to market along the road network (the distance-to-market cost). Gross margin for crop $i$ is calculated as:

$$GM_i = \underbrace{Y_i p_i}_{\text{output}} - \underbrace{f p_f - s p_s - l p_l}_{\text{input costs}} - \underbrace{d_a p_a - d_m p_m}_{\text{location costs}}$$

where $p_m$ is the appropriate price or cost, $f$ is quantity of fertilizer used, $s$ is quantity of seed bought, $l$ is labour hired, $d_a$ is the access distance, and $d_m$ is the distance to market. The land use that maximizes expected gross margin is assigned to that facet, and the results mapped.

To illustrate the use of the model, assume that there are three crops, 'vegetables', 'maize', and 'cotton', and in classical fashion their values decrease in this order, as do their transportation costs. We generated a landscape of 300 facets with slopes varying from 0% to about 9%. Tables 2 and 3 show yields and input levels (all data are purely hypothetical). If we set access costs to be much less than distance-to-market costs, then, with suitable adjustments to relative costs, we can generate standard Von Thünen rings around the location of the market (Figure 2A). The transport costs are such that well-defined crop production frontiers are produced; this is essentially the same as Figure 1, with land use depending principally on the distance of the plot of the local market along the road network. But if we set access costs to be much greater than those of road transport, then land use changes markedly. The high-value crop ('vegetables') is now grown in facets adjacent to the roadside. Because there is still a non-zero distance-to-market cost, the width of the swath of 'vegetables' alongside the road decreases markedly as this cost increases (in a northerly direction along the road network, Figure 2B). As the access distance to the road increases, 'maize' briefly becomes the most profitable crop, but the 'cotton' production frontier is soon reached.

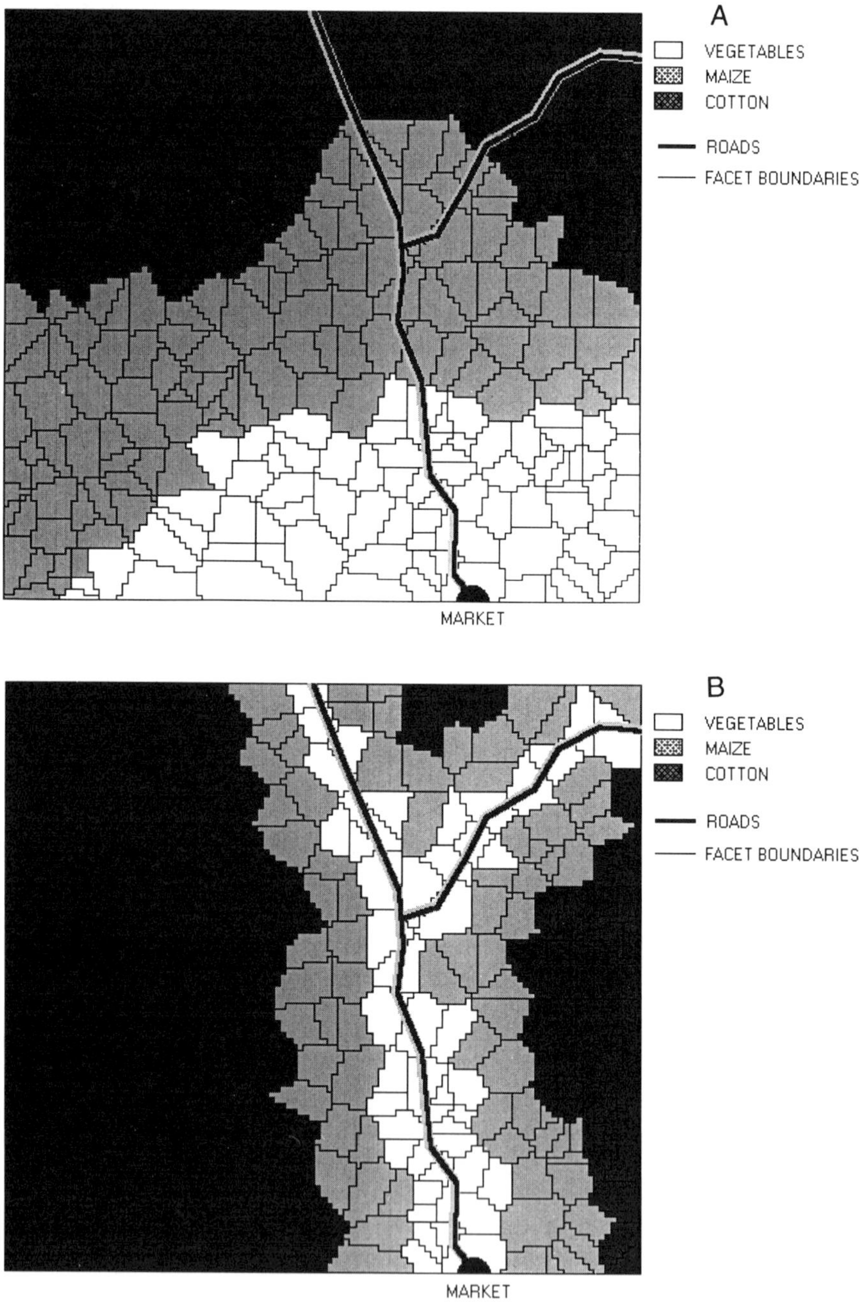


*Figure 2.* Agricultural land use in a randomly generated landscape with different transportation costs (see text for details). (A) Distance-to-market costs outweighing road network access distance costs. (B) Road network access distance costs outweighing distance-to-market costs.

None of these results is new or surprising; the model described above is simply an implementation of the classical economic model that allows the sensitivity of costs and prices to be assessed in a randomly generated landscape. In the next section, we extend the basic model, so that more factors can be taken into account, with concomitant increases in realism.

## A more complex land-use model

The simple model outlined above might be characterized as the first stage in a progression towards a more complete and dynamic representation. The model above has a one-dimensional objective function ('maximize expected gross margin per hectare'), and it is assumed that there are no interactions between facets belonging to the same household or land manager. To construct a slightly more complex model, we amalgamated land facets into managed areas and made the model dynamic rather than static, in the sense that land use in one time period is at least partially dependent on that in previous periods. The farmer's decision structure is also modified by experience. Options are limited by both assets and 'knowledge'.

### *Managed areas*

To amalgamate facets into managed areas, the model requires information on the distribution of the number of facets under the control of any household in the study region. To simulate this, an empirical probability density function (pdf) is constructed of the number of facets that the household manages (from 1 to 5, say). To generate households in the landscape, the facet number pdf is sampled randomly and an integer number of facets assigned to the household. In a landscape, facets may be contiguous or fragmented. The model takes account of the degree of fragmentation in a simple fashion as follows: if the fragmentation index is set to 1, then all facets are assigned to a household on a completely random basis, and may be spread anywhere in the landscape. If the fragmentation index is set to 0, then a household's $n$ facets are contiguous, and in fact will consist of the $n-1$ available facets that are nearest to the household or 'seed' facet. If the fragmentation index is set to 0.5, for instance, then of the total $n$ facets belonging to a household, $0.5\,n$ will be contiguous and the balance fragmented (i.e. assigned randomly). The user may change both the empirical pdf of household size and the fragmentation index in the model data input file, along with the relevant costs, prices, and input levels.

### *Dynamic yields*

To introduce dynamism into the land-use model, we implemented a different type of yield model to that used above. Instead of yields being dependent on land quality as defined by slope and drainage, we related these to soil type in the landscape. For each soil type, we defined a transition matrix giving the effects on yield of moving from one land use to another in successive time periods. The transition matrix is

*Table 4*. Transition matrices for crop yields by soil type, for the more complex land-use model.

| Soil | Input level | Bean | Maize | Fallow |
|---|---|---|---|---|
| 1 | Low | 0.90 | 0.85 | 1.10 |
| 'Best' | Medium | 1.00 | 0.93 | 1.10 |
| | High | 1.10 | 1.10 | 1.10 |
| 2 | Low | 0.85 | 0.80 | 1.15 |
| 'Better' | Medium | 1.00 | 0.90 | 1.15 |
| | High | 1.15 | 1.05 | 1.15 |
| 3 | Low | 0.80 | 0.75 | 1.20 |
| 'Worse' | Medium | 0.90 | 0.85 | 1.20 |
| | High | 1.00 | 1.00 | 1.20 |
| 4 | Low | 0.65 | 0.50 | 1.10 |
| 'Worst' | Medium | 0.75 | 0.70 | 1.10 |
| | High | 0.85 | 0.80 | 1.10 |

defined in terms of three crops and three input levels (low, medium, and high). Consider a plot of land in medium-input maize in time period $t$. To find the yield of maize in time period $t+1$, if the plot should happen to be in maize again, multiply the yield by the factor found in the transition matrix (Table 4), which is 0.93 on soil type 1 (the best) and 0.70 on soil type 4 (the least appropriate for cultivation). The transition matrix is thus an expression of a wide variety of factors related to crop production by soil type. Continuous maize in low-input conditions will gradually reduce yields over time, for example, whereas with high inputs of N, yields may increase over time because of residual effects. Similarly, the residual effects of beans on subsequent maize yields will generally be beneficial in terms of increased N availability to the plant. It would be straightforward to generate these transition yield factors using crop simulation models for the landscape, in view of soil type and representative sets of inputs.

### *Household wealth and preferences*

Because the model has to track managed areas and households with more than one plot of land, a proxy for wealth is cumulated and updated every time period:

$$W_i = W_{i-1} + \sum_{j=1}^{n} gm_j a_j$$

where $W$ is wealth in time period $i$, $gm$ is the gross margin of the $j$th plot of the household's total of $n$ plots, and $a$ is its area.

We attempted to take account of each household's preferences for particular crops in the following way. Each household has a certain predilection for growing a particular crop. This preference will change, depending on the household's perception of its performance over time. For instance, if a household grows beans on a plot, and the gross margin obtained is comparatively large, this will enhance the household's preference for growing the crop; if the gross margin is miserably small or negative, it will be reduced. In the model, these preferences are expressed as a number between

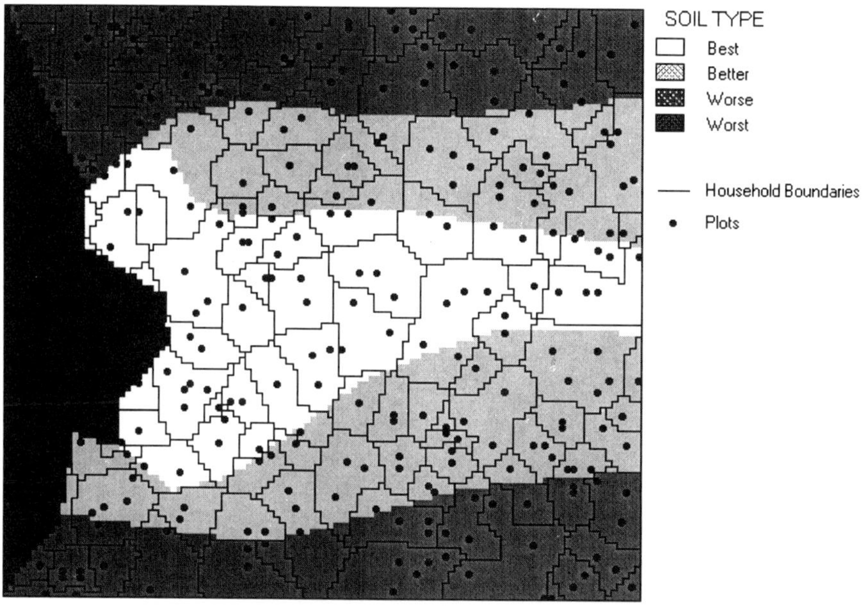


*Figure 3.* Household boundaries and land facet centroids ('plots') in a randomly generated landscape overlaid on soils classified in terms of appropriateness for rainfed agriculture.

0 and 1, and for manipulation these are transformed to normal probabilities. To choose which crop a household will grow on a particular plot in a particular time period, the current preferences are assembled into a cumulative distribution function and sampled randomly. Thus, if the preferences for bean, maize, and cotton at the start of the time step are 0.9, 0.9, and 0.2 respectively, then steps in the cumulative function occur at cumulative probabilities of 0.45 {from 0.9/(0.9+0.9+0.2)}, 0.90, and 1.0; drawing a random number of 0.713 will result in the plot being planted to maize. The input level used will depend on which level maximizes gross margin, provided that the household has enough wealth to purchase the necessary inputs. If not, then the choice is disallowed, and the entire process is repeated until the inputs for the crop of choice can be afforded for that particular plot. This model was set up for three crops, nominally beans, maize, and fallow or pasture. Beans and maize can each be grown with one of three input levels, as for the simpler model described in the previous section. For the fallow or pasture land use, there are neither inputs nor economic returns, so that the household can always 'afford' fallow.

## *Evolution of land use over time*

Despite this model's simplicity, there are many effects that could be studied, to see if it behaves in a plausible fashion. We present just two: land use after a number of iterations (years or seasons) in response to a particular set of costs and prices, and

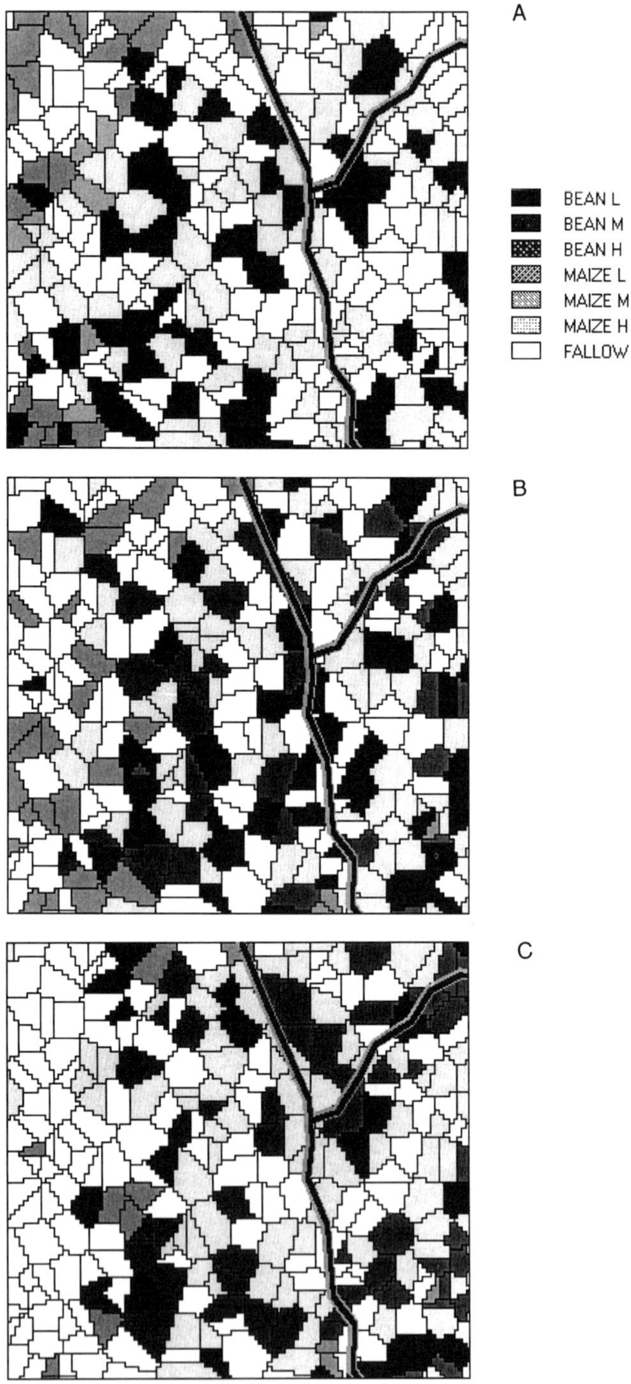


*Figure 4.* Land use in a randomly generated landscape after (A) 1, (B) 5, and (C) 10 iterations of the more complex land-use model.

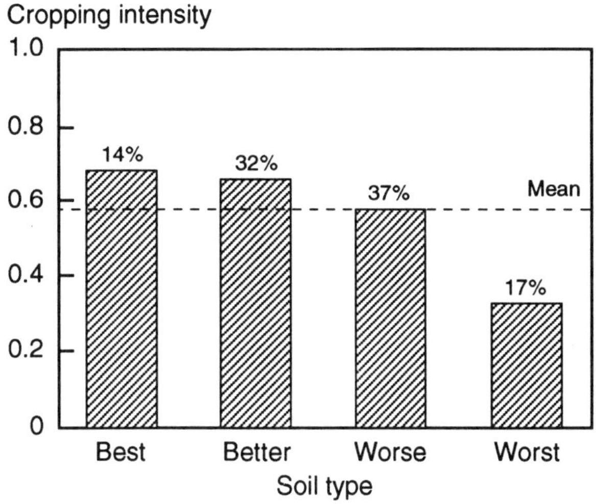


*Figure 5.* Average cropping intensity by soil type over ten iterations of the model; the relative preponderance of each soil type is shown as a percentage.

changes in the distribution of household wealth over time. Figure 3 shows a randomly generated watershed consisting of four soil types classified in terms of their suitability for annual cropping and the 300 land facets of Figure 2 amalgamated into 103 households. Crop yields for the seven land uses (maize and bean by three input levels, and fallow) at time $t=0$ were assigned based on soil type and slope; in subsequent time periods crop yields were calculated using the four transition matrices (Table 4). The wealth of each of the 103 households was initialized at 15 000 monetary units. The road network is the same as that shown in Figure 2.

Figure 4 shows land use after 1, 5, and 10 iterations or seasons of the model, using the rules outlined above. Land use changes considerably over the ten iterations. By the tenth iteration, cropping becomes intermittent on the poorer soil types (a preponderance of facets in fallow); on the better soil types, high-input maize and bean predominate. The impact of transport costs also is still discernible, especially close to the market; land use on the poor soil type around the market (located as in Figure 2) involves somewhat higher inputs compared with land use on the same soil type ('worse' in the classification of Figure 3) in the north of the region. Although this effect is not discernible directly in Figure 4, facet yields by the tenth iteration of the model vary depending on soil type, as might be expected. On the poorer soil types, cropping takes place only once every three years or so, the fallow periods allowing fertility (and yield potential) to build up (see the transition matrices in Table 4); this fertility can then be 'banked' by growing maize or beans and obtaining a good yield and hence good returns. On the better soil types by the tenth iteration, cropping was taking place about two years in three, with much shorter fallow periods. All facets were put into fallow for at least one iteration of the model over the period simulated. These rotational effects are summarized in Figure 5, which shows cropping intensity

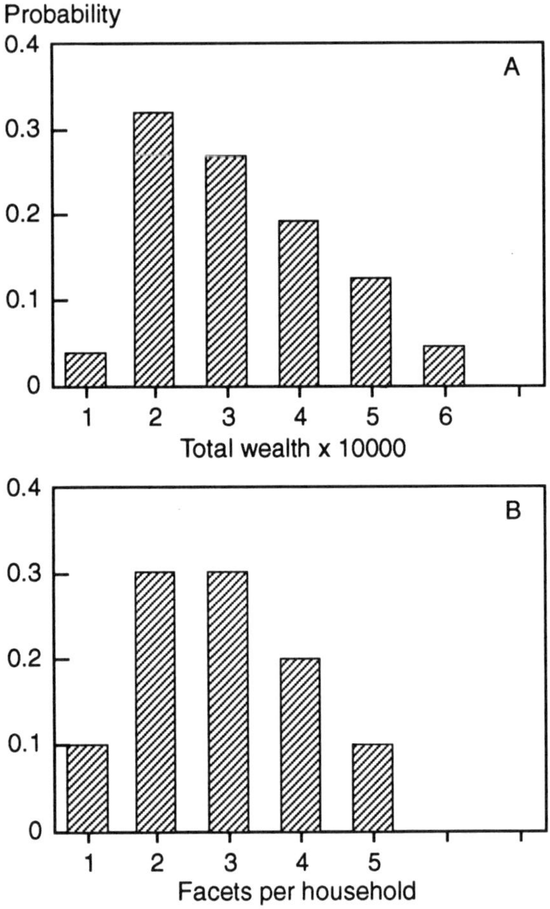


*Figure 6.* (A) Distribution of total household wealth after five iterations of the land-use model. (B) Initial (and final) distribution of facets per household.

as a function of soil type. Cropping intensity is defined here as the proportion of model iterations or seasons in which a plot is planted to maize or beans.

Figure 6A shows the distribution of wealth of the 103 households after five iterations of the model (as noted above, all households started the simulation with 15 000 monetary units). The distribution of number of facets per household (Figure 6B) is a model input, and did not change throughout the simulation (there is as yet no provision for facets to change ownership in the model). As might be expected, the distribution of household wealth is highly correlated with the number of household-owned facets, although this correlation tends to diminish through time because of the intermittent cropping and limited household income that occurs on the poorer soil types. Introducing fragmentation would presumably modify this last effect, if most households had facets on soil types of varying quality.

**Conclusions**

We constructed the model developed above with the sole aim of investigating the feasibility of assembling land-use models based on simple sets of rules, to see if land use evolves in a plausible fashion in response to factors such as changing soil type, particular sets of costs and prices of production, and distance to markets. To date, this activity has proceeded unencumbered with data from the real world; the model is thus conceptual at this stage. There is a temptation to read too much into the results of the simple runs of the previous section; for example, as noted above, crop rotations are already discernible after ten iterations, as the land is rested to allow the household subsequently to 'bank' the fertility that has built up by growing a crop of decently yielding maize. This is certainly plausible, but is it real? This type of modelling activity is similar to that of Edwards-Jones et al. (1996), in the sense that it was conceptual rather than based on real data, but sets of rules were sought that lead to the system's behaving in a plausible fashion. In itself this is not a trivial task. For the land-use model, there are a number of additions to the rules that could be implemented; functions of different householder preference that might necessitate growing maize on at least one household plot, rules to govern the buying and selling of plots in a competitive free market, and so on.

A conceptual model is useful as an aid to setting out what the modellers believe to be the important factors impinging on the system to be modelled, but it is far from a predictive tool. The next step is to consider what might be done to take such work further into the realms of reality, in the search for a useful tool for policy analysis. This requires testing the model in the field. How difficult would this be?

To test the model outlined above in the field, the four major components of the model have to be considered: (1) the landscape, for which real data can often be obtained relatively easily, such as soils maps, digital terrain models, weather data, and infrastructure and cadastral maps; (2) costs of production and produce prices, which again can often be obtained relatively easily for a region; (3) crop yields and input levels, which may be derived using a mixture of farm trial data (if they exist) and crop simulation models; initial yields and the yield transition matrices by soil type could be readily obtained in this fashion; and (4) household preference indices and objectives; this is the category of information that presents most problems in testing a land-use model. Such preferences might have to be derived from a classification of farming systems or household types, with typical preferences assigned for each class, although objective methods for their derivation would seem to be lacking.

We are currently attempting to fit the model to a real-world situation: the forest margins of Brazil. The dynamics of land use on the fringes of the Amazon forest are comparatively straightforward. We also hope to set up the model to run through time with respect to an area where there exist good records of land use over the last 40 or 50 years. This should allow the comparison of modelled land use with observed, in efforts to explain deviations from observation in terms of the relationships in the model. It is only through such comparisons that we will objectively be able to decide if this type of land-use modelling holds out real prospects of being a useful tool to answer important 'what if' questions, or is better abandoned.

## References


Astroth J H (1990) Understanding peasant agriculture. An integrated land-use model of the Punjab. University of Chicago Geography Research Paper No. 223. Chicago, IL, USA.

Carter S E, Bradley P N, Franzel S, Lynam J K (1994) Dealing with spatial variation in research for natural resource management. Paper presented at the workshop on Sustainable Mountain Agriculture, December 1993, International Development Research Center, Nairobi, Kenya.

Eastman J R (1993) IDRISI version 4.1 technical reference manual. Clark University, Worcester, MA, USA.

Edwards-Jones G, Dent J B, Morgan O, McGregor M J, Thornton P K (1996) Incorporating farm household decision making within whole farm models. In Tsuji G Y, Hoogenboom G, Thornton P K (Eds.) The international benchmark sites network for agrotechnology transfer. Kluwer Academic Publishers, Dordrecht, The Netherlands (in press).

Fresco L O (1994) Imaginable futures: a contribution to thinking about and use of planning. Pages 1–8 in Fresco L O, Stroosnijder L, Bouma J, Van Keulen H (Eds.) The future of the land: Mobilising and integrating knowledge for land use options. John Wiley and Sons, Chichester, UK.

Fresco L O (1995) Agroecological knowledge at different scales. Pages 133–141 in Bouma J, Kuyvenhoven A, Bouman B A M, Luyten J C, Zandstra H G (Eds.) Ecoregional approaches for sustainable land use and food production. Kluwer Academic Publishers, Dordrecht, The Netherlands.

Gallagher P, Green R C (1984) A crop land use model: theory and suggestions for estimating planted acreage response. Economic Research Service (ERS) Staff Report No. AGES840410. ERS, United States Department of Agriculture, Washington DC, USA.

Ilbery B W (1979) Decision making in agriculture: A case study of north-east Oxfordshire. Reg. Stud. 13:199–210.

Zuidema G, Van Den Born G J, Alcamo J, Kreileman G J J (1994) Simulating changes in global land cover as affected by economic and climatic factors. Water Air Sci. Pollut. 76:163–198.

# A systems approach to land-use planning in irrigated areas

L. MIRA DA SILVA[1], M. LEMON[2] and J. PARK[1]
[1]*Department of Agriculture, University of Reading, Earley Gate, Whiteknights, P.O. Box 236, Reading RG6 2AT, UK*
[2]*International Ecotechnology Research Centre, Cranfield University, Cranfield, Beds M43 0AL, UK*

**Key words:** integrated method, decision making, irrigation, degradation, Mediterranean, land-use planning

**Abstract**
The introduction of irrigation water into a region may often be accompanied by considerable changes in cropping and land use. In some cases, this takes the form of physical degradation of the soil and water resource or changes in social structures as farms enlarge and markets alter. Alternatively, the take-up of water from irrigation schemes can be limited by a reluctance of local farmers (for a variety of reasons) to utilize the new resource. Two European case studies are used to outline a method for obtaining and combining biophysical and socioeconomic data in a manner which is useful to decision making in land-use planning. It is argued that unless project and model design take account of local issues and the diversity of agendas in their construction, then resultant decision-making tools are unlikely to be capable of analysing the possible effects of change.

## Introduction

The relationship between agriculture and the natural environment is determined by the physical characteristics of that environment and the decisions which affect how it is managed. These decisions in themselves are influenced by a complex set of factors which do not differentiate between physical and social processes (Wolch and Dear 1989; Redclift 1994). While the insights obtained from traditional science about natural processes are of great importance, to be decision relevant they must be related to the socioeconomic and cultural environment in which they occur (Lemon and Park 1993; Butzer 1994). Not only are human and natural processes inseparable, decisions are made upon the perception of their interactions. It is crucial, therefore, that land-use planning incorporates those elements that form the basis for decision making and is not restricted to, or even based upon, those that appear appropriate to the scientist.

Attempts to impose solutions without a comprehensive understanding of the range of local issues will result in unintended consequences (Sperling 1984). Similarly, the variation in local agendas questions any attempt to anticipate farmers' responses. Issues are perceived differently, different issues are perceived to be important, and to further complicate this, perceptions may change over time. The factors that determine the receptivity of local populations to different land-use options must be understood for the effective management of those options (Seaton and Cordey-Hayes 1993). For instance, in many irrigation schemes it is assumed that if water is available



*P.S. Teng* et al. *(eds.): Applications of Systems Approaches at the Farm and Regional Levels, 357–366.*

then farmers will use it. In reality, many schemes are underused because farmers are unable (knowledge or technical constraints) or unwilling (investment cost, reluctance to change) to use the water provided. This suggests that an initial investigative phase may be required to establish the variation in local perception about farming activity and water use. Land-use planning should, therefore, incorporate a set of diagnostic devices that can identify viable future options for locally defined issues.

This paper will develop these ideas through the use of two case studies which focus on the introduction of irrigated agriculture but which have experienced very different transition paths. Both have been subject to outcomes that were not anticipated and that may have been, at least in part, due to a failure to comprehend variety at a local level. One study area, in the Argolid region of southern Greece, has experienced natural resource degradation from the intensive irrigation of citrus fruits over the last 40 years. This study will be used to describe the development of an integrated method for diagnosis which will then be considered in terms of a second study, in the Alentejo region of southern Portugal, which examines the low take-up of irrigation by farmers.

## Background to water degradation in the Argolid (Greece)

The Argolid valley is in the Peloponnese region of Greece, 150 km southwest of Athens. The area is approximately 1000 km$^2$ and consists of a central plain surrounded by foothills. It is dominated by agriculture which consumes about 90% of the groundwater used and is responsible for almost 50% of local employment, although this figure is exaggerated by high levels of multiple activity.

Irrigation has expanded considerably in the last 45 years, from 5500 ha of land using $45 \times 10^6$ m$^3$ of water per year in 1945 to 20 000 ha using $145 \times 10^6$ m$^3$ of water per year in 1990. Water is obtained from personal or collectively owned bore holes or purchased from local water authorities who manage a substantial canal infrastructure for the distribution of spring water to the central plain. There has been a corresponding change in the type of agriculture undertaken with a move away from the rainfed cultivation of cereals and olives alongside some vegetable growing and sheep/goat herding, to the current domination of irrigated citrus fruit production, in particular oranges. Intensive monocropping of citrus has emerged in the central area of the plain with the maintenance of a more diverse production on the periphery.

The intensification of agriculture, with the accompanying increase in technology use, has led to the degradation of water resources (Figure 1). This degradation is both quantitative, with bore holes in the area reaching 400 m in depth (with increased uncertainty about accessing water), and qualitative, with increased salinity due to sea water intrusion as groundwater levels drop. The central areas have suffered primarily from salinity and those on the periphery from the depletion of groundwater resources.

## Integrated method and decision issues

Agricultural systems research is becoming increasingly focused on the integration of biophysical and socioeconomic data (Stomph et al. 1994; Park and Seaton 1996).

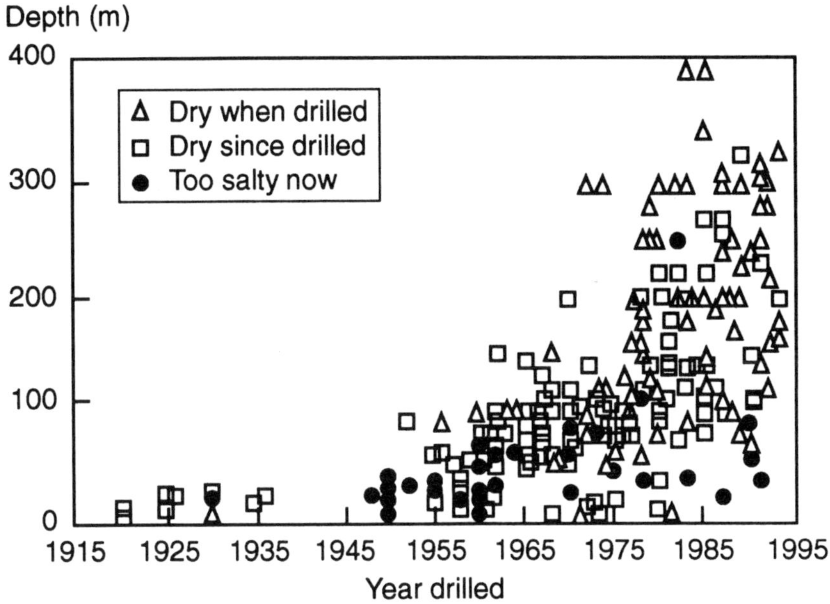


*Figure 1.* Failures (lack of water or excess of salts) in bore-holes drilled in the Argolid (Greece).

This requires the development of methods which are able to collect, interpret and synthesize biophysical, economic and social information and present it in such a way that it is policy relevant. In the following subsections, an integrated method is proposed as a set of research phases which in combination convey a framework to support irrigated land-use planning.

### *Phase 1. Collection of information about the natural environment*

Crop production inevitably interacts with the natural environment and thereby affects agroecological processes. Research which is oriented towards more sustainable land uses must identify key variables which are ecologically relevant, measurable in time and space and can be monitored to provide information to the policy-making process. This may mean that information is collected at a level of definition that would not necessarily be acceptable for more traditional scientific investigation. For instance, Lemon et al. (1994) argue that a comprehensive overview about an agroecological zone can be obtained through social inquiry in situations where more detailed information is not attainable within realistic financial and time constraints.

Information about a range of physical factors (climate, topography, hydrology, soils) provides an initial filter on the choices that are open to decision makers. Certain crops will not grow, or will not yield as much, under unfavourable conditions. However, they might be grown more productively with technological, and related capital, inputs (e.g. irrigation). This introduces the second phase which considers the link between the natural environment and the variability in farmers' decisions.

*Phase 2. Linking the natural environment to decision makers*

The integrated method in the context of this paper is not simply integration of information from a range of different disciplines about phenomena of scientific interest. Rather, it is the representation of decision issues. This distinction between what is issue driven rather than science driven inevitably raises questions about what is appropriate knowledge for the construction of formal linkages between different disciplines and decision issues. For example, in the Argolid most farmers refer to the 'problem' of water when, in quality and quantity, it becomes more difficult to obtain or detrimental to the well-being of their crops. Conversely, hydrologists may refer to a 'problem' which relates demand to the depth and salt content of aquifers. A soil specialist may perceive a 'problem' when the soil deteriorates in its ability to support vegetation and an agronomist with the loss of ability to grow a particular crop. Each interpretation of the issue, or system of interest, has its own boundary conditions and is accompanied by a set of information requirements about the bio-physical environment (water requirements and salt tolerance of different crops, water balance, and salt levels throughout the area) on the one hand, and the capacity of decision makers to act in particular ways (institutional constraints, farm size, labour availability, technical know-how, existence of markets) on the other.

*Phase 3. Linking decision makers to policy issues*

Exploring the relationship between agroecological processes and the actions of farmers is of limited value unless some insight can be gained into the responses of those farmers to a variety of policy instruments and mechanisms aimed at influencing decisions and behaviour. The earlier phases of the method have sought to establish the range of factors which constrain farming decisions and determine the spatial variation in farming activity and natural resources degradation. These constraints incorporate the importance of economic criteria but question the assumption that a change will take place simply because it can be shown to be profitable or financially viable. For example, in the Argolid the European Union (EU) subsidy for uprooting vines was successful in some areas, but in others the crop was retained for the production of wine for domestic consumption.

The third phase, based upon the data collected in the previous two phases, consists of the identification of representative conceptual farmer types, which can provide an insight into the range of possible responses to policy options. If such groupings were scattered evenly across the landscape then such a typology would be of limited interest. However, because it is possible to establish a spatial structure, the implications of policies can be explored at a local level.

*Table 1.* Relevant socioeconomic and physical dimensions in the Argolid (Greece).

| Socioeconomic dimensions | Biophysical dimensions |
|---|---|
| Reliance upon seasonal labour | Soil quality |
| Capital/labour investment | Topography |
| Labour: technology ratio on farm | Microclimate (level of frost) |
| Social acceptability of farming | Availability of water |
| Local, external markets | Quality of water |
| Level of self-sufficiency for farms | Productive level of crops |
| Variety of crops grown | Crop vulnerability |
| **Economic organization** | **Information/public administration** |
| Price support and subsidies | Level of infrastructure |
| Marketing efficiency | Effectiveness of extension service |
| Water and technology prices | Patronage-political involvement |

## Application of the method in Greece

Data to support the first two phases of the integrated method were collected from a mix of secondary sources (e.g. Greek meteorological and agricultural statistical services) and two discrete sets of interviews undertaken in Greek by agronomists who were known and trusted by the farming population. The first of these two sets consisted of semi-structured interviews with individuals involved in activities related to agricultural production in the Argolid valley (i.e. farmers, agronomists, scientists). These focused upon agricultural production and natural resources degradation, in an attempt to map the processes and agencies that were perceived to be, or to have been, influential. While these interviews focused upon production and degradation, the linkages that emerged were not restricted either by issue, time or location. As new actors were mentioned (e.g. private geologists, bore hole drillers) they could also be approached and interviewed.

The second, more structured, set of interviews was carried out with more than 200 farmers (approximately 10% of the farming population). These were designed to provide more extensive data about the qualitative and quantitative attributes that had been identified in the first set of interviews. In addition to information about the use and condition of natural resources (e.g. frequency of irrigation, level of salt in irrigation water), data were collected about the size and dispersal of farms, the use of seasonal and family labour and the distribution of crops and the markets to which they were sold (i.e. local markets, international dealers, domestic consumption). This provided a distribution across the study area that broadly supported the classification of analytical zones.

A number of relevant biophysical and socioeconomic dimensions were identified from the two sets of interviews (Table 1). The term dimension is used to represent a range, and thereby variation, within one attribute (e.g. from high status attached to farming to the work being held in low esteem). It is this representation of different attribute configurations that forms the basis for assessing the options available to farmers in different locations and the range of responses that may result from different policy interventions. Such policies could be directed at the farmer through marketing and technological support, at the crop through price guarantees and subsidies, or at water through price control and rationing. It is possible then to consider the effect of farmers' responses upon the natural resources of the area.

However, to explore the local effects of different policy interventions it is necessary to establish conceptual farmer types (phase 3). In the Argolid, this analysis identified four distinct groups, which are considered to represent the majority of farmers. The first group farm the central and coastal areas. They are predominantly monocroppers of irrigated citrus trees, with high productivity per unit; they sell through local cooperatives and the EU price support system rather than to local markets which are more time consuming. A high level of technology is employed to offset the effects of degradation (e.g. the isolation of bore holes to protect against salt intrusion) which is primarily in the form of groundwater salinity. Primary employment tends to be outside agriculture and the ability to farm in a more diverse way is constrained by limited family and personal commitment to working the land. Due to the existence of second incomes this group can withstand price fluctuations and are less affected by natural resource degradation, but they are not very flexible with regard to labour intensive cropping options.

The second group have larger farms with limited crop variety and insufficient levels of technological investment to combat the effects of degradation. Farms are located primarily in the hills to the north and northeast of the area where productivity is low due to water degradation and poor soils. Farmers are, therefore, constrained by the need to generate additional income outside agriculture. This external income (i.e. through tourism) can become essential to the household budget and thereby lead to a further decrease in the time allocated to farming. These factors restrict the ability of this group to adopt cropping options which require high levels of labour or reliable, good-quality natural resources.

The third group farm predominantly on higher ground in the western part of the valley. They have a greater diversity in the crops they grow and the markets to which they sell, and a high level of agricultural investment both in terms of personal (family) time and capital expenditure on technology. This commitment is manifest through the supportive role played by the extended family (i.e. going to market, working on the farms). There is considerable flexibility in the type of farming undertaken by this group due to the level of commitment and the ability to grow more labour intensive crops alongside others which are produced for domestic consumption.

The final farming type cannot be clearly defined geographically although it tends to be located in the central areas. This group is composed mainly of young farmers, educated in agronomy, who have returned from tertiary education with concerns about soil and groundwater quality and the need to reduce water use and inputs, and to mix crops according to their specific natural characteristics and economic potential. The group also highlight one other important factor which refers to the acquisition of knowledge from outside the area and the use of this information locally.

This typology provides a qualitative picture about the range of responses to different policy options, an insight into their spatial variation and the perceived equity attached to that distribution. However, what it does not do is to generate possible futures which may have to be considered by those involved in policy formulation. It is here that a link needs to be made between the qualitative models that are represented through the elicitation and mapping of attributes perceived to be salient to decision making and quantitative models that are capable of generating a range of possible scenarios. These scenarios might refer to how crop choice is affected by different policies, whether they are focused on the farmer, the crop, or water. The output from such an exercise is then linked to a model which provides an estimate of water requirements and water quality under different policy conditions.

In the Argolid project, the crop choice model was linked to an aquifer model (P. Allen, I. Black, M. Lemon, R. Seaton, Archaeomedes Project, Greece, 1994, unpubl. data) in order to provide information about the effects of various policy interventions on the amount of water used by farmers, its availability and its subsequent quality. The crop choice model estimates the area devoted to different crops and thereby the water requirements. At the same time, crop choice is influenced by water availability and quality. The model contains a deterministic component in which decisions are based upon profitability and a stochastic element which recognizes that other factors also contribute to the final cropping decision.

The sequence of activities described provides one method for combining biophysical and socioeconomic data in a manner which can be used in decision support. This has been achieved by incorporating biophysical and socioeconomic data in a stepwise procedure that initially explores and maps the system of interest, then provides quantitative and qualitative data about the main variables and finally uses models to provide policy relevant information. To further develop this method, and evaluate the possibility of application in other regions, the sequence of phases presented above is being adapted and improved in a new research project in Portugal.

## Application of the method in Portugal

In the Alentejo region of southern Portugal, there are several irrigation schemes in which the level of take-up has often been limited (Table 2). The reasons for these low levels of take-up are sometimes unclear, although some agencies suggest an inadequate consideration during the planning process of the complexity of physical, economic, social, and cultural systems into which irrigation schemes are to be introduced (MAI-CCRA 1982). The tradition from rainfed to irrigated agriculture is a step which in the past may have been underestimated by project decision makers.

The development of an integrated method to apply in irrigated land-use planning is particularly important in the Alentejo region at present because of the proposed Alqueva project which is designed to provide water to irrigate 110000ha. The scheme was initially discussed in the 1940s and has been subject to protracted debate over the last 50 years. If it is to be successful, decision makers must be aware of the problems relating to the potential take-up of irrigation.

The main objective of the research project in Portugal is to develop a method to integrate biophysical and socioeconomic data in a manner which can aid land-use planning. At present, one of the irrigation schemes (Vigia) is being used as a pilot

*Table 2.* Irrigated areas and utilization rates (%) for the main irrigation schemes in Alentejo (Portugal).

| Irrigation scheme | | Camp./ A. Sado | Divor | Caia | Roxo | Mira | Odivelas | Vale do Sado | Vigia | Total |
|---|---|---|---|---|---|---|---|---|---|---|
| Potential irrigation area (ha) | 1981–84 | 5963 | 488 | 7237 | 5040 | 12000 | 6381 | 6171 | | 43280 |
| | 1985–91 | 5963 | 488 | 7237 | 5040 | 12000 | 6381 | 6171 | 1505 | 44785 |
| Utilization rate (actual irrigation area/potential irrigation area | 1981 | 37 | 0 | 51 | 3 | 35 | 37 | 54 | | 37 |
| | 1982 | 37 | 26 | 53 | 21 | 40 | 22 | 85 | | 43 |
| | 1983 | 17 | 1 | 57 | 10 | 47 | 12 | 62 | | 37 |
| | 1984 | 52 | 48 | 49 | 31 | 45 | 30 | 88 | | 49 |
| | 1985 | 57 | 65 | 52 | 29 | 44 | 33 | 90 | 12 | 49 |
| | 1986 | 53 | 66 | 50 | 29 | 43 | 31 | 93 | 18 | 49 |
| | 1987 | 55 | 50 | 51 | 32 | 34 | 34 | 92 | 30 | 47 |
| | 1988 | 64 | 66 | 64 | 39 | 38 | 35 | 91 | 51 | 53 |
| | 1989 | 69 | 44 | 64 | 43 | 43 | 34 | 89 | 52 | 55 |
| | 1990 | 80 | 85 | 65 | 54 | 41 | 41 | 96 | 59 | 60 |
| | 1991 | 68 | 85 | 61 | 49 | 46 | 44 | 97 | 59 | 59 |

study to investigate the take-up of irrigation over the last decade and to qualitatively map the system. Research associated with this process has, in the first instance, involved interviews with farmers and decision makers in the region and the collection of a variety of physical data. It was here that the method developed for the Argolid was most valuable. A similar approach to the one described above was followed in the Vigia scheme and proved to be applicable and able to provide a detailed and comprehensive description of the whole system.

Ongoing research in the Portuguese study includes the development of a decision support model. The model incorporates physical (topography, soil, and climate) and socioeconomic (at the regional and farmer level) variables, together with crop growth requirements, to select crops and estimate their yields, water needs, and economic results. The differences between the model output and real data, obtained in another irrigation scheme, are expected to explain the discrepancies between maximum potential output under irrigated agriculture and actual output under current levels of irrigation.

The final step in this research will be to apply the model to the Alqueva project to aid the selection of areas to irrigate. These decisions have in the past been driven by physical factors such as soil type and topography. It is apparent from other schemes in Alentejo that even where irrigation water is supplied to an area, its final use is governed by the farmers' need for, and ability to make use of, water resources. The model will be run for a number of regional water allocation scenarios to aid in the identification of the most appropriate areas to irrigate. It will take account of both the potential take up by farmers and the physical characteristics of the suggested areas.

## Discussion and implications

Three features are central to this research:

- The need to obtain data of sufficient definition and, at the same time, capable of providing an insight into the variation within the system of study, both in time and space, which is particularly relevant to the measurement of physical phenomena where data are often of a high level of definition but very localized (in time and space).

- The need to improve our understanding about how different social structures affect and are affected by the natural system and how this can provide an insight into the bases for decision making.
- The requirement for these two issues to be represented in an interactive format which can inform about the range of possible futures that might emerge from the influence of different policy instruments.

A major part of the current project in Portugal will inevitably involve the continued exploration of methods for handling large and complicated data sets derived from a range of sources. From a decision-making point of view, there are advantages of constructing simple (or transparent) models which can be used to explore the implications of change, whether these be focused at the farmer or policy level. The increased access to powerful computers has aided the analysis of large amounts of data. However, the construction of complicated quantitative models and their presentation using sophisticated software packages can mean that weaknesses in data and model construction are difficult for the user (the decision maker) to isolate and question. Further, the decision maker may not be able to interpret the perspective from which the system has been analysed and modelled.

There is an interesting difference between the Greek and the Portuguese research projects which enables a clear distinction to be made concerning decision-support modelling. Whereas in Greece the integrated method combined information from the perspective of the sustainability of irrigated production systems, the Portuguese research is attempting to consider the system mainly from a production perspective. This suggests that appropriate decision-support modelling may need to be preceded by smaller projects aimed at selecting appropriate perspectives from which to view a system under study as it is unlikely that a given model can give an accurate representation of the system from several different perspectives. It may also mean that, in some cases, decision-support systems should be made up of suites of relatively simple tools capable of viewing the system from a number of perspectives, rather than a single complicated model.

## Acknowledgements

This work has been undertaken for the European Union DGX11 as part of the Archaeomedes study (EV5V-0021) into land degradation and desertification in the southern Mediterranean and the PRAXIS XXII program which is a research program co-financed by the European Commission and the Portuguese Government.

## References

Butzer K (1994) The Islamic traditions of agroecology: Cross cultural experience, ideas, and innovations. Ecumene 1(1):7–50.

Lemon M, Park J (1993) Elicitation of farming agendas in a complex environment. J. Rural Stud. 9: 405–410.
Lemon M, Seaton R, Park J (1994) Social enquiry and the measurement of natural phenomena: The degradation of irrigation water in the Argolid plain, Greece. Int. J. Sustain. Dev. World Ecol. 1: 206–220.
MAI-CCRA – Ministério da Administração Internaçional – Comissão de Coordenação da Região do Alentejo (1982) Aproveitamento hidroagricola da Vigia. Breves considerações. Evora, Alentejo, Portugal. 45 p.
Park J, Seaton R (1995) Integrative research and sustainable agriculture. Agric. Syst. 50:81–100.
Redclift M (1994) Reflections on the 'sustainable development' debate. Int. J. Sustain. Dev. World Ecol. 1(1):3–21.
Seaton R, Cordey-Hayes M (1993) The development and application of interactive models of industrial technology transfer. Technovation 13(1): 45–53.
Sperling D (1984) Assessment of technological choices using a pathways methodology. Trans-Res. 18a: 343–353.
Stomph T J, Fresco L O, Van Keulen H (1994) Land use system evaluation: Concepts and methodology. Agric. Syst. 44:243–255.
Wolch J, Dear M (1989) The power of geography: How territory shapes social life. Unwin Hyman, Boston, MA, USA.

# Effects of land-use changes on water and nitrogen flows at the scale of West African inland valleys: a conceptual model

N. DE RIDDER, T.J. STOMPH and L.O. FRESCO
*Agronomy Department, Wageningen Agricultural University, Haarweg 333, P.O. Box 341, 6700 AH Wageningen, The Netherlands*

**Key words:** run-off, land use, model, West Africa

**Abstract**
Land use and cover, as influenced by agricultural practices, and the changes in these with increasing pressure on land, are among the factors determining water flows in inland valleys. Changing water flows affect nitrogen flows both at the plot level and at levels higher than plots. We present a conceptual model to show the potential influence of changes in land use on water and nitrogen flows. The model first calculates the three-dimensional redistribution of water at the inland valley level using topographic characteristics as input. For that purpose, the inland valley is divided in columns. Subsequently, the annual one-dimensional water balance is computed for each column using transfer functions. These functions use soil characteristics as input variables. Finally, the annual nitrogen balance is calculated with an equation which uses the calculated infiltration as input variable and differentiates between nitrogen uptake efficiency of annual and perennial vegetation. Both this equation and the transfer functions were tested for West Africa in an earlier study. The model-runs simulating different land-use scenarios illustrate that effects of changes in land use on water and nitrogen flows are not additive, when spatial effects of water redistribution are explicitly modelled. The addition of an explicit time dimension will make the model more dynamic in order to analyse the effects of patterns of land use throughout the year on water and nitrogen flows. The model can be used to target further research on the processes underlying water and nitrogen flows in inland valleys: in particular, the scale dependency of run-off in relation to agricultural practices merits attention in field experiments.

## Introduction

Expanding world population during the last century has induced major changes in land use (Turner et al. 1993). Natural vegetation is increasingly replaced by crops and pastures (Waggoner 1994). These conversions of land covers are particularly important in West Africa (Alexandratos 1988; Becker and Diallo 1992).

Because of its agroecological diversity, West Africa is a suitable area to study the effects of land-use conversion. A characteristic feature of West African landscape are the inland valleys, which are defined as the upper reaches of river systems, inland in respect of main rivers and tributaries, in which alluvial sedimentation processes are absent or of minor importance; inland valleys have a minor floodplain and levee system (Windmeijer and Andriesse 1993; Andriesse et al. 1994). These inland valleys have a size of about one to several square kilometres and occupy between 22 and 52 million hectares of land in West Africa (Windmeijer and Andriesse 1993). Within



*P.S. Teng* et al. *(eds.): Applications of Systems Approaches at the Farm and Regional Levels, 367–381.*

inland valleys several contiguous agroecosystems can be distinguished, primarily based on hydrology: (i) the upland system (crest and slopes), which depends entirely on *in situ* rainfall; (ii) the hydromorphic system, a band at the base of the slopes, which is characterized by the presence of the groundwater table within the rooting zone; and (iii) the temporally flooded lowland system (Andriesse and Fresco 1991). These systems are not distinctly separated units, since boundaries overlap both in space and in time.

Water is the major driving force for interaction between adjacent sections of the toposequence as it moves in various ways along the gradient of the slope from the upper to the lowermost part of the landscape. Rain falling on the upland portion of the landscape may partly leave this agroecosystem as run-off water, moving downslope as surface flows; the rest infiltrates and may be used by the vegetation, may be lost as evaporation or may percolate into deeper layers down to the groundwater table. Percolated water may partly move as subsurface flow to the lowlands and join as seepage the run-off flow in a small stream to leave the inland valley.

The water flows not only contribute to differences in the water regimes between the agroecosystems, but also influence the flows of soluble nutrients, e.g. nitrates, by transporting them downslope via run-off and through leaching by percolation.

In humid West Africa annual crops were originally grown on uplands and upper slopes, but increasing pressure on land has led to a shift of cropping downslope to the lowlands. Changing land use from perennial woody vegetation to crops affects the ratios of run-off to infiltration and of evaporation to transpiration, and thereby the nitrogen flows at plot level as well as at higher levels, e.g. entire slopes and inland valleys.

For land-use management purposes, it would be worthwhile to quantify these effects at higher levels, given the profound influences these changes may have on productivity (Jensen 1992). However, experiments at the level of entire inland valleys would be difficult and expensive, if not impossible. The variation in geomorphology, soil types and land use of inland valleys is enormous and the complexity of the systems is high. Therefore, we have chosen a modelling approach.

Simulation models can be divided into verifiable, explanatory models of recurring systems and speculative or exploratory models of unique systems (De Wit 1993). Models of inland valley behaviour are by their very nature closer to the latter type. Nevertheless, the exploratory nature of such models can be reduced by composing them out of verifiable 'building-blocks' relating to subsystems. Through analysis of system behaviour under varying inputs and assumptions, exploratory models can be used to obtain indications of relations between model input and model output. These same data provide an indication of the priority of the assumptions at which future research should be directed in order to improve understanding of the complex system.

The model presented here is exploratory with the objectives: (i) to conceptualize quantitatively our present understanding of the most important relations between land use and water and nitrogen flows above plot level; (ii) to study quantitatively the effect of changes in land use on changes in the flows given our present under-

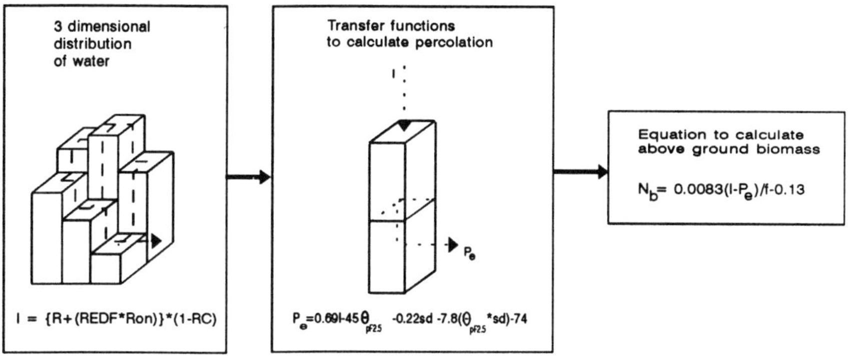


*Figure 1.* The three main characteristics of the conceptual model.

standing; and (iii) to target further research in the field of water and nitrogen flows along slopes as a function of agricultural land use.

## Material and methods

The conceptual model has three main characteristics (Figure 1):

- an approach which describes the three-dimensional redistribution of water at the inland valley level;
- transfer functions to calculate the one-dimensional water balance; and
- an equation for the nitrogen balance.

The model is fully described in De Ridder et al. (1996), but a brief description is given here.

### *Three-dimensional water distribution*

The system is spatially contained by the water divides of an inland valley. An inland valley is divided into *n* columns: columns are of equal length and width of which the magnitude depends on the scale and can be adjusted accordingly. The height of each column is determined by the average altitude of its surface area, derived from topographical data using contour lines. Long-term averages for climatic (rainfall) and hydrological data (groundwater tables) are used as inputs. It is assumed that each year after the rainy season the groundwater table returns to the same lowest average level. The bottom boundary of the system is defined using maximum rooting depth, soil depth and groundwater table as criteria. If the groundwater table is situated deeper in the profile than the maximum rooting depth and no impermeable layer exists creating shallower soils than the maximum rooting depth, maximum rooting depth is used to define the system boundary. If not, groundwater table or depth of a non-permeable layer is used to define the system boundary.

All columns are characterized by texture (percentage of sand). A weighted average for the entire profile is calculated using data from representative soil profile descriptions. Furthermore, an average run-off coefficient per year per column is estimated, based on the assumption that the run-off coefficient for an average rainfall year (average in total amount, distribution and shower intensities) can be related to soil physical parameters (e.g. Withers et al. 1978 in Tauer and Humborg 1992; Stroosnijder and Koné 1982).

Average rainfall per year is applied to each column. The model first sorts the columns from high to low surface altitude. For each column, the average water balance per year is calculated starting with the highest column ($n=1$). The water loss by run-off ($R_{off}$ in mm) from a column is calculated by multiplying the rainfall ($R$ in mm) with the run-off coefficient ($RC$). The remaining water infiltrates ($I$ in mm):

$$I = R \times (1 - RC) \quad (1)$$

Depending on the storage capacity of the soil column and evapo-transpiration, percolation is calculated using a transfer function. The total percolation of the watershed is calculated by adding percolation of all columns.

The run-off water of a column is distributed to adjacent columns with a lower altitude, proportionally to the differences in altitude of the column at issue and these adjacent columns. In this way, slope and its direction are mimicked.

Whenever in previous computing steps run-off water ($R_{on}$ in mm) was passed to the $n$th column as run-on water, it was considered additional water input to the water balance of this column. However, not all run-on water will have the opportunity to infiltrate and, thus, cannot be simply regarded as rainfall. Depending on the slope and the roughness of the surface, the rate of the surface flow will change. At high rates less of the surface flow water will infiltrate than at low rates. Furthermore, the surface flow can concentrate into small channels. The magnitude of concentration in small channels will depend on micro relief and vegetation. This spatial process of infiltration and run-on water is mimicked by the introduction of a reduction factor ($REDF$), which can vary between 0 (all run-on water to a column is considered to leave the column as run-off) and 1 (all run-on water is considered potentially to infiltrate and is added to the rainfall). Thus, infiltration in the $n$th column is:

$$I = \{R + (REDF \times R_{on})\} \times (1 - RC) \quad (2)$$

The run-on water of the $n$th column that does not infiltrate is passed to the adjacent columns and is, again, treated as run-on water for these columns. The run-off of the lowest column, the last column in the computing procedure, is the run-off of the entire inland valley.

### *Transfer functions*

The transfer function used to calculate percolation losses was developed by De Ridder and Van Keulen (1995), using a deterministic model called SAHEL (Van Keulen 1975; Penning de Vries and Van Laar 1982; Van Keulen and Wolf 1986; Penning de Vries et al. 1989; De Ridder et al. 1991). The SAHEL model calculates plant growth

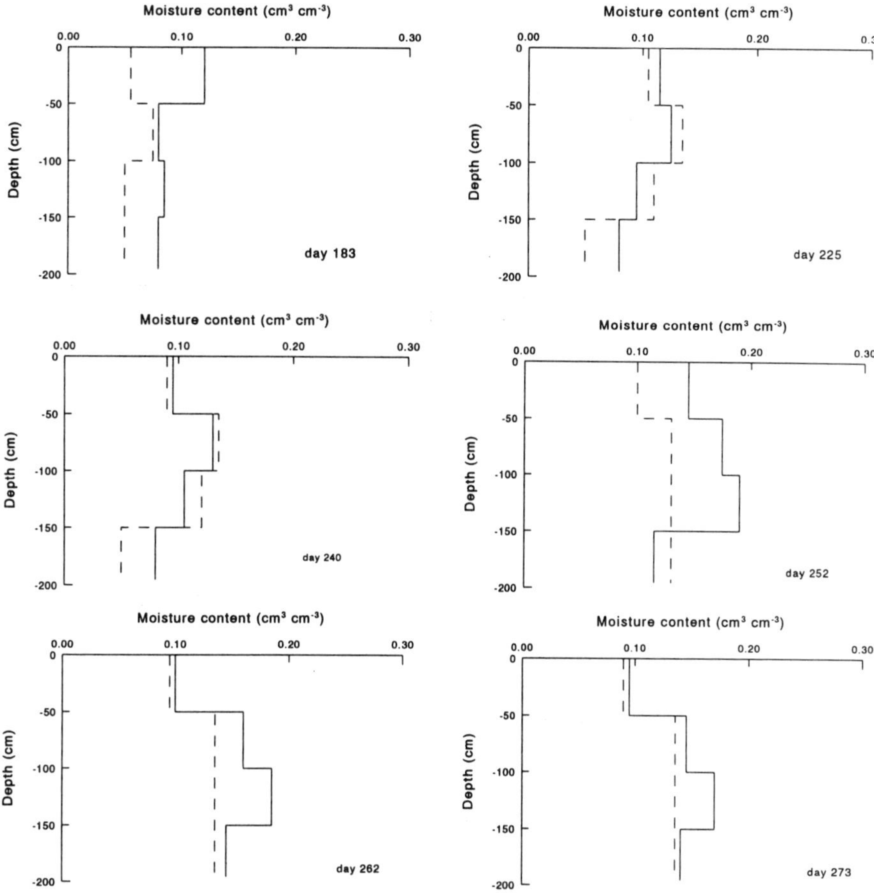


*Figure 2.* Simulated (dashed line) and measured (solid line) water content of a soil profile at six dates within one rainy season at Bambey (Senegal) (*Source:* De Ridder et al. 1991).

and water balances and has been extensively validated on the basis of data from Israel (Migda), Senegal (Bambey) and Mali (Niono) (e.g. Van Keulen et al. 1986; De Ridder et al. 1991). In general, the SAHEL model simulates plant growth and water balances in arid and semi-arid zones correctly. An example of the simulated and measured water content in a soil profile near Bambey (Senegal) is presented in Figure 2.

With this model, annual water balances have been calculated for different circumstances in West Africa: all possible combinations of 50 years of daily rainfall records, 12 soil types varying in texture, a range of run-off coefficients and 6 different soil depths. Cumulative percolation per year at the various soil depths as calculated by the model was correlated with infiltrated water ($I$ in mm), the water content of the soil at pF2.5 ($\Theta_{pF2.5}$ in $cm^3\ cm^{-3}$) and the depth of the soil profile ($sd$ in cm) resulting in a transfer function to calculate the percolation ($P_e$ in mm) (De Ridder and Van Keulen 1995):

$$P_e = 0.69I - 45\Theta_{pF2.5} - 0.22\,sd - 7.8(\Theta_{pF2.5} \times sd) - 74 \qquad (3)$$

The water content of the soil ($cm^3\,cm^{-3}$) at pF2.5 is calculated using a second transfer function, which was developed based on field data from West African soils (De Ridder et al. 1991):

$$\Theta_{pF2.5} = 0.014(37 - 0.35Sp) \qquad (4)$$

in which $Sp$ is the average percentage of sand in the soil profile.

### *Model equation*

To simulate the nitrogen flows, a simple nitrogen balance equation is linked to the water balance per column. The nitrogen per column, which is fixed in the above-ground biomass ($N_b$ in $kg\,ha^{-1}$), is calculated with an equation from De Wit and Krul (1982):

$$N_b = [0.0083\{R + (REDF \times R_{on})\} - R_{off} - P_e]/f - 0.13 \qquad (5)$$

where $f$ represents the yearly losses of nitrogen from the above ground biomass ($kg\,kg^{-1}$). The fraction $f$ represents the losses of nitrogen by volatilization, harvest of products, grazing, bush fires and other losses, e.g. through insects. Nitrogen inputs accounted for are wet deposition through rainfall, dry deposition, and fixation by algae, free living bacteria and legumes.

The equation has been validated for West African circumstances (e.g. Van Keulen et al. 1986; Breman et al. 1991). Correspondence between calculated and measured $N_b$ values is acceptable, although a large spread in individual observations exists. This is mainly due to differences in rainfall distribution within years. Averages over years of calculated and measured $N_b$ values do correspond well. For the present model, this average trend is of prime importance.

Whenever the surface of a column is (mainly) occupied by perennial woody vegetation, we assume that the permanent root system will avoid nitrogen losses through percolation and $P_e$ in equation (5) is set to zero accordingly.

### *Parameterization*

Using data from Hakkeling et al. (1989), we have defined a representative inland valley in West Africa with the following landscape elements and composition: uplands (24%), slopes (56%), bottomlands (13%) and a stream (7%). This inland valley is divided into 362 columns (Figure 3A), of which 27 columns are stream and the

*Figure 3.* (Opposite) (A) The inland valley with upland, slopes, bottomlands and stream as used in the model runs; (B) location of the band with perennial woody vegetation around the bottomlands; and (C) location of the perennial woody vegetation on slope grids which contribute most to reduction of run-on to the bottomlands.

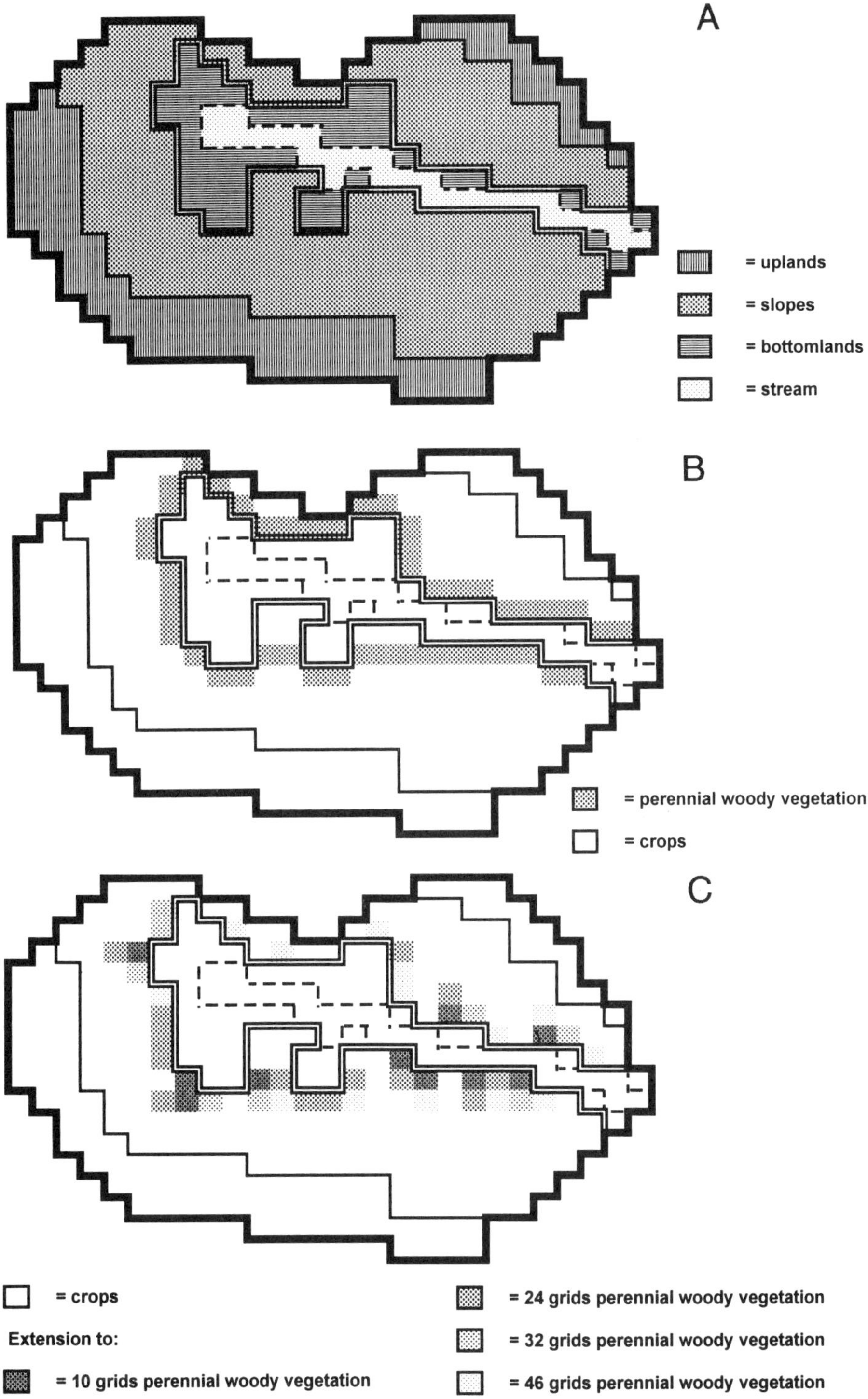
A
= uplands
= slopes
= bottomlands
= stream
B
= perennial woody vegetation
= crops
C
= crops
Extension to:
= 10 grids perennial woody vegetation
= 24 grids perennial woody vegetation
= 32 grids perennial woody vegetation
= 46 grids perennial woody vegetation

remaining 335 are divided in 87 columns of uplands, 202 of slopes, and 46 of bottomlands.

The boundary at the bottom of the system is set at 40 cm in the bottomlands (being the average groundwater table at the lowest point in a year), gradually increasing along the slopes to 200 cm in the uplands (assumed to be the maximum rooting depth).

Percentage of sand is taken as an approximation for soil characteristics and is estimated for the different landscape elements (Luiten and Hakkeling 1990). The run-off coefficient is assumed to increase linearly from 0.12 at a sand percentage of 95 to 0.58 at a sand percentage of 5 (Withers et al. 1978 in Tauer and Humborg 1992; Stroosnijder and Koné 1982). Thus, the landscape elements have the following characteristics: upland soils have 65% sand and a run-off coefficient of 0.27; soils at the slopes have 50% sand with a run-off coefficient of 0.35; and soils in the bottomlands have 35% sand with a run-off coefficient of 0.42. Annual rainfall is set at 1250 mm.

We have distinguished two types of vegetation, a perennial woody vegetation and an annual vegetation (crops). These vegetation types differ in values for $f$ and *REDF*. Nitrogen losses, mainly because of the grain harvest with high nitrogen contents, are supposed to be much larger in crops ($f=0.65$) than in perennial woody vegetation ($f=0.30$) (Breman and De Ridder 1991).

Furthermore, we assumed that a perennial woody vegetation, by the nature of its structure, is more capable of capturing run-on water ($REDF=0.9$) than are crops ($REDF=0.2$). Finally, we assume that vegetation types affect the run-off coefficient differently (e.g. Haylett 1960; Du Plessis and Mostert 1965; Barnes and Franklin 1970). As a first estimate, we assume that the run-off coefficient of a column is divided by two whenever a column is occupied by a perennial woody vegetation, but is not changed when a column is occupied by crops. Water flowing through the stream is assumed to contribute little to infiltration, thus *REDF* is kept at 0 for these columns.

### *Model runs*

Many land-use combinations are possible within the assumed inland valley, e.g. the surface of 1 column (= grid) can be occupied by crops, the rest by a perennial woody vegetation (334 grids). This single cropped grid can be chosen at 335 different locations within the inland valley, each giving a different result in water and nutrient balance. We have chosen various runs of the model to cover a wide range of combinations and to study the effect of location of crops in the inland valley.

In the first runs, the entire inland valley is occupied by either a perennial woody vegetation (WWW) or by crops (CCC), representing two extremes: exploitation is negligible, as it was in the past, and all land is exploited, as it might be expected to be in future.

Subsequently, the model is run with crops only in the bottomlands (WWC-46), only on the slopes (WCW-202) and only on the uplands (CWW-87). The place in the three character combinations of the characters W (perennial woody vegetation) and C (crops), indicates upland (first character), slopes (second character) and bottomlands (last character). The number indicates the total number of grids occupied by crops.

Further combinations investigated are crops on both uplands and slopes (CCW-289), on uplands and in bottomlands (CWC-133) and on slopes and in bottomlands (WCC-248).

In the more humid areas of West Africa bottomlands are traditionally the last part of an inland valley to be reclaimed (e.g. Becker and Diallo 1992). With improved health measures, though, reclamation of bottomlands has become of greater interest, particularly for lowland rice cultivation. With this land-use option one may expect the lowest $N_b$ and highest run-off. It might be of interest to investigate the effects on $N_b$ and run-off, when the perennial woody vegetation, being replaced by crops in the bottomlands (46 grids), is strategically relocated on the slopes. We have followed two strategies. In the first strategy, we have placed a band of 46 grids with perennial woody vegetation around the bottomlands (Figure 3B), aiming at capturing run-off water from the slopes and filtering nitrogen out of percolating water ($CC_bC$-289). In the second strategy, we locate perennial woody vegetation on the slopes. We have run the model with an increasing number of grids occupied with perennial woody vegetation to a maximum of 46 grids (Figure 3C), each time occupying those grids on the slopes through which run-on to the bottomlands was reduced most.

## Results

According to the logic of the model, highest run-off is created when the entire inland valley is occupied with crops (CCC). In Figure 4, the relation between run-off and the number of grids under crops is presented. Run-off is expressed as a fraction of the maximum value for run-off obtained when the entire inland valley is occupied with crops. The dotted line represents the relation between the number of grids under crops and run-off, when spatial effects of water redistribution are not explicitly modelled. The points indicate the calculated run-off of the different model runs, i.e. when spatial effects of water redistribution are explicitly taken into account. The points suggest a solution domain rather than a single line. By decreasing step by step the number of grids occupied by perennial woody vegetation through replacement by crops, we obtain the dashed lines. These lines represent the limits of the domain of possible run-off values, which can be obtained by explicitly modelling spatial effects of water redistribution. The values within the domain depend on the location of grids under crops within the inland valley. With expansion of crops from uplands to slopes and then to bottomlands (WWW → CWW-87 → CCW-289 → CCC), the bottom line of the domain will be followed. When expansion of crops starts from the bottomlands towards the slopes and the uplands (WWW → WWC-46 → WCC-248 → CCC), the upper line of the domain will be followed. The distance between the two lines at any number of grids under crops shows the possible options for changing run-off through reallocation of the cropped land: with the same number of grids under crops, a range of run-off values can be obtained. Thus, it is possible to increase the area of cropped land and, at the same time, decrease run-off by e.g. replacing crops on the uplands and in the bottomlands by perennial woody vegetation, and perennial woody vegetation on the slopes by crops (CWC-133 → WCW-202).

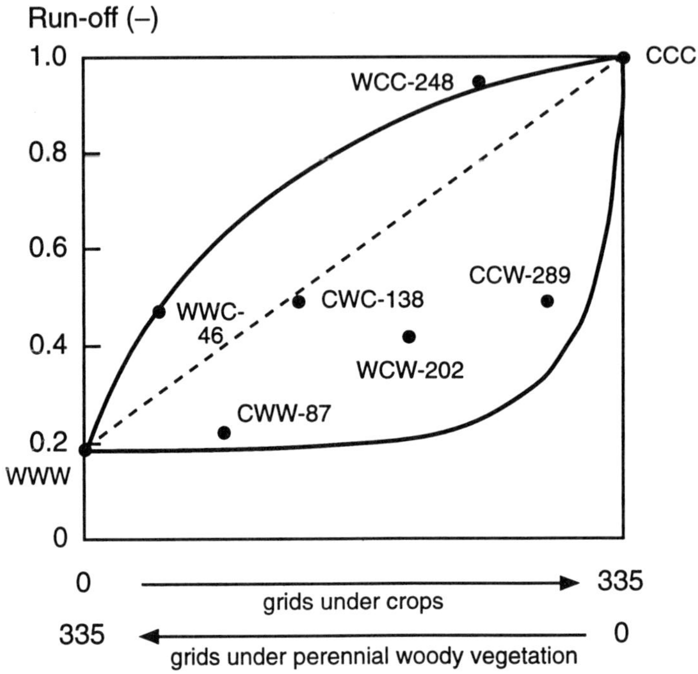


*Figure 4.* The solution domain of run-off values of the inland valley in relation to the number of grids covered with crops. The run-off is expressed as fraction of the maximum run-off obtained when the inland valley is fully covered with crops.

Figure 5 presents the relation between $N_b$ and the number of grids under crops. Here, again according to the logic of the model, the highest $N_b$-value is obtained when the entire inland valley is occupied with perennial woody vegetation. $N_b$-values of the other combinations are expressed as fraction of this highest value. A comparable picture arises as for run-off: a domain of $N_b$-values is created by explicitly modelling spatial effects of water distribution. Relatively high run-off results in lower $N_b$-values, following the bottom line of the domain, when expansion of crops goes from bottomlands to slopes and uplands (WWW$\rightarrow$WWC-46$\rightarrow$WCC-248$\rightarrow$CCC). Relatively low run-off results in higher $N_b$-values, following the upper line of the domain, when expansion of crops goes from uplands to slopes and bottomlands (WWW$\rightarrow$CWW-87$\rightarrow$CCW-289$\rightarrow$CCC). As in Figure 4, the distance between the two lines at any number of grids under crops shows the possible options for changing $N_b$ through reallocation of cropped land: with the same number of grids under crops a range of run-off values and thus $N_b$-values can be obtained. Thus, it is also possible to increase $N_b$, e.g. by replacing crops on the uplands and in the bottomlands by perennial woody vegetation, and perennial woody vegetation on the slopes by crops (CWC-133$\rightarrow$WCW-202). Then, the harvestable $N_b$ (part of $N$ in cropped land) of an inland valley is increased without increasing run-off.

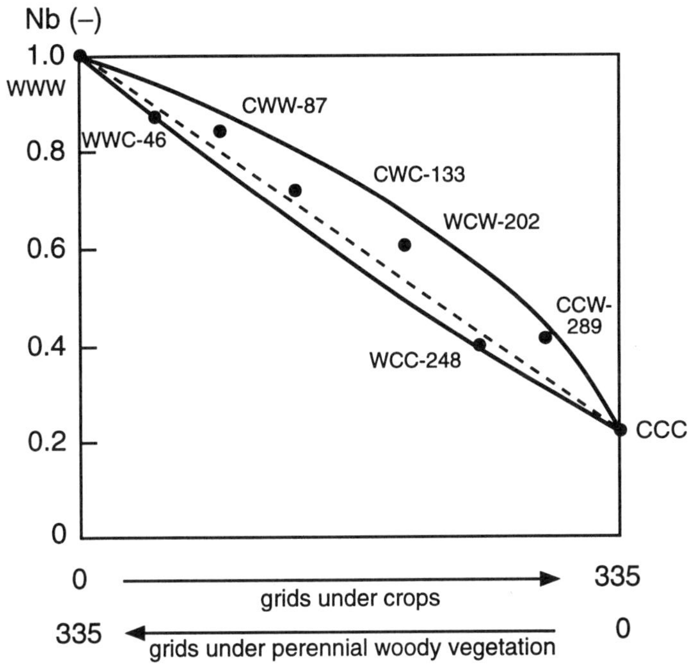


*Figure 5.* The solution domain of $N_b$-values of the inland valley in relation to the number of grids covered with crops. $N_b$ is expressed as fraction of the maximum $N_b$-value obtained when the inland valley is fully covered with perennial woody vegetation.

Figure 6 represents the relation between $N_b$ and number of grids under crops, when perennial woody vegetation, being replaced in the bottomlands by crops (46 grids), is strategically relocated on the slopes. The dotted line represents the relationship when spatial effects of water redistribution are not explicitly modelled. The line with triangles represents the $N_b$-values when grids with perennial woody vegetation are strategically placed at those places on the slopes which contribute most to the run-on to the bottomlands. Equal run-off and $N_b$-values to the situation of perennial woody vegetation occupying the bottomlands (CCW-289) can be obtained when perennial woody vegetation is relocated at strategic places on the slopes (left point of the line). A lower $N_b$-value is obtained when perennial woody vegetation is placed as a band around the bottomlands ($CC_bC$-289). Identical values for run-off and $N_b$ are obtained with less perennial woody vegetation, if perennial woody vegetation is placed strategically on the slopes.

## Discussion and conclusions

The model at issue demonstrates the importance of spatial patterns of land use and their effects on water and nitrogen flows at inland valley level. When surface water flows are spatially conceptualized, it appears that relations between land-use scenarios

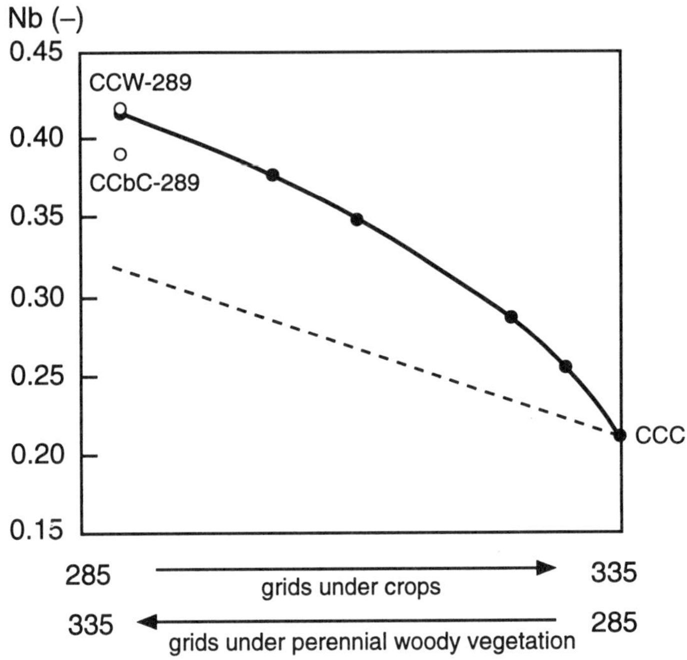


*Figure 6.* The relation between $N_b$-values of the inland valley and the number of grids covered with crops, when perennial woody vegetation is strategically located on slope grids which otherwise would contribute most to run-on to the bottomlands.

(different combinations of perennial woody vegetation and crops) and run-off and $N_b$-values of an inland valley cannot be described by single lines: a domain of values exists, with each value in this domain representing a specific combination of land use (Figures 3–5). The size and shape of the domain will alter with vegetation types, slopes, soil types and size of uplands, slopes and bottomlands (in itself, as well as in relation to each other). This mutual dependency confirms our assumption that optimal combinations are unlikely to be found through experimental work.

The purpose of the present model is by no means to simulate accurately the discharge of an inland valley, nor the daily water balance at any location within an inland valley or erosion and sedimentation rates. To avoid any suggestion in that direction, we have not presented absolute values. The purpose of the model is to illustrate the dependency of water and nitrogen balances of inland valleys on the spatial distribution of surface run-off water. Therefore, we hesitated to complicate the model with detailed relations such as on rainfall intensity and the infiltration/run-off ratio. We have rather chosen to use transfer functions needing only yearly average input values. Nevertheless, inclusion of detailed relations might change the shape and size of the solution domain.

The model represents a concept based on our understanding of water and nitrogen flows. However, while running the model we became aware that some important issues are not addressed in the model.

Evapo-transpiration is not explicitly modelled, but probably differs considerably between perennial woody vegetation and crops on a annual basis (Radersma and De Ridder 1996). Hence, percolation might be considerably lower under perennial woody vegetation than under crops (Stewart 1984). This will not affect the calculated nitrogen balance, since we assumed that nitrogen in the percolation water will be captured by the permanent root system of perennial woody vegetation.

Furthermore, the model equation for $N_b$ (equation (5)) does not take into account losses as a result of denitrification. This model equation was originally developed for the arid and semi-arid Sahelian region in West Africa, where denitrification plays only a minor role (De Wit and Krul 1982). This means that benefits of higher levels of $N_b$ calculated as a result of higher infiltration, at the foot of the slopes and in the bottomlands and through capturing water by introducing perennial woody vegetation, might partly diminish through denitrification.

Finally, nitrogen losses through leaching (i.e. in cropped land) are assumed to leave the system through percolation water. Nevertheless, the percolating water from uplands and upper slopes can return to the surface at the foot of the slopes as seepage, in case a semi-permeable or impermeable layer exists. The leached nitrogen in the percolation water could then contribute to $N_b$-levels in the lower parts of inland valleys. Possible seepage has to be included in the model, because this could affect the run-off of the inland valley as well.

The model results indicate some important research issues to be tackled. The possibility that run-off water might infiltrate lower parts of the slopes, in the model simply introduced by a run-on reduction factor (*RDEF*), appears to be important in the relation between water and N-flows and land use: a domain of solutions instead of single line relations. Experiments are under way to investigate this factor in relation to e.g. vegetation type, micro relief and agricultural practices.

In the model, water and nitrogen flows and relations between the flows are simplified in transfer functions and a model equation. Though by their simplicity useful for a first analysis, these 'building-blocks' are not verifiable. In future models, we will include a time dimension to explain the dynamics of water and nitrogen flows. The processes described in our 'building-blocks' should be validated in field experiments. However, the degree of detail of a 'building-block' will depend on its impact on model behaviour.

Decision makers have to consider possible effects of land-use change on water and nitrogen balances at not just the plot or field level, but increasingly at the level of landscape units such as inland valleys. Models as presented here, provided that the above mentioned issues are included, can give trends of desirable combinations of land use.

## Acknowledgements

Assistance by Annemarie E. Rozema and Daniel W.G. Van Kraalingen in model development is highly appreciated.

## References


Alexandratos N (1988) World agriculture: Toward 2000. Food and Agriculture Organization, Belhaven Press, London, UK. 338 p.

Andriesse W, Fresco L O (1991) A characterization of rice-growing environments in West Africa. Agric. Ecosyst. Environ. 33:377–395.

Andriesse W, Fresco L O, Van Duivenbooden N, Windmeijer P N (1994) Multi-scale characterization of inland valley agro-ecosystems in West Africa. Neth. J. Agric. Sci. 42.159–179.

Barnes D L, Franklin M J (1970) Runoff and soil loss on a sandveld in Rhodesia. Proc. Grassld. Soc. S. Afr. 5:140–144.

Becker L, Diallo R (1992) Characterization and classification of rice agro-ecosystems in Côte d'Ivoire. West Africa Rice Development Association, Bouaké, Ivory Coast. 254 p.

Breman H, De Ridder N (1991) Manuel sur les pâturages des pays sahéliens. Karthala, Paris, France. 485 p.

Breman H, De Ridder N, Van Keulen H (1991) Evaluation semi-détaillée. Pages 131–331 in Breman H, De Ridder N (Eds.) Manuel sur les pâturages des pays sahéliens. Karthala, Paris, France.

De Ridder N, Van Keulen H (1995) Estimating biomass through transfer functions based on simulation results: A case study for the Sahel. Agric. Water Manage. 28(1):57–71.

De Ridder N, Van Keulen H, Breman H, Uithol P W J (1991) La production végétale. Pages 293–355 in Breman H, De Ridder N (Eds.) Manuel sur les pâturages des pays sahéliens. Karthala, Paris, France.

De Ridder N, Stomph T J, Fresco L O (1996) An exploratory multi-scale water balance model to evaluate effects of changing land use and cultivation techniques with reference to West Africa. Catena (submitted).

De Wit C T (1993) Philosophy and terminology. Pages 3–9 in Leffelaar P A (Ed.) On systems analysis and simulation of ecological processes. Kluwer Academic Publishers, Dordrecht, The Netherlands.

De Wit C T, Krul J (1982) La production actuelle dans une situation d'équilibre. Pages 275–283 in Penning de Vries F W T, Djitèye M A (Eds.) La productivité des pâturages sahéliens: Une étude des sols, des végétations et de l'exploitation de cette ressource naturelle. Agr. Res. Rep. 918, Pudoc, Wageningen, The Netherlands.

Du Plessis M C F, Mostert J W C (1965) Afloop en grondverliese by die landbounavorsingsinstituut Glen. S. Afr. Tydskr. Landbouwet. 8:1051–1060.

Hakkeling R T A, Smaling E M A, Diatta S (1989) Detailed soil survey and land evaluation of the WARDA experimental area, Bouaké, Côte d'Ivoire. WSC Report No. 22, Wageningen, The Netherlands. 59 p.

Haylett D G (1960) Run-off and soil erosion studies at Pretoria. S. Afr. J. Agric. Sci. 3:379–394.

Jensen J R (1992) Plant nutrient balance monitoring models for tropical agriculture: A hydrological perspective. Paper presented at the Food and Agriculture Organization Workshop on Development and Testing of Plant Nutrient Balance Monitoring Models at Different Hierarchical Levels in Tropical Agriculture, AGLF, FAO, Rome, Italy. 8 p.

Luiten T, Hakkeling R T A (1990) Fertility parameters of soils in 86 selected West African toposequences. WSC Internal Communications 135, Wageningen, The Netherlands. 20 p.

Penning de Vries F W T, Van Laar H H (Eds.) (1982) Simulation of plant growth and crop production. Simulation Monographs, Pudoc, Wageningen, The Netherlands. 308 p.

Penning de Vries F W T, Jansen D M, Ten Berge H F M, Bakema A (1989) Simulation of ecophysiological processes of growth in several annual crops. Simulation Monographs, Pudoc, Wageningen, The Netherlands. 271 p.

Radersma S, De Ridder N (1996) Computed evapotranspiration of annual and perennial crops at different temporal and spatial scales using published parameter values. Agric. Water Manage. 31:17–34.

Stewart J B (1984) Measurement and prediction of evapotranspiration from forested and agricultural catchments. Agric. Water Manage. 8:1–28.

Stroosnijder L, Koné D (1982) Le bilan d'eau du sol. Pages 133–165 in Penning de Vries F W T, Djitèye, M A (Eds.) La productivité des pâturages sahéliens: Une étude des sols, des végétations et de l'exploitation de cette ressource naturelle. Agr. Res. Rep. 918, Pudoc, Wageningen, The Netherlands.

Tauer W, Humborg G (1992) Runoff irrigation in the Sahel zone: Remote sensing and geographical information systems for determining potential sites. Centre Technique de Coopération Agricole et Rurale, Wageningen, The Netherlands. 192 p.

Turner B L II, Moss R H, Skole D (1993) Relating land use and global land-cover change: A proposal for an IGBP-HDP core-project. IGBP Report No. 24/HDP Report No. 5, Stockholm, Sweden. 65 p.

Van Keulen H (1975) Simulation of water use and herbage growth in arid regions. Simulation Monographs, Pudoc, Wageningen, The Netherlands. 176 p.

Van Keulen H, Wolf J (Eds.) (1986) Modelling of agricultural production: Weather, soils, and crops. Simulation Monographs, Pudoc, Wageningen, The Netherlands. 479 p.
Van Keulen H, Breman H, Van Der Leck J J, Menke J W, Stroosnijder L, Uithol P W J (1986) Prediction of actual primary production under nitrogen limitation. Pages 42–76 in De Ridder N, Van Keulen H, Seligman N G, Neate P J H (Eds.) Modelling of extensive livestock systems. International Livestock Centre for Africa, Addis Ababa, Ethiopia.
Waggoner P E (1994) How much land can ten billion people spare for nature. Task Force Report No. 121, Council for Agricultural Science and Technology, Ames, IA, USA. 126 p.
Windmeijer P N, Andriesse W (1993) Inland valleys in West Africa: An agro-ecological characterization of rice-growing environments. International Institute for Land Reclamation and Improvement Publication 52, Wageningen, The Netherlands. 160 p.

# A model assessing water management strategies for integrated land-use planning in the Mekong Delta of Vietnam

C.T. HOANH[1], H. VAN KEULEN[2] and T.P. TUONG[3]

[1]*International Institute for Aerospace Survey and Earth Sciences, P.O. Box 6, 7500 AA Enschede, The Netherlands; permanent address: Sub-Institute for Water Resources Planning and Management, 253A An Duong Vuong, Q.5, Ho Chi Minh City, Vietnam*

[2]*Section Animal Production Systems, Department of Animal Husbandry, Wageningen Agricultural University and DLO-Research Institute for Agrobiology and Soil Fertility, P.O. Box 14, 6700 AA Wageningen, The Netherlands*

[3]*International Rice Research Institute, P.O. Box 933, 1099 Manila, Philippines*

**Key words:** simulation, multi-goal, multi-criteria evaluation, agriculture, rural development

**Abstract**

Challenges in land-use planning in an agricultural region include balancing food production with resource limitations, balancing economic revenues with environmental protection, integrating sectoral developments and harmonizing farmers' objectives with national goals. This paper describes a PC-based methodology to deal with these challenges in land-use planning at the regional level. The methodology comprises a simulation model consisting of 14 submodels, each representing an analytical step to assess the impact of development and land-use scenarios on water resources, production, socioeconomics, and environment at farm and at regional levels. Interactions among factors are reflected by feedback among submodels. Outputs from the model are linked to GIS for spatial and temporal analysis. Application of the model in the Mekong Delta of Vietnam is discussed.

## Introduction

Land-use planning is essential in any country, because demands for various land uses often exceed the available land resources. Land-use planning involves weighing of trade-offs among conflicting goals resulting from the diverse interests of society (FAO 1993).

FAO (1993) proposed ten steps in land-use planning, based on land evaluation (LE), combining biophysical and socioeconomic factors. Since the 1970s, farming system analysis (FSA) has been applied to take into account the preferences and priorities of the local land users in a bottom-up approach (Huizing 1991). Fresco et al. (1992) suggested LEFSA which combines farming system analysis and land evaluation.

LE, FSA and LEFSA procedures essentially assess the suitability of a land use or farming system based on existing environmental conditions. Some of these procedures have been applied in land evaluation in the Mekong Delta of Vietnam on the basis of present hydrology (Tuong et al. 1991), land uses (Tri et al. 1993) and soil conditions (Van Mensvoort et al. 1993). The environmental conditions, however, can drastically change when an intervention, e.g. a water management measure, is implemented in



*P.S. Teng* et al. *(eds.): Applications of Systems Approaches at the Farm and Regional Levels, 383–397.*

the system. Currently available procedures assess future land use by providing prospective land-use allocation. Not much attention is paid to the question of how that allocation is reached from the present conditions, i.e. the dynamic nature of the changes is not taken into account.

Assessing changes in land-use systems in response to spatial and temporal changes in environmental conditions requires quantification of causal relationships among system components and implies an understanding of ecological and socioeconomic interactions in the system (Fresco et al. 1992). Land-use planning is more complex when different development scenarios are assessed. For each scenario, the balance among goals of agricultural production and 'non-agricultural' land use must be considered dynamically throughout the planning period.

This paper describes a systems approach to evaluating effects of different interventions on temporal and spatial biophysical and socioeconomic parameters and the associated land uses. The approach takes into account the major environmental impacts and conflicting interests of different sectors. A case study is presented for the Quan Lo Phung Hiep area in the Mekong Delta of Vietnam, illustrating an application of the approach for assessing land and water management scenarios for integrated land-use planning at the regional level.

## The approach and tool for integrated land-use planning

### *The approach*

We used multi-level integration of biophysical and socioeconomic factors to assess the effects of interventions through the evaluation of various development scenarios. Each scenario quantifies how key interventions such as water management change the physical conditions. Subsequent biological production and the associated socioeconomic benefits are estimated. Modifications in socioeconomic benefits may lead to changes in the pattern of land and water use which, in their turn, may affect physical conditions. These changes are evaluated in the computations.

To analyse the effects of changes at farm, provincial, regional and national levels, in order to arrive at an acceptable compromise, interests at different levels are integrated, combining top-down and bottom-up sequences. First, effects of national and regional policies (e.g. water management, tax, credit, infrastructure improvement, birth control) on farm production are analysed. These effects are then aggregated to regional and national levels to examine whether the objectives of interventions can be realized and to identify the problems at lower levels in matching their objectives with the policy at higher levels.

Dynamic simulation is used to analyse the interactions among biophysical and socioeconomic factors. When knowledge is not adequate for dynamic simulation, an expert systems approach (Bouma et al. 1993) can be used. This approach makes use of regression functions (e.g. between rice yield and salinity of irrigation water) to simulate system component interactions. The functions are derived from literature, experiments and household surveys. Parametric methods (Driessen and Konijn 1992;

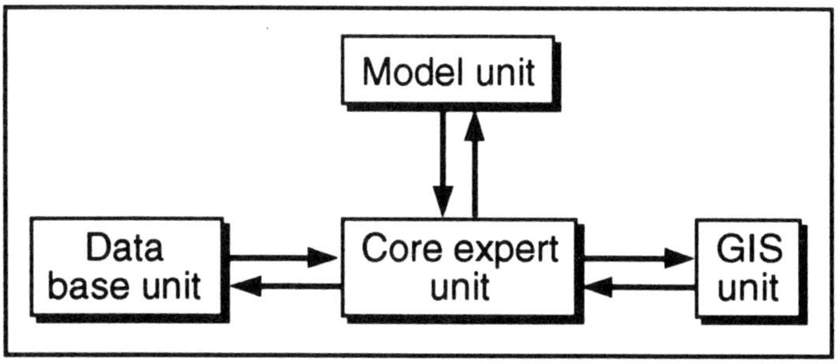


*Figure 1.* Components of CAILUP (Computerized Aid to Integrated Land-Use Planning).

Davidson 1992) are applied when more than one factor interacts with production.

Development goals are set by policy makers. Each goal is expressed quantitatively in terms of 'target values', e.g. rice production is expressed in tons, and net present value in US$. Goals can also be improvement of environmental conditions, e.g. improvement of water quality for domestic consumption.

The concept of 'with' and 'without' intervention cases for evaluating economic indicators such as benefit-cost ratio and internal rate of return (Gittinger 1982) is applied to help decision makers select alternatives among different intervention options. The 'without' case is formulated on the basis of current conditions and used as 'base scenario'. Interventions are evaluated for each goal. Multi-goal impacts of intervention scenarios are ranked by multi-criteria evaluation methods (Nijkamp et al. 1990).

*Computerized Aid to Integrated Land-Use Planning (CAILUP)*

A Computerized Aid to Integrated Land-Use Planning (CAILUP), consisting of four major units (Figure 1), is used to implement the approach. Simulation is carried out in the *model unit* to determine the changes in system parameters by dynamic simulation and expert-based regression functions. The *core expert unit* functions as a 'central processing unit' in managing the system and guiding the analysis process. It comprises a hierarchical menu system that allows the user to work interactively with other units. The *data base unit* handles all input/output data to/from other units; and the *GIS unit* displays spatial data for analysis.

The *model unit* consists on 14 submodels (Figure 2), each representing an analytical step. Inter-sectoral and inter-level integration is implemented by data exchange among submodels and feedback.

The Intervention Generating Submodel [1] generates a biophysical and socio-economic data set, e.g. a construction schedule for an irrigation system, and taxes for the intervention scenarios throughout the planning period, based on the development policies, goals and existing conditions.

The Physical Impact Submodel [2] simulates the dynamic changes in physical conditions of the planning area and over the planning period under the selected

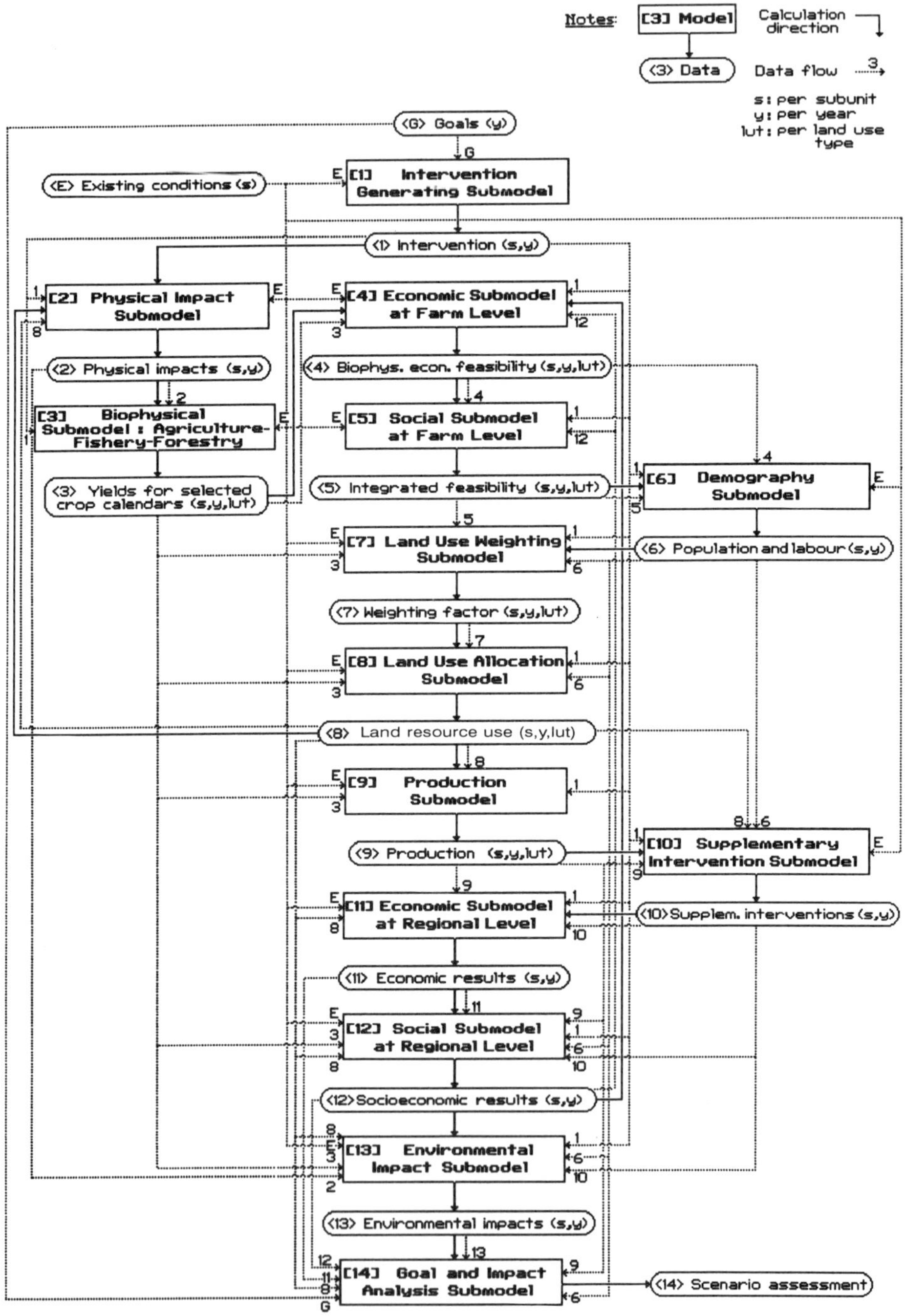


*Figure 2.* Linkages among CAILUP submodels.

intervention scenario. For our work in the Mekong Delta, this submodel comprises (i) the VRSAP (Vietnam River Systems And Plains) hydraulic and salinity dynamic model (Khue 1991; NEDECO 1992) assessing changes in water level, discharge and salinity of the surface water; (ii) expert-based empirical models assessing variations in natural resources, such as degradation of land and water resources in acid sulphate soil areas due to lowering of the water level, or the reduction in water resources due to upstream development. Changes in physical conditions may create changes in land-use patterns which, in turn, affect the physical conditions (e.g. increasing salt water intrusion). This interaction is neglected in the first round of calculations and is considered when land use in all land units is formulated in submodel [8]. Criteria can be set to changes in the physical conditions caused by changes in land-use pattern to determine whether the computation proceeds to submodel [9] or repeats the loop of submodels [3] - [8]. For example, differences in water salinity in all land units between two successive runs must be less than $0.1\,g\,l^{-1}$ before proceeding to submodel [9].

The Biophysical Submodel [3] estimates yields for relevant land-use types under modified physical conditions. For the Mekong Delta, yields were estimated through a series of expert-based regression functions.

The Economic Submodel at Farm Level [4] generates the combined biophysical/ economic feasibility of each land-use type, based on cost-benefit analyses at the farm level. The feasibility is 100% for the land-use type with the highest net income. For other land-use types, the feasibility is expressed as the ratio of their net income and the highest net income, adjusted for different household groups by the ratio of their capital availability and the required capital for the specific land-use type.

The Social Submodel at Farm Level [5] modifies the biophysical/economic feasibility by taking into account social preferences of different household groups, since farmers do not always select the land-use type with the highest economic revenues. For example, traditional rice varieties are preferred to high-yielding varieties for local consumption by high-income household groups. Rice is preferred more than pineapple due to storage flexibility and self-sufficiency considerations when the difference between income from pineapple and rice is below a certain percentage of the income from rice.

The Demography Submodel [6] considers the effects of population growth, rural-urban migration and population redistribution, and generates data on population and both on-farm and off-farm labour availability.

The Land-use Weighting Submodel [7] analyses effects of government policy on farmers' selection. The integrated feasibility from submodel [5] is modified by a factor (between 0 and 1), reflecting the policy of the government, so that the model outcome is compatible with the various goals at levels higher than the farm. The deviation from unity of this factor can be considered as an indication of the required effect by the regional or national government to make the farmers adopt the desired land-use allocations, e.g. through land-use regulations, extension programmes or subsidies.

The Land-use Allocation Submodel [8] generates the area of each land-use type and the water demand in each land unit on the basis of the land-use weighting value from submodel [7] and the water requirement per land-use type. Data on water demand are transferred back to submodel [2] to re-evaluate the physical conditions as discussed in submodel [2].

The Production Submodel [9] generates total production from area [8] and yield [3]. Effects of biophysical risk, such as pests and diseases, can be included in this submodel for risk analysis. A fractional yield reduction is applied for each level of incidence, i.e. high, medium or low.

The Supplementary Intervention Submodel [10] generates requirements for transport, energy and industry associated with each scenario, e.g. products to be transported and processed, fuel to be supplied.

The Economic Submodel at Regional Level [11] calculates the economic indicators – net present value, benefit-cost ratio, internal rates of return – for each land unit and aggregates them to the regional level.

The Socioeconomic Submodel at Regional Level [12] estimates the social indicators such as income distribution and employment generation in the region. The new socioeconomic conditions at the regional level may cause changes in prices of certain commodities or inputs. High labour demands in the region at peak periods may create labour migration. Such changes will affect land-use decisions at farm level. A feedback to submodel [4] is, therefore, required to reflect these effects.

The Environmental Impact Submodel [13] evaluates the environmental impacts of the selected scenario. In some cases, the impacts can be assessed directly from the output of previous submodels, e.g. water quality; in other cases, they are calculated from empirical expressions, e.g. incidence of malaria in relation to the salinity of the canal water (Thu 1991).

The Goal and Impact Analysis Submodel [14] generates a general ranking for the land and water use scenarios, using a multi-criteria rating method (Nijkamp et al. 1990). For each goal $g$ in year $y$ interventions are first evaluated by calculating a relative deviation $Rd(g,y)$ of the realized value $Rv(g,y)$ from the target value $Tv(g,y)$:

$$Rd(g,y) = [Rv(g,y) - Tv(g,y)]/Tv(g,y)$$

The present value of the annual impact of the scenario is evaluated using the 'depreciation' concept. The score for each goal, taking into account the goal priority set by decision makers, is calculated by:

$$Sg(g) = Pv(g) \times \sum_{y=1}^{n} Rd(g,y) \times Df(g,y)$$

where, $Sg(g)$ is the score for single goal $g$, $Pv(g)$ is the priority value of goal $g$, $n$ is number of years in the planning period and $Df(g,y)$ is a discount factor. The overall ranking is based on the final score, obtained by adding all single-goal scores.

## Case study

### *Quan Lo Phung Hiep region*

CAILUP has been used to analyse the effects of different water management and land-use strategies in the Quan Lo Phung Hiep region. The study area is located in

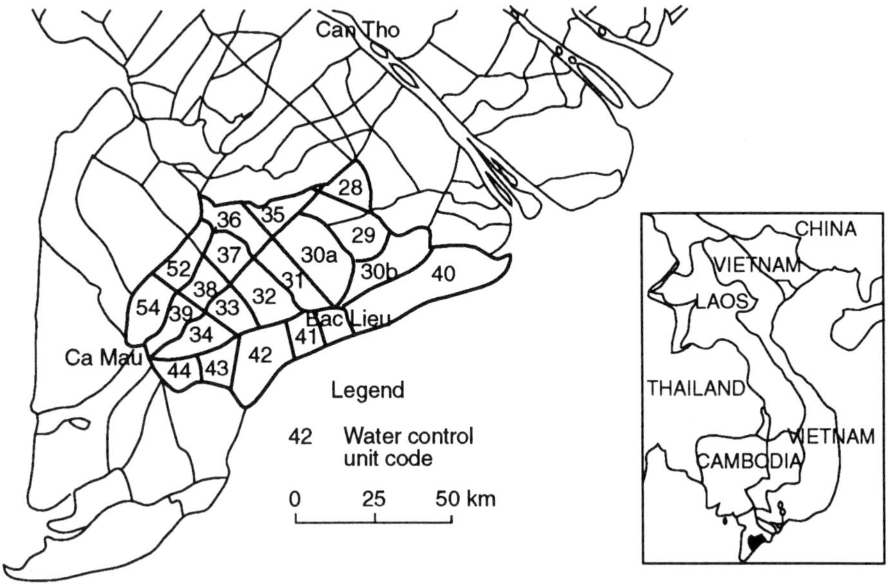


*Figure 3.* Location map of the Quan Lo Phung Hiep region (numbers in the figure indicate water control units).

the Soc Trang and Minh Hai provinces, occupying approximately 500 000 ha in the southern part of the Mekong Delta, Vietnam (Figure 3). The total area comprises 47% saline soils, 14% strongly active acid sulphate soils (ASS), 38% moderately and slightly ASS and 1% sandy soils.

Low rainfall during the dry season (December to April) prevents agricultural production without irrigation. However, salt water intrusion from the sea via two main rivers, My Thanh and Ganh Hao, and an intensive network of canals make water quality in most parts of the region unsuitable for irrigation from January to June. In the early part of the rainy season, leachates from the ASS area contaminate the surface water and reduce its pH to values below 4, which is detrimental to aquaculture production. From August on, about 70% of the region is inundated over 0.5 m for three to four months.

Eighty-five percent of the population in the region is engaged in agricultural, fisheries and forestry activities. In 1990, the region produced 800 000 tons of rice, 6000 tons of brackish and salt water shrimp, 3000 tons of fish and 200 tons of fresh water prawn. Average monthly income per capita was approximately US $35. Over 30% of the total population (1.3 million) consumed less than 1500 kcal (compared to the national target of 2000 kcal) per day, hence food production was a high priority in the development plan.

In the model, the Quan Lo Phung Hiep region was divided into 20 water management units and 181 subunits depending on soil types and administrative boundaries.

*Table 1*. Target values (*Tv*), realized values (*Rv*), and relative deviation of realized value from target ($Rd = (Rv - Tv)/Tv \times 100\%$) of a representative development scenario. For simplicity, data for some years have been omitted.

| Goals | Parameters | Year | | | | | |
|---|---|---|---|---|---|---|---|
| | | 1990 | 1991 | 1995 | 2000 | 2010 | 2020 |
| Rice production | *Tv* (Mtons) | | 0.8 | 1.1 | 1.5 | 1.5 | 1.5 |
| | *Rv* (Mtons) | 0.8 | 0.733 | 1.147 | 1.535 | 1.483 | 1.566 |
| | *Rd* (%) | | −8 | 4 | 2 | −1 | 4 |
| Per capita rice production | *Tv* (kg) | | 630 | 700 | 700 | 700 | 700 |
| | *Rv* (kg) | 630 | 554 | 748 | 860 | 656 | 575 |
| | *Rd* (%) | | −12 | 7 | 23 | −6 | −18 |
| Food distribution[1] | *Tv* (%) | | 20 | 30 | 40 | 45 | 45 |
| | *Rv* (%) | 21 | 21 | 28 | 38 | 42 | 46 |
| | *Rd* (%) | | −3 | 5 | 5 | 6 | −3 |
| Income per capita | *Tv* (US $) | | 40 | 60 | 80 | 80 | 80 |
| | *Rv* (US $) | 35 | 37 | 67 | 82 | 64 | 66 |
| | *Rd* (%) | | −7 | 12 | 2 | −20 | −18 |
| Income distribution[2] | *Tv* (%) | | 30 | 40 | 50 | 50 | 50 |
| | *Rv* (%) | 37 | 37 | 44 | 52 | 55 | 57 |
| | *Rd* (%) | | −23 | −10 | −4 | −9 | −14 |
| Employment | *Tv* (M labour-days) | | 60 | 60 | 70 | 70 | 70 |
| | *Rv* (M labour-days) | 55 | 60.48 | 58.12 | 72.91 | 69.72 | 70.95 |
| | *Rd* (%) | | 1 | −3 | 4 | 0 | 1 |
| Percentage of population with access to fresh water | *Tv* (%) | | 0 | 14 | 14 | 14 | 14 |
| | *Rv* (%) | 0 | 0 | 11 | 16 | 14 | 13 |
| | *Rd* (%) | | 100 | −22 | 11 | 2 | −6 |
| Minimum land cover (%) | *Tv* (%) | | 30 | 35 | 35 | 35 | 35 |
| | *Rv* (%) | 30 | 32 | 35 | 39 | 39 | 39 |
| | *Rd* (%) | | 7 | 1 | 11 | 12 | 13 |
| Total pesticide | *Tv* (tons) | | 500 | 2500 | 5000 | 10500 | 15500 |
| | *Rv* (tons) | 360 | 362 | 2330 | 6064 | 14576 | 22391 |
| | *Rd* (%) | | −28 | −7 | 21 | 39 | 44 |

[1]Percent of population with less than 60% of the average food availability.
[2]Percent of population with less than 60% of the average per capita income.

### *Development objectives and target values*

The national and regional governments have identified five main objectives of the development program for the region: (i) increasing total food production; (ii) increasing total income; (iii) increasing foreign currency earnings; (iv) improving living conditions; and (v) sustaining the economic development rate. These general objectives are translated into particular goals, each with its own target value (Table 1).

Food and income distribution are expressed as the proportion of the population with a food or income availability per capita below 60% of the average in the region. It is likely that areas with favourable soil and water conditions will develop faster than others that are less well endowed, thus increasing inequity. The goal in the model, is therefore, to limit that proportion to a value below 45% (compared to the

*Table 2.* Construction schedules of the new water management system.

| Construction schedule | Main features | Total construction periods (in years) |
|---|---|---|
| A | Main sluices are constructed sequentially from east (Soc Trang) to west (Minh Hai) | 9 |
| B | Main sluices are initiated at the same time in both Soc Trang and Minh Hai provinces | 5 |
| C | As B, but the construction period is longer due to lack of investments | 9 |
| D | As, A for main sluices but, in addition, secondary canals are constructed early in the areas with active acid sulphate soils (ASS) to allow sufficient leaching of acid by salt water prior to the initiation of land-use changes brought about by fresh water availability | 8 |
| E | As B for main sluices, and as D for secondary canals | 5 |
| F | The area protected against salt water is divided into two phases on the basis of two river networks (My Thanh and Ganh Hao). There is also a five-year separation between phases to allow monitoring of the environmental, social, and economic impacts | 12 |
| G | As F, but the area protected against salt water is divided into three phases on the basis of soil types: saline, slightly ASS, and strongly active ASS | 17 |

value of 20% in 1991) after completion of the new water management system.

The regional environmental impact indicators selected were: the proportion of the population supplied with fresh water for domestic use (representing improved water quality); minimum land cover in the dry season (providing a favourable environment for human life, in particular on the land affected by salt water); and total pesticide use (expected to be significantly higher on rice crops after completion of the new water management system).

### *Water management and land-use scenarios*

Previous studies (Sonntag and McNamee 1989; ESSA 1992) have identified water management intervention as the key factor for development of the region. The proposed water management system consists of dams and sluices, primary, secondary and tertiary canals, and on-farm systems to prevent salinity intrusion, and to provide for acid water leaching, water drainage and increased fresh water supply in the dry season.

CAILUP is used to evaluate 28 development scenarios, constructed by combining seven construction schedules (Table 2) with four land-use strategies (Table 3). These scenarios are compared with a 'without case', representing the current water conditions and land-use patterns. However, more land is allocated for specific uses (housing, urbanization, roads, etc.) as a function of population growth.

### *Model calibration*

Hydrology, land use and production in 1985 at the village level were used as initial condition and those in 1990 were used for model calibration.

*Table 3.* Land-use strategies.

| Land-use strategies | Main features |
|---|---|
| 1 | Maximize rice production: rice production is the focus, and the highest-yielding rice cropping system is selected |
| 2 | Maximize income from rice production: rice production is the focus, but the rice cropping system yielding the highest net income is selected |
| 3 | Minimize effects of acid water: acid tolerant crops are cultivated in areas of moderately ASS and no land-use changes are allowed on strongly active ASS |
| 4 | Crop diversification: crops giving the highest net income are selected |

Yields simulated in submodel [3] were calibrated by adapting spatial and temporal coefficients, e.g. that representing the percentage of unfavourable years in rice yield calculations, and that representing the irrigation capacity of small creeks.

The aim of calibrating submodel [4] was to generate values that are proportional with the areas of land-use types in each subunit. Data used in the financial analysis, e.g. the amount of input materials, the number of hired labourers, rate of interest, etc., were revised and adjusted to guarantee that farmers attain a certain net income from the yields simulated in submodel [3].

Calibration of submodel [5] was similar to that of submodel [4]. Data on preferences in the social rules were adjusted to generate the combined feasibility of attaining the same objective.

Submodel [6] was calibrated by adjusting the distribution of emigration across the water management units, migration possibilities and the proportion of the potential migrants who would actually migrate. Calculated migration among subunits is also affected by the calibration of submodel [5]. Thus, calibration should be carried out in series from submodel [3] to submodel [6].

Calibration of submodel [7] was combined with that of submodel [8]. Policy factors were adjusted in combination with the feasibilities from submodel [5] to generate weighting values for land-use allocation with the purpose of matching calculated land-use areas with inventory data.

Submodel [9] was calibrated by modifying the spatial and temporal adjustment coefficients describing the effects of pests and diseases.

Values of demand per capita for local consumption and of factors dealing with raising of pigs and ducks (major 'by-products' from rice production in the region) were adjusted in the calibration of submodel [10], to match the calculated production of pigs and ducks with actual data.

### *Model validation*

Actual data at district level from 1991 to 1994 have been used to validate the model. Constuction of the first three sluices started in 1992 and was completed in 1993, resulting in changes in water conditions and the associated land uses on the eastern side of the region. At the same time, government policies on rice production shifted from a state-controlled economy into a market economy. These modifications, com-

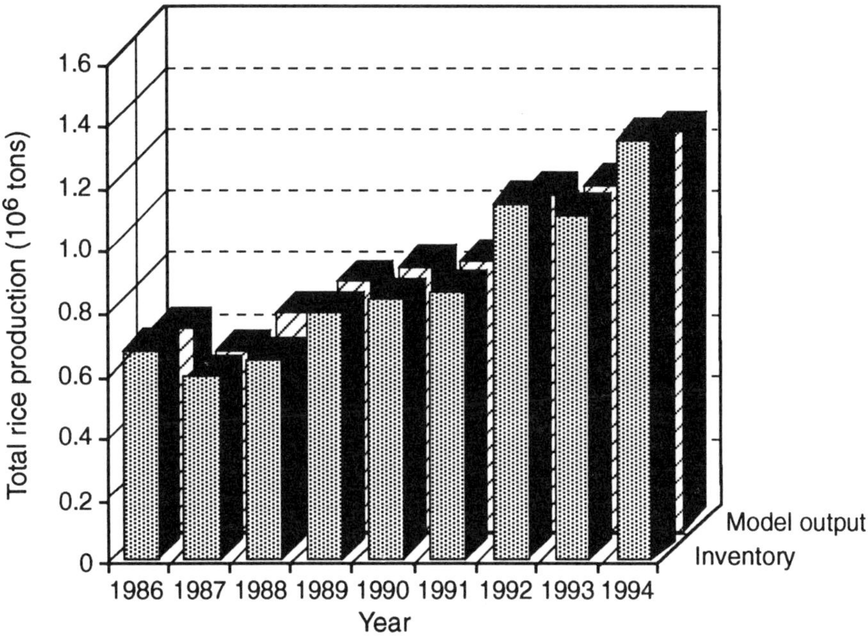


*Figure 4.* Regional rice production simulated by CAILUP and from inventory. The 1986–90 data were used for calibration and 1991–94 for validation purposes.

bined with population and land use in 1990 (as initial conditions), were used in model validation. Actual conditions for the 1990–94 period with respect to rate of migration, policy factors, etc. were introduced in the model.

Average calculated total regional rice production for the period 1991–94 is close to that officially reported (Figure 4). However, differences between calculated and reported yields for the 181 subunits are, in many cases, large (beyond 20%). Generated population distribution over 20 water control units in 1994 is in close agreement with the observations (maximum deviation of 5%). Generated total regional land use is in accordance with the 1994 observations, with maximum deviations of −10% for shrimp ponds and +6% for homestead gardens and specific uses as settlements and roads. The deviation was 61% for fish ponds, since small fish ponds in homestead gardens were included in the inventory while only large commercial fish ponds were taken into account in the model.

### *Assessing development scenarios*

The single scores for each of the goals and the final score for construction schedules A to G with land-use strategy 2 (maximization of income from rice production) and land-use strategies 1–4 for schedule D (Table 4) indicate that short construction periods as in schedule B do not always lead to high economic returns and production. This is because investments in land-use changes by farmers are not in pace

*Table 4.* Single-goal and final scores of representative development scenarios.

| Scenarios | Rice prod. | Rice per capita | Food distrib.[1] | Income per capita | Income distrib.[2] | Employment | % population with access to fresh water | Minimum land cover | Total pesticide | B/C ratio | IRR | Final score[3] |
|---|---|---|---|---|---|---|---|---|---|---|---|---|
| A2 | 1.2 | 0.9 | −0.4 | −1.1 | −2.7 | 0.0 | 3.9 | 2.3 | −4.6 | 0.6 | 1.5 | 1.7 |
| B2 | 0.6 | 0.4 | −1.6 | −1.1 | −3.8 | −0.7 | 4.6 | 2.5 | −3.9 | 0.5 | 1.2 | −1.3 |
| C2 | 1.1 | 0.8 | −0.9 | −1.8 | −2.3 | 0.5 | 4.0 | 2.9 | −4.4 | 0.1 | 0.5 | 0.7 |
| D2 | 1.2 | 0.9 | 0.3 | −1.0 | −2.7 | 0.0 | 3.9 | 2.3 | −4.6 | 0.5 | 1.4 | 1.6 |
| E2 | 0.6 | 0.4 | −1.6 | −1.1 | −3.8 | −0.7 | 4.6 | 2.5 | −3.8 | 0.4 | 1.1 | −1.5 |
| F2 | 0.0 | −0.4 | −0.4 | −2.4 | −3.0 | −0.5 | −1.1 | 2.2 | −3.6 | 0.4 | 0.7 | −8.1 |
| G2 | −0.1 | −0.4 | −0.7 | −3.2 | −2.3 | −0.4 | 2.8 | 2.5 | −2.8 | −0.1 | 0.1 | −4.7 |
| D1 | 2.5 | 2.0 | −0.7 | −0.1 | −3.1 | 0.9 | 3.9 | 2.1 | −5.9 | 0.6 | 0.6 | 1.6 |
| D3 | 0.4 | 0.2 | −1.4 | −2.8 | −1.6 | 0.4 | 3.9 | 3.6 | −3.9 | 0.0 | 0.0 | 0.5 |
| D4 | 1.1 | 0.8 | −0.9 | −1.1 | −2.9 | 0.6 | 3.9 | 3.1 | −4.2 | 0.5 | 1.4 | 2.2 |

[1]Percent of population with less than 60% of the average food availability.
[2]Percent of population with less than 60% of the average per capita income.
[3]All goals are assumed of equal importance in priority setting.

with those for water control by the government. This indicates that integrated planning for land use and water management is essential.

Conflicts among goals are illustrated by scenarios which have a high score for rice production but a low value for economic returns (e.g. D-1). Scenario D-2 has a final score somewhat lower than scenario A-2, due to the costs for secondary canals in the early stages.

For construction schedule D, land-use strategies focusing on rice production such as 1 or 2 lead to higher rice production, but lower economic returns than strategy (4), aiming at crop diversification. However, if the role of the region in producing food for the country is given high priority, the former strategies are more promising.

For some goals, e.g. income/capita, single goal scores are all negative, implying that in none of the scenarios can this goal be attained. In this case, the target value is possibly too high and should be reviewed, or the interventions should be modified to increase those scores.

## Discussion

In the development of land-use patterns, 'biodiversity' leads to a large number of land-use types with various cultivation techniques, while 'sociodiversity' results in a variety of socioeconomic objectives and preferences. Hence, in designing and evaluating alternative land-use patterns, as aimed at in land-use planning, both agro-technical and socioeconomic considerations should be taken into account. Integration in multi-level and multi-sectoral planning, and between biophysical and socio-economic factors can be realized through simulation modelling.

In the development of CAILUP, we attempted to describe the system as 'completely' as possible. The consequence is that, in its current state, the model comprises building-blocks (modules) that are 'weak' in terms of knowledge, information and data, along with modules that are much 'stronger' in this respect. As such, CAILUP also serves to identify knowledge and information gaps as a basis for prioritization of future research.

As an aid in land-use planning, CAILUP has been developed for the purpose of exploration of possibilities rather than for prediction. It is a tool to answer 'what-if' questions in support of planning and management, and not to answer the question 'what is going to happen?'. The 'what-if' questions are related to the impacts of water management and possible supplementary interventions.

To develop and apply CAILUP in a regional study as for the Quan Lo Phung Hiep region, it requires data and information, hardware and software packages. Although these requirements have been taken into account, development and application of CAILUP still presents many challenges. These challenges refer to three main issues: the tension between single discipline and multidisciplinary studies; the tension in modelling between (explanatory) complexity and (descriptive) simplification (Meinke 1996); and the tension between different applications such as research or management, exploratory or prospective.

## Conclusions

Quantitative integration of biophysical and socioeconomic factors may be achieved through applying dynamic simulation modelling techniques. It can support decision-making processes by evaluating the effects of various interventions on the realization of diverse development goals. It also helps to integrate expertise from different disciplines dealing with land-use planning, improve knowledge on interactions among factors and to identify knowledge and information gaps.

In such an integrated context, however, although all submodels are simplified, the model of the whole system is still complicated. A major effort is required, not only for the refinement of each submodel, but also for integration of these submodels.

## References


Bouma J, Wopereis M C S, Wösten J H M, Stein A (1993) Soil data for crop-soil models. Pages 207–220 in Penning de Vries F W T, Teng P S, Metselaar K (Eds.) Systems approaches for agricultural development. Kluwer Academic Publishers, Dordrecht, The Netherlands.

Davidson D A (1992) The evaluation of land resources. Longman Scientific & Technical, Longman Group UK Ltd, Harlow, Essex, UK. 198 p.

Driessen P M, Konijn N T (1992) Land use systems analysis. Wageningen Agricultural University, Wageningen, The Netherlands and INRES (Cooperation between Universitas Brawijaya, Wageningen Agricultural University and State University Leiden), Malang, Indonesia. 230 p.

ESSA – Environmental and Social Systems Analysts Ltd. (1992) Water control project for the Quan Lo Phung Hiep Area, Mekong Delta, Vietnam: A pre-feasibility study. Final technical report. Prepared by ESSA Ltd for Ministry of Water Resources, Ho Chi Minh City, Socialist Republic of Vietnam. Vancouver, BC, Canada. 422 p.

FAO – Food and Agriculture Organization (1993) Guidelines for land use planning. Interdepartmental Working Group on Land Use Planning, FAO of the United Nations, Rome, Italy. 140 p.

Fresco L O, Huizing H, Van Keulen H, Luning H, Schipper R (1992) Land evaluation and farming systems analysis for land use planning. FAO Working Document. Food and Agriculture Organization of the United Nations, Rome, Italy. 209 p.

Gittinger J P (1982) Economic analysis of agricultural projects. The Johns Hopkins University Press, Baltimore, MD, USA. 505 p.

Huizing H (1991) Land evaluation. International Institute for Aerospace Survey and Earth Sciences P.O. Box 6, 7500 AA Enschede, The Netherlands. 172 p.

Khue N N (1991) The Vietnam river system and plain mathematical model for flow and salt concentration. Sub-Institute of Water Resources Planning and Management, Ministry of Water Resources, Ho Chi Minh City, Vietnam. 26 p.

Meinke H (1996) Improving wheat simulation capabilities in Australia from a cropping systems perspective. PhD thesis, Wageningen Agricultural University, Wageningen, The Netherlands. 270 p.

NEDECO (1992) Thematic study on management of water resources, Vol. 4. Hydraulic modelling. Mekong Delta Master Plan Project (VIE/87/031), State Planning Committee-Government of Vietnam, World Bank, Mekong Secretariat, United Nations Development Programme (UNDP). 128 p.

Nijkamp P, Rietveld P, Voogd H (1990) Multicriteria evaluation in physical planning. North-Holland Publisher, Amsterdam, The Netherlands. 702 p.

Sonntag N C, McNamee P J (1989) Eco-development planning for the Quan Lo Phung Hiep area of the Ca Mau Peninsula, Vietnam. ESSA Environmental and Social Systems Analysts Ltd, Vancouver, BC, Canada. 135 p.

Thu L T (1991) Review and assessment of the real situation in public health and social welfare in the Quan Lo Phung Hiep area, Mekong Delta, Vietnam. Institute of Hygiene and Public Health, Ministry of Health, Ho Chi Minh, City, Vietnam. 94 p.

Tri L Q, Nhan N V, Huizing H G J, Van Mensvoort M E F (1993) Present land use as basis for land evaluation in two Mekong Delta districts. Pages 299–320 in Dent D L, Van Mensvoort M E F (Eds.) Selected papers on the Ho Chi Minh City Symposium on Acid Sulphate Soils. International Institute for Land Reclamation and Improvement, Publication No. 53, Wageningen, The Netherlands.

Tuong T P, Hoanh C T, Khiem N T (1991). Agro-hydrological factors as landqualities in land evaluation for rice cropping patterns in the Mekong Delta. Pages 23–30 in Deturck P, Ponnamperuma F N (Eds.) Rice production on acid soils of the tropics. Institute of Fundamental Studies, Kandy, Sri Lanka.

Van Mensvoort M E F, Nhan N V, Tinh T K, Tri L Q (1993) Coarse land evaluation of the acid sulphate soil areas in the Mekong Delta Based on farmers' experience. Pages 321–330 in Dent D L, Van Mensvoort M E F (Eds.) Selected papers on the Ho Chi Minh City Symposium on Acid Sulphate Soils. International Institute for Land Reclamation and Improvement, Publication No. 53, Wageningen, The Netherlands.

# A quantitative tool for regional land-use analysis

D.M. JANSEN, J.J. STOORVOGEL and J.G.P. JANSEN
*Atlantic Zone Program (WAU/CATIE/MAG), Apdo. 224-7210, Guápiles, Costa Rica*

**Key words:** sustainability, biocide, linear programming model, GIS, agricultural policy, resource use, Costa Rica

**Abstract**
An interdisciplinary methodology is being developed to enable analysis of effects of agricultural policies and economic incentives on land use. A linear programming (LP) model that maximizes regional farm income subject to resource and sustainability constraints is used to select the best combination of options for land use in terms of cropping and livestock systems with defined technologies. Constraints can be set at various levels in the agricultural system: field, farm, subregion, or region. By distinguishing different farm types, the LP model explicitly takes into account the variation in resource availability at the farm level. The use of scenarios in the analysis of effects of policy interventions is demonstrated for Guácimo county, a subregion in the Atlantic Zone of Costa Rica. For this county, the methodology is used to analyse possibilities to formulate policy instruments for reducing the use of biocides.

## Introduction

Agricultural policies and economic incentives can be important tools to achieve a more sustainable use of natural resources. Policies are defined by national or regional governments, but the ultimate decision on how to use agricultural land is made on individual farms. Different farms may respond differently to certain policies or incentives. These differences in response are related to differences between farms in physical possibilities (size, soil types, climate), in socioeconomic conditions (market access and prices of inputs and outputs, possibilities to work off-farm), and in objectives and preferences of the persons managing the farm. In many situations, interactions exist between different farms, e.g. through labour exchange. Various types of models exist that enable analysis of farmers' reactions to policy incentives (e.g. Singh et al. 1986; Erenstein and Schipper 1993; Kruseman et al. 1995). Although land use and many of its consequences clearly have spatial dimensions, until recently the available literature did not offer a methodology that allows flexible georeferencing of inputs to, and outputs from, land-use evaluation models. The USTED methodology (*Uso Sostenible de Tierras En el Desarrollo*: Sustainable Land Use in Development), is meant to fill this gap (Jansen and Schipper 1995; Jansen et al. 1995; Schipper et al. 1995; Stoorvogel et al. 1995). It is being developed by the Atlantic Zone Program (AZP), a collaborative project of Wageningen Agricultural University (WAU), The Netherlands, the Tropical Agronomic Research and Higher Education Center (CATIE), Costa Rica and the Costa Rican Ministry of Agricultural and Livestock (MAG). This paper offers a brief description of the USTED methodology, and



*P.S. Teng* et al. *(eds.): Applications of Systems Approaches at the Farm and Regional Levels, 399–411.*

describes the application of USTED to Guácimo county (*cantón*) in the Atlantic Zone of Costa Rica.

## The USTED methodology

A salient feature of the USTED methodology is the explicit inclusion of different levels of analysis and decision making in the agricultural system: field, farm, and region. Various techniques and data bases are integrated in the USTED methodology, while the flow of information between the various components of USTED is managed through a customized data management software called MODUS (Stoorvogel et al. 1995). The modular structure of this software makes updating and upgrading of USTED relatively easy. One set of data bases comprises the options for land use, which are defined as combinations of a land unit (soil group) and a land-use type with a specified technology (Jansen and Schipper 1995). Such combinations are called LUSTs (Land Use Systems at a Defined Technology), where each LUST describes a specific quantitative combination of physical inputs and outputs, thus representing fixed input-output production technologies. To enable calculation of the impact of various policies, different crop and livestock activities need to be described, with a range of technologies that have different income-generating capabilities as well as different sustainability implications. These technologies can be based on prevailing field practices obtained through farm surveys, but should also reflect knowledge of alternatives, e.g. technologies with different input-output relations than those observed in the field. These alternatives can be based on information from literature, field experiments, and expert knowledge, including crop growth simulation models.

LUSTs do not necessarily depict an average crop cycle, but could be described in the form of a long-term cropping sequence, to describe the variability that occurs in the weather, pressure of pests and diseases, as well as the response of farmers to this variability. The capabilities and implications of each LUST are quantified by so-called technical coefficients, which include land and labour requirements, costs and inputs, production, and sustainability indicators. The technical coefficients for each LUST are calculated by combining the physical quantities of relevant inputs and outputs with the required attribute data, such as prices, chemical composition, and toxicity. These attributes are kept outside the LUST descriptions and are stored in separate data bases. To enable comparison of LUSTs with varying duration (as is the case with annuals and perennials), technical coefficients reflecting costs or income are calculated as annuities. The annuity of a range of future cost or income values represents a constant amount per year calculated on the basis of that range's net present value. The net present value is calculated as the sum in year one of all discounted future cost or income values, using a discount rate of 10% per year. Other data bases include the georeferenced information on farms (location, size, soil types, distance to roads, etc.), which is managed by the Geographical Information System (GIS), among others, to facilitate grouping of farms into farm types. Finally, there is a small number of non-georeferenced data on resource availability (on-farm labour force, off-farm work possibilities).

The technical coefficients of the LUSTs are input to a regional linear programming (LP) model (modified from Stoorvogel et al. 1995 and Schipper et al. 1995). The LP model calculates optimal land use at the farm level, by maximizing total regional net farm income, subject to resource constraints at the field, farm and/or regional level. Net farm income is defined as the sum of output value, income from off-farm work on other farms within the region and from other jobs, minus total production costs (including labour).

Limitations in capital availability are not taken into account, since the LP model indicates optimal land use in an equilibrium situation, without revealing the strategy of how to reach this situation. Also, the lack of information on availability of credit to farmers in the Atlantic Zone makes it impossible to include capital as a factor to distinguish farm types and to analyse the effect on land use.

Different farm types are included in the LP model, by describing for each farm type a submatrix with the relevant constraints at field and farm level, and the technical coefficients of the LUSTs that the farm type can choose from. In addition, the methodology allows for the inclusion in the LP model of different subregions (with specific sets of farm types and constraints), to reflect variations in prices and/or resource availability. By treating all prices as fully exogenous, the current version of the model does not yet include demand aspects, thus avoiding the issue of price-supply relations (Hazell and Norton 1986).

Trade-offs between different policy goals can be studied in USTED through scenarios in which the effects on land use are analysed for different (hypothesized) changes in the socioeconomic environment, generally induced by policy instruments. Specific attention is given to the trade-off between economic growth (or regional income) and sustainability. To study the effect of policies or incentives, scenarios are defined as changes relative to a standard situation (the 'base scenario') in attribute data, parameters for the calculation of technical coefficients, constraints or the objective function of the model. The scenario outcomes are in terms of adjustments in land use and choice of technologies, and can be visualized using the GIS through the generation of maps.

## Operationalization of USTED for the Atlantic Zone of Costa Rica

To demonstrate the possibilities of the methodology, USTED was used to analyse land use in the Atlantic Zone (AZ) of Costa Rica. The climate in the region is classified as humid tropical, without dry months (Herrera and Gómez 1993). Average annual rainfall is 3600 mm (1972–88) with an average daily air temperature of about 25°C (1976–88). The various soil series of the original soil maps are aggregated into three main groups: fertile poorly drained (FPD), fertile well drained (FWD), and infertile well drained (IWD) soils. The subregion used in this demonstration, the county of Guácimo, comprises 58 000 ha and 1816 farms (based on the 1984 census by DGEC 1987). Size and location of farms in settlements are known from data from the Institute for Agricultural Development (IDA) that established the settlements. Size and location of plantations and other large farms, as well as of forested areas, are

*Table 1.* Characteristics of the farm types in Guácimo county.

| Type | Average size (ha) | % FWD[1] | % FPD[2] | % IWD[3] | No.of farms |
|---|---|---|---|---|---|
| 1 | 12 | 79 | 19 | 2 | 639 |
| 2 | 12 | 8 | 84 | 8 | 150 |
| 3 | 12 | 0 | 8 | 92 | 274 |
| 4 | 22 | 80 | 16 | 4 | 165 |
| 5 | 22 | 7 | 87 | 6 | 248 |
| 6 | 22 | 0 | 7 | 93 | 286 |
| 7 | 50 | 40 | 30 | 30 | 4 |
| 8 | 50 | 7 | 72 | 21 | 34 |
| 9 | 50 | 0 | 4 | 96 | 16 |
| Region | 17 | 31 | 34 | 35 | 1816 |

[1]% FWD: Fertile Well Drained soil as % of total farm area within group.
[2]% FPD: Fertile Poorly Drained soil as % of total farm area within group.
[3]% IWD: Infertile Well Drained soil as % of total farm area within group.

derived from aerial photographs taken in 1992. Land use in these areas is assumed to be stable, or at least not much affected by the agricultural policies aimed at small- and medium-scale farmers. In the remaining part of the region, the size of individual farms can be estimated from parcel size (Stoorvogel, unpublished data), with the latter determined from the 1992 aerial photographs. In all, nine farm types are distinguished by combining size and soil type distribution (Table 1), where the latter is found using the GIS to make overlays of farm location maps and soil maps.

A constant average labour availability per farm household of 2.0 full-time equivalents is assumed, making each farm type homogeneous in its land to labour ratio. Research in the Atlantic Zone has not resulted in usable classifications of 'farmer types' (Schipper et al. 1995). Also, farms in many areas in the Atlantic Zone (AZ) at present frequently change ownership and, hence, style of farmer. Therefore, no attempt was made to include possible differences between farmer types in the criteria for differentiating farm types.

Seven land-use types are considered important to small and medium size farms in the AZ: cassava, maize, palm heart, pasture with cattle, pineapple, plantain, and tree plantation. In total, 169 LUSTs are described for relevant crop-soil combinations. Differences between LUSTs reflect different sowing dates and/or different input-output relations. Differences in output levels are arrived at by varying the amount of fertilizer (from none to an amount needed for near-potential production) and/or biocides (different strategies in chemical protection against insects, nematodes and diseases, and/or replacing herbicide spraying with manual weeding).

Job opportunities outside the farm in the AZ are mainly on neighbouring farms and in banana plantations. While supply and demand of work on neighbouring farms has to be in equilibrium in the LP model, opportunities to work on banana plantations are assumed non-restricting in view of generally prevailing labour shortages in the AZ.

Prices of all inputs and outputs are in local currency units (Colones) and refer to the year 1990. Outcomes of the model are converted to US dollars at the exchange

*Table 2.* Constraints that can be invoked in the LP model, in relation to the type of analysis.

| Resource | Description | Level | Lower limit | Upper limit | Analysis |
|---|---|---|---|---|---|
| Labour | Monthly use | Farm | 0 | Number of full-time labourers at farm multiplied by number of working hours per month | Always |
| Labour | Monthly use outside farms | Region | 0 | Number of hours required monthly for work outside farming sector | Effect of changing possibilities to work outside farming sector |
| Labour | Monthly hired by other farms | Region | Amount of labour (hours) regionally available minus labour used on own farms and in non-farming sector | See lower limit | Always |
| Land | Monthly used area per soil type | Farm | 0 | Total area per soil type at each farm | Always |
| Soil nutrients | Yearly change in soil N, P, K per unit soil type | Farm | Depends on scenario | Depends on scenario | Limiting soil nutrient losses |
| Biocides | Average yearly $BEII_f$ per ha | Farm | 0 | Depends on scenario | Restrictions to biocide use |
| Biocides | Average yearly $BEII_f$ per ha | Region | 0 | Depends on scenario | Restrictions to biocide use |

rate prevailing in that year, of 120 Colones per US$. Wages are set at US $0.55 per hour for own-household labour, US $0.82 for hired labour, US $0.74 for work on other farms, and US $1.54 for work on a banana plantation. The discount rate for calculating the annuities is set at 10%, reflecting the opportunity cost of capital at the time.

Resource and sustainability constraints in the LP model relate to land, labour, soil nutrients, and biocides (Table 2). Soil nutrients and biocides are used to indicate changes in sustainability of the agricultural system, since soil nutrient depletion and biocide use are considered the principal constraints to the sustainability of agriculture in the AZ (Jansen et al. 1995). Sustainability indicators are quantified by balances for N, P and K in the soil, and by a biocide environmental impact index (*BEII*). N, P and K balances are calculated with an adapted version of the NUTBAL model (Stoorvogel 1993), through separate assessments of nutrient inputs (mineral and organic fertilizer, wet deposition, N-fixation) and nutrient outputs (products, denitrification, leaching). For each commercial biocide formula ($b$), the $BEII_b$ (in units $kg^{-1}$) is calculated according to:

$$BEII_b = \sum_{a=1}^{n} AI_{a,b} \times TOX_a \times DUR_a$$

in which $a$, $n$ is $a^{th}$, and the total number of biocides in commercial product $b$; $AI$ is the fraction active ingredient of biocide $a$; $TOX$ is the indication of toxicity of biocide

$a$, increasing with the toxicity according to its WHO code (Jansen et al. 1995); and $DUR$ is the indication of persistence of biocide $a$ in the environment (here taken as the squared root of its half lifetime).

For each LUST ($l$) the $BEII_l$ (in units ha$^{-1}$ yr$^{-1}$) is the sum of the multiplication of the amount used of each commercial biocide formula ($A_{l,b}$) with its specific $BEII_b$, divided by the duration of the LUST in years ($Y_l$):

$$BEII_l = \frac{1}{Y_l} \times \sum_{b=1}^{m} A_{l,b} \times BEII_b$$

in which $m$ is the total number of commercial biocide formulas used in LUST $l$.

For each farm type ($f$), the $BEII_f$ (units yr$^{-1}$) is the sum of the $BEII_l$ of the LUSTs chosen for this farm type, multiplied by the area under this LUST ($S_{f,l}$). At the regional level ($r$), the $BEII_r$ (units yr$^{-1}$) is calculated as the sum over the farm types of their $BEII_f$:

$$BEII_f = \sum_{l=1}^{k} S_{f,l} \times BEII_l$$

$$BEII_r = \sum_{f=1}^{i} S_{f,l} \times BEII_f$$

with $k$, $i$ being the total number of LUSTs per farm type and farm types per region, respectively.

## Impact analysis of selected policies

To illustrate the potential of the USTED methodology, an example is presented of how USTED can contribute to identification of policies to minimize biocide use. First, optimal land use under actual conditions is calculated in the base scenario. This is followed by an analysis of trade-offs between income and biocide use. Finally, the effects of changes in taxes on biocides are calculated, to indicate the best option for changing biocide use.

### *Base scenario*

In the base scenario, constraints and attributes are set in such a way as to approximate the prevailing attitude of farmers. Thus, no constraints to sustainability indicators are included in the model, since aspects of sustainability do not seem to play a role in land-use decisions by individual farmers in the AZ. Results of the LP model indicate that net income is significantly different across farm types, and is closely correlated to physical potential. Farms with a relatively high percentage of poorly drained soils have lower incomes, a large fraction of which is generated by family members working outside their own farm (Table 3). Optimal land use for all farms consists of a combination of extensive cattle ranging (on fertile poorly drained soils),

*Table 3.* Net income, labour use, and land use under the base scenario (per farm for each farm type and total over the region).

| Farm type | 1 | 2 | 3 | 4 | 5 | 6 | 7 | 8 | 9 | Region |
|---|---|---|---|---|---|---|---|---|---|---|
| Net income (US$ $yr^{-1}$) | 15721 | 8567 | 22780 | 24452 | 10293 | 27228 | 49016 | 24938 | 42868 | 33676805 |
| Total labour use (h $yr^{-1}$) | 2650 | 724 | 4516 | 5059 | 1048 | 6024 | 11569 | 5549 | 9939 | 6250910 |
| Hired labour (h $yr^{-1}$) | 54 | 0 | 1633 | 713 | 0 | 1960 | 6433 | 1767 | 4541 | 1318252 |
| Work on other farms (h $yr^{-1}$) | 944 | 1735 | 12 | 334 | 1021 | 427 | 0 | 385 | 496 | 1318252 |
| Plantation work (h $yr^{-1}$) | 2386 | 3407 | 2893 | 1315 | 3816 | 1474 | 826 | 1717 | 646 | 4485410 |
| $BEII_r$ ($yr^{-1}$ $ha^{-1}$) | 34.7 | 7.4 | 45.8 | 36.0 | 5.7 | 43.0 | 32.0 | 13.6 | 42.2 | 30.0 |
| Use of FPD (ha) extensive cattle | 2.3 | 10.1 | 0.9 | 3.6 | 19.3 | 1.6 | 15.0 | 35.7 | 1.8 | 10337 |
| Use of FWD (ha) palm heart | 9.5 | 1.0 | 0.0 | 17.7 | 1.6 | 0.1 | 20.0 | 3.8 | 0.0 | 9775 |
| Use of IWD (ha) palm heart | 0.0 | 0.0 | 0.0 | 0.0 | 0.0 | 9.4 | 0.0 | 0.0 | 39.9 | 3335 |
| Cassava | 0.2 | 0.9 | 11.1 | 0.8 | 1.2 | 10.9 | 15.0 | 10.4 | 8.2 | 7387 |

*Notes:* See Table 1 for characteristics of each farm type and for meaning of FPD, FWD and IWD.
$BEII_r$: regional Biocide Environmental Impact Index (see text).

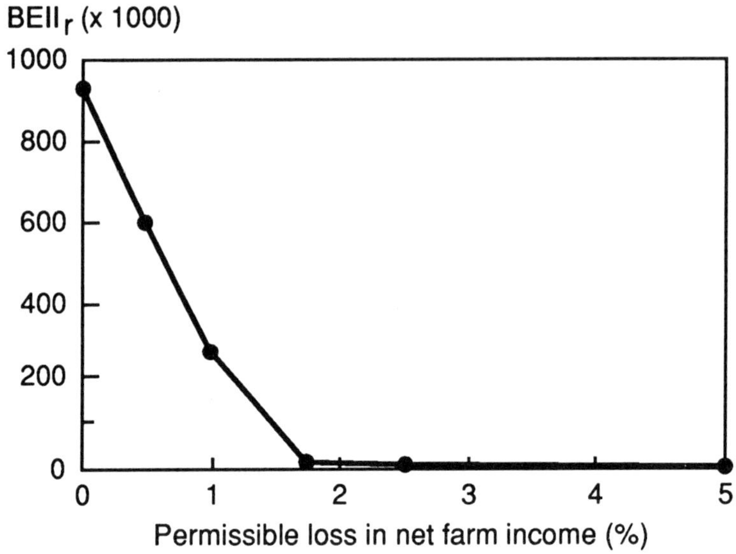


*Figure 1.* Relation between permissible loss in net farm income and minimum regional biocide environmental impact index ($BEII_r$).

palm heart with zero fertilizer input (on fertile well drained and infertile well drained soils) and cassava with high fertilizer input (on infertile well drained soils). The use of herbicides (in both palm heart and cassava), and of insecticides (only in cassava), lead to a regional $BEII_r$ of $9.26 \times 10^5$, while on a hectare basis the $BEII_f$ ranges from 5.7 to 45.8, depending on the farm type (Table 3).

*Possibilities to reduce biocide use*

Policies aimed at changing the behaviour of people can only become effective when the target group has viable alternatives. In the case of biocides, it should be checked whether a reduction in their use does not lead to an unacceptably large reduction in farm income. One option for diminshing biocide use is the replacement of herbicides by labour for manual weeding. The banana plantations in the Atlantic Zone depend to a large extent on the local (farm) population to provide labour. Since these plantations generate large tax revenues, it would be important for the government to check whether a reduction in biocide use does not have unacceptable implications in terms of changes in the availability of labour. This type of checking can be done by analysing the results of the model when its objective is changed from maximization of net income to minimization of the regional $BEII_r$. At the same time, a lower limit has to be set to net income such that each farm type achieves at least a certain percentage of its net income calculated under the base scenario.

In the case of the county of Guácimo, the model results indicate that the average yearly regional $BEII_r$ can be reduced greatly without much effect on either farm

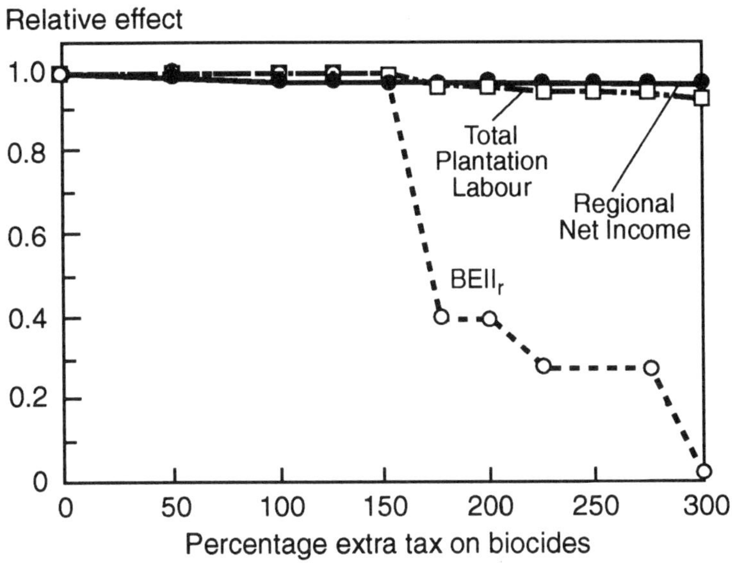


*Figure 2.* Relative effect of extra biocide tax on regional net income, regional biocide environmental impact index ($BEII_r$) and total plantation labour.

income (Figure 1) or the supply of labour to plantations. Since the LUSTs used in this calculation seem realistic, this result is technically feasible, mainly by replacing technologies where herbicides are used by those that use hand weeding. This outcome suggests that it is worthwhile to investigate the effect of various alternative policies aimed at a reduction in the use of biocides.

*Taxing biocide use*

Various alternative tools exist that potentially affect biocide use. For example, the use of (certain types of) biocides may be prohibited; users may be educated better about the undesired effects of biocides; or prices of products grown without biocides may be increased. The analysis in this paper focuses on the effect of increases in taxes on the biocides, since taxation is an instrument that is relatively easy to implement by a government.

When prices of biocides are increased across the board (i.e. a flat tax on all biocides), the model predicts that the $BEII_r$ decreases strongly when the price increase exceeds 150%, while total net farm income and total plantation labour decrease only marginally (Figure 2). The decrease of the $BEII_r$ occurs in steps: at a tax level of 175%, the use of biocides in palm heart is diminished, by replacing herbicide spraying by hand weeding (Figure 3). When the tax rate exceeds 225%, herbicides are also replaced by hand weeding in about a third of the cassava, while at tax rates above 275%, all weeding occurs manually. In all situations, the area under extensive cattle systems remains constant at $10^5$ ha, since on the poorly drained soils there are hardly

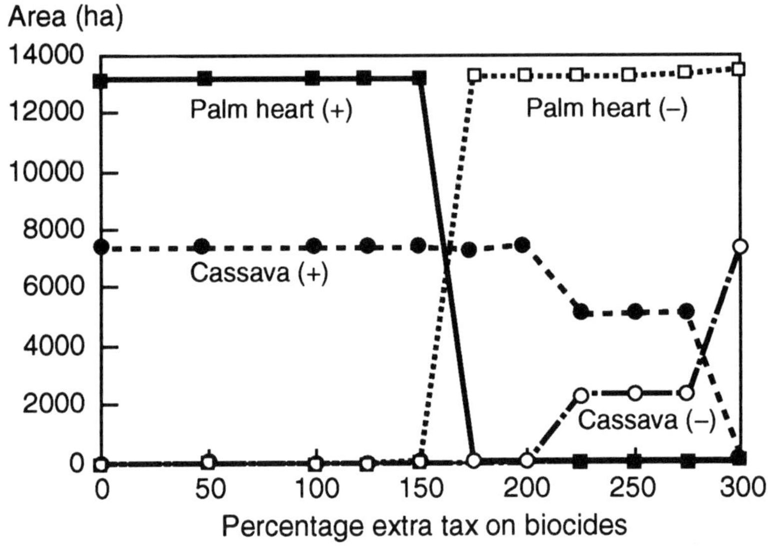


*Figure 3.* Effect of extra biocide tax on total regional area under four groups of land use: palm heart and cassava with (+) or without (–) use of herbicides.

any alternative land uses; cattle ranging requires little labour and does not use biocides.

When increasing prices of all biocides with the same percentage, no difference is made between biocides that have a high environmental impact (or high $BEII_b$) and those that are relatively harmless (low $BEII_b$). To 'punish' more severely the use of the former, a progressive tax is developed, that takes account of the BEII of each individual biocide:

$$Tax_b = \alpha \times BEII_b$$

For $\alpha$ varying between 0.001 and 0.02, the model predicts that such a progressive tax achieves a larger reduction in the $BEII_r$ for each percent of regional income loss than the flat tax (Figure 4). When the use of biocide is to be reduced, such a progessive tax would be preferred over a flat tax, although it would cost a little more effort (and information) to implement it.

## Discussion and conclusions

The USTED methodology for land use analysis links an LP programming model to various data bases and a GIS. The LP model takes account of differences in resource availability at the farm level. However, different farm types are not optimized in isolation, since each type has to take into account the subregional labour supply and demand. Since the methodology explicitly considers both income and sustainability related factors, it can be used to provide the policy maker with information regarding trade-offs between different goals.

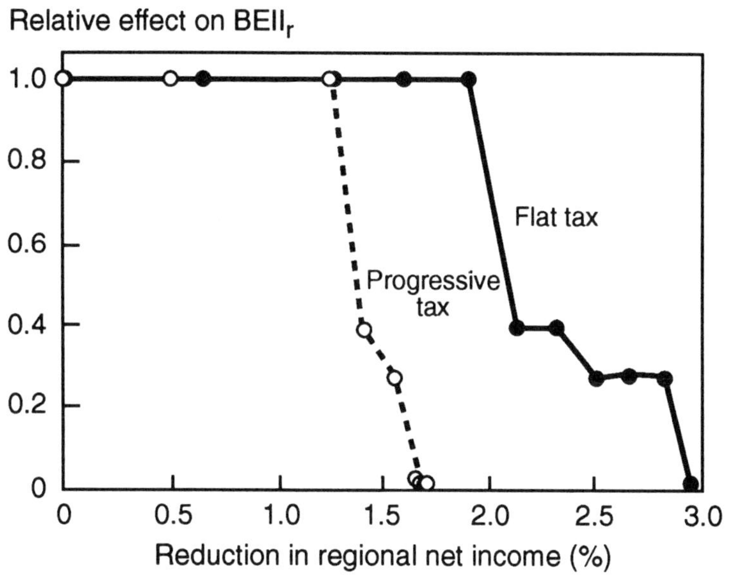


*Figure 4.* Relation between relative effect on regional biocide environmental impact index ($BEII_r$) and reductions in regional net income as affected by two tax systems (see text).

In USTED, farm types are differentiated on the basis of their physical potential (sizes, soil types), whereas subregions are differentiated according to socioeconomic situation (mainly prices) and climatic conditions. Farm types in USTED do not reflect differences between farmers, although it is recognized that these can influence technology adoption decisions (Van Der Ploeg 1990). In fact, here it is assumed that all farmers pursue the same goal: maximization of income. Inclusion of farmer types with different goals (e.g. maximizing income and minimizing income variability) in one model-run is not possible, since LP models can have only one objective function, which should not consist of variables with different units of measurement. Multiple-objective programming, where different goals are optimized in sequential runs, requires subjective measures to weigh the various goals, and as such does not compare goals on the same base. When farmers differ not in goal, but in their preference on how to reach that goal (e.g. cattle farmers versus arable farmers), these differences could be reflected in the USTED methodology, by allowing certain farm types to choose only from a certain set of LUSTs.

Although the area under palm heart is expanding rapidly at present, palm heart and cassava are not as dominant in county Guácimo as calculated by the LP model. Work is in progress to improve the model in terms of its tracking ability of prevailing land-use patterns. Since the emphasis of this paper is on demonstrating the potential use of the methodology, the model results should be regarded as of a preliminary nature. Furthermore, it should be realized that the outcomes of LP models show the exact optimum results, while, at the same time, there are possibly many solutions that deviate only slightly from that optimum. A small change in technical coefficients (e.g. input or output prices) could result in a drastic change in the combination of LUSTs that form the exact optimum. The stepwise shifts in $BEII_r$ (Figure 2) and land-use patterns (Figure 3) illustrate this feature of LP models. The outcome of the LP model should, therefore, be seen as an indicator of change, more than as a precise prediction of how the change will look.

In its present state, USTED can be used to analyse land-use options at the farm level, reflecting the situation where prices are assumed not to be affected by the production at the farm. As such, USTED can be used for an *ex ante* analysis of adoption at the farm level of promising new technologies. It also can be used as an exploratory model, to investigate possibilities and perspectives for land use in the long term (e.g. Rabbinge and Van Latesteijn 1992). Performance of USTED in these types of analysis can be improved by including risk considerations, related to both prices and yields.

At present, the LP model is not yet operational at the regional level, since differences in prices and costs between subregions are not yet included. Also, even for a subregional model, the fixed price assumption is not strictly applicable if the subregion's share of national supply is not negligible. The incorporation of product demand functions will make the model more useful for policy evaluation purposes. In Costa Rica, but also in many other situations, labour moves between regions, while prices of agricultural products are often at least partly influenced by supply from other regions. To further improve USTED such that it becomes an applicable policy tool, it should be extended to include more regions. Although LUSTs have been made for the seven main land-use types in the Atlantic Zone, a larger number of land-use types may need to be included, including various alternative animal production systems.

Adequate incorporation of aspects of sustainability in the model warrants continuing attentions as well. Effects of changes in land use on sustainability can be expressed in changes in the values of the sustainability indicators, or by assigning an economic value to these changes (Painter and Young 1992). For example, costs to replace nutrient losses from the soil nutrient stock can be estimated using actual fertilizer prices and approximations for fertilizer replacement efficiencies (WRI 1991; Ehui and Spencer 1993). The costs of recovering nutrient losses can then be subtracted from the benefits in terms of net farm income, thus assigning equal weights to farm income and sustainability.

Estimation of the costs of the environmental impact of biocides is more difficult. The BEII is a subjective indicator of the damage caused by biocides. Even though it enables an assessment of the trade-off between income and biocide use, it should be realized that the BEII is ordinal, rendering its interpretation difficult. This situation can be improved by assessing the real costs associated with the use of biocides in terms of environmental degradation and human health damage (Pingali et al. 1994). For environmental degradation, however, this is not feasible at present due to lack of data, a problem that is not unique to the Atlantic Zone. Nevertheless, the Dutch Ministry of Agriculture, Nature Conservation and Fishery (MLNV) uses a very simple kind of BEII, i.e. the amount of biocides used, as an indicator for the sustainability of agriculture in The Netherlands. The official policy is to reduce this amount by 50%

in the year 2000 compared to the average use in the period 1984–88 (MLNV 1990), even though the impact of the biocides is not well known.

The cropping history of a field can affect the performance of following crops, e.g. through the build-up of soil-borne pathogens or depletion of soil nutrients. In theory, this interaction can be tackled with multi-period LP models but, in reality, these do not always solve the problem. In USTED, for example, soil nutrient depletion is estimated for each LUST as a flow variable, without taking into account the effects of changing soil fertility. Since each LUST affects the soil nutrient pools differently, after a few periods it becomes practically infeasible to keep track of the different soil types that are formed due to the different cropping histories. Instead, in the USTED methodology a more promising route would be to set limits to the tolerated yearly losses of nutrients from the soil. The latter can be calculated such that no effect of this loss on crop production is expected within a certain number of years. Alternatively, long-term LUSTs can be described that include the interaction between technology, soil nutrient balances, and yields.

## References

DGEC – Directorate General for Statistics and Census (1987) Agrarian census 1984 (in Spanish). Ministry of Economy, Industry and Commerce, San José, Costa Rica. 216 p.

Ehui S K, Spencer D S C (1993) Measuring the sustainability and economic viability of tropical farming systems: A model from sub-Saharan Africa. Agric. Econ. 279–296.

Erenstein O C A, Schipper R A (1993) Linear programming and land use planning: A case study of Matara district, Sri Lanka. Wageningen Economic Studies 28. Pudoc, Wageningen, The Netherlands. 95 p.

Hazell P B R, Norton R D (1986) Mathematical programming for economic analysis in agriculture. Macmillan Publishing Company, New York, USA. 400 p.

Herrera W, Gómez L D, (1993) Map of the biotic units of Cost Rica (in Spanish). Geographic Institute of Costa Rica, San José, Costa Rica.

Jansen D M, Schipper R A (1995) A static, descriptive approach to quantify land use systems. Neth. J. Agric. Sci. 43:31–46.

Jansen D M, Stoorvogel J J, Schipper R A (1995) Using sustainability indicators in agricultural land use analysis: An example from Costa Rica. Neth. J. Agric. Sci. 43:61–82.

Kruseman G, Ruben R, Hensdijk H, Van Ittersum MK (1995) Farm household modeling for estimating the effectiveness of price instruments in land use policy. Neth. J. Agric. Sci. 43:111–123.

MLNV – Ministerie van Landbouw, Natuurbeheer en Visserij (1990) Multi-year plan for crop protection: Policy intentions (in Dutch). MLNV, The Hague, The Netherlands. 273 p.

Painter K, Young D (1992) Technology development versus policy reform: How to best encourage sustainable farming. Am. J. Agric. Econ. 74:1304.

Pingali P L, Marquez C B, Palis F G (1994) Pesticides and Philippine rice farmer health: A medical and economic analysis of impact. Am. J. Agric. Econ. 76:587–592.

Rabbinge R, Van Latesteijn H C (1992) Long-term options for land use in the European Community. Agric. Syst. 40:195–210.

Singh I, Squire L, Strauss J (Eds.) (1986) Agricultural household models: Extensions, applications, and policy. Johns Hopkins University Press for the World Bank, Baltimore, MD, USA. 335 p.

Stoorvogel J J (1993) Optimizing land use distribution to minimize nutrient depletion: A case-study for the Atlantic Zone of Costa Rica. Geoderma 60:277–292.

Stoorvogel J J, Schipper R A, Jansen D M (1995) USTED: A methodology for quantitative analysis of land use scenarios. Neth. J. Agric. Sci. 43:5–18.

Van Der Ploeg J D, (1990) Labor, markets, and agricultural production. Westview Press, Boulder, CO, USA. 313 p.

WRI – World Resources Institute (1991) Accounts overdue: Natural resource depreciation in Costa Rica. WRI, Washington DC, USA and Tropical Science Center, San José, Costa Rica. 110 p.

# Computer aided training on the environmental impacts of agricultural development projects: an Egyptian case study

G. EDWARDS-JONES[1] and A.A. IBRAHIM[2]

[1]*Rural Resource Management Department, Scottish Agricultural College, West Mains Road, Edinburgh, Scotland EH9 3JG, UK*

[2]*Institute of National Planning, Salah-Salem Road, Nasr City, Cairo, Egypt*

**Key words:** environmental impact assessment, training, development planning, knowledge-based systems, irrigation, Egypt

**Abstract**

A knowledge-based system was developed to help a multilateral agency train natural resource planners to be more aware of the full environmental impacts of agricultural development projects. This system, called ECOZONE, has been used successfully in training, but its generic nature limits its ability to capture the true complexity of any given development project. As such its use in training is constrained. A specific case study was developed which used the ECOZONE framework in order to provide a more in-depth representation of the impacts associated with a given development project. This paper describes the ECOZONE software and its uses, and then considers the development of a model of the North Sinai Agricultural Development Project, a large irrigation project in northeast Egypt. The use of this particular case study in training is described, and opportunities for further development of this concept are explored.

## Introduction

As development planning has evolved, there has been a move away from a purely financial analysis of the viability/desirability of projects towards a more holistic view of planning as a multi-objective process of considerable complexity. This increase in complexity has meant that individual planners are less likely to have the skills, time or resources to carry out all the different aspects of a modern, holistic plan. This is not a criticism, as planners are not expected to be experts in their own particular subject and also to have the time, resources or capabilities necessary to fully understand and predict the direct, indirect and cross-sectoral impacts of their planned projects. It is becoming increasingly desirable, however, that development planners are aware of the range of environmental impacts which may arise from projects. It is also desirable that having identified the potential environmental impacts of any given project, modifications should be made in order to minimize adverse environmental effects. Thus, planners need adequate training in environmental science and environmental impact assessment (EIA). This should include as a minimum an appreciation of the overall process of an EIA and the relevant legislative requirements for EIA in the planning process. In addition, the training should relate projects to the impacts they commonly cause, to explain the science behind the causation of these impacts, to suggest relevant mitigating actions and to reinforce the role of good project design in achieving sustainable development.



*P.S. Teng* et al. *(eds.): Applications of Systems Approaches at the Farm and Regional Levels, 413–425.*

*Table 1.* Proposed land allocation and planned plot size per settler in the NSADP.

| Settler type | % of project land allocated | % of population in project area | Size of plots allocated (feddan) | | No. of farms on plots |
|---|---|---|---|---|---|
| Investors | 52 | 7 | 20.0 | (Sand sheet) | 170 |
| | | | 80.0 | (Deep sand) | 1305 |
| Smallholder | 23 | 43 | 4.5–5.2 | (Caly/loam) | 7650 |
| | | | 5.0 | (Sand sheet) | 1610 |
| Graduates | 15 | 20 | 10.0 | (Sand sheet) | 710 |
| | | | 6.6 | (Deep sand) | 3705 |
| Bedouin | 10 | 30 | 3.3 | (Deep sand) | 6470 |

These training requirements are widely recognized by many overseas aid organizations. It was this recognition by The Food and Agriculture Organization of the United Nations (FAO) which provided the impetus for the work described here. This paper provides a brief overview of a generic knowledge-based system (KBS), named ECOZONE, and details a second system which models a specific case-study development project, the North Sinai Agricultural Development Project (NSADP) (Tables 1 and 2). The use of these two types of models in training is described, and the potential for further developments is discussed.

## The development of ECOZONE

The ECOZONE suite of models have been described in detail elsewhere (Edwards-Jones and Gough 1994a,b; Edwards-Jones and Hopkins 1995). A brief overview is provided here.

### *Project aims*

The aim of the project was to develop a computer-based system for use in training development planners, who may possess little knowledge of environmental science, to be more aware of the environmental impacts of projects. As such it was decided that the system should: (a) contain knowledge about a wide range of development activities; (b) be flexible enough to be suitable for training planners from all countries, yet contain some element of geographic specificity (preferably compatible with current FAO designations); (c) recognize (and represent) the complexity of environmental systems and be capable of demonstrating both the higher order and cross-sectoral effects of a project and make explicit the interaction between environmental, economic and social systems; (d) possess extensive explanation facilities; (e) be a suitable vehicle around which to base training courses, and preferably enable interactive group work; and (f) enable specific examples of development projects to be entered for use as case-study material.

*Table 2.* Summary of primary positive and negative impacts likely to arise from the construction and operational phases of the NSAPD.

| Impact type | Project phase | |
|---|---|---|
| | Construction | Operation |
| Positive | Clearing of mines and explosives<br>Creation of employment | Increased employment<br>Improved farming practices (settlers)<br>Improved medical and sanitary conditions<br>Introduction of new fishery resources and new species |
| Negative | Increased wind erosion<br>Unthoughtful dumping of salt crust of Tina Plain<br>Destruction of adjacent desert land due to quarrying<br>Destruction and disturbance of historical and archaeological sites<br>Noise pollution | Crop reduction and salinization due to water shortage<br>Diminishing inflow of Nile water into Lake Manzala<br>Increased seepage of contaminated groundwater in Lake Bardawil<br>Impact of eutrophic irrigation waters on fisheries<br>Contamination of groundwater due to agrochemicals residues<br>Increased health hazards due to wastewater<br>Soil degradation due to waterlogging and salinization<br>Impact of illegal fishing practices<br>Increase in bird harvesting |

### *Knowledge acquisition and engineering*

A knowledge-based approach was adopted in the development of ECOZONE. This was done, first, because there was little quantitative data of a suitable scale in the domain of impact prediction, but much qualitative data and experiential knowledge did exist within the knowledge bases of individual domain experts within FAO. Second, the global scale of the model meant that, even if suitable data had been available, a system based on numerical algorithms would have been extremely complex, large and expensive to develop.

Knowledge acquisition for ECOZONE was initially limited to text analysis. However, subsequent to the construction of an initial rule-base, the rules were checked and amended by domain specialists in FAO.

The rule-base was implemented on the computer as a causal network (Shachter and Kenley 1989). This approach permitted accurate representation of the connectivity of environmental systems, whilst also being easily transcribed into computer code and simple to amend.

### *Software*

Due to the requirement to integrate KBS with hypertext and the necessity of software for use in LDCs to be easily available, inexpensive and robust, it was decided to use the commercially available Toolbook (Asymetrix Corporation, Washington, USA) as the development environment. Toolbook is an object-oriented, card/book-based

*Table 3.* An example of the output of NSADP. Here in the project phase 'Construction' and sector 'Arable' the selected activity is 'Removal of salt crust from Tina plain' which may lead to three primary impacts. All three of these could be analysed further, but for the purposes of clarity only the further impacts arising from 'dumping' of salt crust in sea/lagoons are shown.

| | |
|---|---|
| Project phase: | Construction |
| Sector: | Arable |
| Activities: | Removal of salt crust from Tina plain<br>Disturbance of natural vegetation |
| Primary impacts of: | Removal of salt crust from Tina plain<br>1, dumping of salt crust on land<br>1, dumping of salt crust in sea/lagoons<br>1, increased wind erosion |
| Further impacts of: | 1, dumping of salt crust in sea/lagoons<br>  2, temporary hypersaline environment<br>    3, disruption of natural aquatic ecology<br>      4, reduced biodiversity<br>      4, reduced production in traditional fisheries |

development environment which runs under Windows. Runtime files can be provided with the ECOZONE software, and loading the software is simply a matter of copying discs into an appropriate Directory.

### *Representation of primary and higher-order impacts in the ECOZONE models*

Primary and higher-order impacts are presented to the user through a combination of numerical notation and paragraph indents. In this system, the number signifies the level of impact, i.e. 1 for a primary impact caused directly by the project, and 2 for secondary impacts caused by a primary impact (see Tables 3 and 4).

In reality, the number of higher-order impacts is potentially vast and the knowledge within the system is structured in order to permit realistic simulation of this process. However, when the system is in normal use, no more than five levels of impacts are presented to the user. It was decided to limit the search to five levels after considering the requirement to demonstrate the real complexity of environmental systems and the need to keep search times short.

### *The hypertext information system (HTIS)*

Given that hypertext had been successfully incorporated into other computer systems (Estep et al. 1989; Edwards-Jones et al. 1992), and seemed to meet the requirements of the current system, it was decided to incorporate a free-text data base alongside the knowledge base, with these being connected via hypertext links. In order to render the HTIS more amenable to search by inexperienced users, it was partitioned in to several sections, and upon initial entry into the system the user may choose which section to enter. Available sections include a glossary which contains a brief definition/description of terms and phrases, a text encyclopaedia which contains

*Table 4.* An example of the final impact network for one of 14 primary impacts of introducing housed dairy cattle into the NSADP.

1, use of Friesian/Holstein cattle
  2, potential for high yields
  2, poorly adapted to local conditions
    3, high water requirements
    3, poor disease resistance
      4, decreased yields
        5, decreased income
      4, increased use of veterinary medicines
        5, increased costs of production
        5, need access to qualified vets
    3, poor heat tolerance
  2, need high-quality feed
    3, use of feed concentrates
      4, decreased fodder production
      4, increased cereal production
      4, increased local traffic
        5, increased human health hazard
        5, increased noise disturbance
        5, increased pressure on roads/infrastructure
      4, increased costs of production

textual information on ecozones, sectors, activities and impacts as would be found in a normal book, a section entitled 'Mitigation' which discusses possible methods of avoiding or mitigating impacts of activities, a section named 'Examples' which describes selected case studies of specific projects in various regions and details on their impacts, mitigation procedures, etc. and, finally, a section containing a list of references used during the knowledge acquisition stage of the project, as well as other relevant sources of information. Within each of these sections facilities are available to print, edit or enter further information.

## Specific capabilities of the ECOZONE models

### *ECOZONE-Impacts*

ECOZONE-Impacts incorporates the predictive knowledge base and the hypertext system and serves to collate and relate fragments of knowledge pertaining to development projects. It allows users to build up an understanding of the influence of projects on the environment and how these vary between ecological zones. It recognizes three geographical regions: upland, lowland and coastal; and two climatic regimes: arid and humid. It also recognizes five different sectors within which may occur a varying number of activities. The five sectors are agriculture, forestry, livestock, aquaculture and water resources, and between them they contain 61 activities. However, the activities available for selection within each sector are determined by the combination of geographic region and climatic regime. This sub-sectioning of the total list aims to restrict selection of activities to those most likely to be implemented in a given ecological zone, e.g. seven aquaculture activities are listed for humid coastal

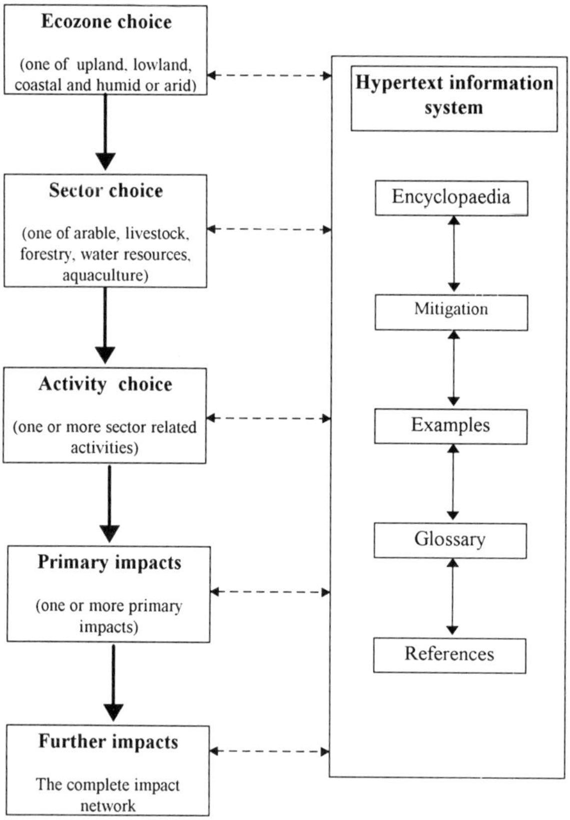


*Figure 1.* Structure of ECOZONE-Impacts. Parentheses indicate user choice necessary to proceed. Solid arrows indicate flow of movement through the software, dashed arrows indicate links available if required.

regions and only two for upland arid regions. The structure of ECOZONE-Impacts is given in Figure 1.

*ECOZONE-Workshop*

Built around the same framework as ECOZONE-Impacts and including full hypertext links, ECOZONE-Workshop is designed to permit different groups of experts to enter rules/facts about the expected environmental impacts related to their specialist sector within a given project. Once completed, these rule-bases may be viewed separately in order to demonstrate to non-specialists the environmental impacts of a given project. Alternatively, the separate rule-bases may be merged into one large rule-base in order to view the overall environmental impacts of the project. Once entered, the rule-base is accessed in a similar manner to that used in ECOZONE-Impacts. The use of a scrolling glossary and a mouse operated rule entry system minimizes keyboard inputs throughout the system. However, should users so desire, edit facilities are available which permit rules to be added/deleted from the glossary, and/or entered into RuleMaker via the keyboard.

*ECOZONE control and EmptyBooks*

Utilizing a similar structure to ECOZONE-Workshop this program allows the user to create empty ECOZONE books and then enter rules into the impact prediction system and information into the HTIS. For example, books could be created for specific case studies, projects, regions or countries. A control book allows the user to access existing books and create new empty books as required. In this way, a library of books containing information specific to certain areas or situations may be developed.

## The use of ECOZONE in training programmes

The ECOZONE models were initially tested in 1992 and early 1993 with university undergraduate and postgraduate students, and participants of a United Nations Development Programme (UNDP) course for planners and administrators from LDCs. Members of these groups completed questionnaires about the system and the training methods employed, and the results of this exercise were subsequently used to improve both elements. During this process, three different types of training programme were evaluated:

(1) Unstructured learning, required students to access either ECOZONE-Impacts or a specific region/project book, scripted using ECOZONE-Emptybook, at will, and by simply playing with the system to learn about the environmental impacts of projects. This type of learning, although suitable for college/university students who have access to the system throughout their academic year, was not suited for 1–2 week training courses.

(2) Task-related training required the students to use the software in order to answer specific questions set individually or as part of a case study. This requires students to use the impact prediction element of ECOZONE to identify environmental impacts and the HTIS to help identify potential mitigating activities. This approach is suitable for short training courses and, by presenting different tasks to different students and getting them to recount their results, leads easily into group discussion. It was apparent, from the pilot testing, that this training method was much more worthwhile when used to analyse case studies of real projects.

(3) When using ECOZONE-Workshop students are treated as an operational project appraisal team and, according to common practice, are split into groups according to discipline, e.g. livestock specialists, economists, soil scientists, foresters, etc. The groups then appraise the case-study project from their discipline's perspective, and by utilizing a combination of their own experience and/or ECOZONE-Impacts, they predict the potential environmental consequences of

the project. They then use the rule-making facility of ECOZONE-Workshop and express their predictions as an impact network. Subsequently, the rule-bases produced by each discipline are amalgamated within ECOZONE-Workshop and the overall network of environmental impacts is produced and viewed. This operation serves to demonstrate the connectivity of environmental systems, the cross-sectoral impacts of many projects and also how different activities within an overall project may have feedback/synergistic effects which may initially impact the natural environment and may subsequently impact managed environments and/or social systems.

The first formal use of the ECOZONE software was in Egypt, in December 1993, in a national training workshop, and subsequently it has been used at national and regional workshops in Jordan (1994), Tunisia (1994), Egypt (1995), Cyprus (1995), Turkey (1995) and will be used in Sudan in 1996. In addition, the original software and manual have been translated into French and Spanish, and several small case studies have been developed using ECOZONE-Emptybook.

Post-course evaluation by the participants suggests they find the approach interesting and useful, and the demand for the courses is continuing. However, as for any training activity, evaluating these training courses in precise objective terms is difficult, and no formal results documenting the relative cost-benefit of utilizing ECOZONE over other training methods are available. To date, the most objective evidence of the value of ECOZONE is that FAO has remained very supportive of the approach and recently commissioned the design of 'Ecozone2'.

## The North Sinai Agricultural Development Project (NSADP)

### *Background*

Sinai is located in the northeast of Egypt between the Gulf of Aqaba in the east, the Gulf of Suez on the west, and in the north it is bordered by the Mediterranean. The project area is located in the northern plains and lies in the so-called 'Mediterranean Foreshore Province' which is characterized by the occurrence of sand dunes, sand sheets, salt flats and low coastal plains. With the exception of a narrow zone of a few kilometres along the coast, which has a more moderate climate similar to other areas in the Eastern Mediterranean Basin, the region has a very arid desert climate. Average rainfall increases in an easterly direction, from 30 mm per year along the Gulf of Suez, to 100 mm in El Arise and 300 mm at Rafah. Sixty percent of the annual rainfall for Sinai is lost to evapotranspiration, 5% runs off to the sea, and 35% percolates deep into the rocks and unconsolidated deposits. Much of the deep percolation recharges groundwater. The average temperatures range from 18°C to 31°C in summer, and from 7°C to 21°C in winter. Currently, the land is wilderness, and is inhabited by nomadic Bedouin, who keep some livestock for subsistence purposes. Several of the coastal lagoons are important areas of natural habitat, particularly for migrating birds, and many undescribed archaeological relics remain in the project area.

*Project aims*

The integrated rural development project aims to (a) save import expenses and to promote export earnings by increasing domestic agricultural output and to increase productivity levels, (b) strengthen infrastructure, including improvement of standards of living, domestic water supply, road networks etc. and (c) disperse a part of the heavily concentrated population in the Nile Valley and Delta to other areas which have agricultural and other industrial development potential.

The project was conceptualized in 1977. Feasibility studies and design of the project took almost 10 years. Phase I of the project, which involved developments west of the Suez canal, started in 1987 and is almost complete. Phase II of the project involves the developments in North Sinai, east of the Suez canal.

*Project description*

The backbone of the NSADP is a 90 km extension of the main El Salam irrigation canal into the project area. This requires the excavation of the canal over a distance of 86.4 km west of the Suez Canal (including a 1350 m siphon to carry the irrigation canal under the Suez Canal), and the construction of necessary control structures and pumping stations. Subsequent to the development of the main irrigation canal, over 400 000 feddans of land will be developed for agriculture (1 feddan is approximately equivalent to 1 acre or 0.4 ha).

Proposed cropping activities include the production of feed and concentrates for livestock, the establishment of fruit and nut plantations and large-scale cultivation of vegetables supplemented by field crops (cereals and oilseeds). It is also intended to expand livestock production so as to reduce the demand for milk and meat imports. Given the importance of fodder as a reclamation crop, livestock activities assume importance in expected future agricultural development. The project focuses on commercial dairy enterprises, and the use of pure-bred Holstein-Friesian cattle has been envisaged in all commercial dairy enterprises. In a target year, about 27 000 head of cattle, 130 000 head of goats and 65 000 head of sheep will be raised in the project area.

Simultaneous with the agricultural development, major infrastructure developments are planned. These include the development of towns, villages, roads, schools, hospitals and light industry. The project will be populated by immigrants from the more populated areas of Egypt of three social classes: small-holders, graduates and investors. The allocation of land, irrigation and credit facilities will differ between these classes (see Table 1). For example, farmers will have access to short-term loans for 90% of reclamation costs at a subsidized interest rate of 3% per annum. Investors will similarly have better access to credit but on less generous terms. A total of LE 112 million (LE = Egyptian pounds) for farm investment will be available in loans over six years.

*Project benefits*

In addition to tangible benefits, the project is expected to realize three further benefits: (a) to become a model for irrigated agriculture in other areas of Egypt; (b) to improve farmers' living standards through an increase in farm income; and (c) to provide employment during construction and through the servicing of the resultant communities.

*Potential environmental impacts of the project*

The project is large and complex, and a wide range of environmental impacts are possible, as summarized in Table 2. Of particular worry are the risks of relatively rapid salinization of the newly irrigated land, the pollution of natural waterways, and the pollution of irrigation channels with agricultural and human waste. In addition, the project will involve huge social change for the native nomadic Bedouin and also for the settlers. However, it is this very change, and the consequent immigration of large numbers of people into the North Sinai region, which provides one of the main benefits of the project, i.e. a reduction in population related pressures in the Nile delta region. Other benefits will accrue from the increased agricultural production the project will bring. Further details of the potential environmental impacts of the project are specified in two Environmental Impact Statements, held by the Egyptian Government, from which most of the information used in model development was derived.

## The NSADP software

On entering the NSADP software, users are required to select the sector (arable, livestock, settlement, water resources or infrastructure) and phase of the project for analysis (construction or operation) (Figure 2). Depending upon this selection, users are presented with a list of potential activities relevant to that sector and project phase. Subsequent to the selection of one or more activities, a list of potential primary impacts is displayed. One or more of these impacts can be selected, and the resulting impact network is presented (Tables 3 and 4). An alternative analysis route at this point is to select one of the primary impacts, and then select the 'Next Impact' option. This facility displays only the next level of impacts, rather than the complete network, and thus permits users to follow a focused path through the impact network, as opposed to being presented with a potentially large complete network.

Users may enter the hypertext information system at any stage of the analysis and search for context-specific information. In the NSADP software the HTIS includes general information on the ecology of arid regions, irrigation schemes, livestock and crop production, and on the social impacts of transmigration and displacement. It also includes specific information on the project, its likely impacts and potential mitigating activities. Users can print information from the textual information system with relative ease.

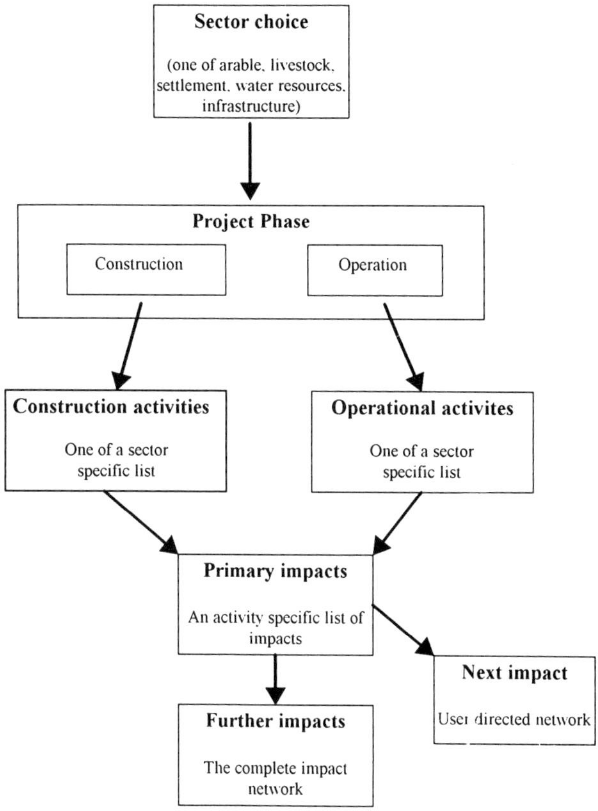


*Figure 2.* A typical user's movement through the NSADP software.

Information on the NSADP project was available in unpublished reports and EIS documents held by the Egyptian Government. Collation of the relevant information from these documents took around 10 person days and structuring the information into the rule-base a further 5 person days. Development of the hypertext took a similar amount of time, while adapting the ECOZONE software to the NSADP format took around 3 days. Given that the programming framework is now complete, the development of further case studies should take less than one person month.

*Using the NSADP software in training*

The NSADP software has been used most in short training courses in Turkey (1995) and Sudan (1996), where its use is integrated with lectures on environmental impact assessment (EIA) and the use of ECOZONE. Currently, these sessions are started with several lectures on the process of, and methodologies for EIA. These sessions are followed by an oral presentation of the NSADP case-study project. Participants are then introduced to the general ECOZONE concept and undertake several familiarization exercises. Once familiar with the software, participants undertake group work where they consider and discuss the institutional and managerial aspects of carrying out an EIA for the NSADP project. They are then asked to use their disciplinary knowledge and ECOZONE-Impacts to identify the likely primary impacts of the project. Each group then presents their results to the entire class and an agreed list of primary impacts is identified. The participants are then split into specialist groups, i.e. crop production specialists, animal production specialists, water resource managers, social scientists and soil scientists, and are asked to consider the higher-order impacts which are likely to derive from the primary impacts relating to their field of discipline. Having identified the impact network for their speciality, they then proceed to represent this in a rule-base on ECOZONE-Workshop. Upon completion of this task, each specialist group views the impact networks of other groups and any overlaps/omissions are discussed. In the penultimate stage of the training, the NSAID-software is presented to the participants and compared with their own impact networks. Again, any mismatches are discussed. Finally, the participants are asked to form groups and use their own work and ther NSADP-software to help complete a formal evaluation of the project. This usually involves completing a Leopold matrix (Leopold et al. 1971) which details the magnitude and importance of impacts arising from the interaction of project activities with specific elements of the environment. The final results are then presented to the class.

These activities are usually undertaken over three full days of a two- or three-week course, and they serve to move participants from a situation of having little formal understanding of EIA, to having a working knowledge of the EIA process, and some predictive and evaluation methodologies. Depending upon the prior skill levels of the participants, other benefits may include an enhanced understanding of the project cycle, the functioning of environmental systems, the management of large-scale irrigation systems and improved computer skills. To date, the workshops have been run by trainers with a good knowledge of the project cycle and EIA, but with varying levels of knowledge about ECOZONE. The minimum requirement for the latter is that trainers should be confident with computers and have a good understanding of ECOZONE functions.

## Discussion

Stone (1992) suggested that knowledge-based systems are well suited for modelling systems for which there is good understanding but little data. As such they seem ideally suited for modelling environmental systems at a relatively large spatial scale, for which good data tend to be lacking. In this situation, they utilize knowledge to make some general predictions which may be accurate, but are unlikely to be precise. While this may be a disadvantage in operational situations, in training situations other criteria are equally, if not more, important: ease of comprehension, ease of use, portability and robustness are all important attributes of training-related software.

Although the lack of precision in knowledge-based systems would seem to be an inherent disadvantage to their operational use, several systems have been developed to aid in EIA (Geraghty 1993). In particular, they have been developed to aid with site selection (Loehle and Osteen 1990), interpreting the legal aspects of EIA (Heikkila and Blewett 1992) and, perhaps, most usefully, in the scoping stage (Fedra et al. 1991). At this stage of the EIA process, all possible impacts of the project are identified and the important ones are selected for further study (Glasson et al. 1994). In this situation, the predictive attributes of knowledge-based systems are useful, while the lack of precision in the output is not a disadvantage as further studies can be undertaken once the most important impacts have been identified.

Given the difficulty of developing generic quantitative models which would be able to predict the multidimensional impacts that typically arise from any development project, it seems that knowledge-based systems have a role to play both in training and operational aspects of environmental impact assessment and management. Indeed, any formalized system which makes apparent the uncertainty and complexity inherent in environmental systems is likely to enhance the planning process.

## Acknowledgements

We gratefully acknowledge the funding and help provided by FAO during this project, and in particular would like to thank Dr. Saad Ahmed El Medini of the FAO Regional Office for the Near East for his continued help and encouragement.

## References

Edwards-Jones G, Gough M (1994a) Use of a simple knowledge-based system for training development planners about the environmental consequences of development projects. Pages 333–337 in Proceedings of the Second International Conference on Expert Systems for Development, 28–31 March 1994, Bangkok, Thailand.

Edwards-Jones G, Gough M (1994b) ECOZONE: A computerized knowledge management system for sensitizing planners to the environmental impacts of development projects. Project Appraisal 9:37–45.

Edwards-Jones G, Hopkins R (1995) A knowledge-based system for predicting the environmental impacts of system level changes to European agricultural systems. Pages 123–130 in Proceedings of the Integrated Crop Protection: Towards Sustainability? 11–14 September 1995, Heriot-Watt University, Edinburgh, UK.

Edwards-Jones G, Mumford J D, Norton G A, Turner R, Proctor G H, May M (1992) A decision support system to aid weed control in sugar beet. Comput. Electron. Agric. 7:35–46.

Estep K W, Hasle A, Omli L, MacIntyre F (1989) Linnaeus: Interactive taxonomy using the Macintosh computer and Hypercard. Bioscience 39:635–638.

Fedra K, Winkelbauer L, Pantulu V R (1991) Expert systems for environmental screening: An application in the Lower Mekong Basin. IIASA, Laxenberg, Austria. 169 p.

Geraghty P J (1993) Environmental assessment and the application of expert systems: An overview. J. Environ. Manage. 39:27–38.

Glasson J, Therivel A, Chadwick A (1994) Introduction to environmental impact assessment. UCL Press, London, UK. 342 P.

Heikkila E J, Blewett E J (1992) Using expert systems to check compliance with municipal building codes. J. Am. Planning Assoc. 58:72–80.

Leopold L B, Clarke F E, Hanshaw B B, Balsley J R (1971) A procedure for evaluating environmental impact. U S Geological Survey Circular 645, Washington DC, USA.

Loehle C, Osteen R (1990) IMPACT: An expert system for environmental impact assessment. AI Appl. Nat. Res. Manage. 4:35–43.

Shachter R D, Kenley R C (1989) Gaussian influence diagrams. Manage. Sci. 35:527–550.

Stone N D (1992) Artificial intelligence approaches to modeling insect systems. Pages 37–52 in Goodenough J L, McKinion J M (Eds.) Basics of insect modelling ASAE, St. Joseph, MI, USA.

# Preservation of the environment on coral islands and intensification of agriculture: applied study on Maré, Loyalty Islands, New Caledonia

C. CORNIAUX[1], T. BECQUER[2], J.P. DANFLOUS[1], C. DUWIG[2], P. VERNIER[1] and D. DULIEU[1]

[1]*Centre de coopération Internationale en Recherche Agronomique pour le Développement (CIRAD), BP 73, Païta, New Caledonia*

[2]*Institut Français de Recherche Scientifique pour le Développement en Coopération (ORSTOM), BP A5, Nouméa, New Caledonia*

**Key words:** cropping systems, preservation of coral environment, subsistence crops, Loyalty Islands

**Abstract**

Elevated Pacific coral islands form a particularly vulnerable ecosystem characterized by fragile soils, usually in thin layers, and by the presence of a fresh water lens that replenishes itself only slowly and is also in balance with sea water. This lens often constitutes the sole supply of drinking water and any pollution affecting it would irremediably jeopardize this vital resource. In the Loyalty Islands, population growth combined with more intensive agricultural methods (use of fertilizers, pesticides, machinery, irrigation), often ill-suited to this type of environment, may cause irreversible damage. Agricultural research must, therefore, now try to obviate the dangers by developing techniques that will protect the environment while allowing essential economic development to proceed. Thus, in 1993, CIRAD New Caledonia, in collaboration with ORSTOM, initiated a new multidisciplinary research project in the Loyalty Islands to investigate problems relating to intensification and stabilization of agriculture on coral islands. The methods involve combinations of different levels of intensification (soil preparation×fertilization rates) and crop rotation. Methods are investigated in conjunction with the major food crops grown in the region (traditional tubers, vegetables), as well as fodder crops. Trials are also under way for monitoring infiltration (nitrates and pesticides). After two years of experiments, the first results have been obtained on crops (yield, labour input), and changes in soil fertility (mineral and organic contents) with an intensive method.

## Introduction

The South Pacific region and the Loyalty Islands (New Caledonia), in particular, are still relatively free of major sources of pollution (Morisson and Brodie 1985; Blanchard 1990; AEEP 1993). However, population growth and the use of more intensive agricultural techniques, often ill-adapted to the specific nature of coral islands, give rise to fears of irreversible damage to the soil and water resources. A selection is needed of proven cultivation methods, which are stable, environmentally sound and socioculturally acceptable, as well as profitable and suitable for development policy makers.

Until the end of the 1980s, studies carried out on the Loyalty Islands about preservation of the environment and intensification of agriculture were essentially made by field projects (Becquer et al. 1993a,b; Bosc and Lhoste 1993; Vernier 1993) which



*P.S. Teng* et al. *(eds.): Applications of Systems Approaches at the Farm and Regional Levels, 427–437.*

had few linkages and could not give comprehensive solutions to pollution problems from agricultural intensification on coral islands. In 1993, we began a multidisciplinary project on Maré, which combines technical (agriculture and pedology) and sociological studies, and involves several research institutions (CIRAD, ORSTOM, BRGM (Bureau de Recherches Géologiques et Minières)). Here we present the results of the first experiments, whose objective was to highlight the advantages and disadvantages of traditional cultivation methods in comparison with those of intensive methods. Propositions relating to an intermediary solution are then made.

## The environment and production methods

### *The physical environment*

New Caledonia lies in the southwest of the South Pacific, between latitudes 18° and 23° South and between longitudes 158° and 172° East, some 1500 km from the east coast of Australia and 1700 km north of New Zealand. The Loyalty Islands comprise a group of elevated coral islands lying on basalt bedrock, which is visible on Maré. The shape of the former atolls is still easily recognizable with the reef crown and the former lagoon, which represents the greater surface area of these islands. Tempered by the trade winds, the climate is described as tropical oceanic. Average annual rainfall varies between 1350 and 1800 mm with minima in the months of August and September. Rainfall is none the less quite regular throughout the year. The temperatures at night are relatively cool (10 to 15°) from June to August.

*The soils.* Acrorthox soils (Latham and Mercky 1983) are the major arable land resource in the Loyalty Islands, covering 19% of the total surface area of Maré, being 130 $km^2$ (Latham and Mercky 1983). These final phases of soil evolution derived from volcanic pumice and ash are mainly composed of iron and aluminium oxides that give the soils their specific agropedological characteristics (Bonzon and Becquer 1994):

- A cation exchange capacity mainly due to the related organic matter.
- A fragile exchange content.
- Limited nutrient content.
- Very rapid infiltration.
- A fragile micro-aggregate structure.

Furthermore, these soils are characterized by extreme variations in depth, which can cause problems in preparation for cultivation and, more particularly, for mechanization (Becquer et al. 1993a). These soils are, therefore, fragile, both chemically (danger of decreased organic matter and mineral content, lixiviation) and physically (wind erosion).

*The fresh water lens.* The Loyalty Islands have no waterways because of the limestone substrata and the absence of relief inside the crowning cliffs. Maré does, however, possess a fresh water lens at a considerable depth. In balance with the sea water, it is fed by rainwater and constitutes almost all the fresh water resource. The lack of depth of the soils, their significant permeability and the cracks in the rocky substrata make it particularly vulnerable to pollution.

*The human factor*

Maré has a surface area of 640 $km^2$ and the census indicates a population of 5600 inhabitants, which gives a relatively low population density (8.4 inhabitants $km^{-2}$). The economic and institutional context of the Loyalty Islands has been marked by a recent, yet far-reaching integration of the population into the market economy and trade exchanges, accompanied by important social changes that are characteristic of economies linked to developed countries. However, contrary to other situations, this integration has not occurred through insular Melanesian agriculture becoming a dominant factor in trade exchanges (Bosc and Lhoste 1993). Although agricultural activities are of little importance in the monetary economy of the islands, they are not negligible, particularly on Maré. Yam cultivation remains significant in native custom and is still fundamental in traditional ceremonies. The agricultural census numbers some 800 plantations with 10% classed as 'market' plantations, integrating production techniques that are more intensive than traditional methods.

*Production methods*

*The traditional method* of production is based on long-term fallow (10 to 20 years) (Jaffré and Veillon 1987). After clearing and burning of the primary and secondary forest, yam cultivation is implanted for one cycle. This is followed by various associated crops that constitute the 'Melanesian garden' (sweet potato, taro, vegetables, banana, manioc, corn). After three years of farming, the land is again left fallow, allowing forest regrowth that generally evolves towards a less dense xerophilous forest.

*The intensive method.* The implantation of new income-earning crops (avocados, litchis, citrus, market gardening) is often synonymous with the permanent exploitation of plots. As is the case when traditional planting (subsistence, market and fruit crops) is intensified, inputs (fertilizer and agricultural chemicals), machinery and irrigation come into use. When the techniques of these elements are not mastered, they represent dangers for the environment on coral islands.

The level of intensification varies from one plantation to the next. Thus, in this article, we will discuss 'the intensive method' for situations in which the growers no longer leave land fallow as part of their production methods (permanent cultivation).

*Themes of research and applied studies*

In view of the concerns expressed above, suitable production methods must be:

- agronomically stable and non-polluting and, therefore, validated over sufficiently long periods to measure long-term effects;

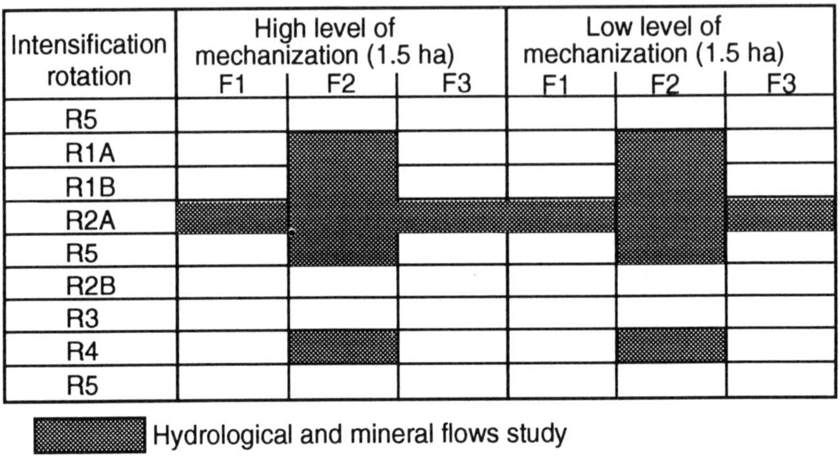


*Figure 1.* Plan of Tawaïnèdre plots (*Source:* Vernier et al. 1993). Fertilizer application: F1 is 300–400 kg $yr^{-2}$; F2 is 2 times F1; F3 is F2 plus organic matter (5 ton $ha^{-1}$). Rotational cropping: R1 is one crop per year (R1A: yam, maize, yam, maize, ...; R2B: maize, yam, maize, yam, ...); R2 is one crop per year plus legume (R2A: yam+leg; maize+leg; yam+leg; maize+leg, ...; R2B: maize+leg; yam+leg; maize+leg; yam+leg; ...); R3 is two crops per year (maize+sweet potato; maize+sweet potato, ...); R4 is three crops per year (with two market garden crops); and R5 is pastures.

- technically adapted to the local situation and growers, yet economically accessible and motivating; and
- diversified to provide producers with a wide variety of contrasting solutions.

Our main experimental program encompasses medium-term (five years) monitoring and comparison of various cultivation methods, which are a combination of three intensification factors: fertilization, crop rotation and the extent to which the land is worked (Vernier et al. 1993; Corniaux and Dulieu 1994; Figure 1). The program includes the major crops that form the basis of the local agricultural economy (yam, sweet potato), as well as the diversification crops (corn, permanent grasslands).

The emphasis is on crop yields and related data (size, weight, etc.), labour input and the evolution of self-propagating flora. In parallel, hydrological and mineral flows are monitored in order to establish a scale of risks for degradation of soil fertility and pollution of the fresh water lens in relation to the different levels of intensification of the planted plots.

Other studies are carried out on various farms. These studies are mainly concerned with soil fertility, the intensification of yam and sweet potato crops, and mechanization.

## Comparison of traditional and intensive methods

### *Maintaining soil fertility: the importance of organic matter*

The soils on Maré are devoid of mineralogical clays (Ternicier 1971). Maintenance of soil fertility (in particular, retention of nutrients) is largely dependent on main-

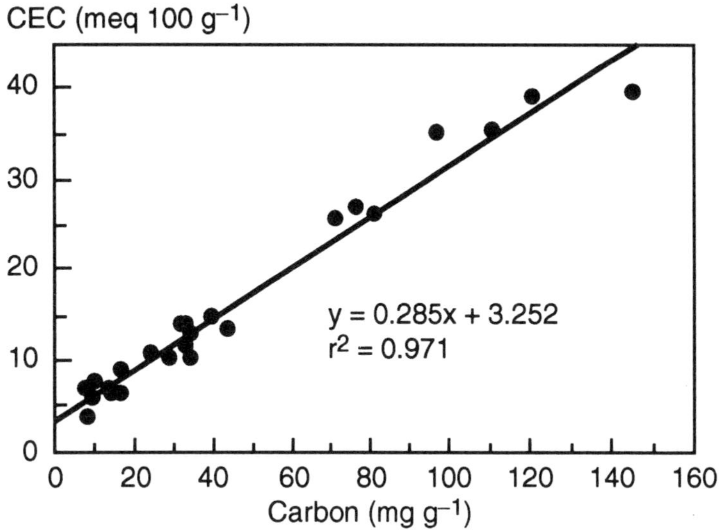


*Figure 2.* Relationship between cation exchange capacity and carbon content for soils on Maré (*Source:* Becquer et al. 1993b).

*Table 1.* Sweet potato yields according to manure and fertilizer supplements on a plot with damaged fertility (yams on a regularly burnt fallow land). Results obtained in 1994 on a plot of 200 $m^2$ in Hnadid (Maré, Loyalty Islands) (*Source:* Polti 1995).

| | T0[1] | T1[2] fertilizers | T2[3] manure | T3[4] fertilizers + manure |
|---|---|---|---|---|
| Total yield (ton $ha^{-1}$) | 9.80 (a) | 10.56 (a) | 17.02 (b) | 10.46 (a) |
| Marketable yield (tubers > 150 g) (ton $ha^{-1}$) | 8.26 (a) | 8.44 (a) | 12.74 (b) | 8.46 (a) |

[1]No fertilizers, no manure.
[2]300 kg $ha^{-1}$ of 0-32-16 + 500 kg $ha^{-1}$ of 13-13-21 (N-P-K).
[3]30 ton $ha^{-1}$ of pig's manure.
[4]T1 + T2.
Statistical significance: $P<0.05$.

taining the organic matter content (Figure 2). A comparison of two fallow plots of different ages and past exploitation has effectively been carried out on two plantations on Maré. The first is a fallow scrub plot where the vegetal ground cover is regularly burnt off; the second plot has been fallow for a very long time and the forest cover has regenerated. Higher levels of organic matter, exchangeable cations and available phosphorus were found on the second plot (Becquer et al. 1993b).

On traditionally cultivated plots that integrate long-term fallow, the soil therefore naturally recovers its fertility. On the other hand, where permanent, intensified farming is practised, soil fertility decreases rapidly if compost and manure are not added. This phenomenon is more prominent in tuber crops (Table 1).

*Table 2.* Labour input (h per plant per 200 m²). Results obtained in 1991 on two yam plots of 200 m² in Hnadid (Maré, Loyalty Islands) (*Source:* Vernier 1993).

| Labour | Intensified cropping method | Traditional cropping method |
|---|---|---|
| Manual ploughing | 0 | 10 |
| Mechanized ploughing | 1 | 0 |
| Manual planting | 8 | 4 |
| Manual support | 2 | 2 |
| Fertilizer supplements | 2 | 0 |
| Agricultural chemical treatments | 5 | 0 |
| Manual maintenance | 20 | 10 |
| Manual harvesting | 20 | 16 |
| Total | 58 | 42 |
| Total/plant | 0.28 | 0.50 |

*Table 3.* Early yam yields (wadrawa: 'chief's yams') according to year and cropping system. Results obtained in Hnadid (Maré, Loyalty Islands) (*Source:* Vernier 1993).

| Plot (200 m²) | Intensified[1] trial 1991–92 | Intensified trial 1990–91 | Traditional trial 1990–91 |
|---|---|---|---|
| Density (plants $ha^{-1}$) | 11 400 (a) | 10 400 (a) | 4000 (b) |
| Plants number | 228 (a) | 208 (a) | 80 (b) |
| Harvested heads number | 228 (a) | 206 (a) | 79 (b) |
| Tubers weight (kg) | 156 (b) | 363 (a) | 104 (c) |
| Marketable tuber weight[2] (kg) | 105 (b) | 294 (a) | 78.5 (c) |
| Average tuber weight (g) | 560 (b) | 650 (a) | 810 (c) |
| Marketable tubers (weight per head (kg)) | 0.47 (b) | 1.42 (a) | 0.99 (c) |
| Yield (ton $ha^{-1}$) | 5.3 (b) | 14.4 (a) | 3.8 (b) |

[1]Harvest was affected by drought in 1991–92.
[2]Marketable yams: tubers ≥ 400 g.
Statistical significance: $P<0.05$.

### *Labour input and tediousness*

Labour input was measured in yam crops grown in both traditional and intensified methods (Table 2). It appears that the labour input for a given surface area is higher in the case of intensive farming. Plantation densities are, however, very different between the two methods. In this experiment, densities were 10 500 plants $ha^{-1}$ for intensive farming as opposed to 4000 using traditional methods. Therefore, the labour input per plant is significantly reduced using intensive methods. The use of machines in intensive farming makes the work less tedious; harvesting of yams is also far easier.

### *Yields*

Yam, sweet potatoes and corn crop yields have been measured for the two production methods (Tables 3 and 4). Intensive production methods provide better yields (per

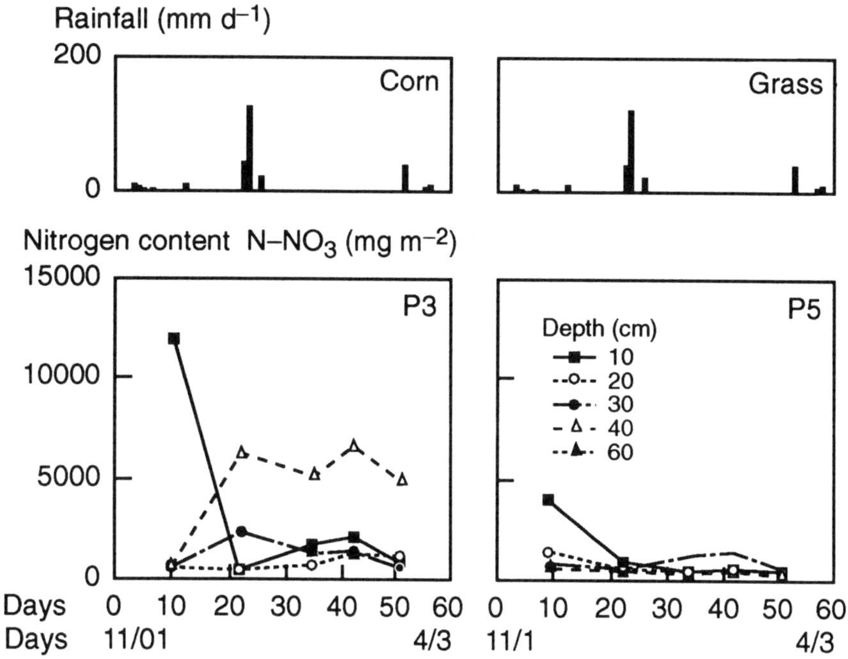


*Figure 3.* Evolution with time and depth of soil nitrogen content according to the crop (corn vs grass) (*Source:* Duwig et al. 1996).

plant and per hectare) for yam and sweet potato, and the tubers are of marketable size. However, traditional methods yield bigger yams that are more suitable for customary ceremonies.

It must be noted that we were unable to identify changes in the yield from the same plot after two years of planting. It would appear that the 'rainfall' factors influenced the rest of the results.

Finally, as is the case for labour input, it must be noted that it is preferable to compare the two methods in terms of yield per plant, rather than yield per hectare. Areas required for traditional planting are generally larger since the plantation densities are lower.

### *Pollution risks*

The pedo-climatic conditions in the Loyalty Islands, and on Maré in particular, are characterized by shallow, permeable soils and sometimes very heavy rainfall. The initial results of a study of nitrate lixiviation through these soils in relation to rainfall clearly show the potential threat of pollution in the fresh water lens (Duwig et al. 1996). An active vegetal cover (pasture) nevertheless reduces the risks (Figure 3). For example, heavy rain (160 mm) was recorded on a plot of corn (shown as P3) a

*Table 4.* Different crop yields according to production method. Results obtained on Tawaïnèdre's experimental station (Maré, Loyalty Islands) in 1994 for sweet potatoes and maize, and in 1995 for yams (*Source:* Vernier et al. 1993; Polti 1995).

| Crops and production method | Yield (ton $ha^{-1}$) | Middle weight per plant (g) | Ears number per plant |
|---|---|---|---|
| Yam (1 crop per year) | | | |
| traditional method | 21 (a) | 990 (a) | – |
| intensive method | 32.5 (b) | 1080 (a) | |
| Maize (1 crop per year) | | | |
| traditional method | 0.07 | – | 0.41 (a) |
| intensive method | 0.14 | – | 0.47 (a) |
| Maize (after sweet potato) | | | |
| traditional method | 0.05 | – | 0.64 (a) |
| intensive method | 0.07 | – | 0.83 (b) |
| Sweet potato | | | |
| traditional method | 2.4 (a) | 258 (a) | – |
| intensive method | 4.0 (b) | 291 (b) | – |

*Note:* Very low corn yields are due to bad establishment because of drought and seed quality. Statistical significance: $P<0.05$.

*Table 5.* Major characteristics of traditional and intensive cropping methods.

| | Traditional planting | Intensive planting |
|---|---|---|
| Pollution risks (fertilizers, pesticides) | 0 | +++ |
| Weed propagation risks | + | ++ |
| Water requirements | 0 | ++ |
| Mechanization requirements | 0 | ++ |
| Surface area requirements | +++ | + |
| Loads (fertilizers, irrigation, pesticides, mechanization) | 0 | ++ |
| Commercial yield | + | ++ |
| Labour input (per production) | ++ | + |
| Remuneration of labour | + | ++ |
| Suitability to the human context | +++ | 0 to + |
| Suitability to the ecosystem | +++ | 0 to + |

fortnight after planting and on a plot that was put into pasture one year earlier (shown as P5). The same amount of fertilizer had been added to the plots at the time the corn was planted (Vernier et al. 1993). Measures of nitrate content provided an estimate of lixiviated nitrate quantities during and after the spell of rain: 5000 mg $m^{-2}$ on P3, which was an 80% loss of the nitrate fertilizer supplement, as opposed to 400 mg $m^{-2}$ on P5.

Table 1 confirms that the plants (sweet potatoes) absorb little fertilizer under intensive growing methods. Supplementing with organic matter is, however, beneficial to their growth. Nevertheless, simultaneous supplements of fertilizer and organic matter (manure or compost) do not apparently give better results (interference in the decomposition of the organic matter).

## Discussion: proposed alternative to traditional and intensive methods

*Advantages and limitations of traditional and intensive methods*

With regard to the intensification of traditional plantations such as subsistence crops (yam, sweet potato, taro, corn), market crops (cabbage, lettuce, carrots, tomatoes) and fruit (banana, pawpaw, pineapple) for commercial purposes, Table 5 summarizes the major advantages and disadvantages of traditional and intensive planting based on the experiment results discussed in the previous section.

*Proposed intermediary method*

Compared to traditional methods, intensive planting appears more profitable and less arduous. On the other hand, traditional methods incorporate long fallow periods and preserve the ecosystem (maintenance of soil fertility, preservation of fresh water lens). Furthermore, traditional methods are perfectly adapted to the human context (Faugère 1994). However, taking into account the large surface areas required for these planting methods, population growth (50% of the Loyalty Islands population is aged under 20) and land ownership questions, management of space can be a significant problem (Herrenschmidt 1995).

Therefore, if one is to propose production methods that are mindful of the environment and human and land ownership issues, but remain attractive to the grower (i.e. profitable and not arduous), one must define a method that is a compromise between traditional and intensive methods.

The advantage of fallow plots has been demonstrated in maintaining soil fertility on traditional plantations. However, fallow scrub lands, which are common in the Loyalty Islands, require long periods, probably more than ten years, in order to be effective. A shorter fallow period leads to a decrease in soil fertility (Becquer et al. 1993b), as well as the development of overbearing and often ligneous species (Jaffré and Veillon 1987). To reconcile the advantages of fallow and the need for sufficient available space, fallow could be replaced by grasslands of both legumes and grasses or of legumes alone.

The advantages offered by grasslands in terms of maintaining soil fertility have already been demonstrated in various situations (Pieri 1989). This type of grassland fallow offers other advantages, which include:

- Reduced fallow period from 10–20 years to 4–6 years.
- Reduced clearing operations which require significant start-up investment, be it done manually (much labour) or mechanically (high mechanization costs). In the case of the grasslands fallow, extensive clearing is only required in the first year of cultivation. For this work, semi-forestry shredders and forestry disc cover-croppers seem well adapted to conditions in the Loyalty Islands (Danflous 1995a,b). Planting would then be possible on unploughed land, or on grasslands that have been ploughed.
- Use of pastures for feeding animals.
- Fixation of atmospheric nitrogen on plots where only legumes have been planted.

At the time of planting, compost or manure should be used rather than mineral fertilizers to maintain soil fertility, provide the nutrients required for plant growth and greatly reduce the risks of pollution of the fresh water lens. Research is nevertheless required to improve plant absorption of and use of fertilizers and to restrict lixiviation, particularly where nitrate is concerned. In this context, it seems that splitting of fertilizer applications is a topic to be immediately investigated.

Furthermore, in sociological terms, production methods to be developed must be as close as possible to traditional cultivation cycles. Yam should be planted in year 1, followed in years 2 and 3 by the subsistence, market and fruit crops found in Melanesian gardens. The methods must also take into consideration the notions of land ownership and the proximity of a water source that the grower can use, as well as the distance of the plots from the grower's home.

Using these methods, it must be underlined that certain crops will be badly suited or even excluded. Such is the case for orchards (citrus, litchi, avocado) which cannot be integrated into a method that incorporates fallow, and are unsuitable when the grower is not the owner of the plot (land unavailable for long periods).

## Conclusions

On coral islands, special care must be taken in perfecting new techniques in order to avoid degrading the already fragile soils and to protect the fresh water lens from chemical pollution through lixiviation of inputs (fertilizers, pesticides). Initial research carried out on Maré since 1991 has enabled us to emphasize the need to develop growing methods that are a compromise between traditional and intensive methods in order to best deal with environmental, technical, economic and sociological issues in the Loyalty Islands.

The proposed method, based on short-term grassland fallow, will require testing over several years in order to validate its reliability. Studies enabling better management of inputs into intensive operations must also be continued, in particular market gardens and fruit crops on irrigated land. Application possibilities on other Pacific Islands where similar problems are encountered must also be studied.

## References

AEEP – Agence de l'Eau et de l'Environnement pour le Pacifique (1993) Elaboration of an exploitation and protection plan for water resource on Lifou and Maré – Qualitative survey of the water resource [in French]. Rapport 1, Chapter 3, Nouméa, New Caledonia. 93 p.

Becquer T, Bourdon E, Nigotte W (1993a) Contribution of the study of soils on Maré: The soils at the agricultural development support centres at Tawaïnèdre and Taodé. 1. Morphological characterization and cartography of soil thickness [in French]. Conv.: Sci. Vie: Agropédol. 17; Institut Français de Recherche Scientifique pour le Développement en Coopération, Paris. France. 37 p.

Becquer T, Bourdon E, Petard J (1993b) Contribution of the study of soils on Maré. The soils at the agricultural development support centres at Tawaïnèdre and Taodé. 2. Physico-chemical characterization of soils [in French]. Conv.: Sci. Vie: Agropédol. 24; Institut Français de Recherche Scientifique pour le Développement en Coopération, Paris, France. 63 p.

Blanchard F (1990) Maré (Loyalty Islands) Management and schematic representation of the fresh water lens [in French]. BRGM, Bureau de Recherches Géologiques et Minières, Paris, France. 75 p.

Bonzon B, Becquer T (1994) Exploitation of Acrorthox soils in the Loyalty Islands. Summary [in French]. ORSTOM. Conv.: Sciences de la vie: Agropédol. 28; Institut Français de Recherche Scientifique pour le Développement en Coopération, Nouméa, New Caledonia. 14 p.

Bosc P M, Lhoste P (1993) Study of the diversity, evolution, and insertion into the market economy of traditional Melanesian production methods in the Loyalty Islands, Province of New Caledonia. Proposals for a socio-economic research programme. Mission report [in French]. Doc CIRAD-SAR no 93/93, Montpellier, France. 36 p.

Corniaux C, Dulieu D (1994) Intensification of agriculture and protection of the environment [in French]. Rapport CORDET, CIRAD-EMVT, Nouméa, New Caledonia. 15 p.

Danflous J P (1995a) Mechanization of planting methods in the Loyalty Islands [in French]. Centre for International Cooperation in Development-oriented Agricultural Research-SAR, Paris, France (in press).

Danflous J P (1995b) Clearing and initial preparation of soil [in French]. Centre for International Cooperation in Development-oriented Agricultural Research-SAR, Paris, France (in press).

Duwig C, Becquer T, Clothier B E, Vauclin M (1996) Nitrate leaching in the Loyalty Islands (New Caledonia) under intensified agricultural practises. In Proceedings of the First International Conference on Contaminants in the Soil Environment in the Australasia-Pacific Region, 18–23 February 1996, Adelaide, Australia (in press).

Faugère E (1994) Rural development and social change on Maré: Initial remarks (assessment of the pre-survey phase) [in French]. Centre for International Cooperation in Development-oriented Agricultural Research, New Caledonia. 8 p.

Herrenschmidt J B (1995) Development in the Loyalty Islands: 1990–1993 [in French]. Centre de Coopération International en Recherche Agronomique pour le Développement-SAR/Institut Français de Recherche Scientifique pour le Développement en Coopération, Paris, France. 30 p.

Jaffré T, Veillon J M (1987) Study of fallow practice and first impressions of secondary succession on Lifou (Loyalty Islands, New Caledonia) [in French]. Rapp. Sci. et Tech.: Sci. Vie: Botanique, 1; Nouméa: Institut Français de Recherche Scientifique pour le Développement en Coopération, Nouméa, New Caledonia. 28 p.

Latham M, Mercky P (1983) Study of the soils in the Loyalty Islands. Pedological map and 1/200 000 map of forestry and cultivation suitability [in French]. Notice explicative No. 99, Institut Français de Recherche Scientifique pour le Développement en Coopération, Paris, France. 45 p.

Morrison R J, Brodie J E (1985) Pollution problems in the South Pacific: Fertilizers, biocides, water supplies and urban wastes. Pages 69–74 in Environment and resources in the pacific. UNEP Regional Seas Reports and Studies, No. 69, Suva, Fiji.

Pieri C (1989) Fertility management of south Sahara savannahs areas. Pages 413–433 in Savannas fertility, 30 years of agricultural research and development balance-sheet in the south of Sahara. Coopération Française/Centre de Coopération International en Recherche Agronomique pour le Développement, Paris, France.

Polti D (1995) Subsistence crop programme. Annual report 1994 [in French]. Centre de Coopération International en Recherche Agronomique pour le Développement-CA, Nouméa, New Caledonia (in press).

Ternicier G (1971) Karst soils on the elevated atoll of Lifou (Loyalty Islands, Territory of New Caledonia) and bauxitization problems. C.R. Acad. Sc. Paris, T. 272:2067–2070.

Vernier P (1993) Subsistence crop programme. Annual report 1992 [in French]. Centre de Coopération International en Recherche Agronomique pour le Développement-CA, Nouméa, New Caledonia. 38 p.

Vernier P, Dulieu D, Danflous J P, Daly P (1993) Intensification of agriculture and preservation of the environment on coral islands. Colloque Institut Français de Recherche Scientifique pour le Développement en Coopération, Paris, France. 12 p.

# Watershed management information systems

J. ADINARAYANA

*Centre of Studies in Resources Engineering, Indian Institute of Technology, Powai, Bombay 400 076, India*

**Key words:** geographic information systems, hilly watershed management, land-use planning, remote sensing, soil erosion, systems approach

**Abstract**

The watershed is the natural planning unit in modelling and monitoring of the effect of watershed management policies. To ensure optimum and sustained productivity through scientific planning, the watershed needs a decision-making information system which involves an appraisal of agroecological characteristics, resource limitations and potential of the watershed for resource development. This complete information helps in generating an information system for watershed management, a Watershed Management Information System (WATMIS). This paper describes an approach to develop a WATMIS for a river basin in the difficult, semi-arid mountainous terrain of the Indian Peninsular region. The methodology includes compilation of various watershed resource data (maps, space and ground-based systems); development of Geographic Information System-based models for soil erosion evaluation; and formulation of a treatment-oriented land-use planning scheme with multidisciplinary knowledge-based rules. A priority sub-watershed delineation survey carried out in the watershed indicates significant variation in sediment yield index values. This calls for conservation planning in cases of high and very high priority sub-watersheds. This systems approach, which is hopefully fast and cost-effective, without individual bias, is related to the optimization of natural resources while minimizing the impact on the watershed environment. This paper envisages some future studies for watershed management strategies.

## Introduction

Watersheds have assumed importance for preserving the ecological balance between natural resource development and conservation, particularly in the fragile and heterogeneous, erosion-susceptible hilly ecosystems. The necessary information for resource planning on a watershed basis includes the agroecological characteristics, resource constraints and the potential of the watershed for sustainable development. This information helps in generating an information system for planning the watershed management programmes of hilly terrain, what may be called a Watershed Management Information System (WATMIS).

An attempt has been made to use remote sensing and Geographic Information Systems (GIS) for developing WATMIS of a hilly watershed. Remote sensing is particularly useful in inaccessible areas. GIS is the most effective tool for integrating the data captured from different sources to produce a picture from which certain decisions can be made. The watershed is chosen as the fundamental unit in both the modelling and monitoring of the effects of watershed management policies.

The approach was tested in a drainage basin of about 100 km$^2$ in the difficult semiarid, mountainous terrain of the Western Ghats (73° 45′–73° 55′ E; 19° 16′–19°



*P.S. Teng* et al. *(eds.): Applications of Systems Approaches at the Farm and Regional Levels, 439–446.*

23′ N). The watershed, which has a basaltic landscape, is located in the Western Plateau and Hill Agroclimatic region of the Indian Peninsular.

Sediments are deposited in the drainage ditches, and ultimately in Yedgaon Reservoir, a terminus of drainage for the watershed. The siltation is largely due to poor soil conservation practices, scantily distributed vegetation cover, steep slopes and overall dry climate with an erratic monsoon rainfall, resulting in a high susceptibility to erosion. Cultivation (cropping systems based on upland rice) is being introduced in the canyons extending into the steeper mountainous area with inadequate soil conservation practices. If erosion is not stopped, further increases in run-off, and sheet and gully erosion on sloping ground will ultimately destroy the productive value of the land. Hence, a systems approach for major planning is necessary for sustainable development of the watershed.

## Data sources

Keeping in mind the integrated approach to be adopted, the following data were used:

- Remotely sensed data: Landsat-TM and IRS LISS-II – digital (Computer Compatible Tapes – CCTs) and visual (False Colour Composites – FCCs) products of monsoon (kharif) and post-monsoon (rabi) agriculture seasons.
- Survey of India (SOI) topographical maps at 1:250 000 and 1:50 000 scales.
- Available literature and maps on various themes.
- Meteorological data (rainfall, temperature, etc.).
- Ground survey information on soil and other terrain characteristics of the sample sites of the watershed.

## Data base construction

The GIS used was the PC-based, indigenously developed package, called GRAM (Devi et al. 1993). The data sources were stored in the GIS in grid-cell format. Digitization, georeferencing and rasterization were done in the input module. The output of this module was used as the base for an analysis module to integrate and process logical combinations and knowledge-based rules. The utilitarian derivative maps and the results were the output of the print module.

Each defined cell (pixel) has an exact location in space, related to the grid orientation and cell size and a list of assigned attributes. The data bases were geometrically corrected, using a SOI topographic map base and resampled to a 30 m resolution (of Landsat-TM). A grid size of 515 × 492 pixels covered the whole watershed.

### *Watershed resources*

Accurate information on the existing land-use pattern and its spatial distribution is a prerequisite for any watershed management program. With the advent of remote sensing, it has been possible to prepare this dynamic resource map at various levels of confidence. However, the application of multi-spectral classification techniques in extracting vegetation classes has been inconsistent because of similarities in spectral responses (Bocco and Valenzuela 1991; Adinarayana et al. 1994a). The problem is more serious with the digital techniques under heterogeneous hilly terrain. The other problem in the collection of land-use data is the presence of clouds and their shadows during the monsoon period. In addition, it is also important to map the agricultural land-use pattern of both monsoon (kharif) and post-monsoon (rabi) seasons together in the rural watersheds. Therefore, multi-seasonal remotely sensed images of Landsat-TM digital (rabi – when most of the crops are matured and harvested) and IRS LISS-II FCC (kharif – cloud-free images showing vegetation in the upland areas) were used. In addition, multi-thematic maps of soil, slope and drainage were used to generate multidisciplinary knowledge-based rules to overcome the above inherent problems in the hilly terrain. These maps were input and stored as separate layers in the GIS for integrated analysis. The above GIS data layers and the 'informed opinions', in the form of knowledge-based rules drawn from multi-disciplinary experts and the knowledge of local land-use patterns plus the field checking, were incorporated in the map to consolidate the agricultural land-use pattern of this heterogeneous highland. To give a few examples:

> **If** Landsat-TM (Kharif class) + Thick Forest and IRS LISS-II (Rabi) = Thick Forest; **Then** class = Thick Forest
> **If** Kharif = Sparse Forest and Rabi = Cropped; **Then** class = Cropped
> **If** Kharif + Rabi = Vegetation, Drainage density = Low, Soil = Typic Chromusterts and Slope = < 5 degrees; **Then** class = Cropped

These *informed opinions*, codified in the form of 'if-then' type of rules, were used to refine the classification accuracy. Initially, the correspondence between the visual and digital classification maps was at best 45%. After applying the rules on the classification output, the accuracy increased to 85–100% (Adinarayana and Rama Krishna 1996). In the process, the visual and digital classification techniques interacted to improve the classification system. Finally, the GIS-assisted improved land-use map consists of five major categories, which were selected from the conservation planning point of view: agriculture, thick forest, sparse forest, degraded pastures and open (with/without scrub) lands. The introduction of multi-disciplinary expert rules provides an extension to the traditional methods of satellite-derived classification which should enable more relevant, and accurate, land-use/cover maps to be prepared.

A collective approach comprising satellite data (Indian Remote Sensing Satellite-IRS), and topographic maps was used for mapping physiographic units. The soil associations of each physiographic unit were established by studying soil profiles in sample strips and by random checking outside the sample strip areas, classifying the soils according to an established soil taxonomy (Soil Survey Staff 1975). Finally, a soil map at 1:50 000 scale was prepared by combining the soil information in each delineated physiographic unit. The developed and moderately deep (50–90 cm) black cultivated soils (Typic Chromusterts) generally occupy the valleys and plains of the watershed. Undulating uplands and subdued hills have degraded pastures or open scrub and partly cultivated lands. These units can be classified into the association of Lithic Ustorthents/Vertic Ustropepts with shallow soil depth (20–50 cm). Under-developed and very shallow (<20 cm) soils (Lithic Ustorthents) cover the hilly regions with steep slopes; these are mainly forest areas.

The contours and high points were digitized from 1:50000 SOI topomaps (20 m interval) to generate the Digital Elevation Model (DEM) and the slope map with the help of the GRAM package. The watershed was divided into three major slope steepness classes for the evaluation of soil erosion and the priority sub-watershed delineation survey: (i) gently sloping lands (0–8 degrees); (ii) moderately sloping lands (8–18 degrees); and (iii) steeply sloping lands (>18 degrees). Steeper slopes promote concentrated run-off and show exposed rocks in some places. Moderate slopes support mostly degraded pastures or open scrub lands and have moderate to high run-off potential. Gently sloping lands covered with moderately deep cultivated black soils have low run-off potential. Once the contours are digitized, GRAM allows one to categorize any number of slope steepness classes as needed.

An isohyetal map was compiled manually from ten years of mean annual rainfall data from six meteorological stations available in and around the watershed area. The watershed was divided into two equal high and low rainfall zones: a <1500 mm low rainfall zone, and a >1500 mm high rainfall zone.

The drainage network was generated from SOI toposheet (of 1975) and the high-order river course systems were mapped to the present conditions from Landsat-TM imagery (of 1989). The drainage density (total length of streams/unit area) was calculated for each sub-watershed and grouped the watershed into low, medium and high drainage density zones. Proximity analysis was done to establish the buffer zone along the drainage lines to demarcate the zones where erosion was likely to be important.

The watershed was systematically divided into 33 sub-watersheds, which are smaller hydrologic units and constitute individual tributaries of the lowest order, or a group of such tributaries. The area of each sub-watershed was planometrically computed and ranged from 1.4 to 7.0 $km^2$, which is considered as a viable working size for implementation of conservation measures and other watershed development programmes.

All the previously described thematic maps were digitized, rasterized and brought to a common coordinate system, which thus acted as data base for GIS analysis to generate the derivative maps.

## Data base analysis

### *Soil erodability assessment*

Soil erosion is the result of interplay between environmental factors, such as soil, topography, drainage, rainfall and land-use pattern, which are generated from different sources/systems. In order to study soil erosion, it is important to integrate these data. GIS is the most effective and viable tool for considering the interaction between the spatially distributed resources. We made a number of useful combinations of the data via overlay analysis and created 'integrated resources units (IRU)'. Each IRU comprises all biophysical resource data on soil, land use, slope, rainfall distribution and drainage buffers for one grid cell. The link between IRU and the soil erosion intensity unit is provided by rules. In the construction of the decision rules, the multidisciplinary expertise and the knowledge of the local terrain parameters, and field observations, were taken into consideration, in addition to the author's expertise in soils, land-use planning, remote sensing and GIS. Two examples are:

> **If** 5° slope, other drainage area, Typic Chromusterts, < 1500 mm rainfall with thick forest; **Then** class = slight to moderate erosion risk
> **If** > 18° slope, 1–2 drainage orders, Typic Ustorthents, > 1500 mm rainfall zone with open areas; **Then** class = very severe erosion risk

Application of these multidisciplinary expert rules permits mapping of the erosion intensity that is likely to occur at any given location, without any subjective bias (Adinarayana et al. 1994b). The soil erosion map was generated with four categories: slight to moderate (< 20 ton $ha^{-1}$ $yr^{-1}$), moderate to severe (20–30 ton $ha^{-1}$ $yr^{-1}$), severe (30–50 ton $ha^{-1}$ $yr^{-1}$) and very severe (> 50 ton $ha^{-1}$ $yr^{-1}$) soil erosion intensity units.

*Conservation management*

A major step in the conservation-oriented planning process is inventorying and classifying the watershed and then judging its capacity to support land use on a sustained basis and also to avoid uses that degrade the land. The erosion causing factors such as slope steepness and soil depth can be yardsticks against which to judge the suitability of alternative land uses and their conservation requirement (El-Swaify et al. 1983). The first step in this process is to create a basic unit by overlay analysis. Each unit comprises the physical resources of soil depth and slope steepness classes, (unlike in IRUs where all the biophysical units are considered) and, hence, were called 'integrated physical land units (IPLU)'. The link between IPLU and the treatment-oriented land-use planning scheme is provided by information-based rules. In the implementation of these rules, the land-use capability classification formulated by Sheng and Stennett (1975) was mainly used, modified according to knowledge of local terrain parameters and field observations, plus the expertise of the author in the subject. These rules and factors were used to generate a sustained land-use system in the watershed with proper soil and water conservation measures (Table 1). Employing this system for land-use planning would be a safeguard against soil erosion and other environmental threats.

Conservation planning in watersheds is expensive and we, therefore, require a selective approach for identifying smaller hydrological units (or sub-watersheds of individual tributaries of the lowest order, or a group of such tributaries) to advise resource management programs more efficiently. Hence, a priority sub-watershed

*Table 1.* Land-use systems with conservation measures.

| Class | Land use | Conservation measures |
|---|---|---|
| CG | Cultivable land, up to 7 degrees | Contour cultivation |
| CM | Cultivable land, on slopes 7–15 degrees | Contour cultivation with soil-conserving crops |
| CS | Cultivable land, on slopes 15–20 degrees | Intensive conservation measures, bench terracing hexagon |
| P | Pastures | Rotational grazing |
| F | Forests | Soils in the watershed are mostly too shallow soils for any soil conservation measures. Contour planting, diversion ditching where the soil depth is up to 50 cm |

*Note:* Gully-dissected lands that prevent normal tillage activities should be pastures/forests; any land that is too stony, preventing tillage and treatment, should be pastures on <20 degrees and forests on >20 degrees.

delineation survey was carried out in the watershed (Adinarayana et al. 1995) using the All India Soil and Land-use Survey's Sediment Yield Index (SYI) model (AISLUS 1991). The methodology to prioritize the watersheds included: (i) generation of information data bases of soil, slope, drainage and density, land-use pattern and delineation of sub-watersheds; (ii) generation of IRUs by overlay analysis; and (iii) assignment of weight and delivery ratio values to IRUs based on combined effects of the environmental factors in terms of the erodability and erosivity potential of IRUs.

On the basis of SYI values, the watershed was divided into four priority classes: low (1000–1099), medium (1100–1199), high (1200–1300) and very high (>1300) priority sub-watersheds. The significant variation in SYI values between sub-watersheds calls for a conservation planning scheme like the one developed with GIS in the present study. The high and very high priority sub-watersheds could be treated with the land-use system suggested here to try to conserve the ecosystem in the watershed. If the appropriate conservation measures are given priority, there would be less erosion, and subsequently massive investments to control soil erosion, or worse, to rehabilitate the affected lands, can be reduced (Adinarayana and Rama Krishna 1995).

## Conclusions

The case study presented here suggests that the evaluation of agroecological characteristics from primary data, soil erosion assessment and aspects of conservation management, could lead to the generation of a WATMIS for a watershed (Figure 1). The development of a WATMIS in the context of local needs and problems, population and infrastructure leads to computation of resource surpluses and shortfalls at the present level of food/fuel demand and supply. Projections for future demands for sustainable development of the watershed can then be formulated based on which thrust areas in each sector can be identified.

The raster-based GIS provides an easy method of integrating the watershed resource data into useful maps. Vector data are transformed to a grid cell by interpolation or polygon conversion. Once in raster form, per pixel logical combinations are simple and could almost be handled in a standard data base.

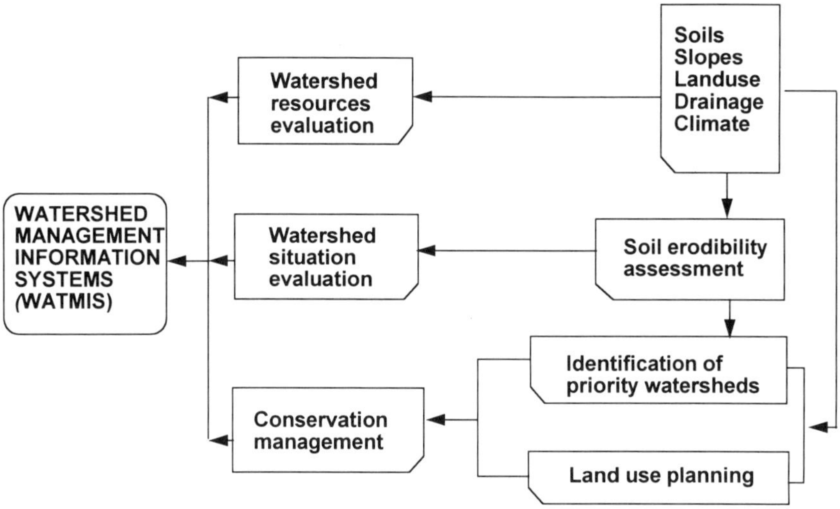


*Figure 1.* Generalized processing flow for watershed management information systems.

IRUs and IPLUs, generated for different purposes in the study, act as the strategic unit for various watershed management aspects as they exhibit strong uniformity in biophysical attributes and are expected to respond similarly to given intensities of human use and management strategies.

The knowledge-based rules used in watershed management planning were expressed as the probability of an item occurring given a particular hypothesis. The rules are the most subjective aspect of the system: they are heuristic, estimated from the knowledge of the pedologist and other resource experts and form a continuum between true and false depending on how sure we are that the rule is true or false (Skidmore et al. 1991). The knowledge bases, with further research and testing on different and larger areas, could lead to future decision-support systems for watershed management programmes.

Sophisticated methodologies, such as image-processing and GIS, are useful tools for planning resource use and management at regional scales. These models could be further improved by using finer resolution terrain analysis, incorporating many more ecological and socioeconomic variables, incorporating *informed opinions* from the experts and local people (the experiential knowledge bases), and employing effective extrapolation of expert system. Eventually, with this improved approach, the local governmental bodies should be able to possess a highly versatile and responsive system that enables their resource managers to assess the breadth and depth of their needs in a geographically precise and consistent manner. It is hoped that the study reported in this paper can illustrate the advantages of 'off-the-shelf' software and GIS technology in providing rapidly the alternative management strategies for large-scale resource planning.

The study also ensures the ability to monitor the dynamic changes of the watershed resources (particularly, the land-use pattern by remote sensing data with periodic intervals) and to redefine the watershed management strategies on the basis of detected change. This systems approach for hilly watersheds with variable rainfall regimes and severe land degradation problems is likely to provide more sustainable and cost-effective resource management strategies.

The methodology attempted in the case study will be further tested in the other areas of hilly watershed ecosystems in the Western Ghat mountainous zone. Although it is an early airing of results obtained from the methodology described, the introduction of multidisciplinary experts' *informed opinion* may provide an extension to the traditional methods of watershed management programs which should enable us to generate and/or evaluate different resource management scenarios.

## Acknowledgements

This research project was financed by a research grant from the Department of Science and Technology, Government of India (through SERC Programme – ES/23/077/89). The author wishes to thank Professor V.S. Chandrasekaran, Head of CSRE, for his constant encouragement and for providing facilities. Thanks are also due to my colleagues in CSRE for useful help, discussions and comments on this paper.

## References

Adinarayana J, Flach J D, Collins W G (1994a) Mapping land use patterns of a river catchment using Geographical Information Systems. J. Environ. Manage. 42(1):55–61.

Adinarayana J, Rama Krishna N, Suri J K, Gopal Rao K, Venkatachalam P V (1994b) Assessment of soil erosion of a watershed using GIS. Int. Arch. Photogramm. Remote Sensing 30(2):1–12.

Adinarayana J, Rama Krishna N (1995) An approach for land use planning in hilly watersheds using Geographical Information Systems. Land Degradation Rehabilitation 6:171–178.

Adinarayana J, Rama Krishna K (1996) Integration of multi-seasonal remotely sensed images for improved landuse classification of a hilly watershed using geographical information systems. Int. J. Remote Sensing (in press).

Adinarayana J, Rama Krishna N, Gopal Rao K (1995) An integrated approach for prioritization of watersheds. J. Environ. Manage. 44:375–384.

AISLUS – All India Soil and Land Use Planning, Ministry of Agriculture, Government of India (1991) Methodology for priority delineation survey. New Delhi, India.

Bocco G, Valenzuela C R (1991) Integration of GIS and Remote Sensing in land use and erosion studies. Pages 192–205 in Valenzuela C R (Ed.) Introduction to GIS. ITC Publication No. LIS-22. Enschede, The Netherlands.

Devi S D, Venkatachalam P, Natarajan C (1993) Geographic Information Systems for environmental impact assessment – a case study. Int. J. Environ. Stud. 43:115–122.

El-Swaify S A, Arsyad S, Krishnarajah P (1983) Soil erosion by water. Pages 99–161 in Carpenter A R (Ed.) Natural systems for development – what planners need to know. Macmillan, New York, USA.

Sheng T C, Stennett H R (1975) Forestry development and watershed management in the upland regions – Jamaica. FAO Working Document No. SF/JAM 505. Food and Agriculture Organization, Rome, Italy.

Skidmore A K, Ryan P J, Dawes W, Short D, O'Loughlin E (1991) Use of expert system to map forest soils from a GIS. Int. J. Geogr. Inf. Syst. 5(4):431–445.

Soil Survey Staff (1975) Soil Taxonomy – a basic system of soil classification for making and interpreting soil survey. USDA Handbook No. 436. United States Department of Agriculture, Washington DC, USA.

# The GCTE Crops Networks: a framework for international collaborative research

P. GREGORY and J.S.I. INGRAM

*Global Change and Terrestrial Ecosystems Core Project (GCTE) Focus 3 Office, NERC Centre for Ecology and Hydrology, Wallingford, UK*

**Key words:** crop growth simulation models, global change, international research networks, GCTE

**Abstract**

The Global Change and Terrestrial Ecosystems Core Project of IGBP has launched a series of international research networks addressing the effects of global change on seven crops of major worldwide importance. The rationale, objectives and *modus operandum* of the GCTE Crop Networks are described. Results from the GCTE Rice Network model comparisons are summarized. It is apparent that the current crop models are able to provide an estimate of yield potential in a wide range of rice growing environments where yield potential ranges from 6 to 15 ton $ha^{-1}$; and that for temperature changes in the range −3 to +3°C, there is general agreement on the yield change of about −4.5% (or −0.3 ton $ha^{-1}$) for each degree of temperature difference from the base climate.

## Introduction

Food production in many parts of the world is already threatened by damage to soil and water resources through poor technology. This situation has usually arisen because increased demands have led to excessive land-use intensification, the consequences of which are now critical in some instances. The general situation may, however, be exacerbated still further by two additional and interactive forces of Global Change – changes in atmospheric composition and the likely associated changes in climate. These further factors will affect these already stressed and rapidly changing systems in ways we cannot yet predict with any accuracy.

While clearly serious in many instances, it must be remembered that not all aspects will be deleterious; global change will affect the world's agricultural production in both beneficial and adverse ways, providing both opportunities and hazards. The ability to capitalize on the former, while avoiding the latter, will require a strong predictive capability, and to develop appropriate adaptive strategies we need to understand the potential impacts of global change on agroecosystems.

## Global Change and Terrestrial Ecosystems: a core project of the International Geosphere-Biosphere Program

To address the major research challenge of how global change will affect terrestrial ecosystems, including agroecosystems, the International Geosphere-Biosphere Program (IGBP) has established a Core Project on Global Change and Terrestrial



*P.S. Teng* et al. *(eds.): Applications of Systems Approaches at the Farm and Regional Levels, 447–455.*

Ecosystems (GCTE). The rationale for this project, and the details of the strategic research plan, are given in the GCTE Operational Plan (Steffen et al. 1992). The GCTE objectives are:

- To predict the effects of changes in climate, atmospheric composition and land use on terrestrial ecosystems, including (i) agriculture, forestry and soils, and (ii) ecological complexity.
- To determine how these effects lead to feedback to the atmosphere and physical climate system.

GCTE's operational definition of 'global change' encompasses far more than just predicted climate change alone. It includes changes in atmospheric composition, such as the concentration of $CO_2$ and other greenhouse gases, which have increased and will continue to increase and which have direct impacts on vegetation with or without climate change. It also includes change in land use, as driven by demographic, economic, technological and social pressures, as mentioned above. Over the next few decades, this human dimension of global change will have a more profound influence on the fate of terrestrial ecosystems than will changes in climate and atmospheric composition, and will be particularly important in tropical regions. Nevertheless, the potential complication of the other drivers of global change warrants serious and urgent attention.

The GCTE plan to meet the research challenge is based on a strategy that is global in scope and consists of a well-integrated set of focused scientific questions. The structure of the programme is designed to ensure cohesiveness by facilitating the flow of people, ideas, expertise and information among its component parts, and is organized into a hierarchy of *Foci, Activities,* and *Tasks.*

## GCTE Focus 3

GCTE includes a major research agenda addressing global change impact on agriculture, managed forestry and soils (Focus 3). While the lack of knowledge about the precise climate in the future makes exact predictions impossible, it is important to be able to predict the consequences of defined scenarios and to identify the most sensitive components of managed systems. To this end, GCTE Focus 3 initiates strategic, interdisciplinary research to improve our general predictive ability for selected agronomic species, grasslands and rangelands; for multi-species cropping systems; for managed forests; for major pests, diseases and weeds; and for soils. The research results will help national programmes to plan site-related management strategies and adaptive research, both for today's rapidly changing environment and for, as and when, climate change occurs. Many nations and international research organizations (e.g. Food and Agriculture Organization (FAO); International Union of Forest Research Organization (IUFRO)) are launching or developing their own programmes in global change issues. The work of this Focus will interact with, and strongly build on, the very large body of existing and ongoing agronomic and forestry related work throughout the world.

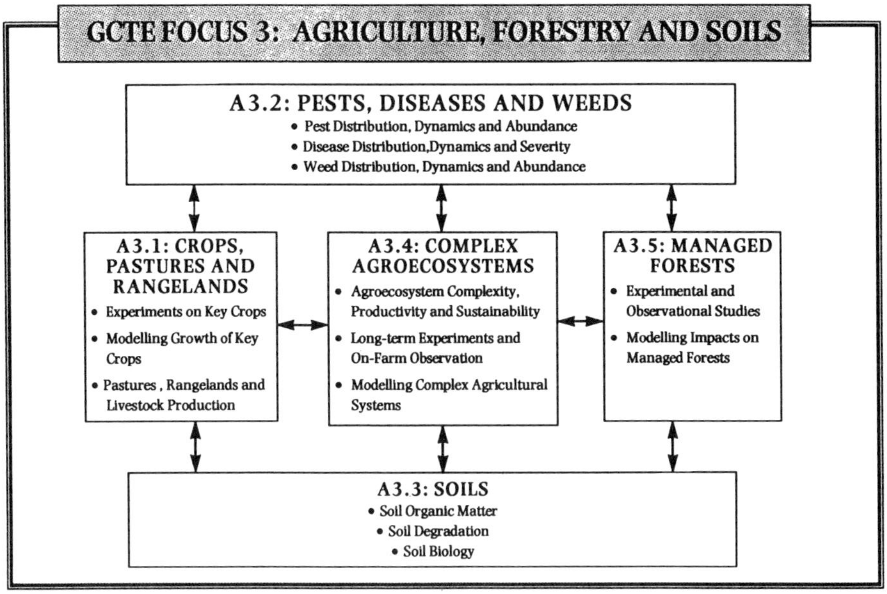


*Figure 1.* The GCTE Focus 3 structure.

Within Focus 3, four major types of production system have been identified: monocrop agriculture, pasture and grazing systems, multi-species cropping (including agroforestry and rotational systems) and forestry. These have been grouped into three Activities:

- *Key agricultural systems:* these include a representative selection of monoculture food crops, together with improved pastures and rangelands; the emphasis in this Activity is on harvestable product, i.e. the usable portion of the crop or animal.
- *Multi-species agroecosystems:* these are the norm in much of the world and may be either spatially or temporally complex mixtures, or both. Modelling such complex agricultural systems is in its infancy, relative to modelling monocrop production or even rotation farming.
- *Managed forests:* these encompass the spectrum from intensively managed plantation crops to more natural forests which are either selectively logged or put to some other use.

Two further Activities cut across these three major areas (Figure 1):

- *Pests, diseases and weeds:* pests, diseases and weeds are all major yield-reducing factors. An increase in their severity and/or incidence may be an early indicator of global change, and a prompt response will be needed to combat further management problems.
- *Soils:* global change will have both direct and indirect impact on the world's soils. Particularly important are changes in soil organic matter dynamics, in erosivity

and in the soil biological processes underpinning nutrient cycling and trace gas emission.

GCTE research will collaborate with, complement and coordinate national and international studies as appropriate. GCTE-initiated research will be general in nature; devoted to principles rather than local issues; devoted to major crops, so that its impact is substantial; and integrative. The collaborative aspect will be especially necessary for linking to the human dimension, particularly apposite when dealing with ecosystems from which a 'harvest' is taken. Global change impact on agriculture and forestry is more than just impact on biology – it is impact on food, wealth, development and people.

## GCTE Crop Networks

Clearly any research programme on 'agriculture' could cover a vast range of topics. It was therefore agreed in early GCTE planning that, due to limited resources, we had to restrict the programme to relatively few issues. Thus, of the approximately 20 main food crops which provide the bulk of the world's food and might form the subject of a research programme in global change, GCTE selected a short-list of seven for initial studies: wheat, rice, potatoes, cassava, sorghum, maize and groundnuts. The crops were selected both in recognition of their importance in global food production and as representatives of a diverse range of crop 'functional types'. While the importance of many other crops is recognized, GCTE believes firmly that it must initially be sharply focused, both to obtain valid results as early as possible, and to test its methodology.

GCTE is launching a series of Crop Networks, one for each of the seven priority crops given above, with linked objectives:

- To refine and adapt current crop production models for use in global change studies in a wide variety of conditions.
- To design and undertake experiments to provide improved mechanistic understanding of global change impact on crop production, to aid in model development.

Robust, reliable crop models are the key to predicting the responses of important agronomic species to changing environmental conditions. Many different crop models have been developed with a wide range of characteristics and initialization requirements. Existing models, however, are frequently constructed and validated using data from one locality, so the identification of modelling approaches that yield generic capability is needed. A set of models will be selected based on relevance to the crop in question, aims of the experiment, and their ability to simulate particular data sets. An initial goal will be the identification of models, or modules, for application to global change field experimental programmes. Once identified, a major aim is to determine (by model comparisons with contrasting data sets) the robustness of existing and new models to environmental changes, and to provide a guide to the application of crop models under global change within national programmes; model validation is an important initial stage. Model comparison will not be a competition, but is designed to stimulate further the development of generic crop models for global change by identifying strengths and weaknesses in existing models. In addition to using pertinent data sets from experiments already conducted, models will be validated using new data sets resulting from the Networks' experimental programmes.

Standardized protocols for the collection and dissemination of data from agricultural experiments have been established, and details are available from the GCTE Focus 3 Office[1]. The structure of input and output data for the models will also be standardized for ease of model comparison and interchange of model components, and the IBSNAT input format (IBSNAT 1986) is proving highly successful.

It is hoped that ultimately the models, which often may have been developed from experimental data from closed chamber and OTC (open top chamber) experiments, will be tested against data sets generated by the GCTE FACE (Free-Air-$CO_2$-Enrichment) (and other field-scale) experiments. Initial comparisons for wheat models have revealed surprising discrepancies, and have highlighted areas where more process understanding is required. Initial comparisons of rice yield models, however, demonstrated much more agreement among the different models (Kropff et al. 1994). A further aim is to strengthen model links to 'yield reducing' factors; working closely with GCTE Activities 3.2 and 3.3, the Crop Networks will further integrate pest, disease, weed and soil components with the current crop models for the priority species, to allow better predictive capabilities for changed environments.

GCTE-initiated research has started with the collation of existing experimental and monitoring data sets and models. Questionnaires soliciting 'metadata' (data describing a piece of research, as distinct from the data or model code itself) have been widely distributed, together with an invitation to apply to join the Network. (Further applications are welcome at any time; please contact the GCTE Focus 3 Office.) All applications are reviewed by an international expert panel (Network Working Group), and the conditions of acceptance are twofold: the quality of science and the pertinence to Network objectives. The collation of metadata serves more than just the establishment of network membership; it helps determine what GCTE-type studies are ongoing, where coordinative effort should be placed, and where new projects should be initiated.

*Network outputs*

As stated previously, the ultimate objective is to develop robust models for use under global change. There are, however, several interim products:

- The metadata (descriptive data) for projects incorporated in the network will be published in the GCTE Report Series, with subsequent editions as necessary; they will also be available on the GCTE WWW HomePage. This compilation will be a unique reference.

[1]GCTE Focus 3 Office, NERC Centre for Ecology and Hydrology, Crowmarsh Gifford, Wallingford OX10 8BB, UK. Tel: +44 1491 692410; Fax +44 1491 692313; email j.ingram@ioh.ac.uk.

- Proposals for synthesis, experimental and model development will be publicized via the network, and further collaboration developed as necessary.
- Because interdisciplinarity is essential if the goals are to be achieved, close links will be developed with branches of crop physiology, soil and pest scientists, and agronomists.
- Results from international workshops to evaluate, compare and validate water and wind erosion models using network data will be distributed widely.
- Network members will be encouraged to prepare multi-authored papers reporting the results of Network activities for publication in the international peer-reviewed literature.

GCTE launched its first Network (for Wheat) in July 1992 and this now comprises 39 formal collaborators from both experimental and modelling backgrounds. Focus 3 Networks for Wheat, Rice and Potato are now formally recognized as GCTE Core Research, and planning for similar Networks for sorghum and cassava is underway.

## The GCTE Rice Network

Given the importance of rice in South and Southeast Asia, coupled with the expected population increase in the region, preparing for global environmental change impact on rice production is crucial. Although the precise climate in the future is unknown, it is important to be able to predict the consequences of defined scenarios and to identify the most sensitive components of rice production systems. As for the other crops, this needs the initiation of strategic, interdisciplinary research to improve our general predictive ability for rice.

The GCTE Rice Research Network was launched at an international workshop in IRRI (International Rice Research Institute, Los Baños, Philippines) in March 1994 (Ingram 1994), where the overall and specific objectives of the network were established as follows:

### *Overall GCTE Rice Network Objective*

To improve understanding of the effects of global change on rice, and thereby develop a predictive capability for determining the impact of these effects on rice production systems worldwide.

### *Specific Objectives*

(1) Develop, improve and validate models of the rice crop/ecosystem. Such models will serve as tools to make reliable assessment of global change projections, and for reducing uncertainties associated with global change including better evaluation of management options.
(2) Conduct experiments that will test scientific hypotheses while providing data that will contribute to development and evaluation of rice crop models.

(3) Foster improved experimental protocols that will support both scientific hypotheses and crop modelling needs.
(4) Promote aggregation and sharing of scientific data and information, and research tools such as models for more effective evaluation of global change and its effects and impacts.
(5) Plan and conduct impact assessments that will contribute to crop management strategies under present conditions and anticipated global change.

The gathering of new data of a suitable type can be costly, so it is essential to use existing data sets as widely and efficiently as possible. Meanwhile, and as for other Networks, it will be important to identify data sets currently available that will help model validation, even if they are not ideal; this exercise will help plan the next generation of experiments. Several suitable rice data sets and models have already been contributed to the Network.

In addition to formalizing the Rice Network and establishing Network protocols, the inaugural workshop at IRRI included a presentation of a rice production model evaluation and comparison (Kropff et al. 1994). This aimed to quantify and explain the yield potential of current varieties in contrasting environments using simulation models which quantitatively describe the processes of yield formation. Experimentally determined yield potential in four environments (a tropical environment at IRRI, Los Baños, Philippines, and in temperate/subtropical environments at the Yanco Agricultural Research Institute in Australia and at Kyoto University in Japan) ranged from 4.7 to 14.7 $ton\,ha^{-1}$.

Four models of varying complexity (ORYZA1 (Kropff et al. 1994), CERES-Rice (Singh et al. 1994), SIMRIW (Horie 1987) and TRYM (Williams et al. 1994)) were tested using these data. Predictions from each model were evaluated and compared with data on rice growth in the environment for which they were developed, and also with data from a range of contrasting environments. The parameters for phenological development were calibrated for all models to ensure that the simulated flowering dates matched the observed data of the different varieties in the different locations.

All four models were run using daily weather data from the four sites and the actual sowing and planting dates. All models were operating at the yield potential level, where N should be non-limiting. In CERES-Rice, a switch was used to switch off N limitation; in the ORYZA1 model the time course of the leaf N content of the highest yielding experiment at IRRI was used; SIMRIW operates at the yield potential level only, and the TRYM model was run for a range of N application levels for all sites (up to 320 $kg\,N\,ha^{-1}$, the maximum level used in Yanco) and the highest-yielding treatment was used in the analysis.

The models simulated yield potential differences among environments very well: the maximum difference between simulated and observed yield ranged from 0.4 to 2.2 $ton\,ha^{-1}$ in the different locations. The results indicate that the current crop models are able to provide an estimate of yield potential in a wide range of rice growing environments where yield potential ranges from 6 to 15 $ton\,ha^{-1}$.

A follow-up workshop in Bangkok (September 1995) designed a multi-site experi-

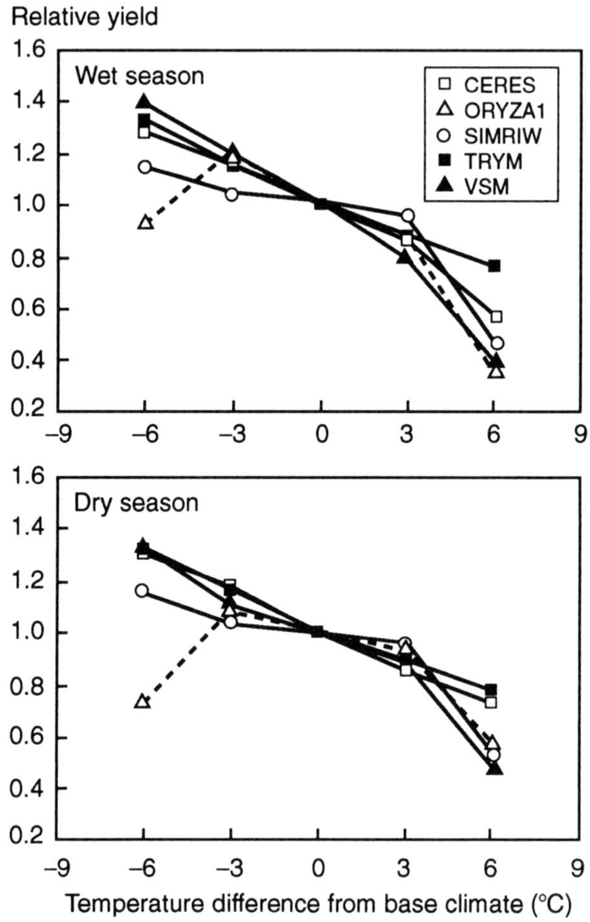


*Figure 2.* The temperature response of five models, shown as the yield relative to that of the base climate, for the wet season (above) and dry season (below). The model VSM was not run with the base climate alone, so the average of yields at −3°C and +3°C was used for the reference value (*Source:* P.L. Mitchell in Ingram 1996).

ment based on temperature gradient tunnels. This experiment will yield data sets which are ideally suited for model validation and development. About four tropical and three temperate sites are envisaged. The workshop also reported results of a refined model sensitivity analysis for five models parameterized with common data. The comparisons involved assessment of the effects of increases in atmospheric $CO_2$ concentration (to twice current levels) and changes in air temperature from −6°C to +6°C. Figure 2 summarizes the temperature response of the models, from which it can be seen that in the temperature range −3°C to +3°C there is general agreement on the yield change of about −4.5% (or −0.3 ton $ha^{-1}$) for each degree of temperature difference from the base climate. Three models (ORYZA1, SIMRIW and VSM (Very

Simple Model by K. Kobayashi, National Institute of Agro-Environmental Sciences, Tsukuba, Japan) consistently predicted a larger decrease in yield for the second step in temperature rise between +3°C and +6°C. Full results of the model sensitivity analysis are presented in Annex 2 (prepared by P.L. Mitchell, University of Sheffield, UK) in GCTE Working Document 19 (Ingram 1996).

## Conclusions

GCTE Crop Networks are designed to maximize the understanding of global change impact on the cropping system. This means on the crops *per se*, and on the crops' pests, weeds and soils (key aspects which are dealt with in GCTE Activities 3.2 and 3.3, respectively), i.e. on the whole agroecosystem. This approach requires the consideration of not only all components of the system, but also the interactive effects of global change drivers. This integrative approach makes GCTE uniquely innovative. Furthermore, by taking the systems approach to crop modelling, these networks cover many aspects of GCTE, and provide a forum for the global change community to undertake and discuss collaborative modelling research. Research results will help national programmes to plan not only long-term strategies, but the improved mechanistic understanding will also assist in short- and medium-term, site-related management. The worldwide Network approach is greatly facilitating the required cross comparison of models with appropriate data sets, and both modellers and experimentalists are essential to a given Network's success.

## References

Horie T (1987) A model for evaluating climatic productivity and water balance of irrigated rice and its application to Southeast Asia. Southeast Asian Stud. 25:62–74.

IBSNAT – International Benchmark Sites Network for Agrotechnology Transfer (1986) Technical Report No. 5: Documentation for IBSNAT crop model input and output files for decision support system for agrotechnology transfer (DSSAT V.2.1). Department of Agronomy and Soil Science, College of Tropical Agriculture and Human Resources, University of Hawaii, Honolulu, HI, USA.

Ingram J S I (1994) Report of the GCTE 'Rice Ecosystems' Workshop, IRRI, Los Baños, 21–22 March 1994. Global Change and Terrestrial Ecosystems Focus 3 Office, Oxford, UK. 29 p.

Ingram J S I (1996) Report of the GCTE Rice Network Experimentation Planning Workshop, Bangkok, Thailand. Global Change and Terrestrial Ecosystems Working Document 19. GCTE Focus 3 Office, Oxford, UK. 29 p.

Kropff M J, Van Laar H H, Matthews R B (Eds.) (1994) ORYZA1: An ecophysiological model for irrigated rice production. SARP Research Proceedings, International Rice Research Institute, P.O. Box 933, Manila, Philippines. 110 p.

Kropff M J, Williams R L, Horie T, Angus J F, Singh U, Centeno H G, Cassman K G (1994) Predicting yield potential of rice in different environments. Pages 657–663 in Humphreys E, Murray E A, Clampett W S, Lewin L G (Eds.) Temperate rice – achievements and potential, Yanco Agricultural Institute, New South Wales Agriculture, NSW 2680, Australia.

Singh U, Ritchie J T, Godwin D C (1993) A user's guide to CERES-RICE – V2.10. International Fertilizer Development Center, Muscle Shoals, AL, USA. 269 p.

Steffen W L, Walker B H, Ingram J S I, Koch G W (Eds.) (1992) Global change and terrestrial ecosystems. The operational plan. IGBP Report No. 21, Stockholm, Sweden. 95 p.

Williams R L, Durkin C O, Stapper M (1994) A simple model of rice yield response to N fertiliser and its use in a decision support system. Pages 703–710 in Humphreys E, Murray E A, Clampett W S, Lewin L G (Eds.) Temperate rice – achievements and potential. Yanco Agricultural Institute, New South Wales Agriculture, NSW 2680, Australia.

# Acronyms used in *Applications of Systems Approaches at the Farm and Regional Levels*, and *Applications of Systems Approaches at the Field Level*

| | |
|---|---|
| AARD | Agency for Agricultural Research and Development |
| AB-DLO | DLO-Research Institute for Agrobiology and Soil Fertility |
| AICCIP | All India Coordinated Crop Improvement Project |
| AIDS | Almost Identical Demand System |
| AISLUS | All India Soil and Land Use Survey |
| AMMI | Additive Main effects and Multiplicative Interaction models |
| APROSC | Agricultural Projects Services Centre |
| APSRU | Agricultural Production Systems Research Unit |
| ARMA | AutoRegressive Moving Average |
| ARO | Advanced Research Organization |
| ASA | American Society of Agronomy |
| ASAE | American Society of Agricultural Engineers |
| AUDPC | Area Under Disease Progress Curve |
| AZP | Atlantic Zone Programme |
| BEII | Biocide Environmental Impact Index |
| BMZ | Federal Ministry for Cooperation |
| BRGM | Bureau de Recherches Géologiques et Minières |
| BULOG | National Logistic Board |
| CAILUP | Computerized Aid to Integrated Land-Use Planning |
| CASER | Center for Agro-Socio Economic Research |
| CATIE | Tropical Agronomic Research and Higher Education Center |
| CCT | Computer Compatible Tapes |
| CEC | Cation Exchange Capacity |
| CERES | Crop Estimation through Resource and Environment Synthesis |
| CGIAR | Consultative Group on International Agricultural Research |
| CGMS | Crop Growth Monitoring System |
| CGPRT | Centre for Coarse Grains, Pulses, Roots and Tuber Crops |
| CIAT | International Center for Tropical Agriculture |
| CIMMYT | Centro Internacional de Mejoramiento de Maiz y Trigo (International Center for the Improvement of Maize and Wheat) |
| CIP | Centro Internacional de la Papa (International Potato Center) |
| CIRAD | Centre de coopération Internationale en Recherche Agronomique pour le Développement (Centre for International Cooperation and Agricultural Development Research) |



*P.S. Teng* et al. *(eds.): Applications of Systems Approaches at the Farm and Regional Levels, 457–462.*

| | |
|---|---|
| CIRAD-URPA | Centre de coopération Internationale en Recherche Agronomique pour le Développement, Unite de Recherche 'Prospective et Politique Agricole' |
| CLIPS | C Language Integrated Production System |
| CLSU | Central Luzon State University |
| CMDT | Compagnie Malienne pour le Développement des Textiles |
| COMAX | CrOp Management eXpert |
| CPAC | Centro de Pesquisa Agropecuária dos Cerrados |
| CR | Crop Residues |
| CRRI | Central Rice Research Institute |
| CSIRO | Commonwealth Scientific and Industrial Research Organization |
| | |
| Danida | Danish International Development Assistance |
| DAS | Days After Sowing |
| DAT | Days After Transplanting |
| DEM | Digital Elevation Model |
| DGIS | Directorate General for International Cooperation |
| DLO | Agricultural Research Department, Ministry of Agriculture, Nature Conservation and Fisheries |
| DNSI | Direction Nationale de la Statistique et de l'Information |
| DQC | Dynamic Quality Control |
| DRR | Directorate of Rice Research |
| DRSPR | Direction Régionale des Systèmes de Production et de la Gestión des Ressources Naturelles |
| DSS | Decision Support System |
| DSSAT | Decision Support System for Agrotechnology Transfer |
| DTM | Digital Terrain Model |
| | |
| EIA | Environmental Impact Assessment |
| EMBRAPA | Empresa Brasileira de Pesquisa Agropecuária |
| EPA | Environmental Protection Agency |
| EPIC | Erosion Productivity Impact Calculator |
| ESPGRN | Equipe Systèmes de Production et Gestión des Ressources Naturelles |
| EU | European Union |
| EUROSTAT | European Statistical Office |
| | |
| FACE | Free-Air-$CO_2$-Enrichment |
| FAO | Food and Agriculture Organization of the United Nations |
| FCC | False Colour Composites |
| FHM | Farm-Household Modelling |
| FPR | Farmer Participatory Research |
| FSA | Farming Systems Analysis |

| | |
|---|---|
| FSR | Farming Systems Research |
| FSR/E | Farming Systems Research and Extension |
| FSS | Farming System Simulator |
| GATT | General Agreement on Tariffs and Trade |
| GBPUAT | Govind Ballabh Plant University of Agriculture and Technology |
| GCM | General Circulation Models |
| GCREC | Gulf Coast Research and Education Center |
| GCTE | Global Change and Terrestrial Ecosystems Core Project |
| GDP | Gross Domestic Product |
| GE | Grain Equivalents |
| GFDL | General Fluid Dynamics Laboratory |
| GIDM | Generic Interactive Dairy Model |
| GIS | Geographic Information System |
| GISS | Goddard Institute for Space Studies |
| GPS | Global Positioning System |
| GRASS | Geographic Resources Analysis Support System |
| GTZ | German Agency for Technical Cooperation |
| G×E | Genotype by Environment |
| HTIS | HyperText Information System |
| IARC | International Agricultural Research Center |
| IARI | Indian Agricultural Research Institute |
| IBSNAT | International Benchmark Sites Network for Agrotechnology Transfer |
| ICAR | Indian Council of Agricultural Research |
| ICARDA | International Center for Agricultural Research in Dry Areas |
| ICASA | International Consortium for Agricultural Systems Applications |
| ICLARM | International Center for Living Aquatic Resources Management |
| ICRAF | International Center for Research in Agroforestry |
| ICRISAT | International Crops Research Institute for the Semi-Arid Tropics |
| IDA | Institute for Agricultural Development |
| IFDC | International Fertilizer Development Center |
| IFPRI | International Food Policy Research Institute |
| IGBP | International Geosphere-Biosphere Programme |
| IIT | Indian Institute of Technology |
| ILCA | International Livestock Center for Africa |
| ILEIA | Information Centre for Low External Input and Sustainable Agriculture |

| | |
|---|---|
| IMGLP | Interactive Multiple Goal Linear Programming |
| INGER | International Network for Genetic Evaluation of Rice |
| INTA | National Institute for Agricultural Research |
| IPLU | Integrated Physical Land Units |
| IRM | Insecticide Resistance Management |
| IRRI | International Rice Research Institute |
| IRS | Indian Remote Sensing Satellite |
| IRU | Integrated Resources Units |
| ITOPE | Interdisciplinary Tool for Optimizing Productivity and Erosion |
| IUFRO | International Union of Forest Research Organization |
| JAU | Jiangxi Agricultural University |
| JRC | Joint Research Centre |
| KBS | Knowledge-Based System |
| KKU | Khon Kaen University |
| LDC | Less Developed Countries |
| LE | Land Evaluation |
| LIS | Livestock Information System |
| LUST | Land-Use Systems at a defined Technology |
| MACROS | Modules of an Annual CRop Simulator |
| MARS | Monitoring Agriculture with Remote Sensing |
| MCDM | Multiple Criteria Decision Models |
| MCO | Multi-Criteria Optimization |
| MDS | Minimum Data Set |
| MEDALUS | Mediterranean Desertification and Land Use |
| MET | Multi-Environment Testing |
| MLNV | Dutch Ministry of Agriculture, Nature Conservation and Fishery |
| MOLP | Multiple Objective Linear Programming |
| MP | Mathematical Programming |
| NARS | National Agricultural Research System |
| NBD | Nutrient Balance model of a Dairy–crop operation |
| NDVI | Normalized Difference Vegetation Index |
| NGO | Non-Governmental Organization |
| NSADP | North Sinai Agricultural Development Project |
| NWO-STW | Netherlands Technology Foundation |
| ORSTOM | Institut Français de Recherche pour le Développement en Coopération |

| | |
|---|---|
| OTC | Open Top Chamber |
| OTELO | Organisation du Travail Et Langage à Objet |
| PAR | Photosynthetically Active Radiation |
| PASW | Plant-Atmosphere-Soil-Water |
| PCA | Principal Component Analysis |
| PDF | Probability Density Function |
| PhilRice | Philippine Rice Research Institute |
| PI | Panicle Initiation |
| PPFD | Photosynthetic Photon Flux Density |
| PRA | Participatory Rapid Appraisal |
| PRODECER | Cerrados Development Program |
| PSS | Project Production Soudano Sahélienne |
| QTL | Quantitative Trait Loci |
| R,D&E | Research, Development and Extension |
| RFLP | Restriction Fragment Length Polymorphism |
| RIDEV | RIce DEVelopment model |
| RMSE | Root Mean Square Error |
| RRA | Rapid Rural Appraisal |
| RSDP | Rice Supply and Demand Project |
| RUE | Radiation Use Efficiency |
| SALT | Sustainable Agriculture Land Technology |
| SANREM CRSP | Sustainable Agriculture and National Resource Management – Collaborative Research Support Program |
| SARP | Simulation and Systems Analysis for Rice Production |
| SC-DLO | DLO-Winand Staring Centre |
| SEBAL | Surface Energy Balance Algorithm for Land |
| SHIELD | Simulation of Hedgerow Intervention against Erosion and Land Degradation |
| SIMRIW | Simulation Model for Rice – Weather Relationships |
| SOI | Survey of India |
| SPAR | Soil Plant Atmosphere Research |
| SSG | Spray Strategy Group |
| SSM | Site Specific Management |
| SUCROSA | Simple and Universal CROp growth Simulator |
| SUSENAS | Survey of Social and Economic National |
| SWAT | Soil and Water Assessment Tool |
| SWRRB | Simulator for Water Resources in Rural Basins |
| SYI | Sediment Yield Index |
| TIGER | Terrestrial Initiative in Global Environmental Research |

| | |
|---|---|
| TLU | Tropical Livestock Units |
| TNAU | Tamil Nadu Agricultural University |
| UKMO | United Kingdom Meteorological Office |
| UN | United Nations |
| UNDP | United Nations Development Programme |
| USDA | United States Department of Agriculture |
| USDA-ARS | United States Department of Agriculture-Agricultural Research Service |
| USTED | Uso Sostenible de Tierras En el Desarrollo (Sustainable Land Use in Development) |
| VAMPP | Veterinary Automated Management and Production control Programme |
| VRSAP | Vietnam River Systems And Plains |
| WARDA | West Africa Rice Development Association |
| WAS | West African Sahel |
| WATMIS | Watershed Management Information System |
| WAU | Wageningen Agricultural University |
| WB | World Bank |
| WCED | World Commission on Environment and Development |
| WDVI | Weighted Difference Vegetation Index |
| WEPS | Wind Erosion Prediction System |
| WRR | Netherlands Scientific Council for Government Policy |

# Index





*P.S. Teng* et al. *(eds.): Applications of Systems Approaches at the Farm and Regional Levels, 463–468.*